BASIC LABORATORY and INDUSTRIAL CHEMICALS

A CRC Quick Reference Handbook

BASIC LABORATORY and INDUSTRIAL CHEMICALS

A CRC Quick Reference Handbook

David R. Lide

Editor-in-Chief, CRC *Handbook of Chemistry and Physics*
and
Former Director, Standard Reference Data
National Institute of Standards and Technology

CRC Press
Boca Raton   Ann Arbor   London   Tokyo

Library of Congress Cataloging-in-Publication Data

Lide, David R., 1928-
  Basic laboratory and industrial chemicals : a CRC quick reference handbook / author,
David R. Lide.
    p. cm.
  Includes bibliographical references and indexes.
  ISBN 0-8493-4498-0
  1. Chemicals--Handbooks, manuals, etc. I. Title.
  QD64.L53 1993
  546--dc20   93-25269
                                    CIP

To Bettijoyce

TABLE OF CONTENTS

INTRODUCTION

Information on the physical, thermodynamic, and transport properties of chemicals is frequently needed by research workers, engineers, and students. The Chemical Abstracts Service Registry System now contains over ten million substances which have been reported in the scientific literature or included on Government regulatory lists. Fortunately, out of this formidable number of known chemical substances, only a small fraction — perhaps a few percent — are ever likely to be encountered in the laboratory or workplace. In fact, most requests for information on chemicals are focused on an even smaller core set of ubiquitous compounds. The objective of this *Handbook* is to present basic property data in a convenient format for such a core set of about 1000 substances which are frequently found in chemical laboratories or industrial facilities.

Compound Selection

The following criteria were used in selecting this core set of basic chemicals:

1. Appearance of the substance in many other widely used reference sources, such as the *CRC Handbook of Chemistry and Physics*, *The Merck Index*, and the *DIPPR Database of Pure Compound Properties*.
2. Occurrence on regulatory lists of chemicals used in commerce which are considered hazardous in one way or another.
3. Availability of reliable property data on the substance (a less important factor than the first two).

As an additional filter, compounds of biochemical interest, such as amino acids and sugars, were generally excluded because the properties tabulated here are not highly pertinent to such compounds. For similar reasons, the number of inorganic salts is limited. However, the more important elements have been included in the list, as well as their most common halides, oxides, and a few other salts.

It is apparent that a certain degree of subjectivity entered into the selection process, and any user of this book will probably find compounds missing that could be considered "basic". However, the author hopes that the coverage is representative of the interests of a broad cross-section of industrial and academic chemists.

Properties Covered

The properties listed are divided into five groups:

1. Identifying information such as name, formula, etc.
2. Melting point, boiling point, and other single-valued physical constants
3. Properties associated with phase transitions
4. Thermodynamic and transport properties at 25°C
5. Comments on any unusual hazards associated with the substance

These data are presented in a standard format with three substances on each page. The details on each property are given below.

The data in this book have been taken primarily from evaluated sources, i.e., from publications and electronic databases in which experts have reviewed the data in the primary research literature and made recommendations of best values. It was not practical in the present format to give a specific reference for each data item which appears (in fact, in some cases the author has selected the value based upon an assessment of the available sources). However, a list of the sources that were utilized, classified by property, appears at the end of this Introduction.

While quantitative uncertainties in the property values are not given, the number of significant figures has been adjusted to reflect the accuracy of the data. An exception was made for the enthalpy, entropy, and heat capacity values, where the number of decimal places has been made uniform in order to facilitate calculations that involve addition and subtraction of these quantities.

Arrangement and Indexes

Substances are listed in alphabetical order by the name considered to be the most easily recognized. As a general rule, systematic names consistent with IUPAC nomenclature recommendations have been chosen. However a number of exceptions are made in cases where a non-systematic name is almost universally used; hence acetone instead of 2-propanone and formic acid instead of methanoic acid. When the primary name is non-systematic, the systematic name is listed as a synonym.

To facilitate location of a substance on the basis of name, a combined name/synonym index appears at the end of the book. Since this index includes both the primary names and the major synonyms, it is probably the most efficient way to locate an entry. Indexes by molecular formula and Chemical Abstracts Service Registry Number are also included.

Description of Entries

Name: The primary name selected for the substance.

Synonyms: One or two other names in common use. If the primary name is non-systematic, the systematic name is given as a synonym. In general, trade names are not included.

Mol. Form.: Molecular formula written in the Hill order, with C first (if present), H second, and the other elements in alphabetical order by symbol.

CAS RN: Chemical Abstracts Service Registry Number.

Merck No.: Number of the monograph describing the substance in the Eleventh Edition of *The Merck Index* (Reference 2).

Mol. Wt.: Molecular weight as calculated with the 1991 IUPAC Standard Atomic Weights. Although the value is given to three decimal places, some of the atomic weights are less accurate. In the case of radioactive elements for which the terrestrial isotopic abundance cannot be precisely defined, the atomic mass of the longest-lived isotope is used.

T_m: Normal melting point temperature at atmospheric pressure, given in both degrees Celsius and kelvins. For a few substances the normal melting point temperature is undefined, because the sublimation pressure of the solid reaches one atmosphere at a temperature for which the liquid phase does not exist. In such a case the triple point temperature T_t, corresponding to the point on the phase diagram where solid, liquid, and gas are in equilibrium, is given instead of T_m, and the sublimation point temperature T_s is listed instead of T_b. This is noted in the COMMENTS line.

T_b: Normal boiling point in °C and K referred to a pressure of 101.325 kPa (760 mmHg). The temperature T_s at which the sublimation pressure of the solid reaches 101.325 kPa (the "sublimation point") is given instead of T_b for a few substances (see discussion under normal melting point).

T_c: Critical temperature in °C and K.

P_c: Critical pressure in megapascals (1 MPa = 1000 kPa = 10 bar = 9.86923 atm).

μ: Electric dipole moment in debye units (1 D = 3.33564 × 10^{-30} C m). Only values measured in the gas phase are given. An entry of 0 indicates that the dipole moment is identically zero because of molecular symmetry.

IP: First ionization potential in electron volts (1 eV/molecule = 96.4853 kJ/mol)

$\Delta_{fus} H(T_m)$: Enthalpy (heat) of fusion at the normal melting point in kJ/mol.

$\Delta_{vap} H(T_b)$: Enthalpy of vaporization at the normal boiling point in kJ/mol.

$\Delta_{vap} H(25°C)$: Enthalpy of vaporization at 25°C in kJ/mol.

Vapor pressure: Values are given, when available, at 0, 25, and 100°C in units of kPa (1 kPa = 7.50062 mmHg = 0.00986923 atm). An entry of N/A indicates either a refractory substance with negligible vapor pressure at ambient temperature or a substance whose critical temperature is lower than the indicated temperature (so that the vapor pressure is not defined.)

$\Delta_f H°$: Standard enthalpy of formation in the state indicated at 25°C and a pressure of 100 kPa, in kJ/mol. An entry of 0 for an element indicates the defined reference state.

$S°$: Standard entropy in the state indicated at 25°C and 100 kPa, in J/mol K.

C_p: Heat capacity at constant pressure at a temperature of 25°C and a nominal pressure of 100 kPa, in J/mol K.

d: Density in the state indicated at top of column. If the stable state at 25°C and 1 kPa is solid or liquid, the density value is given in g/mL (identical to g/cm^3); if the substance is a gas at 25°C and 1 kPa, the entry is the calculated ideal gas value in g/L. All entries are true densities, not specific gravities.

η: Dynamic viscosity in the state indicated. If the state is liquid, the value is given in millipascal seconds (mPa s), which is identical to the centipoise unit (cP). Values for gases are given in micropascal seconds (μPa s). An entry of N/A indicates a solid.

k: Thermal conductivity in units of watts per meter kelvin (W/m K). Most values for solids refer to polycrystalline samples.

COMMENTS: The following notations refer to special hazards of the substance:

TLV — Threshold Limit Value for exposure to airborne concentrations in the workplace, in parts per million by volume (see Reference 28).

carcinogen — substance has been listed by one or more organizations as a confirmed or suspected human carcinogen (see Reference 31).

highly toxic — substance shows unusually high human toxicity, as indicated by its inclusion in regulatory lists such as SARA 302(A), SARA 313, and TSCA 8(A) (see Reference 31) or assignment of a Health Hazard rating of 3 or 4 by the National Fire Protection Association (see Reference 29).

flammable — substance is assigned a Flammability Rating of 3 ("Materials which can be ignited under almost all normal temperature conditions") by the National Fire Protection Association (NFPA) (see Reference 29).

very flammable — substance is assigned an NFPA flammability rating of 4 ("Very flammable gases or very volatile flammable liquids") (see Reference 29).

REFERENCES

Space limitations preclude the citation of a specific reference for each data item. However, the major sources of data utilized in preparing these tables are listed below. In many cases, these references give data over a range of temperature and pressure, not only the values at standard conditions listed here. In addition to these sources, each of which covers a broad range of substances, many articles on individual compounds in the *Journal of Physical and*

Chemical Reference Data were utilized. The cumulative index to that journal provides a useful entry point for property data of the type given in this book.

Multi-property sources:

1. Lide, D. R., Editor, *CRC Handbook of Chemistry and Physics,* 74th Edition, CRC Press, Boca Raton, FL, 1993.
2. Budavari, S., Editor, *The Merck Index, Eleventh Edition,* Merck & Co., Rahway, NJ, 1989.
3. Riddick, J. A., Bunger, W. B., and Sakano, T. K., *Organic Solvents, Fourth Edition*, John Wiley & Sons, New York, 1986.
4. *DIPPR Data Compilation of Pure Compound Properties,* Design Institute for Physical Property Data, American Institute of Chemical Engineers, New York.
5. *TRC Thermodynamic Tables,* Thermodynamic Research Center, Texas A & M University, College Station, TX.
6. *Physical Constants of Hydrocarbon and Non-Hydrocarbon Compounds, 2nd Edition*, ASTM Data Series DS 4B, ASTM, Philadelphia, 1991.

Thermochemical properties:

7. Wagman, D. D., Evans, W. H., Parker, V. B., Schumm, R. H., Halow, I., Bailey, S. M., Churney, K. L., and Nuttall, R. L., *The NBS Tables of Chemical Thermodynamic Properties, J. Phys. Chem. Reference Data,* Vol. 11, Suppl. 2, 1982.
8. Chase, M. W., Davies, C. A., Downey, J. R., Frurip, D. J., McDonald, R. A., and Syverud, A. N., *JANAF Thermochemical Tables, Third Edition, J. Phys. Chem. Reference Data,* Vol. 14, Suppl. 1, 1985.
9. Gurvich, L. V., Veyts, I. V., and Alcock, C. B., *Thermodynamic Properties of Individual Substances, Fourth Edition*, Hemisphere Publishing Corp., New York, 1989.
10. Pedley, J. B., Naylor, R. D., and Kirby, S. P., *Thermochemical Data of Organic Compounds, Second Edition*, Chapman and Hall, London, 1986.
11. Stevenson, R. M., and Malanowski, S., *Handbook of the Thermodynamics of Organic Compounds*, Elsevier, New York, 1987.
12. Majer, V., and Svoboda, V., *Enthalpies of Vaporization of Organic Compounds*, Blackwell Scientific Publications, Oxford, 1985.
13. Dolmalski, E. S., Evans, W. H., and Hearing, E. D., *Heat Capacities and Entropies of Organic Compounds in the Condensed Phase, J. Phys. Chem. Reference Data,* Vol. 13, Suppl. 1, 1984; Vol. 19, No. 4, 881-1047, 1990.
14. Cox, J. D., Wagman, D. D., and Medvedev, V. A., *CODATA Key Values for Thermodynamics*, Hemisphere Publishing Corp. New York, 1989.
15. Dinsdale, A. T., "SGTE Data for Pure Elements", *CALPHAD*, 15, 317-425, 1991.

Vapor Pressure

16. *TRC Thermodynamic Tables* (Reference 5).
17. *DIPPR Data Compilation of Pure Compound Properties* (Reference 4).
18. Riddick, J. A., Bunger, W. B., and Sakano, T. K., *Organic Solvents* (Reference 3).

Viscosity and Thermal Conductivity

19. Viswanath, D. S., and Natarajan, G., *Data Book on the Viscosity of Liquids*, Hemisphere Publishing Corp., New York, 1989.
20. Liley, P. E., Makita, T., and Tanaka, Y., *Properties of Inorganic and Organic Fluids*, Hemisphere Publishing Corp., New York, 1988.
21. Ho, C. Y., Powell, R. W., and Liley, P. E., "Thermal Conductivity of the Elements", *J. Phys. Chem. Reference Data*, 2, 279-421, 1972.
22. Lide, D. R., *CRC Handbook of Chemistry and Physics* (Reference 1).
23. Riddick, J. A., Bunger, W. B., and Sakano, T. K., *Organic Solvents* (Reference 3).
24. *DIPPR Data Compilation of Pure Compound Properties* (Reference 4.).

Dipole Moment and Ionization Potential

25. *Landolt-Börnstein, Numerical Data and Functional Relationships in Science and Technology*, Group II, Vol.6, 1974; Vol. 14, Subvol. a, 1982; Vol. 14, Subvol. b, 1983; Springer-Verlag, Heidelberg.
26. Lias, S. G., Bartmess, J. E., Liebman, J. F., Holmes, J. L., Levin, R. D., and Mallard, W. G., *Gas-Phase Ion and Neutral Thermochemistry, J. Phys. Chem. Reference Data*, Vol. 17, Suppl. 1, 1988.
27. Lide, D. R., *CRC Handbook of Chemistry and Physics* (Reference 1).

Information on Chemical Hazards

28. *Threshold Limit Values for Chemical Substances and Physical Agents 1991-92*, American Conference of Governmental Industrial Hygienists, Cincinnati, OII, 1991.
29. *Fire Hazard Properties of Flammable Liquids, Gases, and Volatile Solids*, NFPA 325M, National Fire Protection Association, Quincy, MA, 1984.
30. Lewis, R. J., *Hazardous Chemicals Desk Reference, Second Edition*, Van Nostrand Reinhold, New York, 1991.
31. *List of Lists of Worldwide Hazardous Chemicals and Pollutants*, J. B. Lippincott Company, Philadelphia, 1990.
32. *Hazardous Substances Data Bank*, National Library of Medicine, Bethesda, MD.
33. Budavari, S., *The Merck Index* (Reference 2).
34. Lide, D. R., *CRC Handbook of Chemistry and Physics* (Reference 1).

Name: Acenaphthene
Synonyms: 1,2-Dihydroacenaphthylene
1,8-Ethylenenaphthalene
Mol. Form.: $C_{12}H_{10}$

CAS RN: 83-32-9
Merck No.: 23
Mol. Wt.: 154.211

PHYSICAL CONSTANTS

T_m: 93.4°C (366.5 K)
T_b: 279°C (552 K)

T_c:
P_c:

μ: 0.85 D
IP: 7.68 eV

TRANSITION PROPERTIES

$\Delta_{fus}H$ (T_m): 21.54 kJ/mol
$\Delta_{vap}H$ (T_b):
$\Delta_{vap}H$ (25°C):

Vapor pressure (0°C):
Vapor pressure (25°C):
Vapor pressure (100°C): 0.281 kPa

PROPERTIES AT 25°C AND 100 kPa

	Solid	Liquid	Gas		Solid
$\Delta_f H°$/kJ mol^{-1}:	70.3		156.0	d:	1.19 g/mL
$S°$/J mol^{-1}K^{-1}:	188.9			η:	N/A
C_p/J mol^{-1}K^{-1}:	190.4			k:	

COMMENTS: Highly toxic

Name: Acetaldehyde
Synonym: Ethanal

Mol. Form.: C_2H_4O

CAS RN: 75-07-0
Merck No.: 32
Mol. Wt.: 44.053

PHYSICAL CONSTANTS

T_m: -123°C (150 K)
T_b: 20.1°C (293.2 K)

T_c: 193°C (466 K)
P_c:

μ: 2.750 D
IP: 10.23 eV

TRANSITION PROPERTIES

$\Delta_{fus}H$ (T_m):
$\Delta_{vap}H$ (T_b): 25.76 kJ/mol
$\Delta_{vap}H$ (25°C): 25.47 kJ/mol

Vapor pressure(0°C): 44.3 kPa
Vapor pressure (25°C): 120 kPa
Vapor pressure (100°C):

PROPERTIES AT 25°C AND 100 kPa

	Solid	Liquid	Gas		Gas
$\Delta_f H°$/kJ mol^{-1}:		-191.8	-166.2	d:	1.801 g/L
$S°$/J mol^{-1}K^{-1}:		160.2	263.7	η:	
C_p/J mol^{-1}K^{-1}:		89.0	55.3	k:	

COMMENTS: TLV=100 ppm; carcinogen; highly toxic; very flammable

---

Name: Acetamide
Synonym: Ethanamide

Mol. Form.: C_2H_5NO

CAS RN: 60-35-5
Merck No.: 36
Mol. Wt.: 59.068

PHYSICAL CONSTANTS

T_m: 81°C (354 K)
T_b: 222.01°C (495.16 K)

T_c:
P_c:

μ: 3.76 D
IP: 9.65 eV

TRANSITION PROPERTIES

$\Delta_{fus}H$ (T_m):
$\Delta_{vap}H$ (T_b):
$\Delta_{vap}H$ (25°C):

Vapor pressure(0°C):
Vapor pressure (25°C):
Vapor pressure (100°C):

PROPERTIES AT 25°C AND 100 kPa

	Solid	Liquid	Gas		Solid
$\Delta_f H°$/kJ mol^{-1}:	-317.0		-238.3	d:	
$S°$/J mol^{-1}K^{-1}:	115.0			η:	N/A
C_p/J mol^{-1}K^{-1}:	91.3			k:	

COMMENTS: Carcinogen

Name: Acetic acid   CAS RN: 64-19-7
Synonym: Ethanoic acid   Merck No.: 47
                                                                                    Mol. Wt.: 60.053
Mol. Form.: $C_2H_4O_2$

PHYSICAL CONSTANTS

T_m: 16.6°C (289.7 K) T_c: 319.56°C (592.71 K)   μ: 1.70 D
T_b: 117.9°C (391.0 K) P_c: 5.786 MPa   IP: 10.66 eV

TRANSITION PROPERTIES

$\Delta_{fus}H$ (T_m): 11.54 kJ/mol   Vapor pressure(0°C):
$\Delta_{vap}H$ (T_b): 23.70 kJ/mol   Vapor pressure (25°C): 2.07 kPa
$\Delta_{vap}H$ (25°C): 23.36 kJ/mol   Vapor pressure (100°C): 57.0 kPa

PROPERTIES AT 25°C AND 100 kPa

	Solid	Liquid	Gas		Liquid
$\Delta_fH°$/kJ mol^{-1}:		-484.5	-432.8	d:	1.0439 g/mL
$S°$/J mol^{-1}K^{-1}:		159.8	282.5	η:	1.06 mPa s
C_p/J mol^{-1}K^{-1}:		123.3	66.5	k:	0.158 W/m K

COMMENTS: TLV=10 ppm

Name: Acetic anhydride   CAS RN: 108-24-7
Synonyms: Acetyl acetate   Merck No.: 48
          Ethanoic anhydride   Mol. Wt.: 102.090
Mol. Form.: $C_4H_6O_3$

PHYSICAL CONSTANTS

T_m: -73°C (200 K) T_c: 333°C (606 K)   μ: 2.8 D
T_b: 139.55°C (412.70 K) P_c: 4.0 MPa   IP: 10.00 eV

TRANSITION PROPERTIES

$\Delta_{fus}H$ (T_m):   Vapor pressure(0°C):
$\Delta_{vap}H$ (T_b): 38.20 kJ/mol   Vapor pressure (25°C): 0.680 kPa
$\Delta_{vap}H$ (25°C): 51.9 kJ/mol   Vapor pressure (100°C): 27.4 kPa

PROPERTIES AT 25°C AND 100 kPa

	Solid	Liquid	Gas		Liquid
$\Delta_fH°$/kJ mol^{-1}:		-624.4	-572.5	d:	1.077 g/mL
$S°$/J mol^{-1}K^{-1}:				η:	0.843 mPa s
C_p/J mol^{-1}K^{-1}:				k:	

COMMENTS: TLV=5 ppm

Name: Acetone   CAS RN: 67-64-1
Synonyms: 2-Propanone   Merck No.: 58
          Dimethyl ketone   Mol. Wt.: 58.080
Mol. Form.: C_3H_6O

PHYSICAL CONSTANTS

T_m: -94.8°C (178.3 K) T_c: 235.0°C (508.1 K)   μ: 2.88 D
T_b: 56.05°C (329.20 K) P_c: 4.700 MPa   IP: 9.71 eV

TRANSITION PROPERTIES

$\Delta_{fus}H$ (T_m): 5.69 kJ/mol   Vapor pressure(0°C): 9.35 kPa
$\Delta_{vap}H$ (T_b): 29.10 kJ/mol   Vapor pressure (25°C): 30.8 kPa
$\Delta_{vap}H$ (25°C): 30.99 kJ/mol   Vapor pressure (100°C):

PROPERTIES AT 25°C AND 100 kPa

	Solid	Liquid	Gas		Liquid
$\Delta_fH°$/kJ mol^{-1}:		-248.1	-217.3	d:	0.7844 g/mL
$S°$/J mol^{-1}K^{-1}:		199.8	297.6	η:	0.306 mPa s
C_p/J mol^{-1}K^{-1}:		126.3	75.0	k:	0.161 W/m K

COMMENTS: TLV=750 ppm; flammable

Name: Acetonitrile

Synonyms: Ethanenitrile
          Methyl cyanide

Mol. Form.: C_2H_3N

CAS RN: 75-05-8

Merck No.: 62

Mol. Wt.: 41.053

PHYSICAL CONSTANTS

T_m: -43.83°C (229.32 K)
T_b: 81.65°C (354.80 K)

T_c: 272.4°C (545.5 K)
P_c: 4.85 MPa

μ: 3.924 D
IP: 12.19 eV

TRANSITION PROPERTIES

$\Delta_{fus}H$ (T_m): 8.17 kJ/mol
$\Delta_{vap}H$ (T_b): 29.75 kJ/mol
$\Delta_{vap}H$ (25°C): 32.94 kJ/mol

Vapor pressure(0°C):
Vapor pressure (25°C): 11.9 kPa
Vapor pressure (100°C):

PROPERTIES AT 25°C AND 100 kPa

	Solid	Liquid	Gas		Liquid
$\Delta_fH°$/kJ mol⁻¹:		31.4	64.3		d: 0.7765 g/mL
$S°$/J mol⁻¹K⁻¹:		149.6	245.1		η: 0.344 mPa s
C_p/J mol⁻¹K⁻¹:		91.4	52.2		k: 0.188 W/m K

COMMENTS: TLV=40 ppm; highly toxic; flammable

Name: Acetophenone

Synonyms: Methyl phenyl ketone
          1-Phenylethanone

Mol. Form.: C_8H_8O

CAS RN: 98-86-2

Merck No.: 65

Mol. Wt.: 120.151

PHYSICAL CONSTANTS

T_m: 20°C (293 K)
T_b: 202°C (475 K)

T_c: 436.4°C (709.5 K)
P_c:

μ: 3.02 D
IP: 9.29 eV

TRANSITION PROPERTIES

$\Delta_{fus}H$ (T_m):
$\Delta_{vap}H$ (T_b): 38.81 kJ/mol
$\Delta_{vap}H$ (25°C): 53.39 kJ/mol

Vapor pressure(0°C):
Vapor pressure (25°C): 0.049 kPa
Vapor pressure (100°C): 3.58 kPa

PROPERTIES AT 25°C AND 100 kPa

	Solid	Liquid	Gas		Liquid
$\Delta_fH°$/kJ mol⁻¹:		-142.5	-86.7		d: 1.023 g/mL
$S°$/J mol⁻¹K⁻¹:					η: 1.68 mPa s
C_p/J mol⁻¹K⁻¹:		204.6			k:

COMMENTS:

Name: Acetyl bromide

Synonym: Ethanoyl bromide

Mol. Form.: C_2H_3BrO

CAS RN: 506-96-7

Merck No.: 76

Mol. Wt.: 122.949

PHYSICAL CONSTANTS

T_m: -96°C (177 K)
T_b: 76°C (349 K)

T_c:
P_c:

μ:
IP:

TRANSITION PROPERTIES

$\Delta_{fus}H$ (T_m):
$\Delta_{vap}H$ (T_b):
$\Delta_{vap}H$ (25°C): 33.0 kJ/mol

Vapor pressure(0°C):
Vapor pressure (25°C): 16.2 kPa
Vapor pressure (100°C):

PROPERTIES AT 25°C AND 100 kPa

	Solid	Liquid	Gas		Liquid
$\Delta_fH°$/kJ mol⁻¹:		-223.4	-190.4		d: 1.65 g/mL
$S°$/J mol⁻¹K⁻¹:					η:
C_p/J mol⁻¹K⁻¹:					k:

COMMENTS:

Name: Acetyl chloride   CAS RN: 75-36-5
Synonym: Ethanoyl chloride   Merck No.: 79
                                                        Mol. Wt.: 78.498

Mol. Form.: C_2H_3ClO

PHYSICAL CONSTANTS

T_m: -112.85°C (160.30 K) T_c: μ: 2.72 D
T_b: 50.75°C (323.90 K) P_c: IP: 10.85 eV

TRANSITION PROPERTIES

$\Delta_{fus}H$ (T_m):   Vapor pressure(0°C): 12.1 kPa
$\Delta_{vap}H$ (T_b):   Vapor pressure (25°C): 38.4 kPa
$\Delta_{vap}H$ (25°C): 30.3 kJ/mol Vapor pressure (100°C):

PROPERTIES AT 25°C AND 100 kPa

	Solid	Liquid	Gas		Liquid
$D_fH°$/kJ mol^{-1}:		-273.8	-243.5		d: 1.100 g/mL
$S°$/J mol^{-1}K^{-1}:		200.8	295.1		h: 0.368 mPa s
C_p/J mol^{-1}K^{-1}:		117.0	67.8		k:

COMMENTS: Highly toxic; flammable

Name: Acetylene   CAS RN: 74-86-2
Synonym: Ethyne   Merck No.: 84
                                                        Mol. Wt.: 26.038

Mol. Form.: C_2H_2

PHYSICAL CONSTANTS

T_t: -80.75°C (192.40 K)* T_c: 35.18°C (308.33 K)   μ: 0 D
T_s: -84.72°C (188.43 K)† P_c: 6.139 MPa   IP: 11.40 eV

TRANSITION PROPERTIES

$\Delta_{fus}H$ (T_m):   Vapor pressure(0°C):
$\Delta_{vap}H$ (T_b):   Vapor pressure (25°C):
$\Delta_{vap}H$ (25°C):   Vapor pressure (100°C): N/A

PROPERTIES AT 25°C AND 100 kPa

	Solid	Liquid	Gas		Gas
$\Delta_fH°$/kJ mol^{-1}:			228.2		d: 1.064 g/L
$S°$/J mol^{-1}K^{-1}:			200.9		η: 10.4 μPa s
C_p/J mol^{-1}K^{-1}:			43.9		k: 0.0214 W/m K

COMMENTS: Very flammable. *Triple point. †Sublimation point.

Name: Adipic acid   CAS RN: 124-04-9
Synonym: Hexanedioic acid   Merck No.: 152
                                                        Mol. Wt.: 146.143

Mol. Form.: $C_6H_{10}O_4$

PHYSICAL CONSTANTS

T_m: 153.2°C (426.3 K) T_c: μ:
T_b: 337.5°C (610.6 K) P_c: IP:

TRANSITION PROPERTIES

$\Delta_{fus}H$ (T_m): 34.85 kJ/mol Vapor pressure(0°C):
$\Delta_{vap}H$ (T_b):   Vapor pressure (25°C):
$\Delta_{vap}H$ (25°C):   Vapor pressure (100°C):

PROPERTIES AT 25°C AND 100 kPa

	Solid	Liquid	Gas		Solid
$\Delta_fH°$/kJ mol^{-1}:	-994.3		-865.0		d: 1.360 g/mL
$S°$/J mol^{-1}K^{-1}:					η: N/A
C_p/J mol^{-1}K^{-1}:					k:

COMMENTS:

Name: Adiponitrile
Synonyms: Hexanedinitrile
      1,4-Dicyanobutane
Mol. Form.: $C_6H_8N_2$

CAS RN: 111-69-3
Merck No.:
Mol. Wt.: 108.143

PHYSICAL CONSTANTS

T_m: 1°C (274 K)
T_b: 295°C (568 K)

T_c:
P_c:

μ:
IP:

TRANSITION PROPERTIES

$\Delta_{fus}H$ (T_m):
$\Delta_{vap}H$ (T_b):
$\Delta_{vap}H$ (25°C): 64.4 kJ/mol

Vapor pressure(0°C):
Vapor pressure (25°C):
Vapor pressure (100°C):

PROPERTIES AT 25°C AND 100 kPa

	Solid	Liquid	Gas		Liquid
$\Delta_f H°$/kJ mol^{-1}:		85.1	149.5		d: 0.9599 g/mL
$S°$/J mol^{-1}K^{-1}:					η:
C_p/J mol^{-1}K^{-1}:		128.7			k:

COMMENTS: Highly toxic

Name: Allene
Synonym: 1,2-Propadiene

Mol. Form.: C_3H_4

CAS RN: 463-49-0
Merck No.:
Mol. Wt.: 40.065

PHYSICAL CONSTANTS

T_m: -136.28°C (136.87 K)
T_b: -34.45°C (238.70 K)

T_c: 120°C (393 K)
P_c:

μ: 0 D
IP: 9.69 eV

TRANSITION PROPERTIES

$\Delta_{fus}H$ (T_m):
$\Delta_{vap}H$ (T_b):
$\Delta_{vap}H$ (25°C):

Vapor pressure(0°C):
Vapor pressure (25°C):
Vapor pressure (100°C):

PROPERTIES AT 25°C AND 100 kPa

	Solid	Liquid	Gas		Gas
$\Delta_f H°$/kJ mol^{-1}:			190.5		d: 1.638 g/L
$S°$/J mol^{-1}K^{-1}:					η:
C_p/J mol^{-1}K^{-1}:					k:

COMMENTS:

Name: Allyl alcohol
Synonyms: 2-Propen-1-ol
      Vinylcarbinol
Mol. Form.: C_3H_6O

CAS RN: 107-18-6
Merck No.: 284
Mol. Wt.: 58.080

PHYSICAL CONSTANTS

T_m: -129°C (144 K)
T_b: 97.08°C (370.23 K)

T_c:
P_c:

μ: 1.60 D
IP: 9.67 eV

TRANSITION PROPERTIES

$\Delta_{fus}H$ (T_m):
$\Delta_{vap}H$ (T_b): 39.96 kJ/mol
$\Delta_{vap}H$ (25°C): 47.3 kJ/mol

Vapor pressure(0°C):
Vapor pressure (25°C): 3.14 kPa
Vapor pressure (100°C): 113 kPa

PROPERTIES AT 25°C AND 100 kPa

	Solid	Liquid	Gas		Liquid
$\Delta_f H°$/kJ mol^{-1}:		-171.8	-124.5		d: 0.850 g/mL
$S°$/J mol^{-1}K^{-1}:					η: 1.22 mPa s
C_p/J mol^{-1}K^{-1}:		138.9			k:

COMMENTS: TLV=2 ppm; highly toxic; flammable

Name: Allylamine
Synonyms: 2-Propen-1-amine
         3-Amino-1-propene
Mol. Form.: C_3H_7N

CAS RN: 107-11-9
Merck No.: 285
Mol. Wt.: 57.095

PHYSICAL CONSTANTS

T_m: -88.2°C (184.9 K) T_c: μ: 1.2 D
T_b: 53.3°C (326.4 K) P_c: IP: 8.76 eV

TRANSITION PROPERTIES

$\Delta_{fus}H$ (T_m):   Vapor pressure(0°C): 9.77 kPa
$\Delta_{vap}H$ (T_b):   Vapor pressure (25°C): 33.1 kPa
$\Delta_{vap}H$ (25°C):   Vapor pressure (100°C):

PROPERTIES AT 25°C AND 100 kPa

	Solid	Liquid	Gas		Liquid
$\Delta_f H°$/kJ mol^{-1}:		-10.0		d:	0.755 g/mL
$S°$/J mol^{-1}K^{-1}:				η:	0.374 mPa s
C_p/J mol^{-1}K^{-1}:				k:	

COMMENTS: Highly toxic; flammable

Name: Allyl ethyl ether
Synonym: 3-Ethoxy-1-propene

Mol. Form.: $C_5H_{10}O$

CAS RN: 557-31-3
Merck No.: 291
Mol. Wt.: 86.134

PHYSICAL CONSTANTS

T_m: T_c: 245°C (518 K) μ:
T_b: 67.65°C (340.80 K) P_c: IP:

TRANSITION PROPERTIES

$\Delta_{fus}H$ (T_m):   Vapor pressure (0°C): 6.00 kPa
$\Delta_{vap}H$ (T_b):   Vapor pressure (25°C): 20.4 kPa
$\Delta_{vap}H$ (25°C):   Vapor pressure (100°C): 259 kPa

PROPERTIES AT 25°C AND 100 kPa

	Solid	Liquid	Gas		Liquid
$\Delta_f H°$/kJ mol^{-1}:				d:	0.761 g/mL
$S°$/J mol^{-1}K^{-1}:				η:	
C_p/J mol^{-1}K^{-1}:				k:	

COMMENTS: Very flammable

Name: Aluminum

Mol. Form.: Al

CAS RN: 7429-90-5
Merck No.: 321
Mol. Wt.: 26.982

PHYSICAL CONSTANTS

T_m: 660.32°C (933.47 K) T_c: μ:
T_b: 2519°C (2792 K) P_c: IP: 5.99 eV

TRANSITION PROPERTIES

$\Delta_{fus}H$ (T_m): 10.71 kJ/mol   Vapor pressure (0°C): N/A
$\Delta_{vap}H$ (T_b): 294.00 kJ/mol   Vapor pressure (25°C): N/A
$\Delta_{vap}H$ (25°C):   Vapor pressure (100°C): N/A

PROPERTIES AT 25°C AND 100 kPa

	Solid	Liquid	Gas		Solid
$\Delta_f H°$/kJ mol^{-1}:	0.0		330.0	d:	2.70 g/mL
$S°$/J mol^{-1}K^{-1}:	28.3		164.6	η:	N/A
C_p/J mol^{-1}K^{-1}:	24.4		21.4	k:	237 W/m K

COMMENTS:

Name: Aluminum borohydride
Synonym: Aluminum trihydride-tris(borane)

CAS RN: 16962-07-5
Merck No.: 331
Mol. Wt.: 71.510

Mol. Form.: AlB_3H_{12}

PHYSICAL CONSTANTS

T_m: -64.5°C (208.6 K)
T_b: 44.5°C (317.6 K)

T_c:
P_c:

μ:
IP:

TRANSITION PROPERTIES

$\Delta_{fus}H$ (T_m):
$\Delta_{vap}H$ (T_b): 30.00 kJ/mol
$\Delta_{vap}H$ (25°C): 29.3 kJ/mol

Vapor pressure (0°C):
Vapor pressure (25°C):
Vapor pressure (100°C):

PROPERTIES AT 25°C AND 100 kPa

	Solid	Liquid	Gas		Liquid
$\Delta_f H°$/kJ mol^{-1}:		-16.3	13.0	d:	
$S°$/J mol^{-1}K^{-1}:		289.1	379.2	η:	
C_p/J mol^{-1}K^{-1}:		194.6		k:	

COMMENTS:

Name: Aluminum oxide (Al_2O_3)
Synonyms: Dialuminum trioxide
          Alumina, corundum
Mol. Form.: Al_2O_3

CAS RN: 1344-28-1
Merck No.: 360
Mol. Wt.: 101.961

PHYSICAL CONSTANTS

T_m: 2054°C (2327 K)
T_b:

T_c:
P_c:

μ:
IP:

TRANSITION PROPERTIES

$\Delta_{fus}H$ (T_m): 111.10 kJ/mol
$\Delta_{vap}H$ (T_b):
$\Delta_{vap}H$ (25°C):

Vapor pressure (0°C): N/A
Vapor pressure (25°C): N/A
Vapor pressure (100°C): N/A

PROPERTIES AT 25°C AND 100 kPa

	Solid	Liquid	Gas		Solid
$\Delta_f H°$/kJ mol^{-1}:	-1675.7			d:	3.97 g/mL
$S°$/J mol^{-1}K^{-1}:				η:	N/A
C_p/J mol^{-1}K^{-1}:	79.0			k:	30 W/m K

COMMENTS: Data refer to corundum

Name: Aluminum tribromide
Synonym: Aluminum(III) bromide

CAS RN: 7727-15-3
Merck No.: 332
Mol. Wt.: 266.694

Mol. Form.: $AlBr_3$

PHYSICAL CONSTANTS

T_m: 97.5°C (370.6 K)
T_b: 255°C (528 K)

T_c: 490°C (763 K)
P_c: 2.89 MPa

μ: 0 D
IP: 10.40 eV

TRANSITION PROPERTIES

$\Delta_{fus}H$ (T_m): 11.25 kJ/mol
$\Delta_{vap}H$ (T_b): 23.50 kJ/mol
$\Delta_{vap}H$ (25°C):

Vapor pressure (0°C):
Vapor pressure (25°C):
Vapor pressure (100°C):

PROPERTIES AT 25°C AND 100 kPa

	Solid	Liquid	Gas		Solid
$\Delta_f H°$/kJ mol^{-1}:	-527.2		-425.1	d:	3.21 g/mL
$S°$/J mol^{-1}K^{-1}:				η:	N/A
C_p/J mol^{-1}K^{-1}:	101.7			k:	

COMMENTS:

Name: Aluminum trichloride   CAS RN: 7446-70-0
Synonym: Aluminum(III) chloride   Merck No.: 338
                                                                 Mol. Wt.: 133.340

Mol. Form.: $AlCl_3$

PHYSICAL CONSTANTS

T_m: 190°C (463 K) T_c: 347°C (620 K) μ: 0 D
T_b: P_c: 2.63 MPa   IP: 12.01 eV

TRANSITION PROPERTIES

$\Delta_{fus}H$ (T_m): 35.40 kJ/mol   Vapor pressure (0°C):
$\Delta_{vap}H$ (T_b):   Vapor pressure (25°C):
$\Delta_{vap}H$ (25°C):   Vapor pressure (100°C):

PROPERTIES AT 25°C AND 100 kPa

	Solid	Liquid	Gas		Solid
$\Delta_f H°$/kJ mol^{-1}:	-704.2		-583.2		d: 2.48 g/mL
$S°$/J mol^{-1}K^{-1}:	110.7				η: N/A
C_p/J mol^{-1}K^{-1}:	91.8				k:
COMMENTS:					

Name: Aluminum triiodide   CAS RN: 7784-23-8
Synonym: Aluminum(III) iodide   Merck No.: 348
                                                                 Mol. Wt.: 407.695

Mol. Form.: AlI_3

PHYSICAL CONSTANTS

T_m: 191°C (464 K) T_c: 710°C (983 K) μ: 0 D
T_b: 382°C (655 K) P_c: IP: 9.10 eV

TRANSITION PROPERTIES

$\Delta_{fus}H$ (T_m): 15.90 kJ/mol   Vapor pressure (0°C):
$\Delta_{vap}H$ (T_b): 32.20 kJ/mol   Vapor pressure (25°C):
$\Delta_{vap}H$ (25°C):   Vapor pressure (100°C):

PROPERTIES AT 25°C AND 100 kPa

	Solid	Liquid	Gas		Solid
$\Delta_f H°$/kJ mol^{-1}:	-313.8		-207.5		d: 3.98 g/mL
$S°$/J mol^{-1}K^{-1}:	159.0				η: N/A
C_p/J mol^{-1}K^{-1}:	98.7				k:
COMMENTS:					

Name: Ammonia   CAS RN: 7664-41-7
                                                                 Merck No.: 510
                                                                 Mol. Wt.: 17.031

Mol. Form.: H_3N

PHYSICAL CONSTANTS

T_m: -77.74°C (195.41 K) T_c: 132.4°C (405.5 K) μ: 1.471 D
T_b: -33.33°C (239.82 K) P_c: 11.35 MPa   IP: 10.16 eV

TRANSITION PROPERTIES

$\Delta_{fus}H$ (T_m): 5.66 kJ/mol   Vapor pressure (0°C): 429 kPa
$\Delta_{vap}H$ (T_b): 23.33 kJ/mol   Vapor pressure (25°C): 1003 kPa
$\Delta_{vap}H$ (25°C): 19.86 kJ/mol   Vapor pressure (100°C): 6253 kPa

PROPERTIES AT 25°C AND 100 kPa

	Solid	Liquid	Gas		Gas
$\Delta_f H°$/kJ mol^{-1}:			-45.9		d: 0.696 g/L
$S°$/J mol^{-1}K^{-1}:			192.8		η: 10.2 μPa s
C_p/J mol^{-1}K^{-1}:			35.1		k: 0.0244 W/m K
COMMENTS: TLV=25 ppm; highly toxic					

Name: Ammonium chloride
Synonym: Sal ammoniac

Mol. Form.: ClH_4N

CAS RN: 12125-02-9
Merck No.: 531
Mol. Wt.: 53.491

PHYSICAL CONSTANTS

T_m: 520°C (793 K)
T_b:

T_c: 882°C (1155 K)
P_c: 163.5 MPa

μ:
IP:

TRANSITION PROPERTIES

$\Delta_{fus}H$ (T_m):
$\Delta_{vap}H$ (T_b):
$\Delta_{vap}H$ (25°C):

Vapor pressure (0°C): N/A
Vapor pressure (25°C): N/A
Vapor pressure (100°C): N/A

PROPERTIES AT 25°C AND 100 kPa

	Solid	Liquid	Gas	Solid
$\Delta_f H°$/kJ mol^{-1}:	-314.4			d: 1.519 g/mL
$S°$/J mol^{-1}K^{-1}:	94.6			η: N/A
C_p/J mol^{-1}K^{-1}:	84.1			k:
COMMENTS:				

Name: Ammonium nitrate

Mol. Form.: $H_4N_2O_3$

CAS RN: 6484-52-2
Merck No.: 561
Mol. Wt.: 80.043

PHYSICAL CONSTANTS

T_m: 169.6°C (442.7 K)
T_b:

T_c:
P_c:

μ:
IP:

TRANSITION PROPERTIES

$\Delta_{fus}H$ (T_m): 6.40 kJ/mol
$\Delta_{vap}H$ (T_b):
$\Delta_{vap}H$ (25°C):

Vapor pressure (0°C):
Vapor pressure (25°C):
Vapor pressure (100°C):

PROPERTIES AT 25°C AND 100 kPa

	Solid	Liquid	Gas	Solid
$\Delta_f H°$/kJ mol^{-1}:	-365.6			d: 1.72 g/mL
$S°$/J mol^{-1}K^{-1}:	151.1			η: N/A
C_p/J mol^{-1}K^{-1}:	139.3			k:
COMMENTS:				

Name: Aniline
Synonyms: Benzenamine
          Phenylamine
Mol. Form.: C_6H_7N

CAS RN: 62-53-3
Merck No.: 687
Mol. Wt.: 93.128

PHYSICAL CONSTANTS

T_m: -6.02°C (267.13 K)
T_b: 184.17°C (457.32 K)

T_c: 426°C (699 K)
P_c: 4.89 MPa

μ: 1.13 D
IP: 7.72 eV

TRANSITION PROPERTIES

$\Delta_{fus}H$ (T_m): 10.56 kJ/mol
$\Delta_{vap}H$ (T_b): 42.44 kJ/mol
$\Delta_{vap}H$ (25°C): 55.83 kJ/mol

Vapor pressure (0°C):
Vapor pressure (25°C): 0.090 kPa
Vapor pressure (100°C): 6.10 kPa

PROPERTIES AT 25°C AND 100 kPa

	Solid	Liquid	Gas	Liquid
$\Delta_f H°$/kJ mol^{-1}:		31.3	87.5	d: 1.0175 g/mL
$S°$/J mol^{-1}K^{-1}:		191.0	317.9	η: 3.85 mPa s
C_p/J mol^{-1}K^{-1}:		191.9	107.9	k:
COMMENTS: TLV=2 ppm; carcinogen; highly toxic				

Name: Anisole
Synonyms: Methyl phenyl ether
          Methoxybenzene
Mol. Form.: C_7H_8O

CAS RN: 100-66-3
Merck No.: 699
Mol. Wt.: 108.140

PHYSICAL CONSTANTS

T_m: -37.5°C (235.6 K)
T_b: 153.7°C (426.8 K)

T_c: 372.5°C (645.6 K)
P_c: 4.25 MPa

μ: 1.38 D
IP: 8.21 eV

TRANSITION PROPERTIES

$\Delta_{fus}H$ (T_m):
$\Delta_{vap}H$ (T_b): 38.97 kJ/mol
$\Delta_{vap}H$ (25°C): 46.90 kJ/mol

Vapor pressure (0°C):
Vapor pressure (25°C): 0.472 kPa
Vapor pressure (100°C): 18.5 kPa

PROPERTIES AT 25°C AND 100 kPa

	Solid	Liquid	Gas	Liquid
$\Delta_f H°$/kJ mol^{-1}:		-114.8	-67.9	d: 0.9893 g/mL
$S°$/J mol^{-1}K^{-1}:				η: 1.06 mPa s
C_p/J mol^{-1}K^{-1}:		199.0		k: 0.156 W/m K
COMMENTS:				

Name: Anthracene

CAS RN: 120-12-7
Merck No.: 712
Mol. Wt.: 178.233

Mol. Form.: $C_{14}H_{10}$

PHYSICAL CONSTANTS

T_m: 215.0°C (488.1 K)
T_b: 339.95°C (613.10 K)

T_c: 596.2°C (869.3 K)
P_c:

μ: 0 D
IP: 7.45 eV

TRANSITION PROPERTIES

$\Delta_{fus}H$ (T_m): 28.83 kJ/mol
$\Delta_{vap}H$ (T_b):
$\Delta_{vap}H$ (25°C):

Vapor pressure (0°C):
Vapor pressure (25°C):
Vapor pressure (100°C):

PROPERTIES AT 25°C AND 100 kPa

	Solid	Liquid	Gas	Solid
$\Delta_f H°$/kJ mol^{-1}:	129.2		230.9	d: 1.28 g/mL
$S°$/J mol^{-1}K^{-1}:	207.5			η: N/A
C_p/J mol^{-1}K^{-1}:	210.5			k:
COMMENTS: Highly toxic				

Name: Antimony

CAS RN: 7440-36-0
Merck No.: 724
Mol. Wt.: 121.757

Mol. Form.: Sb

PHYSICAL CONSTANTS

T_m: 630.63°C (903.78 K)
T_b: 1587°C (1860 K)

T_c:
P_c:

μ:
IP: 8.64 eV

TRANSITION PROPERTIES

$\Delta_{fus}H$ (T_m): 19.87 kJ/mol
$\Delta_{vap}H$ (T_b):
$\Delta_{vap}H$ (25°C):

Vapor pressure (0°C): N/A
Vapor pressure (25°C): N/A
Vapor pressure (100°C): N/A

PROPERTIES AT 25°C AND 100 kPa

	Solid	Liquid	Gas	Solid
$\Delta_f H°$/kJ mol^{-1}:	0.0		262.3	d: 6.68 g/mL
$S°$/J mol^{-1}K^{-1}:	45.7		180.3	η: N/A
C_p/J mol^{-1}K^{-1}:	25.2		20.8	k: 24.3 W/m K
COMMENTS:				

Name: Antimony tribromide

Synonym: Antimony(III) bromide

CAS RN: 7789-61-9

Merck No.: 738

Mol. Wt.: 361.469

Mol. Form.: Br_3Sb

PHYSICAL CONSTANTS

T_m: 96.6°C (369.7 K)

T_b: 280°C (553 K)

T_c: 631°C (904 K)

P_c:

μ:

IP:

TRANSITION PROPERTIES

$\Delta_{fus}H$ (T_m):

$\Delta_{vap}H$ (T_b): 59.00 kJ/mol

$\Delta_{vap}H$ (25°C):

Vapor pressure (0°C):

Vapor pressure (25°C):

Vapor pressure (100°C):

PROPERTIES AT 25°C AND 100 kPa

	Solid	Liquid	Gas		Solid
$\Delta_f H°$/kJ mol^{-1}:	-259.4		-194.6	d:	4.35 g/mL
$S°$/J mol^{-1}K^{-1}:	207.1		372.9	η:	N/A
C_p/J mol^{-1}K^{-1}:			80.2	k:	
COMMENTS:					

Name: Antimony trichloride

Synonym: Antimony(III) chloride

CAS RN: 10025-91-9

Merck No.: 740

Mol. Wt.: 228.115

Mol. Form.: Cl_3Sb

PHYSICAL CONSTANTS

T_m: 73.4°C (346.5 K)

T_b: 220.3°C (493.4 K)

T_c: 521°C (794 K)

P_c:

μ:

IP: 10.10 eV

TRANSITION PROPERTIES

$\Delta_{fus}H$ (T_m): 12.70 kJ/mol

$\Delta_{vap}H$ (T_b): 45.19 kJ/mol

$\Delta_{vap}H$ (25°C):

Vapor pressure (0°C):

Vapor pressure (25°C):

Vapor pressure (100°C):

PROPERTIES AT 25°C AND 100 kPa

	Solid	Liquid	Gas		Solid
$\Delta_f H°$/kJ mol^{-1}:	-382.2			d:	3.14 g/mL
$S°$/J mol^{-1}K^{-1}:	184.1			η:	N/A
C_p/J mol^{-1}K^{-1}:	107.9			k:	
COMMENTS:					

Name: Antimony trifluoride

Synonym: Antimony(III) fluoride

CAS RN: 7783-56-4

Merck No.: 741

Mol. Wt.: 178.752

Mol. Form.: F_3Sb

PHYSICAL CONSTANTS

T_m: 292°C (565 K)

T_b: 376°C (649 K)

T_c:

P_c:

μ:

IP:

TRANSITION PROPERTIES

$\Delta_{fus}H$ (T_m):

$\Delta_{vap}H$ (T_b):

$\Delta_{vap}H$ (25°C):

Vapor pressure (0°C): N/A

Vapor pressure (25°C): N/A

Vapor pressure (100°C): N/A

PROPERTIES AT 25°C AND 100 kPa

	Solid	Liquid	Gas		Solid
$\Delta_f H°$/kJ mol^{-1}:	-915.5			d:	4.38 g/mL
$S°$/J mol^{-1}K^{-1}:				η:	N/A
C_p/J mol^{-1}K^{-1}:				k:	
COMMENTS: Highly toxic					

Name: Antimony triiodide   CAS RN: 7790-44-5
Synonym: Antimony(III) iodide   Merck No.: 742
                                                                            Mol. Wt.: 502.470
Mol. Form.: I_3Sb

PHYSICAL CONSTANTS

T_m: 168°C (441 K) T_c: 829°C (1102 K)   μ:
T_b: 401°C (674 K) P_c: IP:

TRANSITION PROPERTIES

$\Delta_{fus}H$ (T_m):   Vapor pressure (0°C):
$\Delta_{vap}H$ (T_b): 68.60 kJ/mol   Vapor pressure (25°C):
$\Delta_{vap}H$ (25°C):   Vapor pressure (100°C):

PROPERTIES AT 25°C AND 100 kPa

	Solid	Liquid	Gas		Solid
$\Delta_fH°$/kJ mol^{-1}:	-100.4			*d*: 4.92 g/mL	
$S°$/J mol^{-1}K^{-1}:				η: N/A	
C_p/J mol^{-1}K^{-1}:				*k*:	
COMMENTS:					

Name: Argon   CAS RN: 7440-37-1
                                                                            Merck No.: 808
                                                                            Mol. Wt.: 39.948
Mol. Form.: Ar

PHYSICAL CONSTANTS

T_m: -189.35°C (83.80 K) T_c: -122.28°C (150.87 K)   μ: 0 D
T_b: -185.85°C (87.30 K) P_c: 4.898 MPa   IP: 15.76 eV

TRANSITION PROPERTIES

$\Delta_{fus}H$ (T_m): 1.12 kJ/mol   Vapor pressure (0°C): N/A
$\Delta_{vap}H$ (T_b): 6.43 kJ/mol   Vapor pressure (25°C): N/A
$\Delta_{vap}H$ (25°C):   Vapor pressure (100°C): N/A

PROPERTIES AT 25°C AND 100 kPa

	Solid	Liquid	Gas		Gas
$\Delta_fH°$/kJ mol^{-1}:			0.0	*d*: 1.633 g/L	
$S°$/J mol^{-1}K^{-1}:			154.8	η: 22.7 μPa s	
C_p/J mol^{-1}K^{-1}:			20.8	*k*: 0.0179 W/m K	
COMMENTS:					

Name: Arsenic   CAS RN: 7440-38-2
                                                                            Merck No.: 820
                                                                            Mol. Wt.: 74.922
Mol. Form.: As

PHYSICAL CONSTANTS

T_t: 817°C (1090 K)* T_c: 1400°C (1673 K)   μ:
T_s: 614°C (887 K)† P_c: IP: 9.82 eV

TRANSITION PROPERTIES

$\Delta_{fus}H$ (T_m): 24.44 kJ/mol   Vapor pressure (0°C): N/A
$\Delta_{vap}H$ (T_b):   Vapor pressure (25°C): N/A
$\Delta_{vap}H$ (25°C):   Vapor pressure (100°C): N/A

PROPERTIES AT 25°C AND 100 kPa

	Solid	Liquid	Gas		Solid
$\Delta_fH°$/kJ mol^{-1}:	0.0		302.5	*d*: 5.78 g/mL	
$S°$/J mol^{-1}K^{-1}:	35.1		174.2	η: N/A	
C_p/J mol^{-1}K^{-1}:	24.6		20.8	*k*: 50.2 W/m K	
COMMENTS: Carcinogen; highly toxic. *Triple point. †Sublimation point.					

Name: Arsenic pentafluoride

Synonym: Arsenic(V) fluoride

CAS RN: 7784-36-3

Merck No.: 824

Mol. Wt.: 169.914

Mol. Form.: AsF_5

PHYSICAL CONSTANTS

T_m: -79.8°C (193.3 K)	T_c:	μ:
T_b: -53.2°C (219.9 K)	P_c:	IP:

TRANSITION PROPERTIES

$\Delta_{fus}H$ (T_m):

$\Delta_{vap}H$ (T_b): 20.80 kJ/mol

$\Delta_{vap}H$ (25°C):

Vapor pressure (0°C):

Vapor pressure (25°C):

Vapor pressure (100°C):

PROPERTIES AT 25°C AND 100 kPa

	Solid	Liquid	Gas		Gas
$\Delta_f H°$/kJ mol^{-1}:				d: 6.945 g/L
$S°$/J mol^{-1}K^{-1}:				η:
C_p/J mol^{-1}K^{-1}:				k:

COMMENTS: Carcinogen; highly toxic

Name: Arsenic pentoxide (As_2O_5)

Synonym: Arsenic(V) oxide

CAS RN: 1303-28-2

Merck No.: 827

Mol. Wt.: 229.840

Mol. Form.: As_2O_5

PHYSICAL CONSTANTS

T_m:	T_c:	μ:
T_b:	P_c:	IP:

TRANSITION PROPERTIES

$\Delta_{fus}H$ (T_m):

$\Delta_{vap}H$ (T_b):

$\Delta_{vap}H$ (25°C):

Vapor pressure (0°C):

Vapor pressure (25°C):

Vapor pressure (100°C):

PROPERTIES AT 25°C AND 100 kPa

	Solid	Liquid	Gas		Solid
$\Delta_f H°$/kJ mol^{-1}:	-924.9			d:
$S°$/J mol^{-1}K^{-1}:	105.4			η: N/A
C_p/J mol^{-1}K^{-1}:	116.5			k:

COMMENTS: Carcinogen; highly toxic

Name: Arsenic trichloride

Synonym: Arsenic(III) chloride

CAS RN: 7784-34-1

Merck No.: 829

Mol. Wt.: 181.280

Mol. Form.: $AsCl_3$

PHYSICAL CONSTANTS

T_m: -16°C (257 K)	T_c: 381°C (654 K)	μ: 1.59 D
T_b: 130°C (403 K)	P_c:	IP: 10.55 eV

TRANSITION PROPERTIES

$\Delta_{fus}H$ (T_m): 10.10 kJ/mol

$\Delta_{vap}H$ (T_b): 35.01 kJ/mol

$\Delta_{vap}H$ (25°C): 43.5 kJ/mol

Vapor pressure (0°C):

Vapor pressure (25°C): 1.29 kPa

Vapor pressure (100°C): 41.6 kPa

PROPERTIES AT 25°C AND 100 kPa

	Solid	Liquid	Gas		Liquid
$\Delta_f H°$/kJ mol^{-1}:		-305.0	-261.5		d: 2.150 g/mL
$S°$/J mol^{-1}K^{-1}:		216.3	327.2		η:
C_p/J mol^{-1}K^{-1}:			75.7		k:

COMMENTS: Carcinogen; highly toxic

Name: Arsenic trifluoride
Synonym: Arsenic(III) fluoride

Mol. Form.: AsF$_3$

CAS RN: 7784-35-2
Merck No.: 830
Mol. Wt.: 131.917

PHYSICAL CONSTANTS

T_m: -5.9°C (267.2 K)	T_c:	μ: 2.59 D
T_b: 57.8°C (330.9 K)	P_c:	IP: 12.84 eV

TRANSITION PROPERTIES

$\Delta_{fus}H$ (T_m): 10.40 kJ/mol
$\Delta_{vap}H$ (T_b): 29.70 kJ/mol
$\Delta_{vap}H$ (25°C): 35.5 kJ/mol

Vapor pressure (0°C):
Vapor pressure (25°C):
Vapor pressure (100°C):

PROPERTIES AT 25°C AND 100 kPa

	Solid	Liquid	Gas		Liquid
$\Delta_f H°$/kJ mol^{-1}:		-821.3	-785.8	d:	2.7 g/mL
$S°$/J mol^{-1}K^{-1}:		181.2	289.1	η:	
C_p/J mol^{-1}K^{-1}:		126.6	65.6	k:	

COMMENTS: Carcinogen; highly toxic

Name: Arsine
Synonym: Arsane

Mol. Form.: AsH$_3$

CAS RN: 7784-42-1
Merck No.: 837
Mol. Wt.: 77.945

PHYSICAL CONSTANTS

T_m: -116°C (157 K)	T_c: 100.0°C (373.1 K)	μ: 0.20 D
T_b: -62.5°C (210.6 K)	P_c:	IP: 9.89 eV

TRANSITION PROPERTIES

$\Delta_{fus}H$ (T_m):
$\Delta_{vap}H$ (T_b): 16.69 kJ/mol
$\Delta_{vap}H$ (25°C):

Vapor pressure (0°C):
Vapor pressure (25°C):
Vapor pressure (100°C):

PROPERTIES AT 25°C AND 100 kPa

	Solid	Liquid	Gas		Gas
$\Delta_f H°$/kJ mol^{-1}:			66.4	d:	3.186 g/L
$S°$/J mol^{-1}K^{-1}:			222.8	η:	
C_p/J mol^{-1}K^{-1}:			38.1	k:	

COMMENTS: TLV=0.05 ppm; carcinogen; highly toxic

Name: Azobenzene
Synonym: Diphenyldiazene

Mol. Form.: C$_{12}$H$_{10}$N$_2$

CAS RN: 103-33-3
Merck No.: 930
Mol. Wt.: 182.225

PHYSICAL CONSTANTS

T_m: 67.1°C (340.2 K)	T_c:	μ:
T_b: 293°C (566 K)	P_c:	IP:

TRANSITION PROPERTIES

$\Delta_{fus}H$ (T_m): 22.04 kJ/mol
$\Delta_{vap}H$ (T_b):
$\Delta_{vap}H$ (25°C):

Vapor pressure (0°C):
Vapor pressure (25°C):
Vapor pressure (100°C):

PROPERTIES AT 25°C AND 100 kPa

	Solid	Liquid	Gas		Solid
$\Delta_f H°$/kJ mol^{-1}:				d:	1.2 g/mL
$S°$/J mol^{-1}K^{-1}:				η:	N/A
C_p/J mol^{-1}K^{-1}:				k:	

COMMENTS:

Name: Azoxybenzene
Synonym: Diphenyldiazene, 1-oxide

Mol. Form.: $C_{12}H_{10}N_2O$

CAS RN: 495-48-7
Merck No.: 937
Mol. Wt.: 198.224

PHYSICAL CONSTANTS

T_m: 36°C (309 K) T_c: μ:
T_b: P_c: IP: 8.10 eV

TRANSITION PROPERTIES

$\Delta_{fus}H$ (T_m): 17.93 kJ/mol Vapor pressure (0°C):
$\Delta_{vap}H$ (T_b): Vapor pressure (25°C):
$\Delta_{vap}H$ (25°C): Vapor pressure (100°C):

PROPERTIES AT 25°C AND 100 kPa

	Solid	Liquid	Gas		Solid
$\Delta_f H°$/kJ mol^{-1}:				d:	1.159 g/mL
$S°$/J mol^{-1}K^{-1}:				η:	N/A
C_p/J mol^{-1}K^{-1}:				k:	
COMMENTS:					

Name: Azulene
Synonym: Bicyclo[5.3.0]decapentaene

Mol. Form.: $C_{10}H_8$

CAS RN: 275-51-4
Merck No.: 939
Mol. Wt.: 128.174

PHYSICAL CONSTANTS

T_m: 99°C (372 K) T_c: μ: 0.80 D
T_b: P_c: IP: 7.41 eV

TRANSITION PROPERTIES

$\Delta_{fus}H$ (T_m): Vapor pressure (0°C):
$\Delta_{vap}H$ (T_b): Vapor pressure (25°C):
$\Delta_{vap}H$ (25°C): Vapor pressure (100°C):

PROPERTIES AT 25°C AND 100 kPa

	Solid	Liquid	Gas		Solid
$\Delta_f H°$/kJ mol^{-1}:	212.3		289.1	d:	1.175 g/mL
$S°$/J mol^{-1}K^{-1}:				η:	N/A
C_p/J mol^{-1}K^{-1}:				k:	
COMMENTS:					

Name: Barium

Mol. Form.: Ba

CAS RN: 7440-39-3
Merck No.: 974
Mol. Wt.: 137.327

PHYSICAL CONSTANTS

T_m: 727°C (1000 K) T_c: μ:
T_b: 1897°C (2170 K) P_c: IP: 5.21 eV

TRANSITION PROPERTIES

$\Delta_{fus}H$ (T_m): 7.12 kJ/mol Vapor pressure (0°C): N/A
$\Delta_{vap}H$ (T_b): 140.00 kJ/mol Vapor pressure (25°C): N/A
$\Delta_{vap}H$ (25°C): Vapor pressure (100°C): N/A

PROPERTIES AT 25°C AND 100 kPa

	Solid	Liquid	Gas		Solid
$\Delta_f H°$/kJ mol^{-1}:	0.0		180.0	d:	3.62 g/mL
$S°$/J mol^{-1}K^{-1}:	62.8		170.2	η:	N/A
C_p/J mol^{-1}K^{-1}:	28.1		20.8	k:	18.4 W/m K
COMMENTS: Highly toxic					

Name: Barium chloride   CAS RN: 10361-37-2
Synonym: Barium dichloride   Merck No.: 981
                                                                        Mol. Wt.: 208.232
Mol. Form.: BaCl$_2$

PHYSICAL CONSTANTS

T_m: 962°C (1235 K) T_c: μ:
T_b: P_c: IP:

TRANSITION PROPERTIES

$\Delta_{fus}H$ (T_m): 16.00 kJ/mol   Vapor pressure (0°C): N/A
$\Delta_{vap}H$ (T_b):   Vapor pressure (25°C): N/A
$\Delta_{vap}H$ (25°C):   Vapor pressure (100°C): N/A

PROPERTIES AT 25°C AND 100 kPa

	Solid	Liquid	Gas		Solid
$\Delta_f H°$/kJ mol^{-1}:	-858.6			d:	3.9 g/mL
$S°$/J mol^{-1}K^{-1}:	123.7			η:	N/A
C_p/J mol^{-1}K^{-1}:	75.1			k:	

COMMENTS: Highly toxic

Name: Barium fluoride   CAS RN: 7787-32-8
Synonym: Barium difluoride   Merck No.: 986
                                                                        Mol. Wt.: 175.324
Mol. Form.: BaF$_2$

PHYSICAL CONSTANTS

T_m: 1368°C (1641 K) T_c: μ:
T_b: P_c: IP:

TRANSITION PROPERTIES

$\Delta_{fus}H$ (T_m): 23.36 kJ/mol   Vapor pressure (0°C): N/A
$\Delta_{vap}H$ (T_b):   Vapor pressure (25°C): N/A
$\Delta_{vap}H$ (25°C):   Vapor pressure (100°C): N/A

PROPERTIES AT 25°C AND 100 kPa

	Solid	Liquid	Gas		Solid
$\Delta_f H°$/kJ mol^{-1}:	-1207.1			d:	4.893 g/mL
$S°$/J mol^{-1}K^{-1}:	96.4			η:	N/A
C_p/J mol^{-1}K^{-1}:	71.2			k:	

COMMENTS: Highly toxic

Name: Barium sulfate   CAS RN: 7727-43-7
Synonym: Barite   Merck No.: 1006
                                                                        Mol. Wt.: 233.391
Mol. Form.: BaO$_4$S

PHYSICAL CONSTANTS

T_m: 1350°C (1623 K) T_c: μ:
T_b: P_c: IP:

TRANSITION PROPERTIES

$\Delta_{fus}H$ (T_m): 40.60 kJ/mol   Vapor pressure (0°C): N/A
$\Delta_{vap}H$ (T_b):   Vapor pressure (25°C): N/A
$\Delta_{vap}H$ (25°C):   Vapor pressure (100°C): N/A

PROPERTIES AT 25°C AND 100 kPa

	Solid	Liquid	Gas		Solid
$\Delta_f H°$/kJ mol^{-1}:	-1473.2			d:	4.49 g/mL
$S°$/J mol^{-1}K^{-1}:	132.2			η:	N/A
C_p/J mol^{-1}K^{-1}:	101.8			k:	

COMMENTS: Highly toxic

Name: Benzaldehyde
Synonym: Benzenecarboxaldehyde

Mol. Form.: C_7H_6O

CAS RN: 100-52-7
Merck No.: 1065
Mol. Wt.: 106.124

PHYSICAL CONSTANTS

T_m: -26°C (247 K)
T_b: 179.05°C (452.20 K)

T_c: 422°C (695 K)
P_c: 4.65 MPa

μ:
IP: 9.49 eV

TRANSITION PROPERTIES

$\Delta_{fus}H$ (T_m): 9.32 kJ/mol
$\Delta_{vap}H$ (T_b): 42.50 kJ/mol
$\Delta_{vap}H$ (25°C): 50.30 kJ/mol

Vapor pressure (0°C):
Vapor pressure (25°C): 0.169 kPa
Vapor pressure (100°C): 8.37 kPa

PROPERTIES AT 25°C AND 100 kPa

	Solid	Liquid	Gas		Liquid
$\Delta_f H°$/kJ mol^{-1}:		-87.0	-36.7		d: 1.044 g/mL
$S°$/J mol^{-1}K^{-1}:		221.2			η: 1.32 mPa s
C_p/J mol^{-1}K^{-1}:		172.0			k: 0.151 W/m K

COMMENTS:

Name: Benzamide
Synonym: Benzoic acid amide

Mol. Form.: C_7H_7NO

CAS RN: 55-21-0
Merck No.: 1067
Mol. Wt.: 121.139

PHYSICAL CONSTANTS

T_m: 129.1°C (402.2 K)
T_b: 290°C (563 K)

T_c:
P_c:

μ:
IP: 9.45 eV

TRANSITION PROPERTIES

$\Delta_{fus}H$ (T_m): 18.49 kJ/mol
$\Delta_{vap}H$ (T_b):
$\Delta_{vap}H$ (25°C):

Vapor pressure (0°C):
Vapor pressure (25°C):
Vapor pressure (100°C):

PROPERTIES AT 25°C AND 100 kPa

	Solid	Liquid	Gas		Solid
$\Delta_f H°$/kJ mol^{-1}:	-202.6				d: 1.28 g/mL
$S°$/J mol^{-1}K^{-1}:					η: N/A
C_p/J mol^{-1}K^{-1}:					k:

COMMENTS: Highly toxic

Name: Benzene
Synonym: [6]Annulene

Mol. Form.: C_6H_6

CAS RN: 71-43-2
Merck No.: 1074
Mol. Wt.: 78.114

PHYSICAL CONSTANTS

T_m: 5.53°C (278.68 K)
T_b: 80.09°C (353.24 K)

T_c: 289.01°C (562.16 K)
P_c: 4.898 MPa

μ: 0 D
IP: 9.25 eV

TRANSITION PROPERTIES

$\Delta_{fus}H$ (T_m): 9.95 kJ/mol
$\Delta_{vap}H$ (T_b): 30.72 kJ/mol
$\Delta_{vap}H$ (25°C): 33.83 kJ/mol

Vapor pressure (0°C): 3.29 kPa
Vapor pressure (25°C): 12.7 kPa
Vapor pressure (100°C): 179 kPa

PROPERTIES AT 25°C AND 100 kPa

	Solid	Liquid	Gas		Liquid
$\Delta_f H°$/kJ mol^{-1}:		49.0	82.8		d: 0.8736 g/mL
$S°$/J mol^{-1}K^{-1}:			269.2		η: 0.604 mPa s
C_p/J mol^{-1}K^{-1}:		136.3	82.4		k: 0.1411 W/m K

COMMENTS: TLV=0.1 ppm; carcinogen; highly toxic; flammable

Name: Benzenethiol
Synonym: Phenyl mercaptan

Mol. Form.: C_6H_6S

CAS RN: 108-98-5
Merck No.: 9285
Mol. Wt.: 110.180

PHYSICAL CONSTANTS

T_m: -14.9°C (258.2 K)	T_c:	μ:
T_b: 169.1°C (442.2 K)	P_c:	IP: 8.30 eV

TRANSITION PROPERTIES

$\Delta_{fus}H$ (T_m): 11.48 kJ/mol
$\Delta_{vap}H$ (T_b): 39.93 kJ/mol
$\Delta_{vap}H$ (25°C): 47.56 kJ/mol

Vapor pressure (0°C):
Vapor pressure (25°C): 0.397 kPa
Vapor pressure (100°C): 11.6 kPa

PROPERTIES AT 25°C AND 100 kPa

	Solid	Liquid	Gas		Liquid
$\Delta_f H°$/kJ mol^{-1}:		63.7	112.4	d:	1.0727 g/mL
$S°$/J mol^{-1}K^{-1}:		222.8		η:	1.14 mPa s
C_p/J mol^{-1}K^{-1}:		173.2		k:	

COMMENTS: TLV=0.5 ppm; highly toxic

Name: 1,2,3-Benzenetriol
Synonym: Pyrogallol

Mol. Form.: $C_6H_6O_3$

CAS RN: 87-66-1
Merck No.: 8010
Mol. Wt.: 126.112

PHYSICAL CONSTANTS

T_m: 133°C (406 K)	T_c:	μ:
T_b: 309°C (582 K)	P_c:	IP:

TRANSITION PROPERTIES

$\Delta_{fus}H$ (T_m):
$\Delta_{vap}H$ (T_b):
$\Delta_{vap}H$ (25°C):

Vapor pressure (0°C):
Vapor pressure (25°C):
Vapor pressure (100°C):

PROPERTIES AT 25°C AND 100 kPa

	Solid	Liquid	Gas		Solid
$\Delta_f H°$/kJ mol^{-1}:				d:	
$S°$/J mol^{-1}K^{-1}:				η:	N/A
C_p/J mol^{-1}K^{-1}:				k:	

COMMENTS:

Name: *p*-Benzidine
Synonym: 4,4′-Biphenyldiamine

Mol. Form.: $C_{12}H_{12}N_2$

CAS RN: 92-87-5
Merck No.: 1086
Mol. Wt.: 184.241

PHYSICAL CONSTANTS

T_m: 120°C (393 K)	T_c:	μ:
T_b:	P_c:	IP:

TRANSITION PROPERTIES

$\Delta_{fus}H$ (T_m):
$\Delta_{vap}H$ (T_b):
$\Delta_{vap}H$ (25°C):

Vapor pressure (0°C):
Vapor pressure (25°C):
Vapor pressure (100°C):

PROPERTIES AT 25°C AND 100 kPa

	Solid	Liquid	Gas		Solid
$\Delta_f H°$/kJ mol^{-1}:	70.7			d:	
$S°$/J mol^{-1}K^{-1}:				η:	N/A
C_p/J mol^{-1}K^{-1}:				k:	

COMMENTS: Carcinogen; highly toxic

Name: Benzil
Synonyms: Diphenylethanedione
       Dibenzoyl
Mol. Form.: $C_{14}H_{10}O_2$

CAS RN: 134-81-6
Merck No.: 1087
Mol. Wt.: 210.232

PHYSICAL CONSTANTS

T_m: 94.86°C (368.01 K) T_c: μ:
T_b: 347°C (620 K) P_c: IP:

TRANSITION PROPERTIES

$\Delta_{fus}H$ (T_m): 23.54 kJ/mol   Vapor pressure (0°C):
$\Delta_{vap}H$ (T_b):   Vapor pressure (25°C):
$\Delta_{vap}H$ (25°C):   Vapor pressure (100°C):

PROPERTIES AT 25°C AND 100 kPa

	Solid	Liquid	Gas		Solid
$\Delta_f H°$/kJ mol^{-1}:	-153.9		-55.5		d: 1.22 g/mL
$S°$/J mol^{-1}K^{-1}:					η: N/A
C_p/J mol^{-1}K^{-1}:					k:
COMMENTS:					

Name: Benzoic acid
Synonym: Benzenecarboxylic acid

Mol. Form.: $C_7H_6O_2$

CAS RN: 65-85-0
Merck No.: 1101
Mol. Wt.: 122.123

PHYSICAL CONSTANTS

T_m: 122.4°C (395.5 K) T_c: μ:
T_b: 249.2°C (522.3 K) P_c: IP: 9.47 eV

TRANSITION PROPERTIES

$\Delta_{fus}H$ (T_m): 18.06 kJ/mol   Vapor pressure (0°C):
$\Delta_{vap}H$ (T_b):   Vapor pressure (25°C):
$\Delta_{vap}H$ (25°C):   Vapor pressure (100°C):

PROPERTIES AT 25°C AND 100 kPa

	Solid	Liquid	Gas		Solid
$\Delta_f H°$/kJ mol^{-1}:	-385.2		-294.1		d: 1.322 g/mL
$S°$/J mol^{-1}K^{-1}:	167.6				η: N/A
C_p/J mol^{-1}K^{-1}:	146.8				k:
COMMENTS:					

Name: Benzonitrile
Synonyms: Phenyl cyanide
      Cyanobenzene
Mol. Form.: C_7H_5N

CAS RN: 100-47-0
Merck No.: 1107
Mol. Wt.: 103.123

PHYSICAL CONSTANTS

T_m: -12.75°C (260.40 K) T_c: 426.3°C (699.4 K)   μ: 4.18 D
T_b: 191.15°C (464.30 K) P_c: 4.21 MPa   IP: 9.62 eV

TRANSITION PROPERTIES

$\Delta_{fus}H$ (T_m): 10.88 kJ/mol   Vapor pressure (0°C):
$\Delta_{vap}H$ (T_b): 45.94 kJ/mol   Vapor pressure (25°C): 0.110 kPa
$\Delta_{vap}H$ (25°C): 55.48 kJ/mol   Vapor pressure (100°C): 5.41 kPa

PROPERTIES AT 25°C AND 100 kPa

	Solid	Liquid	Gas		Liquid
$\Delta_f H°$/kJ mol^{-1}:		163.2	215.7		d: 1.00 g/mL
$S°$/J mol^{-1}K^{-1}:		209.1			η: 1.27 mPa s
C_p/J mol^{-1}K^{-1}:		165.2			k:
COMMENTS: Highly toxic					

Name: Benzophenone
Synonyms: Diphenyl ketone
        Diphenylmethanone
Mol. Form.: $C_{13}H_{10}O$

CAS RN: 119-61-9
Merck No.: 1108
Mol. Wt.: 182.222

PHYSICAL CONSTANTS

T_m: 47.88°C (321.03 K) T_c: μ:
T_b: 305.4°C (578.5 K) P_c: IP: 9.05 eV

TRANSITION PROPERTIES

$\Delta_{fus}H$ (T_m): 18.19 kJ/mol   Vapor pressure (0°C):
$\Delta_{vap}H$ (T_b):   Vapor pressure (25°C):
$\Delta_{vap}H$ (25°C):   Vapor pressure (100°C):

PROPERTIES AT 25°C AND 100 kPa

	Solid	Liquid	Gas	Solid
$\Delta_f H°$/kJ mol^{-1}:	-34.5		54.9	d: 1.103 g/mL
$S°$/J mol^{-1}K^{-1}:				η: N/A
C_p/J mol^{-1}K^{-1}:	224.8			k:
COMMENTS:				

Name: 1,4-Benzoquinone
Synonyms: 2,5-Cyclohexadiene-1,4-dione
        p-Quinone
Mol. Form.: $C_6H_4O_2$

CAS RN: 106-51-4
Merck No.: 8103
Mol. Wt.: 108.097

PHYSICAL CONSTANTS

T_m: 115.7°C (388.8 K) T_c: μ: 0 D
T_b: P_c: IP: 10.04 eV

TRANSITION PROPERTIES

$\Delta_{fus}H$ (T_m): 18.53 kJ/mol   Vapor pressure (0°C):
$\Delta_{vap}H$ (T_b):   Vapor pressure (25°C):
$\Delta_{vap}H$ (25°C):   Vapor pressure (100°C):

PROPERTIES AT 25°C AND 100 kPa

	Solid	Liquid	Gas	Solid
$\Delta_f H°$/kJ mol^{-1}:	-185.7		-122.9	d: 1.31 g/mL
$S°$/J mol^{-1}K^{-1}:				η: N/A
C_p/J mol^{-1}K^{-1}:	129.0			k:
COMMENTS: TLV=0.1 ppm; highly toxic				

Name: Benzoyl chloride
Synonym: Benzenecarbonyl chloride

Mol. Form.: C_7H_5ClO

CAS RN: 98-88-4
Merck No.: 1124
Mol. Wt.: 140.569

PHYSICAL CONSTANTS

T_m: -1.0°C (272.1 K) T_c: μ:
T_b: 197.2°C (470.3 K) P_c: IP: 9.54 eV

TRANSITION PROPERTIES

$\Delta_{fus}H$ (T_m):   Vapor pressure (0°C):
$\Delta_{vap}H$ (T_b):   Vapor pressure (25°C): 0.084 kPa
$\Delta_{vap}H$ (25°C): 54.8 kJ/mol   Vapor pressure (100°C): 4.53 kPa

PROPERTIES AT 25°C AND 100 kPa

	Solid	Liquid	Gas	Liquid
$\Delta_f H°$/kJ mol^{-1}:		-158.0	-103.2	d: 1.2070 g/mL
$S°$/J mol^{-1}K^{-1}:				η:
C_p/J mol^{-1}K^{-1}:		187.0		k:
COMMENTS: Highly toxic				

Name: Benzyl acetate
Synonyms: (Acetoxymethyl)benzene
        Benzyl ethanoate
Mol. Form.: $C_9H_{10}O_2$

CAS RN: 140-11-4
Merck No.: 1137
Mol. Wt.: 150.177

PHYSICAL CONSTANTS

T_m: -51.3°C (221.8 K)	T_c:	μ:
T_b: 213°C (486 K)	P_c:	IP:

TRANSITION PROPERTIES

$\Delta_{fus}H$ (T_m):   Vapor pressure (0°C):
$\Delta_{vap}H$ (T_b): 49.40 kJ/mol   Vapor pressure (25°C):
$\Delta_{vap}H$ (25°C):   Vapor pressure (100°C):

PROPERTIES AT 25°C AND 100 kPa

	Solid	Liquid	Gas	Liquid
$\Delta_f H°$/kJ mol^{-1}:				d: 1.050 g/mL
$S°$/J mol^{-1}K^{-1}:				η: 8.29 mPa s
C_p/J mol^{-1}K^{-1}:		148.5		k:
COMMENTS:				

Name: Benzyl alcohol
Synonyms: Phenylmethanol
        Benzenecarbinol
Mol. Form.: C_7H_8O

CAS RN: 100-51-6
Merck No.: 1138
Mol. Wt.: 108.140

PHYSICAL CONSTANTS

T_m: -15.2°C (257.9 K)	T_c: 442°C (715 K)	μ: 1.71 D
T_b: 205.31°C (478.46 K)	P_c: 4.3 MPa	IP: 8.50 eV

TRANSITION PROPERTIES

$\Delta_{fus}H$ (T_m): 8.97 kJ/mol   Vapor pressure (0°C):
$\Delta_{vap}H$ (T_b): 50.48 kJ/mol   Vapor pressure (25°C): 0.015 kPa
$\Delta_{vap}H$ (25°C): 60.3 kJ/mol   Vapor pressure (100°C): 2.27 kPa

PROPERTIES AT 25°C AND 100 kPa

	Solid	Liquid	Gas	Liquid
$\Delta_f H°$/kJ mol^{-1}:		-160.7	-100.4	d: 1.044 g/mL
$S°$/J mol^{-1}K^{-1}:		216.7		η: 5.47 mPa s
C_p/J mol^{-1}K^{-1}:		217.9		k:
COMMENTS:				

Name: Benzylamine
Synonym: Benzenemethanamine

Mol. Form.: C_7H_9N

CAS RN: 100-46-9
Merck No.: 1139
Mol. Wt.: 107.155

PHYSICAL CONSTANTS

T_m:	T_c:	μ:
T_b: 185°C (458 K)	P_c:	IP: 8.64 eV

TRANSITION PROPERTIES

$\Delta_{fus}H$ (T_m):   Vapor pressure (0°C):
$\Delta_{vap}H$ (T_b):   Vapor pressure (25°C): 0.096 kPa
$\Delta_{vap}H$ (25°C): 60.16 kJ/mol   Vapor pressure (100°C): 5.97 kPa

PROPERTIES AT 25°C AND 100 kPa

	Solid	Liquid	Gas	Liquid
$\Delta_f H°$/kJ mol^{-1}:		34.2	87.8	d: 0.977 g/mL
$S°$/J mol^{-1}K^{-1}:				η: 1.62 mPa s
C_p/J mol^{-1}K^{-1}:				k:
COMMENTS:				

Name: Benzyl benzoate
Synonym: Phenylmethyl benzoate

Mol. Form.: $C_{14}H_{12}O_2$

CAS RN: 120-51-4
Merck No.: 1141
Mol. Wt.: 212.248

PHYSICAL CONSTANTS

T_m: 21°C (294 K) T_c: μ:
T_b: 323.5°C (596.6 K) P_c: IP:

TRANSITION PROPERTIES

$\Delta_{fus}H$ (T_m):   Vapor pressure (0°C):
$\Delta_{vap}H$ (T_b): 53.60 kJ/mol   Vapor pressure (25°C):
$\Delta_{vap}H$ (25°C):   Vapor pressure (100°C):

PROPERTIES AT 25°C AND 100 kPa

	Solid	Liquid	Gas	Liquid
$\Delta_f H°$/kJ mol^{-1}:				d: 1.1145 g/mL
$S°$/J mol^{-1}K^{-1}:				η: 8.5 mPa s
C_p/J mol^{-1}K^{-1}:				k:
COMMENTS:				

Name: Benzyl chloride
Synonym: (Chloromethyl)benzene

Mol. Form.: C_7H_7Cl

CAS RN: 100-44-7
Merck No.: 1143
Mol. Wt.: 126.585

PHYSICAL CONSTANTS

T_m: -45°C (228 K) T_c: μ:
T_b: 179°C (452 K) P_c: IP:

TRANSITION PROPERTIES

$\Delta_{fus}H$ (T_m):   Vapor pressure (0°C):
$\Delta_{vap}H$ (T_b):   Vapor pressure (25°C): 0.164 kPa
$\Delta_{vap}H$ (25°C): 51.50 kJ/mol   Vapor pressure (100°C): 7.83 kPa

PROPERTIES AT 25°C AND 100 kPa

	Solid	Liquid	Gas	Liquid
$\Delta_f H°$/kJ mol^{-1}:		-32.6	18.9	d: 1.095 g/mL
$S°$/J mol^{-1}K^{-1}:				η: 1.28 mPa s
C_p/J mol^{-1}K^{-1}:		182.4		k:
COMMENTS: TLV=1 ppm; carcinogen; highly toxic				

Name: Benzyl ethyl ether
Synonyms: (Ethoxymethyl)benzene
         Ethyl benzyl ether
Mol. Form.: $C_9H_{12}O$

CAS RN: 539-30-0
Merck No.: 1147
Mol. Wt.: 136.194

PHYSICAL CONSTANTS

T_m: T_c: μ:
T_b: 186°C (459 K) P_c: IP:

TRANSITION PROPERTIES

$\Delta_{fus}H$ (T_m):   Vapor pressure (0°C):
$\Delta_{vap}H$ (T_b):   Vapor pressure (25°C): 0.134 kPa
$\Delta_{vap}H$ (25°C):   Vapor pressure (100°C): 6.38 kPa

PROPERTIES AT 25°C AND 100 kPa

	Solid	Liquid	Gas	Liquid
$\Delta_f H°$/kJ mol^{-1}:				d: 0.9446 g/mL
$S°$/J mol^{-1}K^{-1}:				η:
C_p/J mol^{-1}K^{-1}:				k:
COMMENTS:				

Name: Beryllium

CAS RN: 7440-41-7
Merck No.: 1177
Mol. Wt.: 9.012

Mol. Form.: Be

PHYSICAL CONSTANTS

T_m: 1287°C (1560 K)	T_c:	μ:
T_b: 2471°C (2744 K)	P_c:	IP: 9.32 eV

TRANSITION PROPERTIES

$\Delta_{fus}H$ (T_m): 7.90 kJ/mol	Vapor pressure (0°C): N/A
$\Delta_{vap}H$ (T_b):	Vapor pressure (25°C): N/A
$\Delta_{vap}H$ (25°C):	Vapor pressure (100°C): N/A

PROPERTIES AT 25°C AND 100 kPa

	Solid	Liquid	Gas		Solid
$\Delta_f H°$/kJ mol^{-1}:	0.0		324.0	d:	1.85 g/mL
$S°$/J mol^{-1}K^{-1}:	9.5		136.3	η:	N/A
C_p/J mol^{-1}K^{-1}:	16.4		20.8	k:	201 W/m K

COMMENTS: Carcinogen; highly toxic

Name: Beryllium chloride
Synonym: Beryllium dichloride

CAS RN: 7787-47-5
Merck No.: 1184
Mol. Wt.: 79.918

Mol. Form.: BeCl$_2$

PHYSICAL CONSTANTS

T_m: 415°C (688 K)	T_c:	μ:
T_b: 547°C (820 K)	P_c:	IP:

TRANSITION PROPERTIES

$\Delta_{fus}H$ (T_m): 8.66 kJ/mol	Vapor pressure (0°C): N/A
$\Delta_{vap}H$ (T_b): 105.00 kJ/mol	Vapor pressure (25°C): N/A
$\Delta_{vap}H$ (25°C):	Vapor pressure (100°C): N/A

PROPERTIES AT 25°C AND 100 kPa

	Solid	Liquid	Gas		Solid
$\Delta_f H°$/kJ mol^{-1}:	-490.4			d:	1.90 g/mL
$S°$/J mol^{-1}K^{-1}:	82.7			η:	N/A
C_p/J mol^{-1}K^{-1}:	64.8			k:	

COMMENTS: Carcinogen; highly toxic

Name: Beryllium fluoride
Synonym: Beryllium difluoride

CAS RN: 7787-49-7
Merck No.: 1185
Mol. Wt.: 47.009

Mol. Form.: BeF$_2$

PHYSICAL CONSTANTS

T_m: 552°C (825 K)	T_c:	μ:
T_b: 1159°C (1432 K)	P_c:	IP:

TRANSITION PROPERTIES

$\Delta_{fus}H$ (T_m): 4.76 kJ/mol	Vapor pressure (0°C): N/A
$\Delta_{vap}H$ (T_b):	Vapor pressure (25°C): N/A
$\Delta_{vap}H$ (25°C):	Vapor pressure (100°C): N/A

PROPERTIES AT 25°C AND 100 kPa

	Solid	Liquid	Gas		Solid
$\Delta_f H°$/kJ mol^{-1}:	-1026.8			d:	2.1 g/mL
$S°$/J mol^{-1}K^{-1}:	53.4			η:	N/A
C_p/J mol^{-1}K^{-1}:	51.8			k:	

COMMENTS: Carcinogen; highly toxic

Name: Beryllium oxide
Synonym: Beryllia

Mol. Form.: BeO

CAS RN: 1304-56-9
Merck No.: 1192
Mol. Wt.: 25.012

PHYSICAL CONSTANTS

T_m: 2507°C (2780 K)
T_b:

T_c:
P_c:

μ:
IP: 10.10 eV

TRANSITION PROPERTIES

$\Delta_{fus}H$ (T_m): 85.00 kJ/mol
$\Delta_{vap}H$ (T_b):
$\Delta_{vap}H$ (25°C):

Vapor pressure (0°C): N/A
Vapor pressure (25°C): N/A
Vapor pressure (100°C): N/A

PROPERTIES AT 25°C AND 100 kPa

	Solid	Liquid	Gas	Solid
$\Delta_f H°$/kJ mol^{-1}:	-609.4			d: 3.01 g/mL
$S°$/J mol^{-1}K^{-1}:	13.8			η: N/A
C_p/J mol^{-1}K^{-1}:				k: 270 W/m K

COMMENTS: Carcinogen; highly toxic

Name: Biphenyl
Synonym: Diphenyl

Mol. Form.: $C_{12}H_{10}$

CAS RN: 92-52-4
Merck No.: 3314
Mol. Wt.: 154.211

PHYSICAL CONSTANTS

T_m: 69°C (342 K)
T_b: 256.1°C (529.2 K)

T_c: 516°C (789 K)
P_c: 3.85 MPa

μ: 0 D
IP: 7.95 eV

TRANSITION PROPERTIES

$\Delta_{fus}H$ (T_m): 18.60 kJ/mol
$\Delta_{vap}H$ (T_b):
$\Delta_{vap}H$ (25°C):

Vapor pressure (0°C):
Vapor pressure (25°C):
Vapor pressure (100°C): 0.586 kPa

PROPERTIES AT 25°C AND 100 kPa

	Solid	Liquid	Gas	Solid
$\Delta_f H°$/kJ mol^{-1}:	99.4		181.4	d: 1.04 g/mL
$S°$/J mol^{-1}K^{-1}:	209.4			η: N/A
C_p/J mol^{-1}K^{-1}:	198.4			k:

COMMENTS: TLV=0.2 ppm; highly toxic

Name: Bis(2-amimoethyl)amine
Synonyms: Diethylenetriamine
             2,2′-Diaminodiethylamine
Mol. Form.: $C_4H_{13}N_3$

CAS RN: 111-40-0
Merck No.:
Mol. Wt.: 103.167

PHYSICAL CONSTANTS

T_m: -39°C (234 K)
T_b: 207°C (480 K)

T_c:
P_c:

μ:
IP:

TRANSITION PROPERTIES

$\Delta_{fus}H$ (T_m):
$\Delta_{vap}H$ (T_b):
$\Delta_{vap}H$ (25°C):

Vapor pressure (0°C):
Vapor pressure (25°C): 0.030 kPa
Vapor pressure (100°C): 2.74 kPa

PROPERTIES AT 25°C AND 100 kPa

	Solid	Liquid	Gas	Liquid
$\Delta_f H°$/kJ mol^{-1}:				d: 0.952 g/mL
$S°$/J mol^{-1}K^{-1}:				η:
C_p/J mol^{-1}K^{-1}:				k:

COMMENTS: TLV=1 ppm

Name: Bis(2-chloroethyl) ether
Synonyms: 2,2'-Dichlorethyl ether
        1,1'-Oxybis(2-chloroethane)
Mol. Form.: $C_4H_8Cl_2O$

CAS RN: 111-44-4
Merck No.: 3055
Mol. Wt.: 143.012

PHYSICAL CONSTANTS

T_m: -51.9°C (221.2 K)
T_b: 178.5°C (451.6 K)

T_c:
P_c:

μ:
IP:

TRANSITION PROPERTIES

$\Delta_{fus}H$ (T_m): 8.66 kJ/mol
$\Delta_{vap}H$ (T_b): 45.23 kJ/mol
$\Delta_{vap}H$ (25°C):

Vapor pressure (0°C):
Vapor pressure (25°C): 0.143 kPa
Vapor pressure (100°C): 7.58 kPa

PROPERTIES AT 25°C AND 100 kPa

	Solid	Liquid	Gas		Liquid
$\Delta_f H°$/kJ mol^{-1}:					d: 1.2130 g/mL
$S°$/J mol^{-1}K^{-1}:					η: 2.14 mPa s
C_p/J mol^{-1}K^{-1}:		220.9			k:

COMMENTS: TLV=5 ppm; highly toxic

Name: Bis(chloromethyl) ether
Synonym: Chloromethyl ether

Mol. Form.: $C_2H_4Cl_2O$

CAS RN: 542-88-1
Merck No.: 3058
Mol. Wt.: 114.959

PHYSICAL CONSTANTS

T_m: -41.5°C (231.6 K)
T_b: 106°C (379 K)

T_c:
P_c:

μ:
IP:

TRANSITION PROPERTIES

$\Delta_{fus}H$ (T_m):
$\Delta_{vap}H$ (T_b):
$\Delta_{vap}H$ (25°C):

Vapor pressure (0°C):
Vapor pressure (25°C):
Vapor pressure (100°C):

PROPERTIES AT 25°C AND 100 kPa

	Solid	Liquid	Gas		Liquid
$\Delta_f H°$/kJ mol^{-1}:					d: 1.31 g/mL
$S°$/J mol^{-1}K^{-1}:					η:
C_p/J mol^{-1}K^{-1}:					k:

COMMENTS: TLV=0.001 ppm; carcinogen; highly toxic

Name: Bismuth

Mol. Form.: Bi

CAS RN: 7440-69-9
Merck No.: 1268
Mol. Wt.: 208.980

PHYSICAL CONSTANTS

T_m: 271.40°C (544.55 K)
T_b: 1564°C (1837 K)

T_c:
P_c:

μ:
IP: 7.29 eV

TRANSITION PROPERTIES

$\Delta_{fus}H$ (T_m): 11.30 kJ/mol
$\Delta_{vap}H$ (T_b): 151.00 kJ/mol
$\Delta_{vap}H$ (25°C):

Vapor pressure (0°C): N/A
Vapor pressure (25°C): N/A
Vapor pressure (100°C): N/A

PROPERTIES AT 25°C AND 100 kPa

	Solid	Liquid	Gas		Solid
$\Delta_f H°$/kJ mol^{-1}:	0.0		207.1		d: 9.79 g/mL
$S°$/J mol^{-1}K^{-1}:	56.7		187.0		η: N/A
C_p/J mol^{-1}K^{-1}:	25.5		20.8		k: 7.92 W/m K

COMMENTS:

Name: Bismuth tribromide
Synonym: Bismuth(III) bromide

Mol. Form.: BiBr$_3$

CAS RN: 7787-58-8
Merck No.: 1270
Mol. Wt.: 448.692

PHYSICAL CONSTANTS

T_m: 218°C (491 K) T_c: 947°C (1220 K)   μ:
T_b: 453°C (726 K) P_c: IP:

TRANSITION PROPERTIES

$\Delta_{fus}H$ (T_m):   Vapor pressure (0°C):
$\Delta_{vap}H$ (T_b): 75.40 kJ/mol   Vapor pressure (25°C):
$\Delta_{vap}H$ (25°C):   Vapor pressure (100°C):

PROPERTIES AT 25°C AND 100 kPa

	Solid	Liquid	Gas	Solid
$\Delta_fH°$/kJ mol^{-1}:				d: 5.72 g/mL
$S°$/J mol^{-1}K^{-1}:				η: N/A
C_p/J mol^{-1}K^{-1}:				k:
COMMENTS:				

Name: Bismuth trichloride
Synonym: Bismuth(III) chloride

Mol. Form.: BiCl$_3$

CAS RN: 7787-60-2
Merck No.: 1273
Mol. Wt.: 315.339

PHYSICAL CONSTANTS

T_m: 230°C (503 K) T_c: 906°C (1179 K)   μ:
T_b: 447°C (720 K) P_c: 12.0 MPa   IP: 10.40 eV

TRANSITION PROPERTIES

$\Delta_{fus}H$ (T_m): 10.90 kJ/mol   Vapor pressure (0°C): N/A
$\Delta_{vap}H$ (T_b): 72.61 kJ/mol   Vapor pressure (25°C): N/A
$\Delta_{vap}H$ (25°C):   Vapor pressure (100°C): N/A

PROPERTIES AT 25°C AND 100 kPa

	Solid	Liquid	Gas	Solid
$\Delta_fH°$/kJ mol^{-1}:	-379.1		-265.7	d: 4.75 g/mL
$S°$/J mol^{-1}K^{-1}:	177.0		358.9	η: N/A
C_p/J mol^{-1}K^{-1}:	105.0		79.7	k:
COMMENTS:				

Name: Boric acid (H$_3$BO$_3$)
Synonym: Orthoboric acid

Mol. Form.: BH$_3$O$_3$

CAS RN: 10043-35-3
Merck No.: 1336
Mol. Wt.: 61.833

PHYSICAL CONSTANTS

T_m: 170.9°C (444.0 K) T_c: μ:
T_b: P_c: IP:

TRANSITION PROPERTIES

$\Delta_{fus}H$ (T_m):   Vapor pressure (0°C):
$\Delta_{vap}H$ (T_b):   Vapor pressure (25°C):
$\Delta_{vap}H$ (25°C):   Vapor pressure (100°C):

PROPERTIES AT 25°C AND 100 kPa

	Solid	Liquid	Gas	Solid
$\Delta_fH°$/kJ mol^{-1}:	-1094.3		-994.1	d:
$S°$/J mol^{-1}K^{-1}:	88.8			η: N/A
C_p/J mol^{-1}K^{-1}:	81.4			k:
COMMENTS: Highly toxic				

Name: Boron

CAS RN: 7440-42-8
Merck No.: 1345
Mol. Wt.: 10.811

Mol. Form.: B

PHYSICAL CONSTANTS

T_m: 2075°C (2348 K)	T_c:	μ:
T_b: 4000°C (4273 K)	P_c:	IP: 8.30 eV

TRANSITION PROPERTIES

$\Delta_{fus}H$ (T_m): 50.20 kJ/mol
$\Delta_{vap}H$ (T_b): 480.00 kJ/mol
$\Delta_{vap}H$ (25°C):

Vapor pressure (0°C): N/A
Vapor pressure (25°C): N/A
Vapor pressure (100°C): N/A

PROPERTIES AT 25°C AND 100 kPa

	Solid	Liquid	Gas		Solid
$\Delta_f H°$/kJ mol^{-1}:	0.0		565.0		d: 2.34 g/mL
$S°$/J mol^{-1}K^{-1}:	5.9		153.4		η: N/A
C_p/J mol^{-1}K^{-1}:	11.1		20.8		k: 27.4 W/m K

COMMENTS: Data refer to rhombic form

Name: Boron oxide (B_2O_3)
Synonym: Diboron trioxide

CAS RN: 1303-86-2
Merck No.: 1337
Mol. Wt.: 69.620

Mol. Form.: B_2O_3

PHYSICAL CONSTANTS

T_m: 450°C (723 K)	T_c:	μ:
T_b:	P_c:	IP: 13.50 eV

TRANSITION PROPERTIES

$\Delta_{fus}H$ (T_m): 24.07 kJ/mol
$\Delta_{vap}H$ (T_b):
$\Delta_{vap}H$ (25°C): 417.4 kJ/mol

Vapor pressure (0°C): N/A
Vapor pressure (25°C): N/A
Vapor pressure (100°C): N/A

PROPERTIES AT 25°C AND 100 kPa

	Solid	Liquid	Gas		Solid
$\Delta_f H°$/kJ mol^{-1}:	-1271.9	-1253.4	-836.0		d: 2.46 g/mL
$S°$/J mol^{-1}K^{-1}:	54.0	78.4	283.8		η: N/A
C_p/J mol^{-1}K^{-1}:	62.6	62.8	66.9		k:

COMMENTS:

Name: Boron tribromide
Synonym: Boron(III) bromide

CAS RN: 10294-33-4
Merck No.: 1349
Mol. Wt.: 250.523

Mol. Form.: BBr_3

PHYSICAL CONSTANTS

T_m: -45°C (228 K)	T_c: 308°C (581 K)	μ: 0 D
T_b: 91°C (364 K)	P_c:	IP: 10.51 eV

TRANSITION PROPERTIES

$\Delta_{fus}H$ (T_m):
$\Delta_{vap}H$ (T_b): 30.50 kJ/mol
$\Delta_{vap}H$ (25°C): 34.1 kJ/mol

Vapor pressure (0°C): 2.46 kPa
Vapor pressure (25°C): 8.94 kPa
Vapor pressure (100°C):

PROPERTIES AT 25°C AND 100 kPa

	Solid	Liquid	Gas		Liquid
$\Delta_f H°$/kJ mol^{-1}:		-239.7	-205.6		d: 2.6 g/mL
$S°$/J mol^{-1}K^{-1}:		229.7	324.2		η:
C_p/J mol^{-1}K^{-1}:			67.8		k:

COMMENTS: TLV=1 ppm; highly toxic

Name: Boron trichloride
Synonym: Boron(III) chloride

Mol. Form.: BCl_3

CAS RN: 10294-34-5
Merck No.: 1350
Mol. Wt.: 117.169

PHYSICAL CONSTANTS

T_m: -107°C (166 K)
T_b: 12.65°C (285.80 K)

T_c: 182°C (455 K)
P_c: 3.87 MPa

μ: 0 D
IP: 11.60 eV

TRANSITION PROPERTIES

$\Delta_{fus}H$ (T_m): 2.10 kJ/mol
$\Delta_{vap}H$ (T_b): 23.77 kJ/mol
$\Delta_{vap}H$ (25°C): 23.10 kJ/mol

Vapor pressure (0°C):
Vapor pressure (25°C):
Vapor pressure (100°C):

PROPERTIES AT 25°C AND 100 kPa

	Solid	Liquid	Gas		Gas
$\Delta_f H°$/kJ mol^{-1}:		-427.2	-403.8		d: 4.789 g/L
$S°$/J mol^{-1}K^{-1}:		206.3	290.1		η:
C_p/J mol^{-1}K^{-1}:		106.7	62.7		k:

COMMENTS: Highly toxic

Name: Boron trifluoride
Synonym: Boron(III) fluoride

Mol. Form.: BF_3

CAS RN: 7637-07-2
Merck No.: 1351
Mol. Wt.: 67.806

PHYSICAL CONSTANTS

T_m: -126.8°C (146.3 K)
T_b: -101°C (172 K)

T_c: -12.3°C (260.8 K)
P_c: 4.98 MPa

μ: 0 D
IP: 15.56 eV

TRANSITION PROPERTIES

$\Delta_{fus}H$ (T_m): 4.20 kJ/mol
$\Delta_{vap}H$ (T_b): 19.33 kJ/mol
$\Delta_{vap}H$ (25°C):

Vapor pressure (0°C): N/A
Vapor pressure (25°C): N/A
Vapor pressure (100°C): N/A

PROPERTIES AT 25°C AND 100 kPa

	Solid	Liquid	Gas		Gas
$\Delta_f H°$/kJ mol^{-1}:			-1136.0		d: 2.771 g/L
$S°$/J mol^{-1}K^{-1}:			254.4		η: 17.0 µPa s
C_p/J mol^{-1}K^{-1}:					k: 0.0190 W/m K

COMMENTS: TLV=1 ppm; highly toxic

Name: Bromine (Br_2)
Synonym: Dibromine

Mol. Form.: Br_2

CAS RN: 7726-95-6
Merck No.: 1382
Mol. Wt.: 159.808

PHYSICAL CONSTANTS

T_m: -7.2°C (265.9 K)
T_b: 58.8°C (331.9 K)

T_c: 315°C (588 K)
P_c: 10.34 MPa

μ: 0 D
IP: 10.52 eV

TRANSITION PROPERTIES

$\Delta_{fus}H$ (T_m): 10.57 kJ/mol
$\Delta_{vap}H$ (T_b): 29.96 kJ/mol
$\Delta_{vap}H$ (25°C): 30.91 kJ/mol

Vapor pressure (0°C): 8.80 kPa
Vapor pressure (25°C): 28.7 kPa
Vapor pressure (100°C):

PROPERTIES AT 25°C AND 100 kPa

	Solid	Liquid	Gas		Liquid
$\Delta_f H°$/kJ mol^{-1}:		0.0	30.9		d: 3.1028 g/mL
$S°$/J mol^{-1}K^{-1}:		152.2	245.5		η: 0.944 mPa s
C_p/J mol^{-1}K^{-1}:		75.7	36.0		k: 0.122 W/m K

COMMENTS: TLV=0.1 ppm; highly toxic

Name: Bromine pentafluoride
Synonym: Bromine(V) fluoride

Mol. Form.: BrF_5

CAS RN: 7789-30-2
Merck No.: 1383
Mol. Wt.: 174.896

PHYSICAL CONSTANTS

T_m: -60.5°C (212.6 K)	T_c:	μ: 1.51 D
T_b: 40.76°C (313.91 K)	P_c:	IP: 13.17 eV

TRANSITION PROPERTIES

$\Delta_{fus}H$ (T_m): 5.67 kJ/mol
$\Delta_{vap}H$ (T_b): 30.60 kJ/mol
$\Delta_{vap}H$ (25°C): 29.7 kJ/mol

Vapor pressure (0°C):
Vapor pressure (25°C):
Vapor pressure (100°C):

PROPERTIES AT 25°C AND 100 kPa

	Solid	Liquid	Gas		Liquid
$\Delta_f H°$/kJ mol^{-1}:		-458.6	-428.9	d:	2.460 g/mL
$S°$/J mol^{-1}K^{-1}:		225.1	320.2	η:	
C_p/J mol^{-1}K^{-1}:			99.6	k:	
COMMENTS: TLV=0.1 ppm					

Name: Bromine trifluoride
Synonyms: Bromine fluoride (BrF3)
              Bromine(III) fluoride
Mol. Form.: BrF_3

CAS RN: 7787-71-5
Merck No.: 1384
Mol. Wt.: 136.899

PHYSICAL CONSTANTS

T_m: 8.77°C (281.92 K)	T_c:	μ:
T_b: 125.8°C (398.9 K)	P_c:	IP:

TRANSITION PROPERTIES

$\Delta_{fus}H$ (T_m):
$\Delta_{vap}H$ (T_b): 47.57 kJ/mol
$\Delta_{vap}H$ (25°C): 45.2 kJ/mol

Vapor pressure (0°C):
Vapor pressure (25°C):
Vapor pressure (100°C):

PROPERTIES AT 25°C AND 100 kPa

	Solid	Liquid	Gas		Liquid
$\Delta_f H°$/kJ mol^{-1}:		-300.8	-255.6	d:	2.803 g/mL
$S°$/J mol^{-1}K^{-1}:		178.2	292.5	η:	
C_p/J mol^{-1}K^{-1}:		124.6	66.6	k:	
COMMENTS:					

Name: Bromobenzene
Synonym: Phenyl bromide

Mol. Form.: C_6H_5Br

CAS RN: 108-86-1
Merck No.: 1394
Mol. Wt.: 157.010

PHYSICAL CONSTANTS

T_m: -30.6°C (242.5 K)	T_c: 397°C (670 K)	μ: 1.70 D
T_b: 156.06°C (429.21 K)	P_c: 4.52 MPa	IP: 8.98 eV

TRANSITION PROPERTIES

$\Delta_{fus}H$ (T_m): 10.62 kJ/mol
$\Delta_{vap}H$ (T_b):
$\Delta_{vap}H$ (25°C): 44.54 kJ/mol

Vapor pressure (0°C):
Vapor pressure (25°C): 0.556 kPa
Vapor pressure (100°C): 18.9 kPa

PROPERTIES AT 25°C AND 100 kPa

	Solid	Liquid	Gas		Liquid
$\Delta_f H°$/kJ mol^{-1}:		60.9	105.4	d:	1.4882 g/mL
$S°$/J mol^{-1}K^{-1}:		219.2		η:	1.07 mPa s
C_p/J mol^{-1}K^{-1}:		154.3		k:	
COMMENTS:					

Name: 1-Bromobutane

Synonym: Butyl bromide

Mol. Form.: C_4H_9Br

CAS RN: 109-65-9
Merck No.: 1553
Mol. Wt.: 137.020

PHYSICAL CONSTANTS

T_m: -112.4°C (160.7 K)	T_c:	μ: 2.08 D
T_b: 101.6°C (374.7 K)	P_c:	IP: 10.13 eV

TRANSITION PROPERTIES

$\Delta_{fus}H$ (T_m): 6.69 kJ/mol

$\Delta_{vap}H$ (T_b): 32.51 kJ/mol

$\Delta_{vap}H$ (25°C): 36.64 kJ/mol

Vapor pressure (0°C): 1.41 kPa

Vapor pressure (25°C): 5.26 kPa

Vapor pressure (100°C):

PROPERTIES AT 25°C AND 100 kPa

	Solid	Liquid	Gas		Liquid
$\Delta_f H°$/kJ mol^{-1}:		-143.8	-107.1		d: 1.2686 g/mL
$S°$/J mol^{-1}K^{-1}:					η: 0.606 mPa s
C_p/J mol^{-1}K^{-1}:		109.3			k:

COMMENTS: Flammable

Name: 2-Bromobutane

Synonym: *sec*-Butyl bromide

Mol. Form.: C_4H_9Br

CAS RN: 78-76-2
Merck No.: 1554
Mol. Wt.: 137.020

PHYSICAL CONSTANTS

T_m: -112.7°C (160.4 K)	T_c:	μ: 2.23 D
T_b: 91.4°C (364.5 K)	P_c:	IP: 9.98 eV

TRANSITION PROPERTIES

$\Delta_{fus}H$ (T_m): 6.89 kJ/mol

$\Delta_{vap}H$ (T_b): 30.77 kJ/mol

$\Delta_{vap}H$ (25°C): 34.41 kJ/mol

Vapor pressure (0°C): 2.61 kPa

Vapor pressure (25°C): 9.32 kPa

Vapor pressure (100°C): 129 kPa

PROPERTIES AT 25°C AND 100 kPa

	Solid	Liquid	Gas		Liquid
$\Delta_f H°$/kJ mol^{-1}:		-154.8	-120.3		d: 1.2530 g/mL
$S°$/J mol^{-1}K^{-1}:					η: 0.563 mPa s
C_p/J mol^{-1}K^{-1}:					k:

COMMENTS:

Name: Bromochlorodifluoromethane

Synonyms: Refrigerant 12B1
          Halon 1211

Mol. Form.: $CBrClF_2$

CAS RN: 353-59-3
Merck No.:
Mol. Wt.: 165.365

PHYSICAL CONSTANTS

T_m: -159.5°C (113.6 K)	T_c: 153.73°C (426.88 K)	μ:
T_b: -3.72°C (269.43 K)	P_c: 4.254 MPa	IP: 11.83 eV

TRANSITION PROPERTIES

$\Delta_{fus}H$ (T_m):

$\Delta_{vap}H$ (T_b):

$\Delta_{vap}H$ (25°C):

Vapor pressure (0°C): 118 kPa

Vapor pressure (25°C):

Vapor pressure (100°C):

PROPERTIES AT 25°C AND 100 kPa

	Solid	Liquid	Gas		Gas
$\Delta_f H°$/kJ mol^{-1}:					d: 6.759 g/L
$S°$/J mol^{-1}K^{-1}:			318.5		η:
C_p/J mol^{-1}K^{-1}:			74.6		k:

COMMENTS: Highly toxic

Name: Bromochloromethane

Synonym: Halon 1011

Mol. Form.: CH_2BrCl

CAS RN: 74-97-5

Merck No.:

Mol. Wt.: 129.384

PHYSICAL CONSTANTS

T_m: -87.95°C (185.20 K)

T_b: 68.05°C (341.20 K)

T_c:

P_c:

μ:

IP: 10.77 eV

TRANSITION PROPERTIES

$\Delta_{fus}H$ (T_m):

$\Delta_{vap}H$ (T_b): 30.00 kJ/mol

$\Delta_{vap}H$ (25°C): 32.85 kJ/mol

Vapor pressure (0°C): 5.26 kPa

Vapor pressure (25°C): 19.5 kPa

Vapor pressure (100°C):

PROPERTIES AT 25°C AND 100 kPa

	Solid	Liquid	Gas		Liquid
$\Delta_f H°$/kJ mol^{-1}:				d:	1.925 g/mL
$S°$/J mol^{-1}K^{-1}:			287.6	η:	
C_p/J mol^{-1}K^{-1}:		52.7	52.7	k:	

COMMENTS: TLV=200 ppm

Name: Bromoethane

Synonym: Ethyl bromide

Mol. Form.: C_2H_5Br

CAS RN: 74-96-4

Merck No.: 3730

Mol. Wt.: 108.966

PHYSICAL CONSTANTS

T_m: -118.6°C (154.5 K)

T_b: 38.5°C (311.6 K)

T_c: 230.8°C (503.9 K)

P_c: 6.23 MPa

μ: 2.03 D

IP: 10.28 eV

TRANSITION PROPERTIES

$\Delta_{fus}H$ (T_m): 5.86 kJ/mol

$\Delta_{vap}H$ (T_b): 27.04 kJ/mol

$\Delta_{vap}H$ (25°C): 28.03 kJ/mol

Vapor pressure (0°C): 21.7 kPa

Vapor pressure (25°C): 62.5 kPa

Vapor pressure (100°C): 571 kPa

PROPERTIES AT 25°C AND 100 kPa

	Solid	Liquid	Gas		Liquid
$\Delta_f H°$/kJ mol^{-1}:		-90.1	-61.9	d:	1.4505 g/mL
$S°$/J mol^{-1}K^{-1}:		198.7	286.7	η:	0.374 mPa s
C_p/J mol^{-1}K^{-1}:		100.8	64.5	k:	

COMMENTS: TLV=5 ppm; carcinogen

Name: Bromoethylene

Synonym: Vinyl bromide

Mol. Form.: C_2H_3Br

CAS RN: 593-60-2

Merck No.:

Mol. Wt.: 106.950

PHYSICAL CONSTANTS

T_m: -137.8°C (135.3 K)

T_b: 15.8°C (288.9 K)

T_c:

P_c:

μ: 1.42 D

IP: 9.80 eV

TRANSITION PROPERTIES

$\Delta_{fus}H$ (T_m): 5.12 kJ/mol

$\Delta_{vap}H$ (T_b): 23.43 kJ/mol

$\Delta_{vap}H$ (25°C): 22.60 kJ/mol

Vapor pressure (0°C): 54.5 kPa

Vapor pressure (25°C): 141 kPa

Vapor pressure (100°C):

PROPERTIES AT 25°C AND 100 kPa

	Solid	Liquid	Gas		Gas
$\Delta_f H°$/kJ mol^{-1}:			79.2	d:	4.371 g/L
$S°$/J mol^{-1}K^{-1}:			275.8	η:	
C_p/J mol^{-1}K^{-1}:		55.5	55.5	k:	

COMMENTS:

Name: Bromomethane
Synonyms: Methyl bromide

Mol. Wt.: 94.939
Mol. Form.: CH_3Br

CAS RN: 74-83-9
Merck No.: 5951
Halon 1001

PHYSICAL CONSTANTS

T_m: -93.7°C (179.4 K)	T_c:	μ: 1.822 D
T_b: 3.5°C (276.6 K)	P_c:	IP: 10.54 eV

TRANSITION PROPERTIES

$\Delta_{fus}H$ (T_m): 5.98 kJ/mol
$\Delta_{vap}H$ (T_b): 23.91 kJ/mol
$\Delta_{vap}H$ (25°C): 22.81 kJ/mol

Vapor pressure (0°C): 88.0 kPa
Vapor pressure (25°C): 217 kPa
Vapor pressure (100°C):

PROPERTIES AT 25°C AND 100 kPa

	Solid	Liquid	Gas	Gas
$\Delta_f H°$/kJ mol^{-1}:		-59.4	-35.5	d: 3.881 g/L
$S°$/J mol^{-1}K^{-1}:			246.4	η:
C_p/J mol^{-1}K^{-1}:			42.4	k:

COMMENTS: TLV=5 ppm; carcinogen; highly toxic

Name: 2-Bromo-2-methylpropane
Synonym: *tert*-Butyl bromide

Mol. Form.: C_4H_9Br

CAS RN: 507-19-7
Merck No.: 1555
Mol. Wt.: 137.020

PHYSICAL CONSTANTS

T_m: -16.2°C (256.9 K)	T_c:	μ:
T_b: 73.3°C (346.4 K)	P_c:	IP: 9.92 eV

TRANSITION PROPERTIES

$\Delta_{fus}H$ (T_m): 1.97 kJ/mol
$\Delta_{vap}H$ (T_b): 29.23 kJ/mol
$\Delta_{vap}H$ (25°C): 31.81 kJ/mol

Vapor pressure (0°C): 5.58 kPa
Vapor pressure (25°C): 17.7 kPa
Vapor pressure (100°C):

PROPERTIES AT 25°C AND 100 kPa

	Solid	Liquid	Gas	Liquid
$\Delta_f H°$/kJ mol^{-1}:		-163.8	-132.4	d: 1.4252 g/mL
$S°$/J mol^{-1}K^{-1}:				η: 0.750 mPa s
C_p/J mol^{-1}K^{-1}:		151.0		k:

COMMENTS: Flammable

Name: 1-Bromonaphthalene
Synonym: 1-Naphthyl bromide

Mol. Form.: $C_{10}H_7Br$

CAS RN: 90-11-9
Merck No.: 1413
Mol. Wt.: 207.070

PHYSICAL CONSTANTS

T_m: -1.8°C (271.3 K)	T_c:	μ:
T_b: 281°C (554 K)	P_c:	IP: 8.09 eV

TRANSITION PROPERTIES

$\Delta_{fus}H$ (T_m): 15.16 kJ/mol
$\Delta_{vap}H$ (T_b): 39.30 kJ/mol
$\Delta_{vap}H$ (25°C):

Vapor pressure (0°C):
Vapor pressure (25°C): 0.001 kPa
Vapor pressure (100°C): 0.286 kPa

PROPERTIES AT 25°C AND 100 kPa

	Solid	Liquid	Gas	Liquid
$\Delta_f H°$/kJ mol^{-1}:				d: 1.471 g/mL
$S°$/J mol^{-1}K^{-1}:				η: 4.52 mPa s
C_p/J mol^{-1}K^{-1}:				k:

COMMENTS:

Name: 1-Bromooctane
Synonym: Octyl bromide

Mol. Form.: $C_8H_{17}Br$

CAS RN: 111-83-1
Merck No.: 6684
Mol. Wt.: 193.127

PHYSICAL CONSTANTS

T_m: -55°C (218 K) T_c: μ:
T_b: 200°C (473 K) P_c: IP:

TRANSITION PROPERTIES

$\Delta_{fus}H$ (T_m):   Vapor pressure (0°C):
$\Delta_{vap}H$ (T_b):   Vapor pressure (25°C):
$\Delta_{vap}H$ (25°C): 55.77 kJ/mol   Vapor pressure (100°C):

PROPERTIES AT 25°C AND 100 kPa

	Solid	Liquid	Gas		Liquid
$\Delta_f H°$/kJ mol^{-1}:		-245.1	-189.7		d: 1.108 g/mL
$S°$/J mol^{-1}K^{-1}:					η:
C_p/J mol^{-1}K^{-1}:					k:
COMMENTS:					

Name: 1-Bromopentane
Synonyms: Amyl bromide
              Pentyl bromide
Mol. Form.: $C_5H_{11}Br$

CAS RN: 110-53-2
Merck No.: 637
Mol. Wt.: 151.046

PHYSICAL CONSTANTS

T_m: -95°C (178 K) T_c: μ: 2.20 D
T_b: 129.8°C (402.9 K) P_c: IP: 10.09 eV

TRANSITION PROPERTIES

$\Delta_{fus}H$ (T_m): 11.46 kJ/mol   Vapor pressure (0°C):
$\Delta_{vap}H$ (T_b): 35.01 kJ/mol   Vapor pressure (25°C): 1.68 kPa
$\Delta_{vap}H$ (25°C): 41.28 kJ/mol   Vapor pressure (100°C): 41.9 kPa

PROPERTIES AT 25°C AND 100 kPa

	Solid	Liquid	Gas		Liquid
$\Delta_f H°$/kJ mol^{-1}:		-170.2	-129.0		d: 1.212 g/mL
$S°$/J mol^{-1}K^{-1}:					η: 0.753 mPa s
C_p/J mol^{-1}K^{-1}:		132.2			k:
COMMENTS:					

Name: 1-Bromopropane
Synonym: Propyl bromide

Mol. Form.: C_3H_7Br

CAS RN: 106-94-5
Merck No.: 7857
Mol. Wt.: 122.993

PHYSICAL CONSTANTS

T_m: -110°C (163 K) T_c: μ: 2.18 D
T_b: 71.1°C (344.2 K) P_c: IP: 10.18 eV

TRANSITION PROPERTIES

$\Delta_{fus}H$ (T_m):   Vapor pressure (0°C): 5.56 kPa
$\Delta_{vap}H$ (T_b): 29.84 kJ/mol   Vapor pressure (25°C): 18.6 kPa
$\Delta_{vap}H$ (25°C): 32.01 kJ/mol   Vapor pressure (100°C):

PROPERTIES AT 25°C AND 100 kPa

	Solid	Liquid	Gas		Liquid
$\Delta_f H°$/kJ mol^{-1}:		-121.8	-87.0		d: 1.3452 g/mL
$S°$/J mol^{-1}K^{-1}:					η: 0.489 mPa s
C_p/J mol^{-1}K^{-1}:		86.4			k:
COMMENTS: Flammable					

Name: 2-Bromopropane
Synonym: Isopropyl bromide

Mol. Form.: C_3H_7Br

CAS RN: 75-26-3
Merck No.: 5098
Mol. Wt.: 122.993

PHYSICAL CONSTANTS

T_m: -89°C (184 K)
T_b: 59.5°C (332.6 K)

T_c:
P_c:

μ: 2.21 D
IP: 10.07 eV

TRANSITION PROPERTIES

$\Delta_{fus}H$ (T_m):
$\Delta_{vap}H$ (T_b): 28.33 kJ/mol
$\Delta_{vap}H$ (25°C): 30.17 kJ/mol

Vapor pressure (0°C): 9.23 kPa
Vapor pressure (25°C): 28.9 kPa
Vapor pressure (100°C):

PROPERTIES AT 25°C AND 100 kPa

	Solid	Liquid	Gas		Liquid
$\Delta_f H°$/kJ mol^{-1}:		-130.5	-99.4		d: 1.3060 g/mL
$S°$/J mol^{-1}K^{-1}:					η: 0.458 mPa s
C_p/J mol^{-1}K^{-1}:					k:
COMMENTS: Flammable					

Name: 3-Bromopropene
Synonyms: Allyl bromide
            3-Bromopropylene
Mol. Form.: C_3H_5Br

CAS RN: 106-95-6
Merck No.: 286
Mol. Wt.: 120.977

PHYSICAL CONSTANTS

T_m: -119°C (154 K)
T_b: 70.1°C (343.2 K)

T_c:
P_c:

μ: 1.9 D
IP: 10.06 eV

TRANSITION PROPERTIES

$\Delta_{fus}H$ (T_m):
$\Delta_{vap}H$ (T_b): 30.24 kJ/mol
$\Delta_{vap}H$ (25°C): 32.73 kJ/mol

Vapor pressure (0°C):
Vapor pressure (25°C): 18.6 kPa
Vapor pressure (100°C):

PROPERTIES AT 25°C AND 100 kPa

	Solid	Liquid	Gas		Liquid
$\Delta_f H°$/kJ mol^{-1}:		12.2	45.2		d: 1.391 g/mL
$S°$/J mol^{-1}K^{-1}:					η: 0.471 mPa s
C_p/J mol^{-1}K^{-1}:					k:
COMMENTS: Highly toxic; flammable					

Name: *p*-Bromotoluene
Synonyms: 4-Bromo-1-methylbenzene
            4-Tolyl bromide
Mol. Form.: C_7H_7Br

CAS RN: 106-38-7
Merck No.: 1429
Mol. Wt.: 171.037

PHYSICAL CONSTANTS

T_m: 28.5°C (301.6 K)
T_b: 184.3°C (457.4 K)

T_c:
P_c:

μ:
IP: 8.67 eV

TRANSITION PROPERTIES

$\Delta_{fus}H$ (T_m): 14.90 kJ/mol
$\Delta_{vap}H$ (T_b):
$\Delta_{vap}H$ (25°C):

Vapor pressure (0°C):
Vapor pressure (25°C):
Vapor pressure (100°C): 7.41 kPa

PROPERTIES AT 25°C AND 100 kPa

	Solid	Liquid	Gas		Solid
$\Delta_f H°$/kJ mol^{-1}:					d:
$S°$/J mol^{-1}K^{-1}:					η: N/A
C_p/J mol^{-1}K^{-1}:					k:
COMMENTS:					

Name: Bromotrichloromethane   CAS RN: 75-62-7
Synonym: Carbon bromotrichloride   Merck No.:
                                                                             Mol. Wt.: 198.273
Mol. Form.: $CBrCl_3$

PHYSICAL CONSTANTS

T_m: -5.7°C (267.4 K) T_c: μ:
T_b: 105°C (378 K) P_c: IP: 10.60 eV

TRANSITION PROPERTIES

$\Delta_{fus}H$ (T_m): 2.54 kJ/mol   Vapor pressure (0°C): 1.47 kPa
$\Delta_{vap}H$ (T_b):   Vapor pressure (25°C): 5.35 kPa
$\Delta_{vap}H$ (25°C):   Vapor pressure (100°C): 87.8 kPa

PROPERTIES AT 25°C AND 100 kPa

	Solid	Liquid	Gas		Liquid
$\Delta_f H°$/kJ mol^{-1}:			-41.8	d: 2.012 g/mL	
$S°$/J mol^{-1}K^{-1}:				η:	
C_p/J mol^{-1}K^{-1}:			85.3	k:	

COMMENTS:

Name: Bromotrifluoromethane   CAS RN: 75-63-8
Synonyms: Carbon bromotrifluoride   Merck No.:
          Refrigerant 13B1   Mol. Wt.: 148.910
Mol. Form.: $CBrF_3$

PHYSICAL CONSTANTS

T_m: -172°C (101 K) T_c: 67.1°C (340.2 K) μ: 0.65 D
T_b: -57.89°C (215.26 K) P_c: 3.97 MPa   IP: 11.40 eV

TRANSITION PROPERTIES

$\Delta_{fus}H$ (T_m):   Vapor pressure (0°C):
$\Delta_{vap}H$ (T_b):   Vapor pressure (25°C):
$\Delta_{vap}H$ (25°C):   Vapor pressure (100°C): N/A

PROPERTIES AT 25°C AND 100 kPa

	Solid	Liquid	Gas		Gas
$\Delta_f H°$/kJ mol^{-1}:			-648.3	d: 6.087 g/L	
$S°$/J mol^{-1}K^{-1}:				η:	
C_p/J mol^{-1}K^{-1}:			69.3	k:	

COMMENTS: Highly toxic

Name: 1,2-Butadiene   CAS RN: 590-19-2
Synonym: Methylallene   Merck No.:
                                                                             Mol. Wt.: 54.092
Mol. Form.: C_4H_6

PHYSICAL CONSTANTS

T_m: -136.2°C (136.9 K) T_c: μ: 0.403 D
T_b: 10.9°C (284.0 K) P_c: IP: 9.03 eV

TRANSITION PROPERTIES

$\Delta_{fus}H$ (T_m):   Vapor pressure (0°C): 66.3 kPa
$\Delta_{vap}H$ (T_b): 24.02 kJ/mol   Vapor pressure (25°C): 167 kPa
$\Delta_{vap}H$ (25°C): 23.21 kJ/mol   Vapor pressure (100°C):

PROPERTIES AT 25°C AND 100 kPa

	Solid	Liquid	Gas		Gas
$\Delta_f H°$/kJ mol^{-1}:		139.0	162.3	d: 2.211 g/L	
$S°$/J mol^{-1}K^{-1}:				η:	
C_p/J mol^{-1}K^{-1}:				k:	

COMMENTS:

Name: 1,3-Butadiene
Synonym: Vinylethylene

Mol. Form.: C_4H_6

CAS RN: 106-99-0
Merck No.: 1500
Mol. Wt.: 54.092

PHYSICAL CONSTANTS

T_m: -108.9°C (164.2 K)
T_b: -4.41°C (268.74 K)

T_c: 152°C (425 K)
P_c: 4.33 MPa

μ: 0 D
IP: 9.07 eV

TRANSITION PROPERTIES

$\Delta_{fus}H$ (T_m): 7.98 kJ/mol
$\Delta_{vap}H$ (T_b): 22.47 kJ/mol
$\Delta_{vap}H$ (25°C): 20.86 kJ/mol

Vapor pressure (0°C):
Vapor pressure (25°C):
Vapor pressure (100°C):

PROPERTIES AT 25°C AND 100 kPa

	Solid	Liquid	Gas		Gas
$\Delta_f H°$/kJ mol^{-1}:		87.9	110.0		d: 2.211 g/L
$S°$/J mol^{-1}K^{-1}:		199.0			η:
C_p/J mol^{-1}K^{-1}:		123.6			k:

COMMENTS: TLV=10 ppm; carcinogen; highly toxic; very flammable

---

Name: Butanal
Synonym: Butyraldehyde

Mol. Form.: C_4H_8O

CAS RN: 123-72-8
Merck No.: 1591
Mol. Wt.: 72.107

PHYSICAL CONSTANTS

T_m: -99°C (174 K)
T_b: 74.8°C (347.9 K)

T_c: 264.1°C (537.2 K)
P_c: 4.32 MPa

μ: 2.72 D
IP: 9.84 eV

TRANSITION PROPERTIES

$\Delta_{fus}H$ (T_m): 11.09 kJ/mol
$\Delta_{vap}H$ (T_b): 31.50 kJ/mol
$\Delta_{vap}H$ (25°C): 33.68 kJ/mol

Vapor pressure (0°C):
Vapor pressure (25°C): 15.7 kPa
Vapor pressure (100°C):

PROPERTIES AT 25°C AND 100 kPa

	Solid	Liquid	Gas		Liquid
$\Delta_f H°$/kJ mol^{-1}:		-239.2	-204.8		d: 0.798 g/mL
$S°$/J mol^{-1}K^{-1}:		246.6	343.7		η:
C_p/J mol^{-1}K^{-1}:		163.7	103.4		k:

COMMENTS: Highly toxic; flammable

Name: Butane

Mol. Form.: C_4H_{10}

CAS RN: 106-97-8
Merck No.: 1507
Mol. Wt.: 58.123

PHYSICAL CONSTANTS

T_m: -138.29°C (134.86 K)
T_b: -0.5°C (272.6 K)

T_c: 151.99°C (425.14 K)
P_c: 3.784 MPa

μ:
IP: 10.53 eV

TRANSITION PROPERTIES

$\Delta_{fus}H$ (T_m): 4.66 kJ/mol
$\Delta_{vap}H$ (T_b): 22.44 kJ/mol
$\Delta_{vap}H$ (25°C): 21.02 kJ/mol

Vapor pressure (0°C): 103 kPa
Vapor pressure (25°C): 242 kPa
Vapor pressure (100°C):

PROPERTIES AT 25°C AND 100 kPa

	Solid	Liquid	Gas		Gas
$\Delta_f H°$/kJ mol^{-1}:		-146.6	-125.6		d: 2.376 g/L
$S°$/J mol^{-1}K^{-1}:			310.0		η: 7.5 μPa s
C_p/J mol^{-1}K^{-1}:		142.9	100.7		k: 0.0164 W/m K

COMMENTS: TLV=800 ppm; very flammable

Name: 1,3-Butanediol
Synonym: 1,3-Butylene glycol

Mol. Form.: $C_4H_{10}O_2$

CAS RN: 107-88-0
Merck No.: 1566
Mol. Wt.: 90.122

PHYSICAL CONSTANTS

T_m: <-50°C (<223 K) T_c: μ:
T_b: 207.5°C (480.6 K) P_c: IP:

TRANSITION PROPERTIES

$\Delta_{fus}H$ (T_m):   Vapor pressure (0°C):
$\Delta_{vap}H$ (T_b): 58.45 kJ/mol   Vapor pressure (25°C): 0.008 kPa
$\Delta_{vap}H$ (25°C): 67.80 kJ/mol   Vapor pressure (100°C): 1.37 kPa

PROPERTIES AT 25°C AND 100 kPa

	Solid	Liquid	Gas	Liquid
$\Delta_f H°$/kJ mol^{-1}:		-501.0	-433.2	d: 1.000 g/mL
$S°$/J mol^{-1}K^{-1}:				η: 98.3 mPa s
C_p/J mol^{-1}K^{-1}:				k:
COMMENTS:				

Name: 1,4-Butanediol
Synonyms: Tetramethylene glycol
          Butylene glycol
Mol. Form.: $C_4H_{10}O_2$

CAS RN: 110-63-4
Merck No.:
Mol. Wt.: 90.122

PHYSICAL CONSTANTS

T_m: 20.1°C (293.2 K) T_c: μ:
T_b: 235°C (508 K) P_c: IP:

TRANSITION PROPERTIES

$\Delta_{fus}H$ (T_m):   Vapor pressure (0°C):
$\Delta_{vap}H$ (T_b):   Vapor pressure (25°C):
$\Delta_{vap}H$ (25°C): 76.60 kJ/mol   Vapor pressure (100°C):

PROPERTIES AT 25°C AND 100 kPa

	Solid	Liquid	Gas	Liquid
$\Delta_f H°$/kJ mol^{-1}:		-503.3	-426.7	d: 1.015 g/mL
$S°$/J mol^{-1}K^{-1}:		223.4		η: 71.5 mPa s
C_p/J mol^{-1}K^{-1}:		200.1		k:
COMMENTS:				

Name: Butanenitrile
Synonyms: 1-Cyanopropane
          Propyl cyanide
Mol. Form.: C_4H_7N

CAS RN: 109-74-0
Merck No.: 1597
Mol. Wt.: 69.106

PHYSICAL CONSTANTS

T_m: -111.9°C (161.2 K) T_c: 312.3°C (585.4 K)   μ: 4.07 D
T_b: 117.6°C (390.7 K) P_c: 3.88 MPa   IP: 11.20 eV

TRANSITION PROPERTIES

$\Delta_{fus}H$ (T_m): 5.02 kJ/mol   Vapor pressure (0°C):
$\Delta_{vap}H$ (T_b): 33.68 kJ/mol   Vapor pressure (25°C): 2.55 kPa
$\Delta_{vap}H$ (25°C): 39.33 kJ/mol   Vapor pressure (100°C): 59.5 kPa

PROPERTIES AT 25°C AND 100 kPa

	Solid	Liquid	Gas	Liquid
$\Delta_f H°$/kJ mol^{-1}:		-5.8	33.6	d: 0.7865 g/mL
$S°$/J mol^{-1}K^{-1}:				η: 0.553 mPa s
C_p/J mol^{-1}K^{-1}:				k:
COMMENTS: Highly toxic; flammable				

Name: 1-Butanethiol
Synonyms: Butyl mercaptan
           Thiobutyl alcohol
Mol. Form.: $C_4H_{10}S$

CAS RN: 109-79-5
Merck No.: 1575
Mol. Wt.: 90.189

PHYSICAL CONSTANTS

T_m: -115.7°C (157.4 K)	T_c: 297.0°C (570.1 K)	μ:
T_b: 98.5°C (371.6 K)	P_c:	IP: 9.14 eV

TRANSITION PROPERTIES

$\Delta_{fus}H$ (T_m): 10.46 kJ/mol
$\Delta_{vap}H$ (T_b): 32.23 kJ/mol
$\Delta_{vap}H$ (25°C): 36.63 kJ/mol

Vapor pressure (0°C): 1.51 kPa
Vapor pressure (25°C): 6.07 kPa
Vapor pressure (100°C): 106 kPa

PROPERTIES AT 25°C AND 100 kPa

	Solid	Liquid	Gas		Liquid
$\Delta_f H°$/kJ mol^{-1}:		-124.7	-88.1		d: 0.8367 g/mL
$S°$/J mol^{-1}K^{-1}:					η: 0.512 mPa s
C_p/J mol^{-1}K^{-1}:		171.2			k:

COMMENTS: TLV=0.5 ppm; flammable

Name: 2-Butanethiol
Synonym: *sec*-Butyl mercaptan

Mol. Form.: $C_4H_{10}S$

CAS RN: 513-53-1
Merck No.: 1576
Mol. Wt.: 90.189

PHYSICAL CONSTANTS

T_m: -165°C (108 K)	T_c:	μ:
T_b: 85°C (358 K)	P_c:	IP: 9.10 eV

TRANSITION PROPERTIES

$\Delta_{fus}H$ (T_m):
$\Delta_{vap}H$ (T_b): 30.59 kJ/mol
$\Delta_{vap}H$ (25°C): 33.99 kJ/mol

Vapor pressure (0°C): 2.97 kPa
Vapor pressure (25°C): 10.8 kPa
Vapor pressure (100°C): 156 kPa

PROPERTIES AT 25°C AND 100 kPa

	Solid	Liquid	Gas		Liquid
$\Delta_f H°$/kJ mol^{-1}:		-131.0	-96.9		d: 0.824 g/mL
$S°$/J mol^{-1}K^{-1}:					η:
C_p/J mol^{-1}K^{-1}:					k:

COMMENTS:

Name: Butanoic acid
Synonym: Butyric acid

Mol. Form.: $C_4H_8O_2$

CAS RN: 107-92-6
Merck No.: 1593
Mol. Wt.: 88.106

PHYSICAL CONSTANTS

T_m: -5.7°C (267.4 K)	T_c: 351°C (624 K)	μ:
T_b: 163.75°C (436.90 K)	P_c: 4.03 MPa	IP: 10.17 eV

TRANSITION PROPERTIES

$\Delta_{fus}H$ (T_m): 11.08 kJ/mol
$\Delta_{vap}H$ (T_b):
$\Delta_{vap}H$ (25°C): 40.45 kJ/mol

Vapor pressure (0°C):
Vapor pressure (25°C): 0.102 kPa
Vapor pressure (100°C): 9.39 kPa

PROPERTIES AT 25°C AND 100 kPa

	Solid	Liquid	Gas		Liquid
$\Delta_f H°$/kJ mol^{-1}:		-533.8	-475.8		d: 0.9529 g/mL
$S°$/J mol^{-1}K^{-1}:		222.2			η: 1.43 mPa s
C_p/J mol^{-1}K^{-1}:		178.6			k:

COMMENTS:

Name: Butanoic anhydride

Synonym: Butyric anhydride

Mol. Form.: $C_8H_{14}O_3$

CAS RN: 106-31-0

Merck No.: 1594

Mol. Wt.: 158.197

PHYSICAL CONSTANTS

T_m: -65.7°C (207.4 K)	T_c:	μ:
T_b: 200°C (473 K)	P_c:	IP:

TRANSITION PROPERTIES

$\Delta_{fus}H\ (T_m)$:

$\Delta_{vap}H\ (T_b)$: 50.00 kJ/mol

$\Delta_{vap}H\ (25°C)$:

Vapor pressure (0°C):

Vapor pressure (25°C):

Vapor pressure (100°C):

PROPERTIES AT 25°C AND 100 kPa

	Solid	Liquid	Gas		Liquid
$\Delta_f H°$/kJ mol^{-1}:					d: 0.9620 g/mL
$S°$/J mol^{-1}K^{-1}:					η: 1.49 mPa s
C_p/J mol^{-1}K^{-1}:		283.7			k:
COMMENTS:					

Name: 1-Butanol

Synonyms: Butyl alcohol

Propylcarbinol

Mol. Form.: $C_4H_{10}O$

CAS RN: 71-36-3

Merck No.: 1540

Mol. Wt.: 74.123

PHYSICAL CONSTANTS

T_m: -89.8°C (183.3 K)	T_c: 289.90°C (563.05 K)	μ: 1.66 D
T_b: 117.73°C (390.88 K)	P_c: 4.423 MPa	IP: 10.06 eV

TRANSITION PROPERTIES

$\Delta_{fus}H\ (T_m)$: 9.28 kJ/mol

$\Delta_{vap}H\ (T_b)$: 43.29 kJ/mol

$\Delta_{vap}H\ (25°C)$: 52.35 kJ/mol

Vapor pressure (0°C):

Vapor pressure (25°C): 0.860 kPa

Vapor pressure (100°C): 51.9 kPa

PROPERTIES AT 25°C AND 100 kPa

	Solid	Liquid	Gas		Liquid
$\Delta_f H°$/kJ mol^{-1}:		-327.3	-275.0		d: 0.8058 g/mL
$S°$/J mol^{-1}K^{-1}:		225.8			η: 2.54 mPa s
C_p/J mol^{-1}K^{-1}:		177.2			k: 0.154 W/m K
COMMENTS: TLV=50 ppm; highly toxic; flammable					

Name: 2-Butanol

Synonym: *sec*-Butyl alcohol

Mol. Form.: $C_4H_{10}O$

CAS RN: 78-92-2

Merck No.: 1541

Mol. Wt.: 74.123

PHYSICAL CONSTANTS

T_m: -114.7°C (158.4 K)	T_c: 262.90°C (536.05 K)	μ:
T_b: 99.51°C (372.66 K)	P_c: 4.179 MPa	IP: 9.88 eV

TRANSITION PROPERTIES

$\Delta_{fus}H\ (T_m)$:

$\Delta_{vap}H\ (T_b)$: 40.75 kJ/mol

$\Delta_{vap}H\ (25°C)$: 49.72 kJ/mol

Vapor pressure (0°C):

Vapor pressure (25°C): 2.32 kPa

Vapor pressure (100°C): 103 kPa

PROPERTIES AT 25°C AND 100 kPa

	Solid	Liquid	Gas		Liquid
$\Delta_f H°$/kJ mol^{-1}:		-342.6	-292.9		d: 0.8026 g/mL
$S°$/J mol^{-1}K^{-1}:		214.9	359.5		η: 3.10 mPa s
C_p/J mol^{-1}K^{-1}:		196.9	112.7		k:
COMMENTS: TLV=100 ppm; highly toxic; flammable					

Name: 2-Butanone

Synonyms: Methyl ethyl ketone
Ethyl methyl ketone

Mol. Form.: C_4H_8O

CAS RN: 78-93-3
Merck No.: 5991
Mol. Wt.: 72.107

PHYSICAL CONSTANTS

T_m: -86.67°C (186.48 K)
T_b: 79.59°C (352.74 K)

T_c: 263.63°C (536.78 K)
P_c: 4.207 MPa

μ: 2.78 D
IP: 9.51 eV

TRANSITION PROPERTIES

$\Delta_{fus}H$ (T_m): 8.44 kJ/mol
$\Delta_{vap}H$ (T_b): 31.30 kJ/mol
$\Delta_{vap}H$ (25°C): 34.79 kJ/mol

Vapor pressure (0°C): 3.51 kPa
Vapor pressure (25°C): 12.6 kPa
Vapor pressure (100°C): 184 kPa

PROPERTIES AT 25°C AND 100 kPa

	Solid	Liquid	Gas		Liquid
$\Delta_fH°$/kJ mol^{-1}:		-273.3	-238.7		d: 0.7994 g/mL
$S°$/J mol^{-1}K^{-1}:		239.1	339.9		η: 0.405 mPa s
C_p/J mol^{-1}K^{-1}:		158.7	101.7		k: 0.145 W/m K

COMMENTS: TLV=200 ppm; flammable

Name: *trans*-2-Butenal

Synonym: *trans*-Crotonaldehyde

Mol. Form.: C_4H_6O

CAS RN: 123-73-9
Merck No.: 2599
Mol. Wt.: 70.091

PHYSICAL CONSTANTS

T_m: -76°C (197 K)
T_b: 102.2°C (375.3 K)

T_c:
P_c:

μ: 3.67 D
IP: 9.73 eV

TRANSITION PROPERTIES

$\Delta_{fus}H$ (T_m):
$\Delta_{vap}H$ (T_b):
$\Delta_{vap}H$ (25°C): 38.1 kJ/mol

Vapor pressure (0°C):
Vapor pressure (25°C): 4.92 kPa
Vapor pressure (100°C): 93.2 kPa

PROPERTIES AT 25°C AND 100 kPa

	Solid	Liquid	Gas		Liquid
$\Delta_fH°$/kJ mol^{-1}:		-138.7	-100.6		d: 0.847 g/mL
$S°$/J mol^{-1}K^{-1}:					η:
C_p/J mol^{-1}K^{-1}:		95.4			k:

COMMENTS: Highly toxic; flammable

Name: 1-Butene

Synonym: 1-Butylene

Mol. Form.: C_4H_8

CAS RN: 106-98-9
Merck No.: 1513
Mol. Wt.: 56.107

PHYSICAL CONSTANTS

T_m: -185.35°C (87.80 K)
T_b: -6.26°C (266.89 K)

T_c: 146.42°C (419.57 K)
P_c: 4.023 MPa

μ:
IP: 9.58 eV

TRANSITION PROPERTIES

$\Delta_{fus}H$ (T_m):
$\Delta_{vap}H$ (T_b): 22.07 kJ/mol
$\Delta_{vap}H$ (25°C): 20.22 kJ/mol

Vapor pressure (0°C): 128 kPa
Vapor pressure (25°C): 296 kPa
Vapor pressure (100°C): 1796 kPa

PROPERTIES AT 25°C AND 100 kPa

	Solid	Liquid	Gas		Gas
$\Delta_fH°$/kJ mol^{-1}:		-20.5	0.1		d: 2.293 g/L
$S°$/J mol^{-1}K^{-1}:		227.0			η:
C_p/J mol^{-1}K^{-1}:		118.0			k:

COMMENTS: Flammable

Name: *cis*-2-Butene   CAS RN: 590-18-1
Synonym: *cis*-2-Butylene   Merck No.: 1514
                                                                                       Mol. Wt.: 56.107
Mol. Form.: C_4H_8

PHYSICAL CONSTANTS

T_m: -138.9°C (134.2 K) T_c: 162.43°C (435.58 K)   μ: 0.253 D
T_b: 3.71°C (276.86 K) P_c: 4.197 MPa   IP: 9.11 eV

TRANSITION PROPERTIES

$\Delta_{fus}H$ (T_m): 7.58 kJ/mol   Vapor pressure (0°C): 87.7 kPa
$\Delta_{vap}H$ (T_b): 23.34 kJ/mol   Vapor pressure (25°C): 214 kPa
$\Delta_{vap}H$ (25°C): 22.16 kJ/mol   Vapor pressure (100°C): 1431 kPa

PROPERTIES AT 25°C AND 100 kPa

	Solid	Liquid	Gas		Gas
$\Delta_f H°$/kJ mol^{-1}:		-29.7	-7.1	d:	2.293 g/L
$S°$/J mol^{-1}K^{-1}:		219.9		η:	
C_p/J mol^{-1}K^{-1}:		127.0		k:	

COMMENTS: Flammable

Name: *trans*-2-Butene   CAS RN: 624-64-6
Synonym: *trans*-2-Butylene   Merck No.: 1514
                                                                                       Mol. Wt.: 56.107
Mol. Form.: C_4H_8

PHYSICAL CONSTANTS

T_m: -105.53°C (167.62 K) T_c: 155.48°C (428.63 K)   μ: 0 D
T_b: 0.88°C (274.03 K) P_c: 3.985 MPa   IP: 9.10 eV

TRANSITION PROPERTIES

$\Delta_{fus}H$ (T_m):   Vapor pressure (0°C): 98.0 kPa
$\Delta_{vap}H$ (T_b): 22.72 kJ/mol   Vapor pressure (25°C): 234 kPa
$\Delta_{vap}H$ (25°C): 21.40 kJ/mol   Vapor pressure (100°C): 1528 kPa

PROPERTIES AT 25°C AND 100 kPa

	Solid	Liquid	Gas		Gas
$\Delta_f H°$/kJ mol^{-1}:		-33.0	-11.4	d:	2.293 g/L
$S°$/J mol^{-1}K^{-1}:				η:	
C_p/J mol^{-1}K^{-1}:				k:	

COMMENTS: Flammable

Name: 2-Butoxyethanol   CAS RN: 111-76-2
Synonyms: Ethylene glycol monobutyl ether   Merck No.: 1559
          Butyl cellosolve   Mol. Wt.: 118.176
Mol. Form.: $C_6H_{14}O_2$

PHYSICAL CONSTANTS

T_m: -74.8°C (198.3 K) T_c: 360.8°C (633.9 K)   μ:
T_b: 168.4°C (441.5 K) P_c: IP:

TRANSITION PROPERTIES

$\Delta_{fus}H$ (T_m):   Vapor pressure (0°C):
$\Delta_{vap}H$ (T_b):   Vapor pressure (25°C): 0.150 kPa
$\Delta_{vap}H$ (25°C): 56.59 kJ/mol   Vapor pressure (100°C): 8.71 kPa

PROPERTIES AT 25°C AND 100 kPa

	Solid	Liquid	Gas		Liquid
$\Delta_f H°$/kJ mol^{-1}:				d:	0.8964 g/mL
$S°$/J mol^{-1}K^{-1}:				η:	3.15 mPa s
C_p/J mol^{-1}K^{-1}:		281.0		k:	

COMMENTS: TLV=25 ppm

Name: Butyl acetate
Synonym: Butyl ethanoate

Mol. Form.: $C_6H_{12}O_2$

CAS RN: 123-86-4
Merck No.: 1535
Mol. Wt.: 116.160

PHYSICAL CONSTANTS

T_m: -78°C (195 K) T_c: 306°C (579 K)   μ:
T_b: 126.1°C (399.2 K) P_c: IP: 10.00 eV

TRANSITION PROPERTIES

$\Delta_{fus}H$ (T_m):   Vapor pressure (0°C):
$\Delta_{vap}H$ (T_b): 36.28 kJ/mol   Vapor pressure (25°C): 1.66 kPa
$\Delta_{vap}H$ (25°C): 43.86 kJ/mol   Vapor pressure (100°C): 46.4 kPa

PROPERTIES AT 25°C AND 100 kPa

	Solid	Liquid	Gas		Liquid
$\Delta_f H°$/kJ mol^{-1}:		-529.2	-485.6		d: 0.8761 g/mL
$S°$/J mol^{-1}K^{-1}:					η: 0.685 mPa s
C_p/J mol^{-1}K^{-1}:		227.8			k:

COMMENTS: TLV=150 ppm; flammable

Name: *sec*-Butyl acetate
Synonyms: 1-Methylpropyl acetate
          sec-Butyl ethanoate
Mol. Form.: $C_6H_{12}O_2$

CAS RN: 105-46-4
Merck No.: 1536
Mol. Wt.: 116.160

PHYSICAL CONSTANTS

T_m: -98.9°C (174.2 K) T_c: μ:
T_b: 112°C (385 K) P_c: IP: 9.90 eV

TRANSITION PROPERTIES

$\Delta_{fus}H$ (T_m):   Vapor pressure (0°C):
$\Delta_{vap}H$ (T_b):   Vapor pressure (25°C): 3.20 kPa
$\Delta_{vap}H$ (25°C):   Vapor pressure (100°C):

PROPERTIES AT 25°C AND 100 kPa

	Solid	Liquid	Gas		Liquid
$\Delta_f H°$/kJ mol^{-1}:					d: 0.8694 g/mL
$S°$/J mol^{-1}K^{-1}:					η: 0.65 mPa s
C_p/J mol^{-1}K^{-1}:					k:

COMMENTS: TLV=200 ppm; flammable

Name: Butyl acrylate
Synonym: Butyl-2-propenoate

Mol. Form.: $C_7H_{12}O_2$

CAS RN: 141-32-2
Merck No.: 1539
Mol. Wt.: 128.171

PHYSICAL CONSTANTS

T_m: -64.6°C (208.5 K) T_c: μ:
T_b: 145°C (418 K) P_c: IP:

TRANSITION PROPERTIES

$\Delta_{fus}H$ (T_m):   Vapor pressure (0°C): 0.140 kPa
$\Delta_{vap}H$ (T_b):   Vapor pressure (25°C): 0.731 kPa
$\Delta_{vap}H$ (25°C):   Vapor pressure (100°C): 23.2 kPa

PROPERTIES AT 25°C AND 100 kPa

	Solid	Liquid	Gas		Liquid
$\Delta_f H°$/kJ mol^{-1}:					d: 0.894 g/mL
$S°$/J mol^{-1}K^{-1}:					η:
C_p/J mol^{-1}K^{-1}:					k:

COMMENTS: TLV=10 ppm; highly toxic

Name: Butylamine
Synonyms: 1-Butanamine
         1-Aminobutane
Mol. Form.: $C_4H_{11}N$

CAS RN: 109-73-9
Merck No.: 1543
Mol. Wt.: 73.138

PHYSICAL CONSTANTS

T_m: -49.10°C (224.05 K)
T_b: 77.00°C (350.15 K)

T_c: 258.8°C (531.9 K)
P_c: 4.25 MPa

μ: 1.0 D
IP: 8.71 eV

TRANSITION PROPERTIES

$\Delta_{fus}H$ (T_m):
$\Delta_{vap}H$ (T_b): 31.81 kJ/mol
$\Delta_{vap}H$ (25°C): 35.72 kJ/mol

Vapor pressure (0°C): 3.17 kPa
Vapor pressure (25°C): 12.2 kPa
Vapor pressure (100°C): 200 kPa

PROPERTIES AT 25°C AND 100 kPa

	Solid	Liquid	Gas		Liquid
$\Delta_f H°$/kJ mol^{-1}:		-127.7	-92.0		d: 0.7369 g/mL
$S°$/J mol^{-1}K^{-1}:					η: 0.574 mPa s
C_p/J mol^{-1}K^{-1}:		179.2			k:
COMMENTS: TLV=5 ppm; flammable					

Name: *sec*-Butylamine
Synonyms: 2-Butanamine
         2-Aminobutane
Mol. Form.: $C_4H_{11}N$

CAS RN: 13952-84-6
Merck No.: 1544
Mol. Wt.: 73.138

PHYSICAL CONSTANTS

T_m: -104.5°C (168.6 K)
T_b: 62.73°C (335.88 K)

T_c: 241.2°C (514.3 K)
P_c: 4.20 MPa

μ:
IP: 8.70 eV

TRANSITION PROPERTIES

$\Delta_{fus}H$ (T_m):
$\Delta_{vap}H$ (T_b): 29.92 kJ/mol
$\Delta_{vap}H$ (25°C): 32.85 kJ/mol

Vapor pressure (0°C): 6.62 kPa
Vapor pressure (25°C): 23.0 kPa
Vapor pressure (100°C): 308 kPa

PROPERTIES AT 25°C AND 100 kPa

	Solid	Liquid	Gas		Liquid
$\Delta_f H°$/kJ mol^{-1}:		-137.5	-104.6		d: 0.7200 g/mL
$S°$/J mol^{-1}K^{-1}:					η:
C_p/J mol^{-1}K^{-1}:					k:
COMMENTS: Highly toxic; flammable					

Name: *tert*-Butylamine
Synonyms: 2-Methyl-2-propanamine
         2-Amino-2-methylpropane
Mol. Form.: $C_4H_{11}N$

CAS RN: 75-64-9
Merck No.: 1545
Mol. Wt.: 73.138

PHYSICAL CONSTANTS

T_m: -66.95°C (206.20 K)
T_b: 44.04°C (317.19 K)

T_c: 210.8°C (483.9 K)
P_c: 3.84 MPa

μ:
IP: 8.64 eV

TRANSITION PROPERTIES

$\Delta_{fus}H$ (T_m): 0.88 kJ/mol
$\Delta_{vap}H$ (T_b): 28.27 kJ/mol
$\Delta_{vap}H$ (25°C): 29.64 kJ/mol

Vapor pressure (0°C):
Vapor pressure (25°C): 48.4 kPa
Vapor pressure (100°C):

PROPERTIES AT 25°C AND 100 kPa

	Solid	Liquid	Gas		Liquid
$\Delta_f H°$/kJ mol^{-1}:		-150.6	-120.9		d: 0.6901 g/mL
$S°$/J mol^{-1}K^{-1}:					η:
C_p/J mol^{-1}K^{-1}:		192.1			k:
COMMENTS:					

Name: Butylbenzene

Synonym: 1-Phenylbutane

Mol. Form.: $C_{10}H_{14}$

CAS RN: 104-51-8

Merck No.: 1549

Mol. Wt.: 134.221

PHYSICAL CONSTANTS

T_m: -87.9°C (185.2 K) T_c: 387.4°C (660.5 K)   μ:

T_b: 183.31°C (456.46 K) P_c: 2.887 MPa   IP: 8.69 eV

TRANSITION PROPERTIES

$\Delta_{fus}H$ (T_m): 11.22 kJ/mol   Vapor pressure (0°C):

$\Delta_{vap}H$ (T_b): 38.87 kJ/mol   Vapor pressure (25°C): 0.150 kPa

$\Delta_{vap}H$ (25°C): 51.36 kJ/mol   Vapor pressure (100°C): 7.41 kPa

PROPERTIES AT 25°C AND 100 kPa

	Solid	Liquid	Gas	Liquid
$\Delta_f H°$/kJ mol^{-1}:		-63.2	-13.1	d: 0.856 g/mL
$S°$/J mol^{-1}K^{-1}:		321.2		η: 0.950 mPa s
C_p/J mol^{-1}K^{-1}:		243.4		k:
COMMENTS:				

Name: *sec*-Butylbenzene

Synonyms: (1-Methylpropyl)benzene

          2-Phenylbutane

Mol. Form.: $C_{10}H_{14}$

CAS RN: 135-98-8

Merck No.: 1550

Mol. Wt.: 134.221

PHYSICAL CONSTANTS

T_m: -82.7°C (190.4 K) T_c: μ:

T_b: 173.5°C (446.6 K) P_c: IP: 8.68 eV

TRANSITION PROPERTIES

$\Delta_{fus}H$ (T_m):   Vapor pressure (0°C):

$\Delta_{vap}H$ (T_b):   Vapor pressure (25°C): 0.230 kPa

$\Delta_{vap}H$ (25°C): 47.98 kJ/mol   Vapor pressure (100°C): 10.6 kPa

PROPERTIES AT 25°C AND 100 kPa

	Solid	Liquid	Gas	Liquid
$\Delta_f H°$/kJ mol^{-1}:		-66.4	-17.4	d: 0.8580 g/mL
$S°$/J mol^{-1}K^{-1}:				η:
C_p/J mol^{-1}K^{-1}:				k:
COMMENTS:				

Name: *tert*-Butylbenzene

Synonyms: (1,1-Dimethylethyl)benzene

          2-Methyl-2-phenylpropane

Mol. Form.: $C_{10}H_{14}$

CAS RN: 98-06-6

Merck No.: 1551

Mol. Wt.: 134.221

PHYSICAL CONSTANTS

T_m: -58.1°C (215.0 K) T_c: μ: 0.83 D

T_b: 168.5°C (441.6 K) P_c: IP: 8.64 eV

TRANSITION PROPERTIES

$\Delta_{fus}H$ (T_m):   Vapor pressure (0°C):

$\Delta_{vap}H$ (T_b):   Vapor pressure (25°C): 0.280 kPa

$\Delta_{vap}H$ (25°C): 47.71 kJ/mol   Vapor pressure (100°C): 12.1 kPa

PROPERTIES AT 25°C AND 100 kPa

	Solid	Liquid	Gas	Liquid
$\Delta_f H°$/kJ mol^{-1}:		-70.7	-22.6	d: 0.862 g/mL
$S°$/J mol^{-1}K^{-1}:				η:
C_p/J mol^{-1}K^{-1}:		238.0		k:
COMMENTS:				

Name: Butyl butanoate
Synonym: Butyl butyrate

Mol. Form.: $C_8H_{16}O_2$

CAS RN: 109-21-7
Merck No.: 1556
Mol. Wt.: 144.214

PHYSICAL CONSTANTS

T_m: -91.5°C (181.6 K) T_c: μ:
T_b: 166°C (439 K) P_c: IP:

TRANSITION PROPERTIES

$\Delta_{fus}H\ (T_m)$:   Vapor pressure (0°C):
$\Delta_{vap}H\ (T_b)$:   Vapor pressure (25°C):
$\Delta_{vap}H\ (25°C)$:   Vapor pressure (100°C):

PROPERTIES AT 25°C AND 100 kPa

	Solid	Liquid	Gas	Liquid
$\Delta_f H°$/kJ mol^{-1}:				d: 0.866 g/mL
$S°$/J mol^{-1}K^{-1}:				η: 0.94 mPa s
C_p/J mol^{-1}K^{-1}:				k:
COMMENTS:				

Name: Butyl ethyl ether
Synonyms: 1-Ethoxybutane
          Ethyl butyl ether
Mol. Form.: $C_6H_{14}O$

CAS RN: 628-81-9
Merck No.:
Mol. Wt.: 102.177

PHYSICAL CONSTANTS

T_m: -124°C (149 K) T_c: μ:
T_b: 92.3°C (365.4 K) P_c: IP: 9.36 eV

TRANSITION PROPERTIES

$\Delta_{fus}H\ (T_m)$:   Vapor pressure (0°C):
$\Delta_{vap}H\ (T_b)$: 31.63 kJ/mol   Vapor pressure (25°C): 7.46 kPa
$\Delta_{vap}H\ (25°C)$: 36.32 kJ/mol   Vapor pressure (100°C): 127 kPa

PROPERTIES AT 25°C AND 100 kPa

	Solid	Liquid	Gas	Liquid
$\Delta_f H°$/kJ mol^{-1}:				d: 0.7448 g/mL
$S°$/J mol^{-1}K^{-1}:				η: 0.397 mPa s
C_p/J mol^{-1}K^{-1}:		159.0		k:
COMMENTS: Flammable				

Name: Butyl formate
Synonym: Butyl methanoate

Mol. Form.: $C_5H_{10}O_2$

CAS RN: 592-84-7
Merck No.:
Mol. Wt.: 102.133

PHYSICAL CONSTANTS

T_m: -91.5°C (181.6 K) T_c: μ:
T_b: 106.1°C (379.2 K) P_c: IP: 10.50 eV

TRANSITION PROPERTIES

$\Delta_{fus}H\ (T_m)$:   Vapor pressure (0°C):
$\Delta_{vap}H\ (T_b)$: 36.58 kJ/mol   Vapor pressure (25°C): 3.53 kPa
$\Delta_{vap}H\ (25°C)$: 41.11 kJ/mol   Vapor pressure (100°C): 86.8 kPa

PROPERTIES AT 25°C AND 100 kPa

	Solid	Liquid	Gas	Liquid
$\Delta_f H°$/kJ mol^{-1}:				d: 0.884 g/mL
$S°$/J mol^{-1}K^{-1}:				η: 0.644 mPa s
C_p/J mol^{-1}K^{-1}:				k:
COMMENTS:				

Name: *tert*-Butyl methyl ether
Synonyms: 2-Methyl-2-methoxypropane
             Methyl-*tert*-butyl ether
Mol. Form.: $C_5H_{12}O$

CAS RN: 1634-04-4
Merck No.: 5954
Mol. Wt.: 88.150

PHYSICAL CONSTANTS

T_m: -108.6°C (164.5 K)	T_c: 224.0°C (497.1 K)	μ:
T_b: 55.2°C (328.3 K)	P_c: 3.430 MPa	IP: 9.24 eV

TRANSITION PROPERTIES

$\Delta_{fus}H$ (T_m):   Vapor pressure (0°C):
$\Delta_{vap}H$ (T_b): 27.94 kJ/mol   Vapor pressure (25°C): 33.6 kPa
$\Delta_{vap}H$ (25°C): 29.82 kJ/mol   Vapor pressure (100°C):

PROPERTIES AT 25°C AND 100 kPa

	Solid	Liquid	Gas		Liquid
$\Delta_f H°$/kJ mol^{-1}:		-313.6	-283.5	d:	0.737 g/mL
$S°$/J mol^{-1}K^{-1}:		265.3		η:	
C_p/J mol^{-1}K^{-1}:		187.5		k:	
COMMENTS: Highly toxic					

Name: *p-tert*-Butylphenol
Synonym: 4-(1,1-Dimethylethyl)phenol

Mol. Form.: $C_{10}H_{14}O$

CAS RN: 98-54-4
Merck No.: 1584
Mol. Wt.: 150.221

PHYSICAL CONSTANTS

T_m: 98°C (371 K)	T_c:	μ:
T_b: 237°C (510 K)	P_c:	IP: 7.80 eV

TRANSITION PROPERTIES

$\Delta_{fus}H$ (T_m):   Vapor pressure (0°C):
$\Delta_{vap}H$ (T_b):   Vapor pressure (25°C):
$\Delta_{vap}H$ (25°C):   Vapor pressure (100°C):

PROPERTIES AT 25°C AND 100 kPa

	Solid	Liquid	Gas		Solid
$\Delta_f H°$/kJ mol^{-1}:				d:	
$S°$/J mol^{-1}K^{-1}:				η:	N/A
C_p/J mol^{-1}K^{-1}:				k:	
COMMENTS:					

Name: Butyl propanoate
Synonym: Butyl propionate

Mol. Form.: $C_7H_{14}O_2$

CAS RN: 590-01-2
Merck No.: 1587
Mol. Wt.: 130.187

PHYSICAL CONSTANTS

T_m: -89°C (184 K)	T_c:	μ:
T_b: 146.8°C (419.9 K)	P_c:	IP:

TRANSITION PROPERTIES

$\Delta_{fus}H$ (T_m):   Vapor pressure (0°C):
$\Delta_{vap}H$ (T_b):   Vapor pressure (25°C):
$\Delta_{vap}H$ (25°C):   Vapor pressure (100°C):

PROPERTIES AT 25°C AND 100 kPa

	Solid	Liquid	Gas		Liquid
$\Delta_f H°$/kJ mol^{-1}:				d:	0.871 g/mL
$S°$/J mol^{-1}K^{-1}:				η:	0.78 mPa s
C_p/J mol^{-1}K^{-1}:				k:	
COMMENTS: Flammable					

Name: Butyl stearate

Synonym: Butyl octadecanoate

Mol. Form.: $C_{22}H_{44}O_2$

CAS RN: 123-95-5

Merck No.: 1589

Mol. Wt.: 340.590

PHYSICAL CONSTANTS

T_m: 26.3°C (299.4 K)	T_c:	μ:
T_b: 343°C (616 K)	P_c:	IP:

TRANSITION PROPERTIES

$\Delta_{fus}H$ (T_m): 56.90 kJ/mol	Vapor pressure (0°C):
$\Delta_{vap}H$ (T_b):	Vapor pressure (25°C):
$\Delta_{vap}H$ (25°C):	Vapor pressure (100°C):

PROPERTIES AT 25°C AND 100 kPa

	Solid	Liquid	Gas	Solid
$\Delta_f H°$/kJ mol^{-1}:				d: 0.854 g/mL
$S°$/J mol^{-1}K^{-1}:				η: N/A
C_p/J mol^{-1}K^{-1}:				k:
COMMENTS:				

Name: *p-tert*-Butyltoluene

Synonym: 1-(1,1-Dimethylethyl)-4-methylbenzene

Mol. Form.: $C_{11}H_{16}$

CAS RN: 98-51-1

Merck No.:

Mol. Wt.: 148.248

PHYSICAL CONSTANTS

T_m: -52°C (221 K)	T_c:	μ:
T_b: 190°C (463 K)	P_c:	IP: 8.28 eV

TRANSITION PROPERTIES

$\Delta_{fus}H$ (T_m):	Vapor pressure (0°C):
$\Delta_{vap}H$ (T_b):	Vapor pressure (25°C): 0.090 kPa
$\Delta_{vap}H$ (25°C):	Vapor pressure (100°C): 5.37 kPa

PROPERTIES AT 25°C AND 100 kPa

	Solid	Liquid	Gas	Liquid
$\Delta_f H°$/kJ mol^{-1}:				d: 0.857 g/mL
$S°$/J mol^{-1}K^{-1}:				η:
C_p/J mol^{-1}K^{-1}:				k:
COMMENTS: TLV=10 ppm; highly toxic				

Name: Butyl vinyl ether

Synonym: Vinyl butyl ether

Mol. Form.: $C_6H_{12}O$

CAS RN: 111-34-2

Merck No.:

Mol. Wt.: 100.161

PHYSICAL CONSTANTS

T_m: -92°C (181 K)	T_c:	μ:
T_b: 94°C (367 K)	P_c:	IP:

TRANSITION PROPERTIES

$\Delta_{fus}H$ (T_m):	Vapor pressure (0°C):
$\Delta_{vap}H$ (T_b): 31.58 kJ/mol	Vapor pressure (25°C): 6.80 kPa
$\Delta_{vap}H$ (25°C): 36.17 kJ/mol	Vapor pressure (100°C): 124 kPa

PROPERTIES AT 25°C AND 100 kPa

	Solid	Liquid	Gas	Liquid
$\Delta_f H°$/kJ mol^{-1}:		-218.8	-184.5	d: 0.785 g/mL
$S°$/J mol^{-1}K^{-1}:				η:
C_p/J mol^{-1}K^{-1}:		232.0		k:
COMMENTS: Flammable				

Name: 1-Butyne
Synonym: Ethylacetylene

Mol. Form.: C_4H_6

CAS RN: 107-00-6
Merck No.:
Mol. Wt.: 54.092

PHYSICAL CONSTANTS

T_m: -125.72°C (147.43 K) T_c: 190.6°C (463.7 K) μ: 0.80 D
T_b: 8.08°C (281.23 K) P_c: IP: 10.18 eV

TRANSITION PROPERTIES

$\Delta_{fus}H\ (T_m)$:   Vapor pressure (0°C): 73.1 kPa
$\Delta_{vap}H\ (T_b)$: 24.52 kJ/mol   Vapor pressure (25°C):
$\Delta_{vap}H$ (25°C): 23.35 kJ/mol   Vapor pressure (100°C):

PROPERTIES AT 25°C AND 100 kPa

	Solid	Liquid	Gas		Gas
$\Delta_f H°$/kJ mol^{-1}:		141.9	165.2	d:	2.211 g/L
$S°$/J mol^{-1}K^{-1}:				η:	
C_p/J mol^{-1}K^{-1}:				k:	
COMMENTS:					

Name: 2-Butyne
Synonym: Dimethylacetylene

Mol. Form.: C_4H_6

CAS RN: 503-17-3
Merck No.:
Mol. Wt.: 54.092

PHYSICAL CONSTANTS

T_m: -32.36°C (240.79 K) T_c: 215.6°C (488.7 K) μ: 0 D
T_b: 26.97°C (300.12 K) P_c: IP: 9.56 eV

TRANSITION PROPERTIES

$\Delta_{fus}H\ (T_m)$: 9.23 kJ/mol   Vapor pressure (0°C): 33.7 kPa
$\Delta_{vap}H\ (T_b)$:   Vapor pressure (25°C): 94.3 kPa
$\Delta_{vap}H$ (25°C): 26.6 kJ/mol   Vapor pressure (100°C):

PROPERTIES AT 25°C AND 100 kPa

	Solid	Liquid	Gas		Liquid
$\Delta_f H°$/kJ mol^{-1}:		119.1	145.7	d:	0.688 g/mL
$S°$/J mol^{-1}K^{-1}:				η:	
C_p/J mol^{-1}K^{-1}:				k:	0.121 W/m K
COMMENTS:					

Name: γ-Butyrolactone
Synonyms: Oxolan-2-one
　　　　　Dihydro-2(3H)-furanone
Mol. Form.: $C_4H_6O_2$

CAS RN: 96-48-0
Merck No.: 1596
Mol. Wt.: 86.090

PHYSICAL CONSTANTS

T_m: -43.37°C (229.78 K) T_c: μ: 4.27 D
T_b: 204°C (477 K) P_c: IP:

TRANSITION PROPERTIES

$\Delta_{fus}H\ (T_m)$: 9.57 kJ/mol   Vapor pressure (0°C):
$\Delta_{vap}H\ (T_b)$: 52.22 kJ/mol   Vapor pressure (25°C): 0.430 kPa
$\Delta_{vap}H$ (25°C):   Vapor pressure (100°C):

PROPERTIES AT 25°C AND 100 kPa

	Solid	Liquid	Gas		Liquid
$\Delta_f H°$/kJ mol^{-1}:				d:	1.40 g/mL
$S°$/J mol^{-1}K^{-1}:				η:	1.7 mPa s
C_p/J mol^{-1}K^{-1}:		141.4		k:	
COMMENTS:					

Name: Cadmium

Mol. Form.: Cd

CAS RN: 7440-43-9
Merck No.: 1611
Mol. Wt.: 112.411

PHYSICAL CONSTANTS

T_m: 321.07°C (594.22 K)
T_b: 767°C (1040 K)

T_c:
P_c:

μ:
IP: 8.99 eV

TRANSITION PROPERTIES

$\Delta_{fus}H$ (T_m): 6.19 kJ/mol
$\Delta_{vap}H$ (T_b): 99.87 kJ/mol
$\Delta_{vap}H$ (25°C):

Vapor pressure (0°C): N/A
Vapor pressure (25°C): N/A
Vapor pressure (100°C): N/A

PROPERTIES AT 25°C AND 100 kPa

	Solid	Liquid	Gas		Solid
$\Delta_f H°$/kJ mol^{-1}:	0.0		111.8		d: 8.69 g/mL
$S°$/J mol^{-1}K^{-1}:	51.8		167.7		η: N/A
C_p/J mol^{-1}K^{-1}:	26.0		20.8		k: 96.9 W/m K

COMMENTS: Carcinogen; highly toxic

Name: Cadmium bromide
Synonym: Cadmium dibromide

Mol. Form.: Br$_2$Cd

CAS RN: 7789-42-6
Merck No.: 1613
Mol. Wt.: 272.219

PHYSICAL CONSTANTS

T_m: 568°C (841 K)
T_b: 844°C (1117 K)

T_c:
P_c:

μ:
IP:

TRANSITION PROPERTIES

$\Delta_{fus}H$ (T_m): 20.90 kJ/mol
$\Delta_{vap}H$ (T_b): 115.00 kJ/mol
$\Delta_{vap}H$ (25°C):

Vapor pressure (0°C): N/A
Vapor pressure (25°C): N/A
Vapor pressure (100°C): N/A

PROPERTIES AT 25°C AND 100 kPa

	Solid	Liquid	Gas	Solid
$\Delta_f H°$/kJ mol^{-1}:	-316.2			d: 5.19 g/mL
$S°$/J mol^{-1}K^{-1}:	137.2			η: N/A
C_p/J mol^{-1}K^{-1}:	76.7			k:

COMMENTS: Carcinogen

Name: Cadmium chloride
Synonym: Cadmium dichloride

Mol. Form.: CdCl$_2$

CAS RN: 10108-64-2
Merck No.: 1615
Mol. Wt.: 183.316

PHYSICAL CONSTANTS

T_m: 564°C (837 K)
T_b: 960°C (1233 K)

T_c:
P_c:

μ:
IP:

TRANSITION PROPERTIES

$\Delta_{fus}H$ (T_m): 48.58 kJ/mol
$\Delta_{vap}H$ (T_b): 124.30 kJ/mol
$\Delta_{vap}H$ (25°C):

Vapor pressure (0°C): N/A
Vapor pressure (25°C): N/A
Vapor pressure (100°C): N/A

PROPERTIES AT 25°C AND 100 kPa

	Solid	Liquid	Gas	Solid
$\Delta_f H°$/kJ mol^{-1}:	-391.5			d: 4.08 g/mL
$S°$/J mol^{-1}K^{-1}:	115.3			η: N/A
C_p/J mol^{-1}K^{-1}:	74.7			k:

COMMENTS: Carcinogen; highly toxic

Name: Cadmium fluoride

Synonym: Cadmium difluoride

Mol. Form.: CdF_2

CAS RN: 7790-79-6

Merck No.: 1617

Mol. Wt.: 150.408

PHYSICAL CONSTANTS

T_m: 1110°C (1383 K)	T_c:	μ:
T_b: 1748°C (2021 K)	P_c:	IP:

TRANSITION PROPERTIES

$\Delta_{fus}H$ (T_m): 22.60 kJ/mol

$\Delta_{vap}H$ (T_b): 214.00 kJ/mol

$\Delta_{vap}H$ (25°C):

Vapor pressure (0°C): N/A

Vapor pressure (25°C): N/A

Vapor pressure (100°C): N/A

PROPERTIES AT 25°C AND 100 kPa

	Solid	Liquid	Gas		Solid
$\Delta_f H°$/kJ mol^{-1}:	-700.4			d:	6.33 g/mL
$S°$/J mol^{-1}K^{-1}:	77.4			η:	N/A
C_p/J mol^{-1}K^{-1}:				k:	

COMMENTS: Carcinogen; highly toxic

Name: Calcium

CAS RN: 7440-70-2

Merck No.: 1642

Mol. Wt.: 40.078

Mol. Form.: Ca

PHYSICAL CONSTANTS

T_m: 842°C (1115 K)	T_c:	μ:
T_b: 1484°C (1757 K)	P_c:	IP: 6.11 eV

TRANSITION PROPERTIES

$\Delta_{fus}H$ (T_m): 8.54 kJ/mol

$\Delta_{vap}H$ (T_b):

$\Delta_{vap}H$ (25°C):

Vapor pressure (0°C): N/A

Vapor pressure (25°C): N/A

Vapor pressure (100°C): N/A

PROPERTIES AT 25°C AND 100 kPa

	Solid	Liquid	Gas		Solid
$\Delta_f H°$/kJ mol^{-1}:	0.0		177.8	d:	1.54 g/mL
$S°$/J mol^{-1}K^{-1}:	41.6		154.9	η:	N/A
C_p/J mol^{-1}K^{-1}:	25.9		20.8	k:	200 W/m K

COMMENTS:

Name: Calcium bromide

Synonym: Calcium dibromide

Mol. Form.: Br_2Ca

CAS RN: 7789-41-5

Merck No.: 1653

Mol. Wt.: 199.886

PHYSICAL CONSTANTS

T_m: 742°C (1015 K)	T_c:	μ:
T_b:	P_c:	IP:

TRANSITION PROPERTIES

$\Delta_{fus}H$ (T_m): 29.08 kJ/mol

$\Delta_{vap}H$ (T_b):

$\Delta_{vap}H$ (25°C):

Vapor pressure (0°C): N/A

Vapor pressure (25°C): N/A

Vapor pressure (100°C): N/A

PROPERTIES AT 25°C AND 100 kPa

	Solid	Liquid	Gas		Solid
$\Delta_f H°$/kJ mol^{-1}:	-682.8			d:	3.38 g/mL
$S°$/J mol^{-1}K^{-1}:	130.0			η:	N/A
C_p/J mol^{-1}K^{-1}:				k:	

COMMENTS:

Name: Calcium carbide

CAS RN: 75-20-7
Merck No.: 1656
Mol. Wt.: 64.100

Mol. Form.: C_2Ca

PHYSICAL CONSTANTS

T_m: 2300°C (2573 K) T_c: μ:
T_b: P_c: IP:

TRANSITION PROPERTIES

$\Delta_{fus}H$ (T_m): Vapor pressure (0°C): N/A
$\Delta_{vap}H$ (T_b): Vapor pressure (25°C): N/A
$\Delta_{vap}H$ (25°C): Vapor pressure (100°C): N/A

PROPERTIES AT 25°C AND 100 kPa

	Solid	Liquid	Gas		Solid
$\Delta_f H°$/kJ mol^{-1}:	-59.8			d:	2.22 g/mL
$S°$/J mol^{-1}K^{-1}:	70.0			η:	N/A
C_p/J mol^{-1}K^{-1}:	62.7			k:	
COMMENTS: Very flammable					

Name: Calcium carbonate

CAS RN: 471-34-1
Merck No.: 1657
Mol. Wt.: 100.087

Mol. Form.: $CCaO_3$

PHYSICAL CONSTANTS

T_m: 1339.1°C (1612.2 K) T_c: μ:
T_b: P_c: IP:

TRANSITION PROPERTIES

$\Delta_{fus}H$ (T_m): 53.10 kJ/mol Vapor pressure (0°C): N/A
$\Delta_{vap}H$ (T_b): Vapor pressure (25°C): N/A
$\Delta_{vap}H$ (25°C): Vapor pressure (100°C): N/A

PROPERTIES AT 25°C AND 100 kPa

	Solid	Liquid	Gas		Solid
$\Delta_f H°$/kJ mol^{-1}:	-1207.6			d:	2.711 g/mL
$S°$/J mol^{-1}K^{-1}:	91.7			η:	N/A
C_p/J mol^{-1}K^{-1}:	83.5			k:	
COMMENTS:					

Name: Calcium chloride
Synonym: Calcium dichloride

CAS RN: 10043-52-4
Merck No.: 1659
Mol. Wt.: 110.983

Mol. Form.: $CaCl_2$

PHYSICAL CONSTANTS

T_m: 772°C (1045 K) T_c: μ:
T_b: 1935.5°C (2208.6 K) P_c: IP:

TRANSITION PROPERTIES

$\Delta_{fus}H$ (T_m): 28.54 kJ/mol Vapor pressure (0°C): N/A
$\Delta_{vap}H$ (T_b): Vapor pressure (25°C): N/A
$\Delta_{vap}H$ (25°C): Vapor pressure (100°C): N/A

PROPERTIES AT 25°C AND 100 kPa

	Solid	Liquid	Gas		Solid
$\Delta_f H°$/kJ mol^{-1}:	-795.4			d:	2.22 g/mL
$S°$/J mol^{-1}K^{-1}:	108.4			η:	N/A
C_p/J mol^{-1}K^{-1}:	72.9			k:	
COMMENTS:					

Name: Calcium fluoride

Synonym: Calcium difluoride

Mol. Form.: CaF_2

CAS RN: 7789-75-5

Merck No.: 1669

Mol. Wt.: 78.075

PHYSICAL CONSTANTS

T_m: 1418°C (1691 K)	T_c:	μ:
T_b: 2533.4°C (2806.5 K)	P_c:	IP:

TRANSITION PROPERTIES

$\Delta_{fus}H$ (T_m): 29.71 kJ/mol

$\Delta_{vap}H$ (T_b):

$\Delta_{vap}H$ (25°C):

Vapor pressure (0°C): N/A

Vapor pressure (25°C): N/A

Vapor pressure (100°C): N/A

PROPERTIES AT 25°C AND 100 kPa

	Solid	Liquid	Gas		Solid
$\Delta_f H°$/kJ mol^{-1}:	-1228.0			d:	3.18 g/mL
$S°$/J mol^{-1}K^{-1}:	68.5			η:	N/A
C_p/J mol^{-1}K^{-1}:	67.0			k:	9.5 W/m K
COMMENTS: Highly toxic					

Name: Calcium hydroxide

Synonym: Calcium dihydroxide

Mol. Form.: CaH_2O_2

CAS RN: 1305-62-0

Merck No.: 1676

Mol. Wt.: 74.093

PHYSICAL CONSTANTS

T_m:	T_c:	μ:
T_b:	P_c:	IP:

TRANSITION PROPERTIES

$\Delta_{fus}H$ (T_m):

$\Delta_{vap}H$ (T_b):

$\Delta_{vap}H$ (25°C):

Vapor pressure (0°C):

Vapor pressure (25°C):

Vapor pressure (100°C):

PROPERTIES AT 25°C AND 100 kPa

	Solid	Liquid	Gas		Solid
$\Delta_f H°$/kJ mol^{-1}:	-985.2			d:	
$S°$/J mol^{-1}K^{-1}:	83.4			η:	N/A
C_p/J mol^{-1}K^{-1}:	87.5			k:	
COMMENTS:					

Name: Calcium iodide

Synonym: Calcium diiodide

Mol. Form.: CaI_2

CAS RN: 10102-68-8

Merck No.: 1680

Mol. Wt.: 293.887

PHYSICAL CONSTANTS

T_m: 779°C (1052 K)	T_c:	μ:
T_b:	P_c:	IP:

TRANSITION PROPERTIES

$\Delta_{fus}H$ (T_m): 41.80 kJ/mol

$\Delta_{vap}H$ (T_b):

$\Delta_{vap}H$ (25°C):

Vapor pressure (0°C): N/A

Vapor pressure (25°C): N/A

Vapor pressure (100°C): N/A

PROPERTIES AT 25°C AND 100 kPa

	Solid	Liquid	Gas		Solid
$\Delta_f H°$/kJ mol^{-1}:	-533.5			d:	3.96 g/mL
$S°$/J mol^{-1}K^{-1}:	142.0			η:	N/A
C_p/J mol^{-1}K^{-1}:				k:	
COMMENTS:					

Name: Calcium oxide
Synonym: Lime

Mol. Form.: CaO

CAS RN: 1305-78-8
Merck No.: 1692
Mol. Wt.: 56.077

PHYSICAL CONSTANTS

T_m: 2927°C (3200 K)
T_b:

T_c:
P_c:

μ:
IP: 6.90 eV

TRANSITION PROPERTIES

$\Delta_{fus}H$ (T_m): 59.00 kJ/mol
$\Delta_{vap}H$ (T_b):
$\Delta_{vap}H$ (25°C):

Vapor pressure (0°C): N/A
Vapor pressure (25°C): N/A
Vapor pressure (100°C): N/A

PROPERTIES AT 25°C AND 100 kPa

	Solid	Liquid	Gas		Solid
$\Delta_f H°$/kJ mol^{-1}:	-634.9			d:	3.34 g/mL
$S°$/J mol^{-1}K^{-1}:	38.1			η:	N/A
C_p/J mol^{-1}K^{-1}:	42.0			k:	
COMMENTS:					

Name: Calcium sulfate
Synonym: Anhydrite

Mol. Form.: CaO$_4$S

CAS RN: 7778-18-9
Merck No.: 1713
Mol. Wt.: 136.142

PHYSICAL CONSTANTS

T_m: 1450°C (1723 K)
T_b:

T_c:
P_c:

μ:
IP:

TRANSITION PROPERTIES

$\Delta_{fus}H$ (T_m): 28.03 kJ/mol
$\Delta_{vap}H$ (T_b):
$\Delta_{vap}H$ (25°C):

Vapor pressure (0°C): N/A
Vapor pressure (25°C): N/A
Vapor pressure (100°C): N/A

PROPERTIES AT 25°C AND 100 kPa

	Solid	Liquid	Gas		Solid
$\Delta_f H°$/kJ mol^{-1}:	-1434.5			d:	2.96 g/mL
$S°$/J mol^{-1}K^{-1}:	106.5			η:	N/A
C_p/J mol^{-1}K^{-1}:	99.7			k:	
COMMENTS:					

Name: Camphor
Synonym: 1,7,7-Trimethylbicyclo[2.2.1]hepten-2-one

Mol. Form.: C$_{10}$H$_{16}$O

CAS RN: 76-22-2
Merck No.: 1738
Mol. Wt.: 152.236

PHYSICAL CONSTANTS

T_m: 180°C (453 K)
T_b: 207°C (480 K)

T_c:
P_c:

μ:
IP: 8.76 eV

TRANSITION PROPERTIES

$\Delta_{fus}H$ (T_m): 6.84 kJ/mol
$\Delta_{vap}H$ (T_b): 59.50 kJ/mol
$\Delta_{vap}H$ (25°C):

Vapor pressure (0°C):
Vapor pressure (25°C): 0.050 kPa
Vapor pressure (100°C):

PROPERTIES AT 25°C AND 100 kPa

	Solid	Liquid	Gas		Solid
$\Delta_f H°$/kJ mol^{-1}:	-319.4		-267.5		d:	0.992 g/mL
$S°$/J mol^{-1}K^{-1}:					η:	N/A
C_p/J mol^{-1}K^{-1}:	271.2				k:	
COMMENTS: TLV=2 ppm					

Name: 6-Caprolactam   CAS RN: 105-60-2
Synonyms: Hexahydro-2-azepinone   Merck No.: 1762
    6-Hexanelactam   Mol. Wt.: 113.160
Mol. Form.: $C_6H_{11}NO$

PHYSICAL CONSTANTS

T_m: 69.3°C (342.4 K)	T_c:	μ:
T_b: 270°C (543 K)	P_c:	IP: 9.07 eV

TRANSITION PROPERTIES

$\Delta_{fus}H$ (T_m):	Vapor pressure (0°C):
$\Delta_{vap}H$ (T_b):	Vapor pressure (25°C):
$\Delta_{vap}H$ (25°C):	Vapor pressure (100°C):

PROPERTIES AT 25°C AND 100 kPa

	Solid	Liquid	Gas		Solid
$\Delta_fH°$/kJ mol^{-1}:	-329.4		-246.2	d:	
$S°$/J mol^{-1}K^{-1}:				η: N/A	
C_p/J mol^{-1}K^{-1}:	156.8			k:	
COMMENTS: TLV=5 ppm					

Name: ε-Caprolactone   CAS RN: 502-44-3
Synonyms: 2-Oxepanone   Merck No.:
    1,6-Hexanolide   Mol. Wt.: 114.144
Mol. Form.: $C_6H_{10}O_2$

PHYSICAL CONSTANTS

T_m: -18°C (255 K)	T_c:	μ:
T_b: 215°C (488 K)	P_c:	IP:

TRANSITION PROPERTIES

$\Delta_{fus}H$ (T_m):	Vapor pressure (0°C):
$\Delta_{vap}H$ (T_b):	Vapor pressure (25°C):
$\Delta_{vap}H$ (25°C):	Vapor pressure (100°C):

PROPERTIES AT 25°C AND 100 kPa

	Solid	Liquid	Gas		Liquid
$\Delta_fH°$/kJ mol^{-1}:				d: 1.064 g/mL	
$S°$/J mol^{-1}K^{-1}:				η:	
C_p/J mol^{-1}K^{-1}:				k:	
COMMENTS:					

Name: Carbazole   CAS RN: 86-74-8
Synonyms: Dibenzopyrolle   Merck No.: 1792
    9-Azafluorene   Mol. Wt.: 167.210
Mol. Form.: $C_{12}H_9N$

PHYSICAL CONSTANTS

T_m: 246.2°C (519.3 K)	T_c: 628.7°C (901.8 K)	μ:
T_b: 354.70°C (627.85 K)	P_c: 3.93 MPa	IP: 7.57 eV

TRANSITION PROPERTIES

$\Delta_{fus}H$ (T_m): 26.90 kJ/mol	Vapor pressure (0°C): N/A
$\Delta_{vap}H$ (T_b):	Vapor pressure (25°C): N/A
$\Delta_{vap}H$ (25°C):	Vapor pressure (100°C): N/A

PROPERTIES AT 25°C AND 100 kPa

	Solid	Liquid	Gas		Solid
$\Delta_fH°$/kJ mol^{-1}:	101.7		200.7	d: 1.1 g/mL	
$S°$/J mol^{-1}K^{-1}:				η: N/A	
C_p/J mol^{-1}K^{-1}:				k:	
COMMENTS:					

Name: Carbon

Mol. Form.: C

CAS RN: 7440-44-0
Merck No.: 1814
Mol. Wt.: 12.011

PHYSICAL CONSTANTS

T_t: 3974°C (4247 K)*	T_c:	μ:
T_s: 3825°C (4098 K)†	P_c:	IP: 11.26 eV

TRANSITION PROPERTIES

$\Delta_{fus}H$ (T_m): 104.60 kJ/mol   Vapor pressure (0°C): N/A
$\Delta_{vap}H$ (T_b):   Vapor pressure (25°C): N/A
$\Delta_{vap}H$ (25°C):   Vapor pressure (100°C): N/A

PROPERTIES AT 25°C AND 100 kPa

	Solid	Liquid	Gas		Solid
$\Delta_f H°$/kJ mol^{-1}:	0.0		716.7	d:	3.51 g/mL
$S°$/J mol^{-1}K^{-1}:	5.7		158.1	η:	N/A
C_p/J mol^{-1}K^{-1}:	8.5		20.8	k:	1960 W/m K

COMMENTS: Data refer to graphite. *Triple point. †Sublimation point.

Name: Carbon dioxide
Synonym: Carbonic anhydride

Mol. Form.: CO$_2$

CAS RN: 124-38-9
Merck No.: 1816
Mol. Wt.: 44.010

PHYSICAL CONSTANTS

T_t: -56.57°C (216.58 K)*	T_c: 30.99°C (304.14 K)	μ: 0 D
T_s: -78.4°C (194.7 K)†	P_c: 7.375 MPa	IP: 13.77 eV

TRANSITION PROPERTIES

$\Delta_{fus}H$ (T_m): 9.02 kJ/mol   Vapor pressure (0°C): 3483 kPa
$\Delta_{vap}H$ (T_b):   Vapor pressure (25°C):
$\Delta_{vap}H$ (25°C):   Vapor pressure (100°C): N/A

PROPERTIES AT 25°C AND 100 kPa

	Solid	Liquid	Gas		Gas
$\Delta_f H°$/kJ mol^{-1}:			-393.5	d:	1.799 g/L
$S°$/J mol^{-1}K^{-1}:			213.8	η:	14.9 µPa s
C_p/J mol^{-1}K^{-1}:			37.1	k:	0.0168 W/m K

COMMENTS: TLV=5000 ppm. *Triple point. †Sublimation point.

Name: Carbon disulfide
Synonym: Carbon bisulfide

Mol. Form.: CS$_2$

CAS RN: 75-15-0
Merck No.: 1818
Mol. Wt.: 76.143

PHYSICAL CONSTANTS

T_m: -111.5°C (161.6 K)	T_c: 279°C (552 K)	μ: 0 D
T_b: 46°C (319 K)	P_c: 7.90 MPa	IP: 10.07 eV

TRANSITION PROPERTIES

$\Delta_{fus}H$ (T_m): 4.40 kJ/mol   Vapor pressure (0°C): 16.9 kPa
$\Delta_{vap}H$ (T_b): 26.74 kJ/mol   Vapor pressure (25°C): 48.2 kPa
$\Delta_{vap}H$ (25°C): 27.51 kJ/mol   Vapor pressure (100°C):

PROPERTIES AT 25°C AND 100 kPa

	Solid	Liquid	Gas		Liquid
$\Delta_f H°$/kJ mol^{-1}:		89.0	116.6	d:	1.2555 g/mL
$S°$/J mol^{-1}K^{-1}:		151.3	237.8	η:	0.352 mPa s
C_p/J mol^{-1}K^{-1}:		76.4	45.4	k:	0.149 W/m K

COMMENTS: TLV=10 ppm; highly toxic; flammable

Name: Carbon monoxide

CAS RN: 630-08-0
Merck No.: 1820
Mol. Wt.: 28.010

Mol. Form.: CO

PHYSICAL CONSTANTS

T_m: -205°C (68 K) T_c: -140.24°C (132.91 K) μ: 0.110 D
T_b: -191.5°C (81.6 K) P_c: 3.499 MPa   IP: 14.01 eV

TRANSITION PROPERTIES

$\Delta_{fus}H$ (T_m): 0.83 kJ/mol   Vapor pressure (0°C): N/A
$\Delta_{vap}H$ (T_b): 6.04 kJ/mol   Vapor pressure (25°C): N/A
$\Delta_{vap}H$ (25°C):   Vapor pressure (100°C): N/A

PROPERTIES AT 25°C AND 100 kPa

	Solid	Liquid	Gas		Gas
$\Delta_f H°$/kJ mol^{-1}:			-110.5		d: 1.145 g/L
$S°$/J mol^{-1}K^{-1}:			197.7		η: 17.8 µPa s
C_p/J mol^{-1}K^{-1}:			29.1		k: 0.0250 W/m K

COMMENTS: TLV=25 ppm; highly toxic; very flammable

Name: Carbon oxysulfide
Synonym: Carbonyl sulfide

CAS RN: 463-58-1
Merck No.:
Mol. Wt.: 60.076

Mol. Form.: COS

PHYSICAL CONSTANTS

T_m: -138.8°C (134.3 K) T_c: 102°C (375 K) μ: 0.715 D
T_b: -50°C (223 K) P_c: 5.88 MPa   IP: 11.17 eV

TRANSITION PROPERTIES

$\Delta_{fus}H$ (T_m):   Vapor pressure (0°C):
$\Delta_{vap}H$ (T_b):   Vapor pressure (25°C):
$\Delta_{vap}H$ (25°C):   Vapor pressure (100°C):

PROPERTIES AT 25°C AND 100 kPa

	Solid	Liquid	Gas		Gas
$\Delta_f H°$/kJ mol^{-1}:			-142.0		d: 2.456 g/L
$S°$/J mol^{-1}K^{-1}:			231.6		η:
C_p/J mol^{-1}K^{-1}:			41.5		k:

COMMENTS: Highly toxic

Name: Carbonyl chloride
Synonyms: Phosgene
          Carbonic dichloride
Mol. Form.: CCl$_2$O

CAS RN: 75-44-5
Merck No.: 7310
Mol. Wt.: 98.916

PHYSICAL CONSTANTS

T_m: -127.9°C (145.2 K) T_c: 182°C (455 K) μ: 1.17 D
T_b: 8°C (281 K) P_c: 5.67 MPa   IP: 11.40 eV

TRANSITION PROPERTIES

$\Delta_{fus}H$ (T_m): 5.74 kJ/mol   Vapor pressure (0°C): 75.1 kPa
$\Delta_{vap}H$ (T_b):   Vapor pressure (25°C):
$\Delta_{vap}H$ (25°C):   Vapor pressure (100°C):

PROPERTIES AT 25°C AND 100 kPa

	Solid	Liquid	Gas		Gas
$\Delta_f H°$/kJ mol^{-1}:			-219.1		d: 4.043 g/L
$S°$/J mol^{-1}K^{-1}:			283.5		η:
C_p/J mol^{-1}K^{-1}:			57.7		k:

COMMENTS: TLV=0.1 ppm; highly toxic

Name: Carbonyl fluoride
Synonym: Carbonic difluoride

Mol. Form.: CF$_2$O

CAS RN: 353-50-4
Merck No.: 1826
Mol. Wt.: 66.007

PHYSICAL CONSTANTS

T_m: -111.26°C (161.89 K)	T_c:	μ: 0.95 D
T_b: -84.57°C (188.58 K)	P_c:	IP: 13.03 eV

TRANSITION PROPERTIES

$\Delta_{fus}H$ (T_m):	Vapor pressure (0°C):
$\Delta_{vap}H$ (T_b):	Vapor pressure (25°C):
$\Delta_{vap}H$ (25°C):	Vapor pressure (100°C):

PROPERTIES AT 25°C AND 100 kPa

	Solid	Liquid	Gas		Gas
$\Delta_f H°$/kJ mol^{-1}:			-639.8	d:	2.698 g/L
$S°$/J mol^{-1}K^{-1}:				η:	
C_p/J mol^{-1}K^{-1}:			46.8	k:	

COMMENTS: TLV=2 ppm

Name: Cerium

Mol. Form.: Ce

CAS RN: 7440-45-1
Merck No.: 1988
Mol. Wt.: 140.115

PHYSICAL CONSTANTS

T_m: 799°C (1072 K)	T_c:	μ:
T_b: 3424°C (3697 K)	P_c:	IP: 5.54 eV

TRANSITION PROPERTIES

$\Delta_{fus}H$ (T_m): 5.46 kJ/mol	Vapor pressure (0°C): N/A
$\Delta_{vap}H$ (T_b):	Vapor pressure (25°C): N/A
$\Delta_{vap}H$ (25°C):	Vapor pressure (100°C): N/A

PROPERTIES AT 25°C AND 100 kPa

	Solid	Liquid	Gas		Solid
$\Delta_f H°$/kJ mol^{-1}:	0.0		423.0	d:	8.16 g/mL
$S°$/J mol^{-1}K^{-1}:	72.0		191.8	η:	N/A
C_p/J mol^{-1}K^{-1}:	26.9		23.1	k:	11.3 W/m K

COMMENTS:

Name: Cesium

Mol. Form.: Cs

CAS RN: 7440-46-2
Merck No.: 2001
Mol. Wt.: 132.905

PHYSICAL CONSTANTS

T_m: 28.44°C (301.59 K)	T_c:	μ:
T_b: 671°C (944 K)	P_c:	IP: 3.89 eV

TRANSITION PROPERTIES

$\Delta_{fus}H$ (T_m): 2.10 kJ/mol	Vapor pressure (0°C):
$\Delta_{vap}H$ (T_b):	Vapor pressure (25°C):
$\Delta_{vap}H$ (25°C):	Vapor pressure (100°C):

PROPERTIES AT 25°C AND 100 kPa

	Solid	Liquid	Gas		Solid
$\Delta_f H°$/kJ mol^{-1}:	0.0		76.5	d:	1.93 g/mL
$S°$/J mol^{-1}K^{-1}:	85.2		175.6	η:	N/A
C_p/J mol^{-1}K^{-1}:	32.2		20.8	k:	35.9 W/m K

COMMENTS: Flammable

Name: Cesium chloride

Mol. Form.: ClCs

CAS RN: 7647-17-8
Merck No.: 2004
Mol. Wt.: 168.358

PHYSICAL CONSTANTS

T_m: 645°C (918 K)
T_b: 1297°C (1570 K)

T_c:
P_c:

μ: 10.387 D
IP: 7.84 eV

TRANSITION PROPERTIES

$\Delta_{fus}H$ (T_m): 15.90 kJ/mol
$\Delta_{vap}H$ (T_b):
$\Delta_{vap}H$ (25°C):

Vapor pressure (0°C): N/A
Vapor pressure (25°C): N/A
Vapor pressure (100°C): N/A

PROPERTIES AT 25°C AND 100 kPa

	Solid	Liquid	Gas		Solid
$\Delta_f H°$/kJ mol^{-1}:	-443.0			d:	3.988 g/mL
$S°$/J mol^{-1}K^{-1}:	101.2			η:	N/A
C_p/J mol^{-1}K^{-1}:	52.5			k:	
COMMENTS:					

Name: Chlorine (Cl$_2$)
Synonym: Dichlorine

Mol. Form.: Cl$_2$

CAS RN: 7782-50-5
Merck No.: 2095
Mol. Wt.: 70.905

PHYSICAL CONSTANTS

T_m: -101.5°C (171.6 K)
T_b: -34.04°C (239.11 K)

T_c: 143.8°C (416.9 K)
P_c: 7.991 MPa

μ: 0 D
IP: 11.48 eV

TRANSITION PROPERTIES

$\Delta_{fus}H$ (T_m): 6.40 kJ/mol
$\Delta_{vap}H$ (T_b): 20.41 kJ/mol
$\Delta_{vap}H$ (25°C): 17.65 kJ/mol

Vapor pressure (0°C): 370 kPa
Vapor pressure (25°C): 780 kPa
Vapor pressure (100°C): 3927 kPa

PROPERTIES AT 25°C AND 100 kPa

	Solid	Liquid	Gas		Gas
$\Delta_f H°$/kJ mol^{-1}:			0.0	d:	2.898 g/L
$S°$/J mol^{-1}K^{-1}:			223.1	η:	
C_p/J mol^{-1}K^{-1}:			33.9	k:	0.0088 W/m K
COMMENTS: TLV=0.5 ppm; highly toxic					

Name: Chlorine dioxide (ClO$_2$)
Synonym: Chlorine peroxide

Mol. Form.: ClO$_2$

CAS RN: 10049-04-4
Merck No.: 2096
Mol. Wt.: 67.452

PHYSICAL CONSTANTS

T_m: -59°C (214 K)
T_b: 11°C (284 K)

T_c:
P_c:

μ:
IP: 10.36 eV

TRANSITION PROPERTIES

$\Delta_{fus}H$ (T_m):
$\Delta_{vap}H$ (T_b): 30.00 kJ/mol
$\Delta_{vap}H$ (25°C):

Vapor pressure (0°C): 64.7 kPa
Vapor pressure (25°C): 169 kPa
Vapor pressure (100°C):

PROPERTIES AT 25°C AND 100 kPa

	Solid	Liquid	Gas		Gas
$\Delta_f H°$/kJ mol^{-1}:			102.5	d:	2.757 g/L
$S°$/J mol^{-1}K^{-1}:			256.8	η:	
C_p/J mol^{-1}K^{-1}:			42.0	k:	
COMMENTS: TLV=0.1 ppm; highly toxic					

Name: Chlorine fluoride
Synonym: Chlorine monofluoride

Mol. Form.: ClF

CAS RN: 7790-89-8
Merck No.: 2098
Mol. Wt.: 54.451

PHYSICAL CONSTANTS

T_m: -155.6°C (117.5 K)
T_b: -101.1°C (172.0 K)

T_c:
P_c:

μ: 0.888 D
IP: 12.65 eV

TRANSITION PROPERTIES

$\Delta_{fus}H$ (T_m):
$\Delta_{vap}H$ (T_b): 24.00 kJ/mol
$\Delta_{vap}H$ (25°C):

Vapor pressure (0°C):
Vapor pressure (25°C):
Vapor pressure (100°C):

PROPERTIES AT 25°C AND 100 kPa

	Solid	Liquid	Gas	Gas
$\Delta_f H°$/kJ mol^{-1}:			-50.3	d: 2.226 g/L
$S°$/J mol^{-1}K^{-1}:			217.9	η:
C_p/J mol^{-1}K^{-1}:			32.1	k:
COMMENTS:				

Name: Chlorine trifluoride
Synonym: Chlorine(III) fluoride

Mol. Form.: ClF$_3$

CAS RN: 7790-91-2
Merck No.: 2100
Mol. Wt.: 92.448

PHYSICAL CONSTANTS

T_m: -76.34°C (196.81 K)
T_b: 11.75°C (284.90 K)

T_c:
P_c:

μ: 0.6 D
IP: 12.65 eV

TRANSITION PROPERTIES

$\Delta_{fus}H$ (T_m):
$\Delta_{vap}H$ (T_b): 27.53 kJ/mol
$\Delta_{vap}H$ (25°C): 26.3 kJ/mol

Vapor pressure (0°C):
Vapor pressure (25°C):
Vapor pressure (100°C):

PROPERTIES AT 25°C AND 100 kPa

	Solid	Liquid	Gas	Gas
$\Delta_f H°$/kJ mol^{-1}:		-189.5	-163.2	d: 3.779 g/L
$S°$/J mol^{-1}K^{-1}:			281.6	η:
C_p/J mol^{-1}K^{-1}:			63.9	k:
COMMENTS: TLV=0.1 ppm				

Name: Chloroacetaldehyde
Synonym: 2-Chloro-1-ethanal

Mol. Form.: C$_2$H$_3$ClO

CAS RN: 107-20-0
Merck No.: 2108
Mol. Wt.: 78.498

PHYSICAL CONSTANTS

T_m: -16.3°C (256.8 K)
T_b: 85.5°C (358.6 K)

T_c:
P_c:

μ:
IP: 10.48 eV

TRANSITION PROPERTIES

$\Delta_{fus}H$ (T_m):
$\Delta_{vap}H$ (T_b):
$\Delta_{vap}H$ (25°C):

Vapor pressure (0°C):
Vapor pressure (25°C):
Vapor pressure (100°C):

PROPERTIES AT 25°C AND 100 kPa

	Solid	Liquid	Gas	Liquid
$\Delta_f H°$/kJ mol^{-1}:				d:
$S°$/J mol^{-1}K^{-1}:				η:
C_p/J mol^{-1}K^{-1}:				k:
COMMENTS: TLV=1 ppm				

Name: Chloroacetic acid
Synonym: Chloroethanoic acid

Mol. Form.: $C_2H_3ClO_2$

CAS RN: 79-11-8
Merck No.: 2111
Mol. Wt.: 94.497

PHYSICAL CONSTANTS

T_m: 61.3°C (334.4 K)
T_b: 189.35°C (462.50 K)

T_c:
P_c:

μ:
IP: 10.70 eV

TRANSITION PROPERTIES

$\Delta_{fus}H$ (T_m): 12.28 kJ/mol
$\Delta_{vap}H$ (T_b):
$\Delta_{vap}H$ (25°C):

Vapor pressure (0°C):
Vapor pressure (25°C):
Vapor pressure (100°C): 3.31 kPa

PROPERTIES AT 25°C AND 100 kPa

	Solid	Liquid	Gas		Solid
$\Delta_f H°$/kJ mol^{-1}:	-510.5		-435.2	d:	
$S°$/J mol^{-1}K^{-1}:				η: N/A	
C_p/J mol^{-1}K^{-1}:				k:	

COMMENTS: Highly toxic

Name: Chloroacetyl chloride
Synonym: Chloroacetic acid chloride

Mol. Form.: $C_2H_2Cl_2O$

CAS RN: 79-04-9
Merck No.: 2054
Mol. Wt.: 112.943

PHYSICAL CONSTANTS

T_m: -22°C (251 K)
T_b: 106°C (379 K)

T_c:
P_c:

μ: 2.23 D
IP: 11.00 eV

TRANSITION PROPERTIES

$\Delta_{fus}H$ (T_m):
$\Delta_{vap}H$ (T_b):
$\Delta_{vap}H$ (25°C): 38.9 kJ/mol

Vapor pressure (0°C): 0.676 kPa
Vapor pressure (25°C): 3.33 kPa
Vapor pressure (100°C): 83.8 kPa

PROPERTIES AT 25°C AND 100 kPa

	Solid	Liquid	Gas		Liquid
$\Delta_f H°$/kJ mol^{-1}:		-283.7	-244.8	d:	1.413 g/mL
$S°$/J mol^{-1}K^{-1}:				η:	
C_p/J mol^{-1}K^{-1}:				k:	

COMMENTS: TLV=0.05 ppm

Name: *o*-Chloroaniline
Synonym: 2-Chlorobenzenamine

Mol. Form.: C_6H_6ClN

CAS RN: 95-51-2
Merck No.: 2118
Mol. Wt.: 127.573

PHYSICAL CONSTANTS

T_m: -14°C (259 K)
T_b: 208.84°C (481.99 K)

T_c:
P_c:

μ:
IP: 8.50 eV

TRANSITION PROPERTIES

$\Delta_{fus}H$ (T_m): 11.88 kJ/mol
$\Delta_{vap}H$ (T_b): 44.35 kJ/mol
$\Delta_{vap}H$ (25°C): 56.75 kJ/mol

Vapor pressure (0°C):
Vapor pressure (25°C): 0.034 kPa
Vapor pressure (100°C):

PROPERTIES AT 25°C AND 100 kPa

	Solid	Liquid	Gas		Liquid
$\Delta_f H°$/kJ mol^{-1}:				d:	1.206 g/mL
$S°$/J mol^{-1}K^{-1}:				η:	3.32 mPa s
C_p/J mol^{-1}K^{-1}:				k:	

COMMENTS:

Name: *m*-Chloroaniline
Synonym: 3-Chlorobenzenamine

Mol. Form.: C_6H_6ClN

CAS RN: 108-42-9
Merck No.: 2118
Mol. Wt.: 127.573

PHYSICAL CONSTANTS

T_m: -10.4°C (262.7 K)	T_c:	μ:
T_b: 230.5°C (503.6 K)	P_c:	IP: 8.09 eV

TRANSITION PROPERTIES

$\Delta_{fus}H$ (T_m):	Vapor pressure (0°C):
$\Delta_{vap}H$ (T_b):	Vapor pressure (25°C):
$\Delta_{vap}H$ (25°C):	Vapor pressure (100°C):

PROPERTIES AT 25°C AND 100 kPa

	Solid	Liquid	Gas		Liquid
$\Delta_f H°$/kJ mol^{-1}:					d: 1.210 g/mL
$S°$/J mol^{-1}K^{-1}:					η:
C_p/J mol^{-1}K^{-1}:					k:
COMMENTS:					

Name: *p*-Chloroaniline
Synonym: 4-Chlorobenzenamine

Mol. Form.: C_6H_6ClN

CAS RN: 106-47-8
Merck No.: 2118
Mol. Wt.: 127.573

PHYSICAL CONSTANTS

T_m: 72.5°C (345.6 K)	T_c:	μ:
T_b: 232°C (505 K)	P_c:	IP: 8.18 eV

TRANSITION PROPERTIES

$\Delta_{fus}H$ (T_m):	Vapor pressure (0°C):
$\Delta_{vap}H$ (T_b):	Vapor pressure (25°C):
$\Delta_{vap}H$ (25°C):	Vapor pressure (100°C):

PROPERTIES AT 25°C AND 100 kPa

	Solid	Liquid	Gas		Solid
$\Delta_f H°$/kJ mol^{-1}:					d: 1.41 g/mL
$S°$/J mol^{-1}K^{-1}:					η: N/A
C_p/J mol^{-1}K^{-1}:					k:
COMMENTS:					

Name: Chlorobenzene
Synonym: Phenyl chloride

Mol. Form.: C_6H_5Cl

CAS RN: 108-90-7
Merck No.: 2121
Mol. Wt.: 112.558

PHYSICAL CONSTANTS

T_m: -45.2°C (227.9 K)	T_c: 359.3°C (632.4 K)	μ: 1.69 D
T_b: 131.72°C (404.87 K)	P_c: 4.52 MPa	IP: 9.06 eV

TRANSITION PROPERTIES

$\Delta_{fus}H$ (T_m): 9.61 kJ/mol	Vapor pressure (0°C):
$\Delta_{vap}H$ (T_b): 35.19 kJ/mol	Vapor pressure (25°C): 1.60 kPa
$\Delta_{vap}H$ (25°C): 40.97 kJ/mol	Vapor pressure (100°C): 39.5 kPa

PROPERTIES AT 25°C AND 100 kPa

	Solid	Liquid	Gas		Liquid
$\Delta_f H°$/kJ mol^{-1}:		11.0	52.0		d: 1.1009 g/mL
$S°$/J mol^{-1}K^{-1}:					η: 0.755 mPa s
C_p/J mol^{-1}K^{-1}:		150.1			k: 0.127 W/m K
COMMENTS: TLV=10 ppm; highly toxic; flammable					

Name: *o*-Chlorobenzoic acid   CAS RN: 118-91-2
Synonym: 2-Chlorobenzoic acid   Merck No.: 2125
                                                                Mol. Wt.: 156.568
Mol. Form.: $C_7H_5ClO_2$

PHYSICAL CONSTANTS

T_m: 140.2°C (413.3 K) T_c: μ:
T_b: P_c: IP:

TRANSITION PROPERTIES

$\Delta_{fus}H$ (T_m): 25.73 kJ/mol   Vapor pressure (0°C):
$\Delta_{vap}H$ (T_b):   Vapor pressure (25°C):
$\Delta_{vap}H$ (25°C):   Vapor pressure (100°C):

PROPERTIES AT 25°C AND 100 kPa

	Solid	Liquid	Gas		Solid
$\Delta_f H°$/kJ mol^{-1}:	-428.9		-341.0	d:	
$S°$/J mol^{-1}K^{-1}:				η: N/A	
C_p/J mol^{-1}K^{-1}:	163.2			k:	
COMMENTS:					

Name: 2-Chloro-1,3-butadiene   CAS RN: 126-99-8
Synonym: Chloroprene   Merck No.:
                                                                Mol. Wt.: 88.536
Mol. Form.: C_4H_5Cl

PHYSICAL CONSTANTS

T_m: T_c: μ:
T_b: 59.4°C (332.5 K) P_c: IP:

TRANSITION PROPERTIES

$\Delta_{fus}H$ (T_m):   Vapor pressure (0°C): 9.88 kPa
$\Delta_{vap}H$ (T_b):   Vapor pressure (25°C): 29.5 kPa
$\Delta_{vap}H$ (25°C):   Vapor pressure (100°C):

PROPERTIES AT 25°C AND 100 kPa

	Solid	Liquid	Gas	Liquid
$\Delta_f H°$/kJ mol^{-1}:				d: 0.952 g/mL
$S°$/J mol^{-1}K^{-1}:				η:
C_p/J mol^{-1}K^{-1}:				k:
COMMENTS: TLV=10 ppm; carcinogen; highly toxic; flammable				

Name: 1-Chlorobutane   CAS RN: 109-69-3
Synonym: Butyl chloride   Merck No.: 1560
                                                                Mol. Wt.: 92.568
Mol. Form.: C_4H_9Cl

PHYSICAL CONSTANTS

T_m: -123.1°C (150.0 K) T_c: μ: 2.05 D
T_b: 78.6°C (351.7 K) P_c: IP: 10.67 eV

TRANSITION PROPERTIES

$\Delta_{fus}H$ (T_m):   Vapor pressure (0°C): 3.82 kPa
$\Delta_{vap}H$ (T_b): 30.39 kJ/mol   Vapor pressure (25°C): 13.7 kPa
$\Delta_{vap}H$ (25°C): 33.51 kJ/mol   Vapor pressure (100°C): 188 kPa

PROPERTIES AT 25°C AND 100 kPa

	Solid	Liquid	Gas	Liquid
$\Delta_f H°$/kJ mol^{-1}:		-188.1	-154.6	d: 0.8810 g/mL
$S°$/J mol^{-1}K^{-1}:				η: 0.422 mPa s
C_p/J mol^{-1}K^{-1}:		175.0		k:
COMMENTS: Flammable				

Name: 2-Chlorobutane
Synonym: *sec*-Butyl chloride

Mol. Form.: C_4H_9Cl

CAS RN: 78-86-4
Merck No.: 1561
Mol. Wt.: 92.568

PHYSICAL CONSTANTS

T_m: -131.3°C (141.8 K)	T_c:	μ: 2.04 D
T_b: 68.3°C (341.4 K)	P_c:	IP: 10.53 eV

TRANSITION PROPERTIES

$\Delta_{fus}H\ (T_m)$:
$\Delta_{vap}H\ (T_b)$: 29.17 kJ/mol
$\Delta_{vap}H\ (25°C)$: 31.53 kJ/mol

Vapor pressure (0°C): 6.51 kPa
Vapor pressure (25°C): 21.2 kPa
Vapor pressure (100°C):

PROPERTIES AT 25°C AND 100 kPa

	Solid	Liquid	Gas		Liquid
$\Delta_f H°$/kJ mol^{-1}:		-192.8	-161.2		*d*: 0.866 g/mL
$S°$/J mol^{-1}K^{-1}:					η:
C_p/J mol^{-1}K^{-1}:					*k*:
COMMENTS:					

Name: Chlorocyclohexane
Synonym: Cyclohexyl chloride

Mol. Form.: $C_6H_{11}Cl$

CAS RN: 542-18-7
Merck No.: 2738
Mol. Wt.: 118.606

PHYSICAL CONSTANTS

T_m: -44°C (229 K)	T_c:	μ:
T_b: 142°C (415 K)	P_c:	IP:

TRANSITION PROPERTIES

$\Delta_{fus}H\ (T_m)$:
$\Delta_{vap}H\ (T_b)$:
$\Delta_{vap}H\ (25°C)$: 43.5 kJ/mol

Vapor pressure (0°C):
Vapor pressure (25°C):
Vapor pressure (100°C): 28.7 kPa

PROPERTIES AT 25°C AND 100 kPa

	Solid	Liquid	Gas		Liquid
$\Delta_f H°$/kJ mol^{-1}:		-207.2	-163.7		*d*: 0.995 g/mL
$S°$/J mol^{-1}K^{-1}:					η:
C_p/J mol^{-1}K^{-1}:					*k*:
COMMENTS:					

Name: 1-Chloro-1,1-difluoroethane

Mol. Form.: $C_2H_3ClF_2$

CAS RN: 75-68-3
Merck No.:
Mol. Wt.: 100.495

PHYSICAL CONSTANTS

T_m: -130.8°C (142.3 K)	T_c: 137.14°C (410.29 K)	μ: 2.14 D
T_b: -9.75°C (263.40 K)	P_c: 4.041 MPa	IP: 11.98 eV

TRANSITION PROPERTIES

$\Delta_{fus}H\ (T_m)$: 2.69 kJ/mol
$\Delta_{vap}H\ (T_b)$:
$\Delta_{vap}H\ (25°C)$:

Vapor pressure (0°C): 52.9 kPa
Vapor pressure (25°C): 351 kPa
Vapor pressure (100°C): 2192 kPa

PROPERTIES AT 25°C AND 100 kPa

	Solid	Liquid	Gas		Gas
$\Delta_f H°$/kJ mol^{-1}:					*d*: 4.108 g/L
$S°$/J mol^{-1}K^{-1}:			307.2		η:
C_p/J mol^{-1}K^{-1}:		130.1	82.5		*k*:
COMMENTS:					

Name: 1-Chloro-2,2-difluoroethylene
Synonym: 2-Chloro-1,1-difluoroethene

Mol. Form.: C_2HClF_2

CAS RN: 359-10-4
Merck No.:
Mol. Wt.: 98.479

PHYSICAL CONSTANTS

T_m: -138.5°C (134.6 K)
T_b: -18.55°C (254.60 K)

T_c: 127.5°C (400.6 K)
P_c: 4.46 MPa

μ:
IP: 9.80 eV

TRANSITION PROPERTIES

$\Delta_{fus}H$ (T_m):
$\Delta_{vap}H$ (T_b):
$\Delta_{vap}H$ (25°C):

Vapor pressure (0°C):
Vapor pressure (25°C):
Vapor pressure (100°C):

PROPERTIES AT 25°C AND 100 kPa

	Solid	Liquid	Gas		Gas
$\Delta_fH°$/kJ mol^{-1}:			-315.5		d: 4.025 g/L
$S°$/J mol^{-1}K^{-1}:			303.0		η:
C_p/J mol^{-1}K^{-1}:			72.1		k:
COMMENTS:					

Name: Chlorodifluoromethane
Synonyms: Refrigerant 22
         CFC-22
Mol. Form.: $CHClF_2$

CAS RN: 75-45-6
Merck No.:
Mol. Wt.: 86.468

PHYSICAL CONSTANTS

T_m: -157.42°C (115.73 K)
T_b: -40.75°C (232.40 K)

T_c: 96.2°C (369.3 K)
P_c: 4.99 MPa

μ: 1.42 D
IP: 12.20 eV

TRANSITION PROPERTIES

$\Delta_{fus}H$ (T_m): 4.12 kJ/mol
$\Delta_{vap}H$ (T_b): 20.24 kJ/mol
$\Delta_{vap}H$ (25°C):

Vapor pressure (0°C): 498 kPa
Vapor pressure (25°C): 1044 kPa
Vapor pressure (100°C): N/A

PROPERTIES AT 25°C AND 100 kPa

	Solid	Liquid	Gas		Gas
$\Delta_fH°$/kJ mol^{-1}:			-482.6		d: 3.534 g/L
$S°$/J mol^{-1}K^{-1}:			280.9		η:
C_p/J mol^{-1}K^{-1}:			55.9		k:
COMMENTS: TLV=1000 ppm					

Name: Chloroethane
Synonym: Ethyl chloride

Mol. Form.: C_2H_5Cl

CAS RN: 75-00-3
Merck No.: 3740
Mol. Wt.: 64.514

PHYSICAL CONSTANTS

T_m: -138.7°C (134.4 K)
T_b: 13.1°C (286.2 K)

T_c: 187.3°C (460.4 K)
P_c: 5.3 MPa

μ: 2.05 D
IP: 10.97 eV

TRANSITION PROPERTIES

$\Delta_{fus}H$ (T_m): 4.45 kJ/mol
$\Delta_{vap}H$ (T_b): 24.65 kJ/mol
$\Delta_{vap}H$ (25°C): 24.3 kJ/mol

Vapor pressure (0°C): 62.3 kPa
Vapor pressure (25°C): 160 kPa
Vapor pressure (100°C):

PROPERTIES AT 25°C AND 100 kPa

	Solid	Liquid	Gas		Gas
$\Delta_fH°$/kJ mol^{-1}:		-136.5	-112.2		d: 2.637 g/L
$S°$/J mol^{-1}K^{-1}:		190.8	276.0		η:
C_p/J mol^{-1}K^{-1}:		104.3	62.8		k:
COMMENTS: TLV=1000 ppm; highly toxic; very flammable					

Name: 2-Chloroethanol
Synonym: Ethylene chlorohydrin

Mol. Form.: C_2H_5ClO

CAS RN: 107-07-3
Merck No.: 3750
Mol. Wt.: 80.514

PHYSICAL CONSTANTS

T_m: -67.5°C (205.6 K)
T_b: 128.6°C (401.7 K)

T_c:
P_c:

μ: 1.78 D
IP: 10.52 eV

TRANSITION PROPERTIES

$\Delta_{fus}H$ (T_m):
$\Delta_{vap}H$ (T_b): 41.43 kJ/mol
$\Delta_{vap}H$ (25°C):

Vapor pressure (0°C):
Vapor pressure (25°C):
Vapor pressure (100°C): 38.6 kPa

PROPERTIES AT 25°C AND 100 kPa

	Solid	Liquid	Gas		Liquid
$\Delta_f H°$/kJ mol^{-1}:		-295.4			d: 1.1965 g/mL
$S°$/J mol^{-1}K^{-1}:					η: 3.1 mPa s
C_p/J mol^{-1}K^{-1}:					k:
COMMENTS: TLV=1 ppm; highly toxic					

Name: Chloroethylene
Synonym: Vinyl chloride

Mol. Form.: C_2H_3Cl

CAS RN: 75-01-4
Merck No.: 9898
Mol. Wt.: 62.499

PHYSICAL CONSTANTS

T_m: -153.79°C (119.36 K)
T_b: -13.37°C (259.78 K)

T_c:
P_c:

μ: 1.45 D
IP: 9.99 eV

TRANSITION PROPERTIES

$\Delta_{fus}H$ (T_m): 4.75 kJ/mol
$\Delta_{vap}H$ (T_b): 20.80 kJ/mol
$\Delta_{vap}H$ (25°C): 18.64 kJ/mol

Vapor pressure (0°C): 170 kPa
Vapor pressure (25°C): 355 kPa
Vapor pressure (100°C):

PROPERTIES AT 25°C AND 100 kPa

	Solid	Liquid	Gas		Gas
$\Delta_f H°$/kJ mol^{-1}:		14.6	37.3		d: 2.555 g/L
$S°$/J mol^{-1}K^{-1}:			264.0		η:
C_p/J mol^{-1}K^{-1}:			53.7		k:
COMMENTS: TLV=5 ppm; carcinogen; very flammable					

Name: 1-Chlorohexane
Synonym: Hexyl chloride

Mol. Form.: $C_6H_{13}Cl$

CAS RN: 544-10-5
Merck No.: 2144
Mol. Wt.: 120.622

PHYSICAL CONSTANTS

T_m: -94°C (179 K)
T_b: 135°C (408 K)

T_c:
P_c:

μ:
IP:

TRANSITION PROPERTIES

$\Delta_{fus}H$ (T_m):
$\Delta_{vap}H$ (T_b): 35.67 kJ/mol
$\Delta_{vap}H$ (25°C): 42.83 kJ/mol

Vapor pressure (0°C):
Vapor pressure (25°C): 1.25 kPa
Vapor pressure (100°C): 35.1 kPa

PROPERTIES AT 25°C AND 100 kPa

	Solid	Liquid	Gas		Liquid
$\Delta_f H°$/kJ mol^{-1}:					d: 0.874 g/mL
$S°$/J mol^{-1}K^{-1}:					η:
C_p/J mol^{-1}K^{-1}:					k:
COMMENTS:					

Name: Chloromethane   CAS RN: 74-87-3
Synonyms: Methyl chloride   Merck No.: 5964
       Refrigerant 40   Mol. Wt.: 50.488
Mol. Form.: CH_3Cl

PHYSICAL CONSTANTS

T_m: -97.7°C (175.4 K) T_c: 143.10°C (416.25 K)   μ: 1.892 D
T_b: -24.09°C (249.06 K) P_c: 6.679 MPa   IP: 11.22 eV

TRANSITION PROPERTIES

$\Delta_{fus}H$ (T_m): 6.43 kJ/mol   Vapor pressure (0°C): 259 kPa
$\Delta_{vap}H$ (T_b): 21.40 kJ/mol   Vapor pressure (25°C): 574 kPa
$\Delta_{vap}H$ (25°C): 18.92 kJ/mol   Vapor pressure (100°C): 3219 kPa

PROPERTIES AT 25°C AND 100 kPa

	Solid	Liquid	Gas		Gas
$\Delta_f H°$/kJ mol^{-1}:			-81.9		d: 2.064 g/L
$S°$/J mol^{-1}K^{-1}:			234.6		η:
C_p/J mol^{-1}K^{-1}:			40.8		k:

COMMENTS: TLV=50 ppm; carcinogen; very flammable

Name: 1-Chloro-3-methylbutane   CAS RN: 107-84-6
Synonym: Isoamyl chloride   Merck No.: 4998
                                                                  Mol. Wt.: 106.595
Mol. Form.: $C_5H_{11}Cl$

PHYSICAL CONSTANTS

T_m: -104.4°C (168.7 K) T_c: μ:
T_b: 98.9°C (372.0 K) P_c: IP:

TRANSITION PROPERTIES

$\Delta_{fus}H$ (T_m):   Vapor pressure (0°C):
$\Delta_{vap}H$ (T_b): 32.02 kJ/mol   Vapor pressure (25°C):
$\Delta_{vap}H$ (25°C): 36.24 kJ/mol   Vapor pressure (100°C):

PROPERTIES AT 25°C AND 100 kPa

	Solid	Liquid	Gas		Liquid
$\Delta_f H°$/kJ mol^{-1}:		-216.0	-179.2		d: 0.8700 g/mL
$S°$/J mol^{-1}K^{-1}:					η: 5.0 mPa s
C_p/J mol^{-1}K^{-1}:		175.1			k:

COMMENTS:

Name: 1-Chloro-2-methylpropane   CAS RN: 513-36-0
Synonym: Isobutyl chloride   Merck No.: 5022
                                                                  Mol. Wt.: 92.568
Mol. Form.: C_4H_9Cl

PHYSICAL CONSTANTS

T_m: -130.3°C (142.8 K) T_c: μ: 2.00 D
T_b: 68.5°C (341.6 K) P_c: IP: 10.66 eV

TRANSITION PROPERTIES

$\Delta_{fus}H$ (T_m):   Vapor pressure (0°C):
$\Delta_{vap}H$ (T_b): 29.22 kJ/mol   Vapor pressure (25°C):
$\Delta_{vap}H$ (25°C): 31.67 kJ/mol   Vapor pressure (100°C):

PROPERTIES AT 25°C AND 100 kPa

	Solid	Liquid	Gas		Liquid
$\Delta_f H°$/kJ mol^{-1}:		-191.1	-159.4		d: 0.8717 g/mL
$S°$/J mol^{-1}K^{-1}:					η: 0.4 mPa s
C_p/J mol^{-1}K^{-1}:					k:

COMMENTS:

Name: 2-Chloro-2-methylpropane
Synonym: *tert*-Butyl chloride

Mol. Form.: C_4H_9Cl

CAS RN: 507-20-0
Merck No.: 1562
Mol. Wt.: 92.568

PHYSICAL CONSTANTS

T_m: -26°C (247 K)	T_c:	μ: 2.13 D
T_b: 50.9°C (324.0 K)	P_c:	IP: 10.61 eV

TRANSITION PROPERTIES

$\Delta_{fus}H$ (T_m): 2.09 kJ/mol
$\Delta_{vap}H$ (T_b): 27.55 kJ/mol
$\Delta_{vap}H$ (25°C): 28.98 kJ/mol

Vapor pressure (0°C):
Vapor pressure (25°C): 42.7 kPa
Vapor pressure (100°C):

PROPERTIES AT 25°C AND 100 kPa

	Solid	Liquid	Gas		Liquid
$\Delta_f H°$/kJ mol^{-1}:		-211.2	-182.2	d:	0.8361 g/mL
$S°$/J mol^{-1}K^{-1}:				η:	0.43 mPa s
C_p/J mol^{-1}K^{-1}:				k:	
COMMENTS:					

Name: 1-Chloronaphthalene
Synonym: 1-Naphthyl chloride

Mol. Form.: $C_{10}H_7Cl$

CAS RN: 90-13-1
Merck No.: 2149
Mol. Wt.: 162.618

PHYSICAL CONSTANTS

T_m: -2.5°C (270.6 K)	T_c:	μ:
T_b: 259.3°C (532.4 K)	P_c:	IP: 8.13 eV

TRANSITION PROPERTIES

$\Delta_{fus}H$ (T_m): 12.90 kJ/mol
$\Delta_{vap}H$ (T_b): 52.05 kJ/mol
$\Delta_{vap}H$ (25°C): 65.2 kJ/mol

Vapor pressure (0°C):
Vapor pressure (25°C):
Vapor pressure (100°C):

PROPERTIES AT 25°C AND 100 kPa

	Solid	Liquid	Gas		Liquid
$\Delta_f H°$/kJ mol^{-1}:		54.6	119.8	d:	1.189 g/mL
$S°$/J mol^{-1}K^{-1}:				η:	2.94 mPa s
C_p/J mol^{-1}K^{-1}:		212.6		k:	
COMMENTS: Highly toxic					

Name: *o*-Chloronitrobenzene
Synonym: 1-Chloro-2-nitrobenzene

Mol. Form.: $C_6H_4ClNO_2$

CAS RN: 88-73-3
Merck No.: 2151
Mol. Wt.: 157.556

PHYSICAL CONSTANTS

T_m: 32.5°C (305.6 K)	T_c:	μ: 4.64 D
T_b: 245.5°C (518.6 K)	P_c:	IP:

TRANSITION PROPERTIES

$\Delta_{fus}H$ (T_m):
$\Delta_{vap}H$ (T_b):
$\Delta_{vap}H$ (25°C):

Vapor pressure (0°C):
Vapor pressure (25°C):
Vapor pressure (100°C):

PROPERTIES AT 25°C AND 100 kPa

	Solid	Liquid	Gas		Solid
$\Delta_f H°$/kJ mol^{-1}:				d:	
$S°$/J mol^{-1}K^{-1}:				η:	N/A
C_p/J mol^{-1}K^{-1}:				k:	
COMMENTS:					

Name: *m*-Chloronitrobenzene
Synonym: 1-Chloro-3-nitrobenzene

Mol. Form.: $C_6H_4ClNO_2$

CAS RN: 121-73-3
Merck No.: 2151
Mol. Wt.: 157.556

PHYSICAL CONSTANTS

T_m: 44.4°C (317.5 K) T_c: μ: 3.73 D
T_b: 235.5°C (508.6 K) P_c: IP: 9.92 eV

TRANSITION PROPERTIES

$\Delta_{fus}H$ (T_m): 19.37 kJ/mol
$\Delta_{vap}H$ (T_b):
$\Delta_{vap}H$ (25°C):

Vapor pressure (0°C):
Vapor pressure (25°C):
Vapor pressure (100°C):

PROPERTIES AT 25°C AND 100 kPa

	Solid	Liquid	Gas		Solid
$\Delta_fH°$/kJ mol^{-1}:					*d*: 1.53 g/mL
$S°$/J mol^{-1}K^{-1}:					η: N/A
C_p/J mol^{-1}K^{-1}:					*k*:
COMMENTS:					

Name: *p*-Chloronitrobenzene
Synonym: 1-Chloro-4-nitrobenzene

Mol. Form.: $C_6H_4ClNO_2$

CAS RN: 100-00-5
Merck No.: 2151
Mol. Wt.: 157.556

PHYSICAL CONSTANTS

T_m: 83.5°C (356.6 K) T_c: μ: 2.83 D
T_b: 242°C (515 K) P_c: IP: 9.96 eV

TRANSITION PROPERTIES

$\Delta_{fus}H$ (T_m): 20.77 kJ/mol
$\Delta_{vap}H$ (T_b):
$\Delta_{vap}H$ (25°C):

Vapor pressure (0°C):
Vapor pressure (25°C): 0.003 kPa
Vapor pressure (100°C):

PROPERTIES AT 25°C AND 100 kPa

	Solid	Liquid	Gas		Solid
$\Delta_fH°$/kJ mol^{-1}:					*d*: 1.520 g/mL
$S°$/J mol^{-1}K^{-1}:					η: N/A
C_p/J mol^{-1}K^{-1}:					*k*:
COMMENTS:					

Name: Chloropentafluoroethane
Synonym: Refrigerant 115

Mol. Form.: C_2ClF_5

CAS RN: 76-15-3
Merck No.:
Mol. Wt.: 154.467

PHYSICAL CONSTANTS

T_m: -99.44°C (173.71 K) T_c: 80.1°C (353.2 K)   μ: 0.52 D
T_b: -37.95°C (235.20 K) P_c: 3.229 MPa   IP: 12.60 eV

TRANSITION PROPERTIES

$\Delta_{fus}H$ (T_m): 1.88 kJ/mol
$\Delta_{vap}H$ (T_b): 19.41 kJ/mol
$\Delta_{vap}H$ (25°C):

Vapor pressure (0°C):
Vapor pressure (25°C): 912 kPa
Vapor pressure (100°C): N/A

PROPERTIES AT 25°C AND 100 kPa

	Solid	Liquid	Gas		Gas
$\Delta_fH°$/kJ mol^{-1}:					*d*: 6.314 g/L
$S°$/J mol^{-1}K^{-1}:					η:
C_p/J mol^{-1}K^{-1}:		184.2			*k*:
COMMENTS: TLV=1000 ppm; highly toxic					

Name: 1-Chloropentane

Synonyms: Amyl chloride

            Pentyl chloride

Mol. Form.: $C_5H_{11}Cl$

CAS RN: 543-59-9

Merck No.: 642

Mol. Wt.: 106.595

PHYSICAL CONSTANTS

T_m: -99°C (174 K)

T_b: 107.8°C (380.9 K)

T_c:

P_c:

μ: 2.16 D

IP:

TRANSITION PROPERTIES

$\Delta_{fus}H$ (T_m):

$\Delta_{vap}H$ (T_b): 33.15 kJ/mol

$\Delta_{vap}H$ (25°C): 38.24 kJ/mol

Vapor pressure (0°C):

Vapor pressure (25°C): 4.36 kPa

Vapor pressure (100°C): 79.6 kPa

PROPERTIES AT 25°C AND 100 kPa

	Solid	Liquid	Gas	Liquid
$\Delta_f H°$/kJ mol^{-1}:		-213.2	-175.0	d: 0.877 g/mL
$S°$/J mol^{-1}K^{-1}:				η: 0.55 mPa s
C_p/J mol^{-1}K^{-1}:				k:
COMMENTS: Flammable				

Name: *o*-Chlorophenol

Synonym: 2-Chlorophenol

Mol. Form.: C_6H_5ClO

CAS RN: 95-57-8

Merck No.: 2154

Mol. Wt.: 128.558

PHYSICAL CONSTANTS

T_m: 9.8°C (282.9 K)

T_b: 174.9°C (448.0 K)

T_c:

P_c:

μ:

IP:

TRANSITION PROPERTIES

$\Delta_{fus}H$ (T_m): 12.52 kJ/mol

$\Delta_{vap}H$ (T_b):

$\Delta_{vap}H$ (25°C):

Vapor pressure (0°C):

Vapor pressure (25°C): 0.308 kPa

Vapor pressure (100°C): 10.8 kPa

PROPERTIES AT 25°C AND 100 kPa

	Solid	Liquid	Gas	Liquid
$\Delta_f H°$/kJ mol^{-1}:				d: 1.257 g/mL
$S°$/J mol^{-1}K^{-1}:				η: 3.59 mPa s
C_p/J mol^{-1}K^{-1}:		189.0		k:
COMMENTS: Highly toxic				

Name: *m*-Chlorophenol

Synonym: 3-Chlorophenol

Mol. Form.: C_6H_5ClO

CAS RN: 108-43-0

Merck No.: 2154

Mol. Wt.: 128.558

PHYSICAL CONSTANTS

T_m: 32.6°C (305.7 K)

T_b: 214°C (487 K)

T_c:

P_c:

μ:

IP: 8.65 eV

TRANSITION PROPERTIES

$\Delta_{fus}H$ (T_m): 14.91 kJ/mol

$\Delta_{vap}H$ (T_b):

$\Delta_{vap}H$ (25°C): 36.0 kJ/mol

Vapor pressure (0°C):

Vapor pressure (25°C):

Vapor pressure (100°C):

PROPERTIES AT 25°C AND 100 kPa

	Solid	Liquid	Gas	Solid
$\Delta_f H°$/kJ mol^{-1}:	-206.4	-189.3	-153.3	d: 1.218 g/mL
$S°$/J mol^{-1}K^{-1}:				η: N/A
C_p/J mol^{-1}K^{-1}:				k:
COMMENTS: Highly toxic				

Name: *p*-Chlorophenol
Synonym: 4-Chlorophenol

Mol. Form.: C_6H_5ClO

CAS RN: 106-48-9
Merck No.: 2154
Mol. Wt.: 128.558

PHYSICAL CONSTANTS

T_m: 42.7°C (315.8 K)
T_b: 220°C (493 K)

T_c:
P_c:

μ: 2.11 D
IP: 8.69 eV

TRANSITION PROPERTIES

$\Delta_{fus}H$ (T_m): 14.07 kJ/mol
$\Delta_{vap}H$ (T_b):
$\Delta_{vap}H$ (25°C): 35.5 kJ/mol

Vapor pressure (0°C):
Vapor pressure (25°C):
Vapor pressure (100°C):

PROPERTIES AT 25°C AND 100 kPa

	Solid	Liquid	Gas		Solid
$\Delta_f H°$/kJ mol^{-1}:	-197.7	-181.3	-145.8		*d*: 1.40 g/mL
$S°$/J mol^{-1}K^{-1}:					η: N/A
C_p/J mol^{-1}K^{-1}:					*k*:
COMMENTS: Highly toxic					

Name: 1-Chloropropane
Synonym: Propyl chloride

Mol. Form.: C_3H_7Cl

CAS RN: 540-54-5
Merck No.: 7859
Mol. Wt.: 78.541

PHYSICAL CONSTANTS

T_m: -122.8°C (150.3 K)
T_b: 46.5°C (319.6 K)

T_c: 230°C (503 K)
P_c: 4.58 MPa

μ: 2.05 D
IP: 10.82 eV

TRANSITION PROPERTIES

$\Delta_{fus}H$ (T_m): 5.54 kJ/mol
$\Delta_{vap}H$ (T_b): 27.18 kJ/mol
$\Delta_{vap}H$ (25°C): 28.35 kJ/mol

Vapor pressure (0°C): 15.1 kPa
Vapor pressure (25°C): 45.8 kPa
Vapor pressure (100°C):

PROPERTIES AT 25°C AND 100 kPa

	Solid	Liquid	Gas		Liquid
$\Delta_f H°$/kJ mol^{-1}:		-160.6	-131.9		*d*: 0.8830 g/mL
$S°$/J mol^{-1}K^{-1}:					η: 0.334 mPa s
C_p/J mol^{-1}K^{-1}:		132.2			*k*:
COMMENTS:					

Name: 2-Chloropropane
Synonym: Isopropyl chloride

Mol. Form.: C_3H_7Cl

CAS RN: 75-29-6
Merck No.: 5099
Mol. Wt.: 78.541

PHYSICAL CONSTANTS

T_m: -117.2°C (155.9 K)
T_b: 35.7°C (308.8 K)

T_c:
P_c:

μ: 2.17 D
IP: 10.78 eV

TRANSITION PROPERTIES

$\Delta_{fus}H$ (T_m): 7.39 kJ/mol
$\Delta_{vap}H$ (T_b): 26.30 kJ/mol
$\Delta_{vap}H$ (25°C): 26.90 kJ/mol

Vapor pressure (0°C): 25.1 kPa
Vapor pressure (25°C): 68.9 kPa
Vapor pressure (100°C):

PROPERTIES AT 25°C AND 100 kPa

	Solid	Liquid	Gas		Liquid
$\Delta_f H°$/kJ mol^{-1}:		-172.1	-144.9		*d*: 0.8563 g/mL
$S°$/J mol^{-1}K^{-1}:					η: 0.303 mPa s
C_p/J mol^{-1}K^{-1}:					*k*:
COMMENTS:					

Name: 2-Chloropropene
Synonym: Isopropenyl chloride

Mol. Form.: C_3H_5Cl

CAS RN: 557-98-2
Merck No.:
Mol. Wt.: 76.525

PHYSICAL CONSTANTS

T_m: -137.4°C (135.7 K)
T_b: 22.65°C (295.80 K)

T_c:
P_c:

μ: 1.647 D
IP:

TRANSITION PROPERTIES

$\Delta_{fus}H$ (T_m):
$\Delta_{vap}H$ (T_b):
$\Delta_{vap}H$ (25°C):

Vapor pressure (0°C): 41.4 kPa
Vapor pressure (25°C): 110 kPa
Vapor pressure (100°C):

PROPERTIES AT 25°C AND 100 kPa

	Solid	Liquid	Gas	Gas
$\Delta_f H°$/kJ mol^{-1}:		-21.0		d: 3.128 g/L
$S°$/J mol^{-1}K^{-1}:				η:
C_p/J mol^{-1}K^{-1}:				k:
COMMENTS:				

Name: 3-Chloropropene
Synonyms: Allyl chloride
           3-Chloropropylene
Mol. Form.: C_3H_5Cl

CAS RN: 107-05-1
Merck No.: 287
Mol. Wt.: 76.525

PHYSICAL CONSTANTS

T_m: -134.5°C (138.6 K)
T_b: 45.15°C (318.30 K)

T_c: 241°C (514 K)
P_c:

μ: 1.94 D
IP: 9.90 eV

TRANSITION PROPERTIES

$\Delta_{fus}H$ (T_m):
$\Delta_{vap}H$ (T_b): 29.04 kJ/mol
$\Delta_{vap}H$ (25°C):

Vapor pressure (0°C): 16.4 kPa
Vapor pressure (25°C): 48.9 kPa
Vapor pressure (100°C):

PROPERTIES AT 25°C AND 100 kPa

	Solid	Liquid	Gas	Liquid
$\Delta_f H°$/kJ mol^{-1}:				d: 0.933 g/mL
$S°$/J mol^{-1}K^{-1}:				η: 0.314 mPa s
C_p/J mol^{-1}K^{-1}:				k:
COMMENTS: TLV=1 ppm; carcinogen; highly toxic; flammable				

Name: *o*-Chlorotoluene
Synonyms: 1-Chloro-2-methylbenzene
           2-Tolyl chloride
Mol. Form.: C_7H_7Cl

CAS RN: 95-49-8
Merck No.: 2172
Mol. Wt.: 126.585

PHYSICAL CONSTANTS

T_m: -35.6°C (237.5 K)
T_b: 159.0°C (432.1 K)

T_c:
P_c:

μ: 1.56 D
IP: 8.83 eV

TRANSITION PROPERTIES

$\Delta_{fus}H$ (T_m): 8.37 kJ/mol
$\Delta_{vap}H$ (T_b): 37.53 kJ/mol
$\Delta_{vap}H$ (25°C): 45.63 kJ/mol

Vapor pressure (0°C):
Vapor pressure (25°C): 0.482 kPa
Vapor pressure (100°C): 16.8 kPa

PROPERTIES AT 25°C AND 100 kPa

	Solid	Liquid	Gas	Liquid
$\Delta_f H°$/kJ mol^{-1}:				d: 1.077 g/mL
$S°$/J mol^{-1}K^{-1}:				η: 0.964 mPa s
C_p/J mol^{-1}K^{-1}:		166.8		k:
COMMENTS: TLV=50 ppm				

Name: *m*-Chlorotoluene   CAS RN: 108-41-8
Synonyms: 3-Tolyl chloride   Merck No.: 2172
              1-Chloro-3-methylbenzene   Mol. Wt.: 126.585
Mol. Form.: C_7H_7Cl

PHYSICAL CONSTANTS

T_m: -47.8°C (225.3 K)	T_c:	μ:
T_b: 161.8°C (434.9 K)	P_c:	IP: 8.83 eV

TRANSITION PROPERTIES

$\Delta_{fus}H$ (T_m): 10.46 kJ/mol	Vapor pressure (0°C):
$\Delta_{vap}H$ (T_b):	Vapor pressure (25°C):
$\Delta_{vap}H$ (25°C):	Vapor pressure (100°C): 15.1 kPa

PROPERTIES AT 25°C AND 100 kPa

	Solid	Liquid	Gas		Liquid
$\Delta_fH°$/kJ mol^{-1}:				d:	1.070 g/mL
$S°$/J mol^{-1}K^{-1}:				η:	0.823 mPa s
C_p/J mol^{-1}K^{-1}:				k:	
COMMENTS:					

Name: *p*-Chlorotoluene   CAS RN: 106-43-4
Synonyms: 1-Chloro-4-methylbenzene   Merck No.: 2172
              4-Tolyl chloride   Mol. Wt.: 126.585
Mol. Form.: C_7H_7Cl

PHYSICAL CONSTANTS

T_m: 7.5°C (280.6 K)	T_c:	μ: 2.21 D
T_b: 162.4°C (435.5 K)	P_c:	IP: 8.69 eV

TRANSITION PROPERTIES

$\Delta_{fus}H$ (T_m):	Vapor pressure (0°C):
$\Delta_{vap}H$ (T_b): 38.74 kJ/mol	Vapor pressure (25°C):
$\Delta_{vap}H$ (25°C):	Vapor pressure (100°C): 15.3 kPa

PROPERTIES AT 25°C AND 100 kPa

	Solid	Liquid	Gas		Liquid
$\Delta_fH°$/kJ mol^{-1}:				d:	1.064 g/mL
$S°$/J mol^{-1}K^{-1}:				η:	0.837 mPa s
C_p/J mol^{-1}K^{-1}:				k:	
COMMENTS:					

Name: Chlorotrifluoroethylene   CAS RN: 79-38-9
Synonym: Chlorotrifluoroethene   Merck No.:
                                                                Mol. Wt.: 116.470
Mol. Form.: C_2ClF_3

PHYSICAL CONSTANTS

T_m: -158°C (115 K)	T_c: 106°C (379 K)	μ: 0.40 D
T_b: -27.85°C (245.30 K)	P_c: 4.05 MPa	IP: 9.81 eV

TRANSITION PROPERTIES

$\Delta_{fus}H$ (T_m):	Vapor pressure (0°C):
$\Delta_{vap}H$ (T_b):	Vapor pressure (25°C):
$\Delta_{vap}H$ (25°C):	Vapor pressure (100°C):

PROPERTIES AT 25°C AND 100 kPa

	Solid	Liquid	Gas		Gas
$\Delta_fH°$/kJ mol^{-1}:			-555.2	d:	4.761 g/L
$S°$/J mol^{-1}K^{-1}:			322.1	η:	
C_p/J mol^{-1}K^{-1}:			83.9	k:	
COMMENTS: Very flammable					

Name: Chlorotrifluoromethane
Synonyms: Refrigerant 13
        CFC-13
Mol. Form.: $CClF_3$

CAS RN: 75-72-9
Merck No.:
Mol. Wt.: 104.459

PHYSICAL CONSTANTS

T_m: -181°C (92 K) T_c: 29°C (302 K)   μ: 0.50 D
T_b: -81.4°C (191.7 K) P_c: 3.870 MPa   IP: 12.39 eV

TRANSITION PROPERTIES

$\Delta_{fus}H$ (T_m):   Vapor pressure (0°C):
$\Delta_{vap}H$ (T_b): 15.75 kJ/mol   Vapor pressure (25°C): 3600 kPa
$\Delta_{vap}H$ (25°C):   Vapor pressure (100°C): N/A

PROPERTIES AT 25°C AND 100 kPa

	Solid	Liquid	Gas		Gas
$\Delta_f H°$/kJ mol^{-1}:			-707.8		d: 4.270 g/L
$S°$/J mol^{-1}K^{-1}:			285.4		η:
C_p/J mol^{-1}K^{-1}:			66.9		k:
COMMENTS:					

Name: Chromium

CAS RN: 7440-47-3
Merck No.: 2234
Mol. Wt.: 51.996

Mol. Form.: Cr

PHYSICAL CONSTANTS

T_m: 1907°C (2180 K) T_c: μ:
T_b: 2671°C (2944 K) P_c: IP: 6.77 eV

TRANSITION PROPERTIES

$\Delta_{fus}H$ (T_m): 21.00 kJ/mol   Vapor pressure (0°C): N/A
$\Delta_{vap}H$ (T_b):   Vapor pressure (25°C): N/A
$\Delta_{vap}H$ (25°C):   Vapor pressure (100°C): N/A

PROPERTIES AT 25°C AND 100 kPa

	Solid	Liquid	Gas		Solid
$\Delta_f H°$/kJ mol^{-1}:	0.0		396.6		d: 7.15 g/mL
$S°$/J mol^{-1}K^{-1}:	23.8		174.5		η: N/A
C_p/J mol^{-1}K^{-1}:	23.4		20.8		k: 93.9 W/m K
COMMENTS: Carcinogen; highly toxic					

Name: Chromium chloride ($CrCl_2$)
Synonyms: Chromium dichloride
        Chromium(II) chloride
Mol. Form.: Cl_2Cr

CAS RN: 10049-05-5
Merck No.: 2246
Mol. Wt.: 122.902

PHYSICAL CONSTANTS

T_m: 814°C (1087 K) T_c: μ:
T_b: 1300°C (1573 K) P_c: IP:

TRANSITION PROPERTIES

$\Delta_{fus}H$ (T_m): 32.20 kJ/mol   Vapor pressure (0°C): N/A
$\Delta_{vap}H$ (T_b): 197.00 kJ/mol   Vapor pressure (25°C): N/A
$\Delta_{vap}H$ (25°C):   Vapor pressure (100°C): N/A

PROPERTIES AT 25°C AND 100 kPa

	Solid	Liquid	Gas		Solid
$\Delta_f H°$/kJ mol^{-1}:	-395.4				d: 2.88 g/mL
$S°$/J mol^{-1}K^{-1}:	115.3				η: N/A
C_p/J mol^{-1}K^{-1}:	71.2				k:
COMMENTS: Carcinogen					

Name: Chromium oxide (Cr_2O_3)   CAS RN: 1308-38-9
Synonyms: Dichromium trioxide   Merck No.: 2229
        Chromium(III) oxide   Mol. Wt.: 151.990
Mol. Form.: Cr_2O_3

PHYSICAL CONSTANTS

T_m: 2330°C (2603 K) T_c: μ:
T_b: P_c: IP:

TRANSITION PROPERTIES

$\Delta_{fus}H$ (T_m): 130.00 kJ/mol   Vapor pressure (0°C): N/A
$\Delta_{vap}H$ (T_b):   Vapor pressure (25°C): N/A
$\Delta_{vap}H$ (25°C):   Vapor pressure (100°C): N/A

PROPERTIES AT 25°C AND 100 kPa

	Solid	Liquid	Gas		Solid
$\Delta_f H°$/kJ mol^{-1}:	-1139.7			d:	5.22 g/mL
$S°$/J mol^{-1}K^{-1}:	81.2			η:	N/A
C_p/J mol^{-1}K^{-1}:	118.7			k:	

COMMENTS: Carcinogen; highly toxic

Name: Chromyl chloride (CrO_2Cl_2)   CAS RN: 14977-61-8
Synonym: Chromyl dichloride   Merck No.: 2251
                                                                  Mol. Wt.: 154.900
Mol. Form.: Cl_2CrO_2

PHYSICAL CONSTANTS

T_m: -96.5°C (176.6 K) T_c: μ:
T_b: 117°C (390 K) P_c: IP: 11.60 eV

TRANSITION PROPERTIES

$\Delta_{fus}H$ (T_m):   Vapor pressure (0°C):
$\Delta_{vap}H$ (T_b): 35.10 kJ/mol   Vapor pressure (25°C):
$\Delta_{vap}H$ (25°C): 41.4 kJ/mol   Vapor pressure (100°C):

PROPERTIES AT 25°C AND 100 kPa

	Solid	Liquid	Gas		Liquid
$\Delta_f H°$/kJ mol^{-1}:		-579.5	-538.1	d:	1.91 g/mL
$S°$/J mol^{-1}K^{-1}:		221.8	329.8	η:	
C_p/J mol^{-1}K^{-1}:			84.5	k:	

COMMENTS: TLV=0.025 ppm; carcinogen

Name: Chrysene   CAS RN: 218-01-9
Synonym: 1,2-Benzophenanthrene   Merck No.: 2259
                                                                  Mol. Wt.: 228.293
Mol. Form.: $C_{18}H_{12}$

PHYSICAL CONSTANTS

T_m: 258.2°C (531.3 K) T_c: μ:
T_b: 448°C (721 K) P_c: IP: 7.59 eV

TRANSITION PROPERTIES

$\Delta_{fus}H$ (T_m): 26.15 kJ/mol   Vapor pressure (0°C): N/A
$\Delta_{vap}H$ (T_b):   Vapor pressure (25°C): N/A
$\Delta_{vap}H$ (25°C):   Vapor pressure (100°C): N/A

PROPERTIES AT 25°C AND 100 kPa

	Solid	Liquid	Gas		Solid
$\Delta_f H°$/kJ mol^{-1}:	145.3		269.8	d:	1.268 g/mL
$S°$/J mol^{-1}K^{-1}:				η:	N/A
C_p/J mol^{-1}K^{-1}:				k:	

COMMENTS: Carcinogen; highly toxic

Name: Cobalt

CAS RN: 7440-48-4
Merck No.: 2421
Mol. Wt.: 58.933

Mol. Form.: Co

PHYSICAL CONSTANTS

T_m: 1495°C (1768 K) T_c: μ:
T_b: 2927°C (3200 K) P_c: IP: 7.88 eV

TRANSITION PROPERTIES

$\Delta_{fus}H$ (T_m): 16.20 kJ/mol Vapor pressure (0°C): N/A
$\Delta_{vap}H$ (T_b): Vapor pressure (25°C): N/A
$\Delta_{vap}H$ (25°C): Vapor pressure (100°C): N/A

PROPERTIES AT 25°C AND 100 kPa

	Solid	Liquid	Gas		Solid
$\Delta_f H°$/kJ mol^{-1}:	0.0		424.7		d: 8.86 g/mL
$S°$/J mol^{-1}K^{-1}:	30.0		179.5		η: N/A
C_p/J mol^{-1}K^{-1}:	24.8		23.0		k: 100 W/m K

COMMENTS: Highly toxic

Name: Cobalt oxide (CoO)
Synonym: Cobalt(II) oxide

CAS RN: 1307-96-6
Merck No.: 2440
Mol. Wt.: 74.933

Mol. Form.: CoO

PHYSICAL CONSTANTS

T_m: 1830°C (2103 K) T_c: μ:
T_b: P_c: IP:

TRANSITION PROPERTIES

$\Delta_{fus}H$ (T_m): Vapor pressure (0°C): N/A
$\Delta_{vap}H$ (T_b): Vapor pressure (25°C): N/A
$\Delta_{vap}H$ (25°C): Vapor pressure (100°C): N/A

PROPERTIES AT 25°C AND 100 kPa

	Solid	Liquid	Gas		Solid
$\Delta_f H°$/kJ mol^{-1}:	-237.9				d: 6.5 g/mL
$S°$/J mol^{-1}K^{-1}:	53.0				η: N/A
C_p/J mol^{-1}K^{-1}:	55.2				k:

COMMENTS: Highly toxic

Name: Copper

CAS RN: 7440-50-8
Merck No.: 2514
Mol. Wt.: 63.546

Mol. Form.: Cu

PHYSICAL CONSTANTS

T_m: 1084.62°C (1357.77 K) T_c: μ:
T_b: 2562°C (2835 K) P_c: IP: 7.73 eV

TRANSITION PROPERTIES

$\Delta_{fus}H$ (T_m): 13.26 kJ/mol Vapor pressure (0°C): N/A
$\Delta_{vap}H$ (T_b): Vapor pressure (25°C): N/A
$\Delta_{vap}H$ (25°C): Vapor pressure (100°C): N/A

PROPERTIES AT 25°C AND 100 kPa

	Solid	Liquid	Gas		Solid
$\Delta_f H°$/kJ mol^{-1}:	0.0		337.4		d: 8.96 g/mL
$S°$/J mol^{-1}K^{-1}:	33.2		166.4		η: N/A
C_p/J mol^{-1}K^{-1}:	24.4		20.8		k: 401 W/m K

COMMENTS:

Name: Copper chloride (CuCl$_2$)
Synonyms: Cupric chloride
         Copper(II) chloride
Mol. Form.: Cl$_2$Cu

CAS RN: 7447-39-4
Merck No.: 2636
Mol. Wt.: 134.451

PHYSICAL CONSTANTS

| T_m: 430°C (703 K) | T_c: | μ: |
| T_b: | P_c: | IP: |

TRANSITION PROPERTIES

$\Delta_{fus}H$ (T_m): 20.40 kJ/mol
$\Delta_{vap}H$ (T_b):
$\Delta_{vap}H$ (25°C):

Vapor pressure (0°C): N/A
Vapor pressure (25°C): N/A
Vapor pressure (100°C): N/A

PROPERTIES AT 25°C AND 100 kPa

	Solid	Liquid	Gas		Solid
$\Delta_f H°$/kJ mol^{-1}:	-220.1			d:	3.4 g/mL
$S°$/J mol^{-1}K^{-1}:	108.1			η:	N/A
C_p/J mol^{-1}K^{-1}:	71.9			k:	
COMMENTS:					

Name: Copper oxide (CuO)
Synonyms: Cupric oxide
         Copper(II) oxide
Mol. Form.: CuO

CAS RN: 1317-38-0
Merck No.: 2650
Mol. Wt.: 79.545

PHYSICAL CONSTANTS

| T_m: 1446°C (1719 K) | T_c: | μ: |
| T_b: | P_c: | IP: |

TRANSITION PROPERTIES

$\Delta_{fus}H$ (T_m): 11.80 kJ/mol
$\Delta_{vap}H$ (T_b):
$\Delta_{vap}H$ (25°C):

Vapor pressure (0°C): N/A
Vapor pressure (25°C): N/A
Vapor pressure (100°C): N/A

PROPERTIES AT 25°C AND 100 kPa

	Solid	Liquid	Gas		Solid
$\Delta_f H°$/kJ mol^{-1}:	-157.3			d:	6.5 g/mL
$S°$/J mol^{-1}K^{-1}:	42.6			η:	N/A
C_p/J mol^{-1}K^{-1}:	42.3			k:	
COMMENTS:					

Name: Copper sulfate (CuSO$_4$)
Synonyms: Cupric sulfate
         Copper(II) sulfate
Mol. Form.: CuO$_4$S

CAS RN: 7758-98-7
Merck No.: 2659
Mol. Wt.: 159.610

PHYSICAL CONSTANTS

| T_m: 200°C (473 K) | T_c: | μ: |
| T_b: | P_c: | IP: |

TRANSITION PROPERTIES

$\Delta_{fus}H$ (T_m):
$\Delta_{vap}H$ (T_b):
$\Delta_{vap}H$ (25°C):

Vapor pressure (0°C):
Vapor pressure (25°C):
Vapor pressure (100°C):

PROPERTIES AT 25°C AND 100 kPa

	Solid	Liquid	Gas		Solid
$\Delta_f H°$/kJ mol^{-1}:	-771.4			d:	3.6 g/mL
$S°$/J mol^{-1}K^{-1}:	109.2			η:	N/A
C_p/J mol^{-1}K^{-1}:	98.9			k:	
COMMENTS: Highly toxic					

Name: *o*-Cresol
Synonyms: 2-Methylphenol
        o-Cresylic acid
Mol. Form.: C_7H_8O

CAS RN: 95-48-7
Merck No.: 2580
Mol. Wt.: 108.140

PHYSICAL CONSTANTS

T_m: 29.8°C (302.9 K) T_c: 424.5°C (697.6 K)   μ:
T_b: 191.04°C (464.19 K) P_c: 5.01 MPa   IP: 8.14 eV

TRANSITION PROPERTIES

$\Delta_{fus}H$ (T_m): 13.94 kJ/mol   Vapor pressure (0°C):
$\Delta_{vap}H$ (T_b): 45.19 kJ/mol   Vapor pressure (25°C): 0.041 kPa
$\Delta_{vap}H$ (25°C):   Vapor pressure (100°C): 4.23 kPa

PROPERTIES AT 25°C AND 100 kPa

	Solid	Liquid	Gas		Solid
$\Delta_f H°$/kJ mol⁻¹:	-204.6		-128.6		*d*: 1.135 g/mL
$S°$/J mol⁻¹K⁻¹:	165.4				η: N/A
C_p/J mol⁻¹K⁻¹:	154.6	154.5			*k*:

COMMENTS: TLV=5 ppm; highly toxic

Name: *m*-Cresol
Synonyms: 3-Methylphenol
        m-Cresylic acid
Mol. Form.: C_7H_8O

CAS RN: 108-39-4
Merck No.: 2579
Mol. Wt.: 108.140

PHYSICAL CONSTANTS

T_m: 11.8°C (284.9 K) T_c: 432.7°C (705.8 K)   μ:
T_b: 202.27°C (475.42 K) P_c: 4.56 MPa   IP: 8.29 eV

TRANSITION PROPERTIES

$\Delta_{fus}H$ (T_m): 9.41 kJ/mol   Vapor pressure (0°C):
$\Delta_{vap}H$ (T_b): 47.40 kJ/mol   Vapor pressure (25°C): 0.019 kPa
$\Delta_{vap}H$ (25°C): 61.71 kJ/mol   Vapor pressure (100°C): 2.50 kPa

PROPERTIES AT 25°C AND 100 kPa

	Solid	Liquid	Gas		Liquid
$\Delta_f H°$/kJ mol⁻¹:		-194.0	-132.3		*d*: 1.0302 g/mL
$S°$/J mol⁻¹K⁻¹:		212.6			η: 12.9 mPa s
C_p/J mol⁻¹K⁻¹:		224.9			*k*:

COMMENTS: TLV=5 ppm; highly toxic

Name: *p*-Cresol
Synonyms: 4-Methylphenol
        p-Cresylic acid
Mol. Form.: C_7H_8O

CAS RN: 106-44-5
Merck No.: 2581
Mol. Wt.: 108.140

PHYSICAL CONSTANTS

T_m: 34.7°C (307.8 K) T_c: 431.5°C (704.6 K)   μ:
T_b: 201.98°C (475.13 K) P_c: 5.15 MPa   IP: 8.13 eV

TRANSITION PROPERTIES

$\Delta_{fus}H$ (T_m): 11.89 kJ/mol   Vapor pressure (0°C):
$\Delta_{vap}H$ (T_b): 47.45 kJ/mol   Vapor pressure (25°C): 0.017 kPa
$\Delta_{vap}H$ (25°C):   Vapor pressure (100°C): 2.45 kPa

PROPERTIES AT 25°C AND 100 kPa

	Solid	Liquid	Gas		Solid
$\Delta_f H°$/kJ mol⁻¹:	-199.3		-125.4		*d*: 1.154 g/mL
$S°$/J mol⁻¹K⁻¹:	167.3				η: N/A
C_p/J mol⁻¹K⁻¹:	150.2	150.2	125.0		*k*:

COMMENTS: TLV=5 ppm; highly toxic

Name: *cis*-Crotonic acid
Synonym: *cis*-2-Butenoic acid

Mol. Form.: $C_4H_6O_2$

CAS RN: 503-64-0
Merck No.: 5048
Mol. Wt.: 86.090

PHYSICAL CONSTANTS

T_m: 15°C (288 K) T_c: μ:
T_b: 169°C (442 K) P_c: IP: 10.08 eV

TRANSITION PROPERTIES

$\Delta_{fus}H$ (T_m): 12.57 kJ/mol   Vapor pressure (0°C):
$\Delta_{vap}H$ (T_b):   Vapor pressure (25°C):
$\Delta_{vap}H$ (25°C):   Vapor pressure (100°C): 7.36 kPa

PROPERTIES AT 25°C AND 100 kPa

	Solid	Liquid	Gas	Liquid
$\Delta_f H°$/kJ mol^{-1}:				d: 1.022 g/mL
$S°$/J mol^{-1}K^{-1}:				η:
C_p/J mol^{-1}K^{-1}:				k:

COMMENTS: Highly toxic

Name: *trans*-Crotonic acid
Synonym: *trans*-2-Butenoic acid

Mol. Form.: $C_4H_6O_2$

CAS RN: 107-93-7
Merck No.: 2600
Mol. Wt.: 86.090

PHYSICAL CONSTANTS

T_m: 72°C (345 K) T_c: μ:
T_b: 184.7°C (457.8 K) P_c: IP: 9.90 eV

TRANSITION PROPERTIES

$\Delta_{fus}H$ (T_m): 12.98 kJ/mol   Vapor pressure (0°C):
$\Delta_{vap}H$ (T_b):   Vapor pressure (25°C):
$\Delta_{vap}H$ (25°C):   Vapor pressure (100°C): 3.88 kPa

PROPERTIES AT 25°C AND 100 kPa

	Solid	Liquid	Gas	Solid
$\Delta_f H°$/kJ mol^{-1}:				d:
$S°$/J mol^{-1}K^{-1}:				η: N/A
C_p/J mol^{-1}K^{-1}:				k:

COMMENTS: Highly toxic

Name: Cumene
Synonyms: Isopropylbenzene
            (1-Methylethyl)benzene
Mol. Form.: C_9H_{12}

CAS RN: 98-82-8
Merck No.: 2619
Mol. Wt.: 120.194

PHYSICAL CONSTANTS

T_m: -96.01°C (177.14 K) T_c: 358.0°C (631.1 K)   μ: 0.79 D
T_b: 152.41°C (425.56 K) P_c: 3.209 MPa   IP: 8.73 eV

TRANSITION PROPERTIES

$\Delta_{fus}H$ (T_m): 7.79 kJ/mol   Vapor pressure (0°C):
$\Delta_{vap}H$ (T_b): 37.53 kJ/mol   Vapor pressure (25°C): 0.610 kPa
$\Delta_{vap}H$ (25°C): 45.13 kJ/mol   Vapor pressure (100°C): 20.7 kPa

PROPERTIES AT 25°C AND 100 kPa

	Solid	Liquid	Gas	Liquid
$\Delta_f H°$/kJ mol^{-1}:		-41.1	4.0	d: 0.8574 g/mL
$S°$/J mol^{-1}K^{-1}:				η: 0.737 mPa s
C_p/J mol^{-1}K^{-1}:		210.7		k: 0.128 W/m K

COMMENTS: TLV=50 ppm; highly toxic; flammable

Name: Cyanamide
Synonym: Cyanogenamide

Mol. Form.: CH_2N_2

CAS RN: 420-04-2
Merck No.: 2691
Mol. Wt.: 42.040

PHYSICAL CONSTANTS

T_m: 44°C (317 K)
T_b:

T_c:
P_c:

μ: 4.27 D
IP: 10.40 eV

TRANSITION PROPERTIES

$\Delta_{fus}H$ (T_m): 8.76 kJ/mol
$\Delta_{vap}H$ (T_b):
$\Delta_{vap}H$ (25°C):

Vapor pressure (0°C):
Vapor pressure (25°C):
Vapor pressure (100°C):

PROPERTIES AT 25°C AND 100 kPa

	Solid	Liquid	Gas		Solid
$\Delta_fH°$/kJ mol^{-1}:	58.8			d:	1.25 g/mL
$S°$/J mol^{-1}K^{-1}:				η:	N/A
C_p/J mol^{-1}K^{-1}:				k:	
COMMENTS: Highly toxic					

Name: Cyanogen
Synonym: Ethanedinitrile

Mol. Form.: C_2N_2

CAS RN: 460-19-5
Merck No.: 2698
Mol. Wt.: 52.036

PHYSICAL CONSTANTS

T_m: -27.9°C (245.2 K)
T_b: -21.1°C (252.0 K)

T_c: 127°C (400 K)
P_c: 5.98 MPa

μ: 0 D
IP: 13.37 eV

TRANSITION PROPERTIES

$\Delta_{fus}H$ (T_m):
$\Delta_{vap}H$ (T_b): 23.33 kJ/mol
$\Delta_{vap}H$ (25°C): 19.75 kJ/mol

Vapor pressure (0°C):
Vapor pressure (25°C):
Vapor pressure (100°C):

PROPERTIES AT 25°C AND 100 kPa

	Solid	Liquid	Gas		Gas
$\Delta_fH°$/kJ mol^{-1}:		285.9	306.7	d:	2.127 g/L
$S°$/J mol^{-1}K^{-1}:			241.9	η:	
C_p/J mol^{-1}K^{-1}:			56.8	k:	
COMMENTS: TLV=10 ppm; highly toxic; very flammable					

Name: Cyanogen bromide
Synonym: Bromine cyanide

Mol. Form.: CBrN

CAS RN: 506-68-3
Merck No.: 2700
Mol. Wt.: 105.922

PHYSICAL CONSTANTS

T_m: 52°C (325 K)
T_b: 61.5°C (334.6 K)

T_c:
P_c:

μ:
IP:

TRANSITION PROPERTIES

$\Delta_{fus}H$ (T_m):
$\Delta_{vap}H$ (T_b):
$\Delta_{vap}H$ (25°C):

Vapor pressure (0°C): 2.90 kPa
Vapor pressure (25°C): 15.9 kPa
Vapor pressure (100°C):

PROPERTIES AT 25°C AND 100 kPa

	Solid	Liquid	Gas		Solid
$\Delta_fH°$/kJ mol^{-1}:	140.5		186.2	d:	2.005 g/mL
$S°$/J mol^{-1}K^{-1}:			248.3	η:	N/A
C_p/J mol^{-1}K^{-1}:			46.9	k:	
COMMENTS: Highly toxic					

Name: Cyanogen chloride
Synonym: Chlorine cyanide

Mol. Form.: CClN

CAS RN: 506-77-4
Merck No.: 2701
Mol. Wt.: 61.470

PHYSICAL CONSTANTS

T_m: -6.55°C (266.60 K) T_c: μ: 2.833 D
T_b: 13°C (286 K) P_c: IP: 12.34 eV

TRANSITION PROPERTIES

$\Delta_{fus}H$ (T_m):   Vapor pressure (0°C):
$\Delta_{vap}H$ (T_b):   Vapor pressure (25°C):
$\Delta_{vap}H$ (25°C): 25.9 kJ/mol   Vapor pressure (100°C):

PROPERTIES AT 25°C AND 100 kPa

	Solid	Liquid	Gas		Gas
$\Delta_f H°$/kJ mol^{-1}:		112.1	138.0	d:	2.513 g/L
$S°$/J mol^{-1}K^{-1}:			236.2	η:	
C_p/J mol^{-1}K^{-1}:			45.0	k:	

COMMENTS: TLV=0.3 ppm; highly toxic

Name: Cyclobutane
Synonym: Tetramethylene

Mol. Form.: C$_4$H$_8$

CAS RN: 287-23-0
Merck No.: 2720
Mol. Wt.: 56.107

PHYSICAL CONSTANTS

T_m: -90.67°C (182.48 K) T_c: 186.9°C (460.0 K)   μ: 0 D
T_b: 12.6°C (285.7 K) P_c: 4.98 MPa   IP: 9.92 eV

TRANSITION PROPERTIES

$\Delta_{fus}H$ (T_m):   Vapor pressure (0°C): 62.6 kPa
$\Delta_{vap}H$ (T_b): 24.19 kJ/mol   Vapor pressure (25°C): 157 kPa
$\Delta_{vap}H$ (25°C): 23.51 kJ/mol   Vapor pressure (100°C): 1123 kPa

PROPERTIES AT 25°C AND 100 kPa

	Solid	Liquid	Gas		Gas
$\Delta_f H°$/kJ mol^{-1}:		3.7	28.4	d:	2.293 g/L
$S°$/J mol^{-1}K^{-1}:				η:	
C_p/J mol^{-1}K^{-1}:				k:	

COMMENTS: Very flammable

Name: Cycloheptane

Mol. Form.: C$_7$H$_{14}$

CAS RN: 291-64-5
Merck No.:
Mol. Wt.: 98.188

PHYSICAL CONSTANTS

T_m: -8.03°C (265.12 K) T_c: 331.1°C (604.2 K)   μ:
T_b: 118.48°C (391.63 K) P_c: 3.81 MPa   IP: 9.97 eV

TRANSITION PROPERTIES

$\Delta_{fus}H$ (T_m): 1.88 kJ/mol   Vapor pressure (0°C):
$\Delta_{vap}H$ (T_b):   Vapor pressure (25°C): 2.90 kPa
$\Delta_{vap}H$ (25°C): 38.5 kJ/mol   Vapor pressure (100°C): 58.8 kPa

PROPERTIES AT 25°C AND 100 kPa

	Solid	Liquid	Gas		Liquid
$\Delta_f H°$/kJ mol^{-1}:		-156.6	-118.1	d:	0.8066 g/mL
$S°$/J mol^{-1}K^{-1}:				η:	
C_p/J mol^{-1}K^{-1}:				k:	

COMMENTS: Flammable

Name: Cyclohexane
Synonym: Hexahydrobenzene

Mol. Form.: C_6H_{12}

CAS RN: 110-82-7
Merck No.: 2729
Mol. Wt.: 84.161

PHYSICAL CONSTANTS

T_m: 6.6°C (279.7 K)
T_b: 80.73°C (353.88 K)

T_c: 280.4°C (553.5 K)
P_c: 4.07 MPa

μ:
IP: 9.86 eV

TRANSITION PROPERTIES

$\Delta_{fus}H$ (T_m): 2.63 kJ/mol
$\Delta_{vap}H$ (T_b): 29.97 kJ/mol
$\Delta_{vap}H$ (25°C): 33.01 kJ/mol

Vapor pressure (0°C):
Vapor pressure (25°C): 13.0 kPa
Vapor pressure (100°C): 174 kPa

PROPERTIES AT 25°C AND 100 kPa

	Solid	Liquid	Gas		Liquid
$\Delta_f H°$/kJ mol^{-1}:		-156.4	-123.4		d: 0.7739 g/mL
$S°$/J mol^{-1}K^{-1}:					η: 0.894 mPa s
C_p/J mol^{-1}K^{-1}:		154.9			k: 0.123 W/m K

COMMENTS: TLV=300 ppm; highly toxic; flammable

Name: Cyclohexanol
Synonym: Cyclohexyl alcohol

Mol. Form.: $C_6H_{12}O$

CAS RN: 108-93-0
Merck No.: 2731
Mol. Wt.: 100.161

PHYSICAL CONSTANTS

T_m: 25.46°C (298.61 K)
T_b: 160.84°C (433.99 K)

T_c: 376.9°C (650.0 K)
P_c: 4.26 MPa

μ:
IP: 9.75 eV

TRANSITION PROPERTIES

$\Delta_{fus}H$ (T_m): 1.76 kJ/mol
$\Delta_{vap}H$ (T_b):
$\Delta_{vap}H$ (25°C): 62.01 kJ/mol

Vapor pressure (0°C):
Vapor pressure (25°C): 0.100 kPa
Vapor pressure (100°C): 10.4 kPa

PROPERTIES AT 25°C AND 100 kPa

	Solid	Liquid	Gas		Solid
$\Delta_f H°$/kJ mol^{-1}:		-348.2	-286.2		d: 0.9604 g/mL
$S°$/J mol^{-1}K^{-1}:					η: N/A
C_p/J mol^{-1}K^{-1}:		208.2			k:

COMMENTS: TLV=50 ppm

Name: Cyclohexanone
Synonym: Pimelic ketone

Mol. Form.: $C_6H_{10}O$

CAS RN: 108-94-1
Merck No.: 2732
Mol. Wt.: 98.145

PHYSICAL CONSTANTS

T_m: -31°C (242 K)
T_b: 155.43°C (428.58 K)

T_c: 379.9°C (653.0 K)
P_c: 4.0 MPa

μ: 2.87 D
IP: 9.14 eV

TRANSITION PROPERTIES

$\Delta_{fus}H$ (T_m):
$\Delta_{vap}H$ (T_b): 40.25 kJ/mol
$\Delta_{vap}H$ (25°C): 45.06 kJ/mol

Vapor pressure (0°C):
Vapor pressure (25°C): 0.530 kPa
Vapor pressure (100°C): 18.6 kPa

PROPERTIES AT 25°C AND 100 kPa

	Solid	Liquid	Gas		Liquid
$\Delta_f H°$/kJ mol^{-1}:		-271.2	-226.1		d: 0.9425 g/mL
$S°$/J mol^{-1}K^{-1}:					η: 2.02 mPa s
C_p/J mol^{-1}K^{-1}:		182.2			k:

COMMENTS: TLV=25 ppm

Name: Cyclohexene   CAS RN: 110-83-8
Synonym: Tetrahydrobenzene   Merck No.: 2733
                                                                                                          Mol. Wt.: 82.145

Mol. Form.: C_6H_{10}

PHYSICAL CONSTANTS

T_m: -103.5°C (169.6 K) T_c: 287.33°C (560.48 K)   μ: 0.332 D
T_b: 82.98°C (356.13 K) P_c: IP: 8.95 eV

TRANSITION PROPERTIES

$\Delta_{fus}H$ (T_m): 3.29 kJ/mol   Vapor pressure (0°C):
$\Delta_{vap}H$ (T_b): 30.46 kJ/mol   Vapor pressure (25°C): 11.8 kPa
$\Delta_{vap}H$ (25°C): 33.47 kJ/mol   Vapor pressure (100°C): 165 kPa

PROPERTIES AT 25°C AND 100 kPa

	Solid	Liquid	Gas		Liquid
$\Delta_f H°$/kJ mol^{-1}:		-38.5	-5.0		d: 0.806 g/mL
$S°$/J mol^{-1}K^{-1}:		214.6			η: 0.625 mPa s
C_p/J mol^{-1}K^{-1}:		148.3			k: 0.130 W/m K

COMMENTS: TLV=300 ppm; flammable

Name: Cyclohexylamine   CAS RN: 108-91-8
Synonym: Aminocyclohexane   Merck No.: 2735
                                                                                                          Mol. Wt.: 99.176

Mol. Form.: $C_6H_{13}N$

PHYSICAL CONSTANTS

T_m: -17.7°C (255.4 K) T_c: μ:
T_b: 134°C (407 K) P_c: IP: 8.62 eV

TRANSITION PROPERTIES

$\Delta_{fus}H$ (T_m):   Vapor pressure (0°C):
$\Delta_{vap}H$ (T_b): 36.14 kJ/mol   Vapor pressure (25°C): 1.20 kPa
$\Delta_{vap}H$ (25°C): 43.67 kJ/mol   Vapor pressure (100°C): 36.1 kPa

PROPERTIES AT 25°C AND 100 kPa

	Solid	Liquid	Gas		Liquid
$\Delta_f H°$/kJ mol^{-1}:		-147.7	-104.9		d: 0.8627 g/mL
$S°$/J mol^{-1}K^{-1}:					η: 1.94 mPa s
C_p/J mol^{-1}K^{-1}:					k:

COMMENTS: TLV=10 ppm; highly toxic; flammable

Name: Cyclohexylbenzene   CAS RN: 827-52-1
Synonym: Phenylcyclohexane   Merck No.:
                                                                                                          Mol. Wt.: 160.259

Mol. Form.: $C_{12}H_{16}$

PHYSICAL CONSTANTS

T_m: 7.3°C (280.4 K) T_c: μ:
T_b: 240.1°C (513.2 K) P_c: IP:

TRANSITION PROPERTIES

$\Delta_{fus}H$ (T_m): 15.30 kJ/mol   Vapor pressure (0°C):
$\Delta_{vap}H$ (T_b):   Vapor pressure (25°C):
$\Delta_{vap}H$ (25°C): 59.94 kJ/mol   Vapor pressure (100°C):

PROPERTIES AT 25°C AND 100 kPa

	Solid	Liquid	Gas		Liquid
$\Delta_f H°$/kJ mol^{-1}:		-76.6	-16.7		d: 0.9387 g/mL
$S°$/J mol^{-1}K^{-1}:					η:
C_p/J mol^{-1}K^{-1}:		261.3			k:

COMMENTS:

Name: 1,3-Cyclopentadiene

Synonym: Pyropentylene

Mol. Form.: C_5H_6

CAS RN: 542-92-7

Merck No.: 2744

Mol. Wt.: 66.103

PHYSICAL CONSTANTS

T_m: -85°C (188 K)	T_c:	μ: 0.419 D
T_b: 41°C (314 K)	P_c:	IP: 8.56 eV

TRANSITION PROPERTIES

$\Delta_{fus}H$ (T_m):

$\Delta_{vap}H$ (T_b):

$\Delta_{vap}H$ (25°C): 28.4 kJ/mol

Vapor pressure (0°C):

Vapor pressure (25°C): 58.5 kPa

Vapor pressure (100°C):

PROPERTIES AT 25°C AND 100 kPa

	Solid	Liquid	Gas	Liquid
$\Delta_f H°$/kJ mol^{-1}:		105.9	134.3	d: 0.7966 g/mL
$S°$/J mol^{-1}K^{-1}:				η:
C_p/J mol^{-1}K^{-1}:				k:

COMMENTS: TLV=75 ppm

Name: Cyclopentane

Synonym: Pentamethylene

Mol. Form.: C_5H_{10}

CAS RN: 287-92-3

Merck No.: 2746

Mol. Wt.: 70.134

PHYSICAL CONSTANTS

T_m: -93.8°C (179.3 K)	T_c: 238.6°C (511.7 K)	μ:
T_b: 49.3°C (322.4 K)	P_c: 4.508 MPa	IP: 10.51 eV

TRANSITION PROPERTIES

$\Delta_{fus}H$ (T_m): 0.61 kJ/mol

$\Delta_{vap}H$ (T_b): 27.30 kJ/mol

$\Delta_{vap}H$ (25°C): 28.52 kJ/mol

Vapor pressure (0°C): 14.2 kPa

Vapor pressure (25°C): 42.3 kPa

Vapor pressure (100°C): 418 kPa

PROPERTIES AT 25°C AND 100 kPa

	Solid	Liquid	Gas	Liquid
$\Delta_f H°$/kJ mol^{-1}:		-105.1	-76.4	d: 0.7405 g/mL
$S°$/J mol^{-1}K^{-1}:		204.5		η: 0.413 mPa s
C_p/J mol^{-1}K^{-1}:		128.8		k: 0.126 W/m K

COMMENTS: TLV=600 ppm; flammable

Name: Cyclopentanol

Synonym: Cyclopentyl alcohol

Mol. Form.: $C_5H_{10}O$

CAS RN: 96-41-3

Merck No.: 2747

Mol. Wt.: 86.134

PHYSICAL CONSTANTS

T_m: -19°C (254 K)	T_c: 346.4°C (619.5 K)	μ:
T_b: 140.42°C (413.57 K)	P_c: 4.90 MPa	IP: 9.72 eV

TRANSITION PROPERTIES

$\Delta_{fus}H$ (T_m):

$\Delta_{vap}H$ (T_b):

$\Delta_{vap}H$ (25°C): 57.60 kJ/mol

Vapor pressure (0°C): 0.035 kPa

Vapor pressure (25°C): 0.294 kPa

Vapor pressure (100°C):

PROPERTIES AT 25°C AND 100 kPa

	Solid	Liquid	Gas	Liquid
$\Delta_f H°$/kJ mol^{-1}:		-300.1	-242.6	d: 0.943 g/mL
$S°$/J mol^{-1}K^{-1}:		206.3		η:
C_p/J mol^{-1}K^{-1}:		184.1		k:

COMMENTS:

Name: Cyclopentanone   CAS RN: 120-92-3
Synonym: Adipic ketone   Merck No.: 2748
                                                                  Mol. Wt.: 84.118
Mol. Form.: C_5H_8O

PHYSICAL CONSTANTS

T_m: -51.3°C (221.8 K) T_c: 351.4°C (624.5 K)   μ: 3.3 D
T_b: 130.57°C (403.72 K) P_c: 4.60 MPa   IP: 9.25 eV

TRANSITION PROPERTIES

$\Delta_{fus}H$ (T_m):   Vapor pressure (0°C):
$\Delta_{vap}H$ (T_b): 36.35 kJ/mol   Vapor pressure (25°C): 1.54 kPa
$\Delta_{vap}H$ (25°C): 42.72 kJ/mol   Vapor pressure (100°C): 39.6 kPa

PROPERTIES AT 25°C AND 100 kPa

	Solid	Liquid	Gas		Liquid
$\Delta_f H°$/kJ mol^{-1}:		-235.7	-192.1	*d*:	0.944 g/mL
$S°$/J mol^{-1}K^{-1}:				η:	1.4 mPa s
C_p/J mol^{-1}K^{-1}:		154.5		*k*:	
COMMENTS: Flammable					

Name: Cyclopentene   CAS RN: 142-29-0
                                                                  Merck No.:
                                                                  Mol. Wt.: 68.119
Mol. Form.: C_5H_8

PHYSICAL CONSTANTS

T_m: -135.1°C (138.0 K) T_c: 233.9°C (507.0 K)   μ: 0.20 D
T_b: 44.24°C (317.39 K) P_c: 4.802 MPa   IP: 9.01 eV

TRANSITION PROPERTIES

$\Delta_{fus}H$ (T_m): 3.36 kJ/mol   Vapor pressure (0°C): 17.4 kPa
$\Delta_{vap}H$ (T_b):   Vapor pressure (25°C): 50.7 kPa
$\Delta_{vap}H$ (25°C): 29.5 kJ/mol   Vapor pressure (100°C):

PROPERTIES AT 25°C AND 100 kPa

	Solid	Liquid	Gas		Liquid
$\Delta_f H°$/kJ mol^{-1}:		4.4	33.9	*d*:	0.768 g/mL
$S°$/J mol^{-1}K^{-1}:		201.2		η:	
C_p/J mol^{-1}K^{-1}:		122.4		*k*:	0.129 W/m K
COMMENTS: Flammable					

Name: Cyclopropane   CAS RN: 75-19-4
Synonym: Trimethylene   Merck No.: 2755
                                                                  Mol. Wt.: 42.081
Mol. Form.: C_3H_6

PHYSICAL CONSTANTS

T_m: -127.4°C (145.7 K) T_c: 125.2°C (398.3 K)   μ: 0 D
T_b: -32.81°C (240.34 K) P_c: 5.579 MPa   IP: 9.86 eV

TRANSITION PROPERTIES

$\Delta_{fus}H$ (T_m): 5.44 kJ/mol   Vapor pressure (0°C):
$\Delta_{vap}H$ (T_b): 20.05 kJ/mol   Vapor pressure (25°C):
$\Delta_{vap}H$ (25°C): 16.93 kJ/mol   Vapor pressure (100°C):

PROPERTIES AT 25°C AND 100 kPa

	Solid	Liquid	Gas		Gas
$\Delta_f H°$/kJ mol^{-1}:			53.3	*d*:	1.720 g/L
$S°$/J mol^{-1}K^{-1}:			237.5	η:	
C_p/J mol^{-1}K^{-1}:			55.6	*k*:	
COMMENTS: Very flammable					

Name: *o*-Cymene
Synonyms: 1-Methyl-2-isopropylbenzene
          2-Isopropyltoluene
Mol. Form.: $C_{10}H_{14}$

CAS RN: 527-84-4
Merck No.: 2770
Mol. Wt.: 134.221

PHYSICAL CONSTANTS

T_m: -71.5°C (201.6 K) T_c: μ:
T_b: 178.1°C (451.2 K) P_c: IP:

TRANSITION PROPERTIES

$\Delta_{fus}H$ (T_m):   Vapor pressure (0°C):
$\Delta_{vap}H$ (T_b):   Vapor pressure (25°C):
$\Delta_{vap}H$ (25°C):   Vapor pressure (100°C): 8.90 kPa

PROPERTIES AT 25°C AND 100 kPa

	Solid	Liquid	Gas	Liquid
$\Delta_f H°$/kJ mol^{-1}:		-73.3		d: 0.8726 g/mL
$S°$/J mol^{-1}K^{-1}:				η:
C_p/J mol^{-1}K^{-1}:				k:
COMMENTS:				

Name: *m*-Cymene
Synonyms: 1-Methyl-3-isopropylbenzene
          3-Isopropyltoluene
Mol. Form.: $C_{10}H_{14}$

CAS RN: 535-77-3
Merck No.: 2770
Mol. Wt.: 134.221

PHYSICAL CONSTANTS

T_m: -63.7°C (209.4 K) T_c: μ:
T_b: 175.1°C (448.2 K) P_c: IP:

TRANSITION PROPERTIES

$\Delta_{fus}H$ (T_m):   Vapor pressure (0°C):
$\Delta_{vap}H$ (T_b):   Vapor pressure (25°C):
$\Delta_{vap}H$ (25°C):   Vapor pressure (100°C): 10.0 kPa

PROPERTIES AT 25°C AND 100 kPa

	Solid	Liquid	Gas	Liquid
$\Delta_f H°$/kJ mol^{-1}:		-78.6		d: 0.8570 g/mL
$S°$/J mol^{-1}K^{-1}:				η:
C_p/J mol^{-1}K^{-1}:				k:
COMMENTS:				

Name: *p*-Cymene
Synonyms: 1-Methyl-4-isopropylbenzene
          4-Isopropyltoluene
Mol. Form.: $C_{10}H_{14}$

CAS RN: 99-87-6
Merck No.: 2770
Mol. Wt.: 134.221

PHYSICAL CONSTANTS

T_m: -68.9°C (204.2 K) T_c: 378°C (651 K) μ:
T_b: 177.13°C (450.28 K) P_c: 2.73 MPa   IP: 8.29 eV

TRANSITION PROPERTIES

$\Delta_{fus}H$ (T_m): 9.60 kJ/mol   Vapor pressure (0°C):
$\Delta_{vap}H$ (T_b): 38.16 kJ/mol   Vapor pressure (25°C): 0.190 kPa
$\Delta_{vap}H$ (25°C): 50.29 kJ/mol   Vapor pressure (100°C): 9.15 kPa

PROPERTIES AT 25°C AND 100 kPa

	Solid	Liquid	Gas	Liquid
$\Delta_f H°$/kJ mol^{-1}:		-78.0		d: 0.8525 g/mL
$S°$/J mol^{-1}K^{-1}:				η: 2.2 mPa s
C_p/J mol^{-1}K^{-1}:		236.4		k: 0.122 W/m K
COMMENTS:				

Name: *cis*-Decahydronaphthalene
Synonyms: *cis*-Decalin
      cis-Bicyclo[4.4.0]decane
Mol. Form.: $C_{10}H_{18}$

CAS RN: 493-01-6
Merck No.: 2839
Mol. Wt.: 138.253

PHYSICAL CONSTANTS

T_m: -42.95°C (230.20 K)	T_c: 429.2°C (702.3 K)	μ:
T_b: 195.81°C (468.96 K)	P_c: 3.20 MPa	IP: 9.26 eV

TRANSITION PROPERTIES

$\Delta_{fus}H$ (T_m): 9.49 kJ/mol
$\Delta_{vap}H$ (T_b): 41.00 kJ/mol
$\Delta_{vap}H$ (25°C): 51.34 kJ/mol

Vapor pressure (0°C):
Vapor pressure (25°C): 0.100 kPa
Vapor pressure (100°C): 5.56 kPa

PROPERTIES AT 25°C AND 100 kPa

	Solid	Liquid	Gas		Liquid
$\Delta_f H°$/kJ mol^{-1}:		-219.4	-169.2	d:	0.893 g/mL
$S°$/J mol^{-1}K^{-1}:		265.0		η:	3.04 mPa s
C_p/J mol^{-1}K^{-1}:		232.0		k:	
COMMENTS:					

Name: *trans*-Decahydronaphthalene
Synonyms: *trans*-Decalin
      trans-Bicyclo[4.4.0]decane
Mol. Form.: $C_{10}H_{18}$

CAS RN: 493-02-7
Merck No.: 2839
Mol. Wt.: 138.253

PHYSICAL CONSTANTS

T_m: -30.36°C (242.79 K)	T_c: 414.0°C (687.1 K)	μ:
T_b: 187.31°C (460.46 K)	P_c:	IP: 9.24 eV

TRANSITION PROPERTIES

$\Delta_{fus}H$ (T_m): 14.41 kJ/mol
$\Delta_{vap}H$ (T_b): 40.23 kJ/mol
$\Delta_{vap}H$ (25°C): 49.87 kJ/mol

Vapor pressure (0°C):
Vapor pressure (25°C): 0.164 kPa
Vapor pressure (100°C): 7.46 kPa

PROPERTIES AT 25°C AND 100 kPa

	Solid	Liquid	Gas		Liquid
$\Delta_f H°$/kJ mol^{-1}:		-230.6	-182.1	d:	0.866 g/mL
$S°$/J mol^{-1}K^{-1}:		264.9		η:	1.95 mPa s
C_p/J mol^{-1}K^{-1}:		228.5		k:	
COMMENTS:					

Name: Decanal
Synonyms: Capraldehyde
      Decanaldehyde
Mol. Form.: $C_{10}H_{20}O$

CAS RN: 112-31-2
Merck No.:
Mol. Wt.: 156.268

PHYSICAL CONSTANTS

T_m: -5°C (268 K)	T_c: 401.1°C (674.2 K)	μ:
T_b: 208.5°C (481.6 K)	P_c:	IP:

TRANSITION PROPERTIES

$\Delta_{fus}H$ (T_m):
$\Delta_{vap}H$ (T_b):
$\Delta_{vap}H$ (25°C):

Vapor pressure (0°C):
Vapor pressure (25°C):
Vapor pressure (100°C): 1.97 kPa

PROPERTIES AT 25°C AND 100 kPa

	Solid	Liquid	Gas		Liquid
$\Delta_f H°$/kJ mol^{-1}:				d:	
$S°$/J mol^{-1}K^{-1}:				η:	
C_p/J mol^{-1}K^{-1}:				k:	
COMMENTS:					

Name: Decane   CAS RN: 124-18-5
                                                                    Merck No.:
                                                                    Mol. Wt.: 142.285

Mol. Form.: $C_{10}H_{22}$

PHYSICAL CONSTANTS

T_m: -29.7°C (243.4 K) T_c: 344.50°C (617.65 K)   μ:
T_b: 174.15°C (447.30 K) P_c: 2.104 MPa   IP: 9.65 eV

TRANSITION PROPERTIES

$\Delta_{fus}H$ (T_m): 28.78 kJ/mol   Vapor pressure (0°C):
$\Delta_{vap}H$ (T_b): 38.75 kJ/mol   Vapor pressure (25°C): 0.170 kPa
$\Delta_{vap}H$ (25°C): 51.38 kJ/mol   Vapor pressure (100°C): 9.56 kPa

PROPERTIES AT 25°C AND 100 kPa

	Solid	Liquid	Gas		Liquid
$\Delta_f H°$/kJ mol^{-1}:		-300.9	-249.5	d:	0.7264 g/mL
$S°$/J mol^{-1}K^{-1}:				η:	0.838 mPa s
C_p/J mol^{-1}K^{-1}:		314.4		k:	0.132 W/m K
COMMENTS:					

Name: Decanoic acid   CAS RN: 334-48-5
Synonyms: Capric acid   Merck No.: 1759
          Decylic acid   Mol. Wt.: 172.268
Mol. Form.: $C_{10}H_{20}O_2$

PHYSICAL CONSTANTS

T_m: 31.99°C (305.14 K) T_c: 453°C (726 K)   μ:
T_b: 268.75°C (541.90 K) P_c: 2.23 MPa   IP:

TRANSITION PROPERTIES

$\Delta_{fus}H$ (T_m): 28.02 kJ/mol   Vapor pressure (0°C):
$\Delta_{vap}H$ (T_b):   Vapor pressure (25°C):
$\Delta_{vap}H$ (25°C): 89.4 kJ/mol   Vapor pressure (100°C):

PROPERTIES AT 25°C AND 100 kPa

	Solid	Liquid	Gas		Solid
$\Delta_f H°$/kJ mol^{-1}:	-713.7	-684.3	-594.9	d:	
$S°$/J mol^{-1}K^{-1}:				η:	N/A
C_p/J mol^{-1}K^{-1}:				k:	
COMMENTS:					

Name: 1-Decanol   CAS RN: 112-30-1
Synonyms: Capric alcohol   Merck No.: 2847
          Decyl alcohol   Mol. Wt.: 158.284
Mol. Form.: $C_{10}H_{22}O$

PHYSICAL CONSTANTS

T_m: 6.9°C (280.0 K) T_c: 416°C (689 K)   μ:
T_b: 231.1°C (504.2 K) P_c: 2.41 MPa   IP:

TRANSITION PROPERTIES

$\Delta_{fus}H$ (T_m):   Vapor pressure (0°C):
$\Delta_{vap}H$ (T_b):   Vapor pressure (25°C): 0.009 kPa
$\Delta_{vap}H$ (25°C): 81.50 kJ/mol   Vapor pressure (100°C): 0.584 kPa

PROPERTIES AT 25°C AND 100 kPa

	Solid	Liquid	Gas		Liquid
$\Delta_f H°$/kJ mol^{-1}:		-478.1	-396.4	d:	0.8263 g/mL
$S°$/J mol^{-1}K^{-1}:				η:	10.9 mPa s
C_p/J mol^{-1}K^{-1}:		370.6		k:	0.162 W/m K
COMMENTS:					

Name: 1-Decene

Mol. Form.: $C_{10}H_{20}$

CAS RN: 872-05-9
Merck No.:
Mol. Wt.: 140.269

PHYSICAL CONSTANTS

T_m: -66.3°C (206.8 K)	T_c: 343.3°C (616.4 K)	μ:
T_b: 170.5°C (443.6 K)	P_c: 2.218 MPa	IP: 9.42 eV

TRANSITION PROPERTIES

$\Delta_{fus}H$ (T_m): 21.10 kJ/mol   Vapor pressure (0°C):
$\Delta_{vap}H$ (T_b): 38.66 kJ/mol   Vapor pressure (25°C): 0.210 kPa
$\Delta_{vap}H$ (25°C): 50.43 kJ/mol   Vapor pressure (100°C): 10.9 kPa

PROPERTIES AT 25°C AND 100 kPa

	Solid	Liquid	Gas	Liquid
$\Delta_fH°$/kJ mol^{-1}:		-173.8	-123.4	d: 0.737 g/mL
$S°$/J mol^{-1}K^{-1}:		425.0		η: 0.756 mPa s
C_p/J mol^{-1}K^{-1}:		300.8		k:
COMMENTS:				

Name: Diacetone alcohol
Synonym: 4-Hydroxy-4-methyl-2-pentanone

Mol. Form.: $C_6H_{12}O_2$

CAS RN: 123-42-2
Merck No.: 2944
Mol. Wt.: 116.160

PHYSICAL CONSTANTS

T_m: -44°C (229 K)	T_c:	μ:
T_b: 167.9°C (441.0 K)	P_c:	IP:

TRANSITION PROPERTIES

$\Delta_{fus}H$ (T_m):   Vapor pressure (0°C):
$\Delta_{vap}H$ (T_b):   Vapor pressure (25°C): 0.220 kPa
$\Delta_{vap}H$ (25°C):   Vapor pressure (100°C): 10.5 kPa

PROPERTIES AT 25°C AND 100 kPa

	Solid	Liquid	Gas	Liquid
$\Delta_fH°$/kJ mol^{-1}:				d: 0.9342 g/mL
$S°$/J mol^{-1}K^{-1}:				η: 2.80 mPa s
C_p/J mol^{-1}K^{-1}:		221.3		k:
COMMENTS: TLV=50 ppm				

Name: Diallylsulfide
Synonyms: Bis(2-propenyl) sulfide
        4-Thia-1,6-heptadiene
Mol. Form.: $C_6H_{10}S$

CAS RN: 592-88-1
Merck No.: 295
Mol. Wt.: 114.211

PHYSICAL CONSTANTS

T_m: -85°C (188 K)	T_c: 380°C (653 K)	μ:
T_b: 138.6°C (411.7 K)	P_c:	IP:

TRANSITION PROPERTIES

$\Delta_{fus}H$ (T_m):   Vapor pressure (0°C):
$\Delta_{vap}H$ (T_b):   Vapor pressure (25°C):
$\Delta_{vap}H$ (25°C):   Vapor pressure (100°C):

PROPERTIES AT 25°C AND 100 kPa

	Solid	Liquid	Gas	Liquid
$\Delta_fH°$/kJ mol^{-1}:				d: 0.89 g/mL
$S°$/J mol^{-1}K^{-1}:				η:
C_p/J mol^{-1}K^{-1}:				k:
COMMENTS:				

Name: Diazomethane   CAS RN: 334-88-3
Synonym: Diazirine   Merck No.: 2983
         Mol. Wt.: 42.040

Mol. Form.: CH_2N_2

PHYSICAL CONSTANTS

T_m: -145°C (128 K)	T_c:	μ: 1.50 D
T_b: -23°C (250 K)	P_c:	IP: 9.00 eV

TRANSITION PROPERTIES

$\Delta_{fus}H$ (T_m):	Vapor pressure (0°C):
$\Delta_{vap}H$ (T_b):	Vapor pressure (25°C):
$\Delta_{vap}H$ (25°C):	Vapor pressure (100°C):

PROPERTIES AT 25°C AND 100 kPa

	Solid	Liquid	Gas		Gas
$\Delta_fH°$/kJ mol^{-1}:				d:	1.718 g/L
$S°$/J mol^{-1}K^{-1}:			242.9	η:	
C_p/J mol^{-1}K^{-1}:			52.5	k:	

COMMENTS: TLV=0.2 ppm; carcinogen; highly toxic

Name: Diborane   CAS RN: 19287-45-7
Synonym: Diborane(6)   Merck No.: 2997
         Mol. Wt.: 27.670

Mol. Form.: B_2H_6

PHYSICAL CONSTANTS

T_m: -165.5°C (107.6 K)	T_c: 16.7°C (289.8 K)	μ:
T_b: -87.55°C (185.60 K)	P_c: 4.05 MPa	IP: 11.38 eV

TRANSITION PROPERTIES

$\Delta_{fus}H$ (T_m):	Vapor pressure (0°C):
$\Delta_{vap}H$ (T_b): 14.28 kJ/mol	Vapor pressure (25°C): N/A
$\Delta_{vap}H$ (25°C):	Vapor pressure (100°C): N/A

PROPERTIES AT 25°C AND 100 kPa

	Solid	Liquid	Gas		Gas
$\Delta_fH°$/kJ mol^{-1}:			35.6	d:	1.131 g/L
$S°$/J mol^{-1}K^{-1}:			232.1	η:	
C_p/J mol^{-1}K^{-1}:			56.9	k:	

COMMENTS: TLV=0.1 ppm; highly toxic; very flammable

Name: Dibromodifluoromethane   CAS RN: 75-61-6
Synonyms: Refrigerant 12B2   Merck No.:
         Halon 1202   Mol. Wt.: 209.816
Mol. Form.: CBr_2F_2

PHYSICAL CONSTANTS

T_m: -110.1°C (163.0 K)	T_c: 198.2°C (471.3 K)	μ: 0.66 D
T_b: 25°C (298 K)	P_c:	IP: 11.07 eV

TRANSITION PROPERTIES

$\Delta_{fus}H$ (T_m):	Vapor pressure (0°C): 41.9 kPa
$\Delta_{vap}H$ (T_b):	Vapor pressure (25°C): 110 kPa
$\Delta_{vap}H$ (25°C):	Vapor pressure (100°C):

PROPERTIES AT 25°C AND 100 kPa

	Solid	Liquid	Gas		Gas
$\Delta_fH°$/kJ mol^{-1}:				d:	8.576 g/L
$S°$/J mol^{-1}K^{-1}:			325.3	η:	
C_p/J mol^{-1}K^{-1}:			77.0	k:	

COMMENTS: TLV=100 ppm

Name: 1,2-Dibromoethane
Synonym: Ethylene dibromide

CAS RN: 106-93-4
Merck No.: 3753
Mol. Wt.: 187.862

Mol. Form.: $C_2H_4Br_2$

PHYSICAL CONSTANTS

T_m: 9.93°C (283.08 K)
T_b: 131.6°C (404.7 K)

T_c: 309.9°C (583.0 K)
P_c: 7.2 MPa

μ:
IP: 10.37 eV

TRANSITION PROPERTIES

$\Delta_{fus}H$ (T_m): 10.84 kJ/mol
$\Delta_{vap}H$ (T_b): 34.77 kJ/mol
$\Delta_{vap}H$ (25°C): 41.73 kJ/mol

Vapor pressure (0°C):
Vapor pressure (25°C): 1.55 kPa
Vapor pressure (100°C): 40.7 kPa

PROPERTIES AT 25°C AND 100 kPa

	Solid	Liquid	Gas		Liquid
$\Delta_f H°$/kJ mol^{-1}:		-79.2	-37.5		d: 2.1687 g/mL
$S°$/J mol^{-1}K^{-1}:		223.3			η: 1.60 mPa s
C_p/J mol^{-1}K^{-1}:		136.0			k:

COMMENTS: Carcinogen; highly toxic

Name: Dibromomethane
Synonym: Methylene bromide

CAS RN: 74-95-3
Merck No.: 5980
Mol. Wt.: 173.835

Mol. Form.: CH_2Br_2

PHYSICAL CONSTANTS

T_m: -52.55°C (220.60 K)
T_b: 97°C (370 K)

T_c:
P_c:

μ: 1.43 D
IP: 10.50 eV

TRANSITION PROPERTIES

$\Delta_{fus}H$ (T_m):
$\Delta_{vap}H$ (T_b): 32.92 kJ/mol
$\Delta_{vap}H$ (25°C): 36.97 kJ/mol

Vapor pressure (0°C): 1.53 kPa
Vapor pressure (25°C): 6.12 kPa
Vapor pressure (100°C): 111 kPa

PROPERTIES AT 25°C AND 100 kPa

	Solid	Liquid	Gas		Liquid
$\Delta_f H°$/kJ mol^{-1}:					d: 2.48 g/mL
$S°$/J mol^{-1}K^{-1}:			293.2		η: 0.980 mPa s
C_p/J mol^{-1}K^{-1}:			54.7		k: 0.108 W/m K

COMMENTS: Highly toxic

Name: 1,2-Dibromopropane
Synonym: Propylene dibromide

CAS RN: 78-75-1
Merck No.: 7866
Mol. Wt.: 201.889

Mol. Form.: $C_3H_6Br_2$

PHYSICAL CONSTANTS

T_m: -55.2°C (217.9 K)
T_b: 141.9°C (415.0 K)

T_c:
P_c:

μ:
IP: 10.10 eV

TRANSITION PROPERTIES

$\Delta_{fus}H$ (T_m): 8.94 kJ/mol
$\Delta_{vap}H$ (T_b): 35.61 kJ/mol
$\Delta_{vap}H$ (25°C): 41.67 kJ/mol

Vapor pressure (0°C):
Vapor pressure (25°C): 1.07 kPa
Vapor pressure (100°C): 27.1 kPa

PROPERTIES AT 25°C AND 100 kPa

	Solid	Liquid	Gas		Liquid
$\Delta_f H°$/kJ mol^{-1}:			-71.5		d: 1.9234 g/mL
$S°$/J mol^{-1}K^{-1}:					η: 1.55 mPa s
C_p/J mol^{-1}K^{-1}:		160.0			k:

COMMENTS:

Name: 1,2-Dibromotetrafluoroethane
Synonyms: Refrigerant 114B2
      Halon 2402
Mol. Form.: $C_2Br_2F_4$

CAS RN: 124-73-2
Merck No.:
Mol. Wt.: 259.824

PHYSICAL CONSTANTS

T_m: -110.4°C (162.7 K)
T_b: 47.35°C (320.50 K)

T_c: 214.7°C (487.8 K)
P_c: 3.393 MPa

μ:
IP: 11.10 eV

TRANSITION PROPERTIES

$\Delta_{fus}H$ (T_m): 7.04 kJ/mol
$\Delta_{vap}H$ (T_b): 27.03 kJ/mol
$\Delta_{vap}H$ (25°C): 28.39 kJ/mol

Vapor pressure (0°C): 14.3 kPa
Vapor pressure (25°C): 43.4 kPa
Vapor pressure (100°C):

PROPERTIES AT 25°C AND 100 kPa

	Solid	Liquid	Gas	Liquid
$\Delta_f H°$/kJ mol^{-1}:			-789.1	d: 2.163 g/mL
$S°$/J mol^{-1}K^{-1}:				η: 0.72 mPa s
C_p/J mol^{-1}K^{-1}:		180.3		k:

COMMENTS: Highly toxic

Name: Dibutylamine
Synonym: *N*-Butyl-1-butanamine

Mol. Form.: $C_8H_{19}N$

CAS RN: 111-92-2
Merck No.: 3019
Mol. Wt.: 129.246

PHYSICAL CONSTANTS

T_m: -62°C (211 K)
T_b: 159.6°C (432.7 K)

T_c: 334.4°C (607.5 K)
P_c: 3.11 MPa

μ:
IP: 7.69 eV

TRANSITION PROPERTIES

$\Delta_{fus}H$ (T_m):
$\Delta_{vap}H$ (T_b): 38.44 kJ/mol
$\Delta_{vap}H$ (25°C): 49.45 kJ/mol

Vapor pressure (0°C):
Vapor pressure (25°C): 0.340 kPa
Vapor pressure (100°C): 14.4 kPa

PROPERTIES AT 25°C AND 100 kPa

	Solid	Liquid	Gas	Liquid
$\Delta_f H°$/kJ mol^{-1}:		-206.0	-156.6	d: 0.7571 g/mL
$S°$/J mol^{-1}K^{-1}:				η: 0.918 mPa s
C_p/J mol^{-1}K^{-1}:		292.9		k:

COMMENTS: Highly toxic

Name: Dibutyl ether
Synonyms: 1,1′-Oxybisbutane
      Butyl ether
Mol. Form.: $C_8H_{18}O$

CAS RN: 142-96-1
Merck No.: 1568
Mol. Wt.: 130.230

PHYSICAL CONSTANTS

T_m: -95.2°C (177.9 K)
T_b: 140.28°C (413.43 K)

T_c: 311.0°C (584.1 K)
P_c: 3.01 MPa

μ: 1.17 D
IP: 9.43 eV

TRANSITION PROPERTIES

$\Delta_{fus}H$ (T_m):
$\Delta_{vap}H$ (T_b): 36.49 kJ/mol
$\Delta_{vap}H$ (25°C): 44.97 kJ/mol

Vapor pressure (0°C):
Vapor pressure (25°C): 0.898 kPa
Vapor pressure (100°C): 29.1 kPa

PROPERTIES AT 25°C AND 100 kPa

	Solid	Liquid	Gas	Liquid
$\Delta_f H°$/kJ mol^{-1}:		-377.9	-333.4	d: 0.7641 g/mL
$S°$/J mol^{-1}K^{-1}:				η: 0.637 mPa s
C_p/J mol^{-1}K^{-1}:		278.2		k:

COMMENTS: Flammable

Name: Di-*tert*-butyl ether   CAS RN: 6163-66-2
Synonyms: 2,2′-Oxybis(2-methyl)propane   Merck No.: 3020
      tert-butyl ether   Mol. Wt.: 130.230
Mol. Form.: $C_8H_{18}O$

PHYSICAL CONSTANTS

T_m:	T_c: 277°C (550 K)	μ:
T_b: 107.23°C (380.38 K)	P_c:	IP: 8.81 eV

TRANSITION PROPERTIES

$\Delta_{fus}H$ (T_m):   Vapor pressure (0°C):
$\Delta_{vap}H$ (T_b): 32.15 kJ/mol   Vapor pressure (25°C): 4.34 kPa
$\Delta_{vap}H$ (25°C): 37.61 kJ/mol   Vapor pressure (100°C): 78.2 kPa

PROPERTIES AT 25°C AND 100 kPa

	Solid	Liquid	Gas		Liquid
$\Delta_f H°$/kJ mol^{-1}:		-399.6	-362.0		d: 0.762 g/mL
$S°$/J mol^{-1}K^{-1}:					η:
C_p/J mol^{-1}K^{-1}:		276.1			k:
COMMENTS: Flammable					

Name: Dibutyl phthalate   CAS RN: 84-74-2
Synonym: Dibutyl 1,2-benzenedicarboxylate   Merck No.: 1586
                                                                 Mol. Wt.: 278.348
Mol. Form.: $C_{16}H_{22}O_4$

PHYSICAL CONSTANTS

T_m: -35°C (238 K)	T_c:	μ:
T_b: 340°C (613 K)	P_c:	IP:

TRANSITION PROPERTIES

$\Delta_{fus}H$ (T_m):   Vapor pressure (0°C):
$\Delta_{vap}H$ (T_b): 79.20 kJ/mol   Vapor pressure (25°C):
$\Delta_{vap}H$ (25°C): 91.7 kJ/mol   Vapor pressure (100°C):

PROPERTIES AT 25°C AND 100 kPa

	Solid	Liquid	Gas		Liquid
$\Delta_f H°$/kJ mol^{-1}:		-842.6	-750.9		d: 1.0426 g/mL
$S°$/J mol^{-1}K^{-1}:					η: 16.6 mPa s
C_p/J mol^{-1}K^{-1}:		498.0			k: 0.136 W/m K
COMMENTS:					

Name: Dibutyl sulfide   CAS RN: 544-40-1
Synonym: 5-Thianonane   Merck No.: 1590
                                                                 Mol. Wt.: 146.297
Mol. Form.: $C_8H_{18}S$

PHYSICAL CONSTANTS

T_m: -75.0°C (198.1 K)	T_c:	μ:
T_b: 188.9°C (462.0 K)	P_c:	IP: 8.20 eV

TRANSITION PROPERTIES

$\Delta_{fus}H$ (T_m):   Vapor pressure (0°C):
$\Delta_{vap}H$ (T_b): 41.28 kJ/mol   Vapor pressure (25°C):
$\Delta_{vap}H$ (25°C): 52.96 kJ/mol   Vapor pressure (100°C):

PROPERTIES AT 25°C AND 100 kPa

	Solid	Liquid	Gas		Liquid
$\Delta_f H°$/kJ mol^{-1}:		-220.7	-167.4		d: 0.834 g/mL
$S°$/J mol^{-1}K^{-1}:		405.1			η: 0.98 mPa s
C_p/J mol^{-1}K^{-1}:		284.3			k:
COMMENTS:					

Name: Di-*tert*-butyl sulfide
Synonym: *tert*-Butyl sulfide

CAS RN: 107-47-1
Merck No.:
Mol. Wt.: 146.297

Mol. Form.: $C_8H_{18}S$

PHYSICAL CONSTANTS

T_m: -9.0°C (264.1 K)
T_b: 149.1°C (422.2 K)

T_c:
P_c:

μ:
IP: 8.00 eV

TRANSITION PROPERTIES

$\Delta_{fus}H$ (T_m):
$\Delta_{vap}H$ (T_b): 33.26 kJ/mol
$\Delta_{vap}H$ (25°C): 43.76 kJ/mol

Vapor pressure (0°C):
Vapor pressure (25°C):
Vapor pressure (100°C):

PROPERTIES AT 25°C AND 100 kPa

	Solid	Liquid	Gas		Liquid
$\Delta_fH°$/kJ mol^{-1}:		-232.6	-188.9		d: 0.815 g/mL
$S°$/J mol^{-1}K^{-1}:					η:
C_p/J mol^{-1}K^{-1}:					k:
COMMENTS:					

Name: *o*-Dichlorobenzene
Synonym: 1,2-Dichlorobenzene

CAS RN: 95-50-1
Merck No.: 3044
Mol. Wt.: 147.003

Mol. Form.: $C_6H_4Cl_2$

PHYSICAL CONSTANTS

T_m: -16.7°C (256.4 K)
T_b: 180°C (453 K)

T_c:
P_c:

μ: 2.50 D
IP: 9.08 eV

TRANSITION PROPERTIES

$\Delta_{fus}H$ (T_m): 12.93 kJ/mol
$\Delta_{vap}H$ (T_b): 39.66 kJ/mol
$\Delta_{vap}H$ (25°C): 50.21 kJ/mol

Vapor pressure (0°C):
Vapor pressure (25°C): 0.180 kPa
Vapor pressure (100°C): 8.39 kPa

PROPERTIES AT 25°C AND 100 kPa

	Solid	Liquid	Gas		Liquid
$\Delta_fH°$/kJ mol^{-1}:		-17.5	30.2		d: 1.3003 g/mL
$S°$/J mol^{-1}K^{-1}:					η: 1.32 mPa s
C_p/J mol^{-1}K^{-1}:		162.4			k:
COMMENTS: TLV=25 ppm; highly toxic					

Name: *m*-Dichlorobenzene
Synonym: 1,3-Dichlorobenzene

CAS RN: 541-73-1
Merck No.: 3043
Mol. Wt.: 147.003

Mol. Form.: $C_6H_4Cl_2$

PHYSICAL CONSTANTS

T_m: -24.8°C (248.3 K)
T_b: 173°C (446 K)

T_c:
P_c:

μ: 1.72 D
IP: 9.11 eV

TRANSITION PROPERTIES

$\Delta_{fus}H$ (T_m): 12.64 kJ/mol
$\Delta_{vap}H$ (T_b): 38.62 kJ/mol
$\Delta_{vap}H$ (25°C): 48.58 kJ/mol

Vapor pressure (0°C):
Vapor pressure (25°C): 0.252 kPa
Vapor pressure (100°C): 10.9 kPa

PROPERTIES AT 25°C AND 100 kPa

	Solid	Liquid	Gas		Liquid
$\Delta_fH°$/kJ mol^{-1}:		-20.7	25.7		d: 1.2828 g/mL
$S°$/J mol^{-1}K^{-1}:					η: 1.04 mPa s
C_p/J mol^{-1}K^{-1}:					k:
COMMENTS: Highly toxic					

Name: *p*-Dichlorobenzene
Synonym: 1,4-Dichlorobenzene

Mol. Form.: $C_6H_4Cl_2$

CAS RN: 106-46-7
Merck No.: 3045
Mol. Wt.: 147.003

PHYSICAL CONSTANTS

T_m: 52.7°C (325.8 K)	T_c:	μ: 0 D
T_b: 174°C (447 K)	P_c:	IP: 8.89 eV

TRANSITION PROPERTIES

$\Delta_{fus}H$ (T_m): 17.15 kJ/mol
$\Delta_{vap}H$ (T_b): 38.79 kJ/mol
$\Delta_{vap}H$ (25°C): 49.00 kJ/mol

Vapor pressure (0°C):
Vapor pressure (25°C): 0.235 kPa
Vapor pressure (100°C): 10.4 kPa

PROPERTIES AT 25°C AND 100 kPa

	Solid	Liquid	Gas		Solid
$\Delta_f H°$/kJ mol^{-1}:	-42.3		22.5	d:	1.211 g/mL
$S°$/J mol^{-1}K^{-1}:	175.4			η:	N/A
C_p/J mol^{-1}K^{-1}:	147.8			k:	

COMMENTS: TLV=10 ppm; carcinogen; highly toxic

Name: Dichlorodifluoromethane
Synonyms: Refrigerant 12
CFC-12
Mol. Form.: CCl_2F_2

CAS RN: 75-71-8
Merck No.: 3053
Mol. Wt.: 120.913

PHYSICAL CONSTANTS

T_m: -158°C (115 K)	T_c: 111.80°C (384.95 K)	μ: 0.51 D
T_b: -29.8°C (243.3 K)	P_c: 4.136 MPa	IP: 11.75 eV

TRANSITION PROPERTIES

$\Delta_{fus}H$ (T_m): 4.14 kJ/mol
$\Delta_{vap}H$ (T_b): 20.11 kJ/mol
$\Delta_{vap}H$ (25°C):

Vapor pressure (0°C): 308 kPa
Vapor pressure (25°C): 651 kPa
Vapor pressure (100°C): 3332 kPa

PROPERTIES AT 25°C AND 100 kPa

	Solid	Liquid	Gas		Gas
$\Delta_f H°$/kJ mol^{-1}:			-477.4	d:	4.942 g/L
$S°$/J mol^{-1}K^{-1}:			300.8	η:	
C_p/J mol^{-1}K^{-1}:		117.2	72.3	k:	0.0099 W/m K

COMMENTS: TLV=1000 ppm; highly toxic

Name: 1,1-Dichloroethane
Synonym: Ethylidene dichloride

Mol. Form.: $C_2H_4Cl_2$

CAS RN: 75-34-3
Merck No.: 3766
Mol. Wt.: 98.959

PHYSICAL CONSTANTS

T_m: -96.96°C (176.19 K)	T_c: 250°C (523 K)	μ: 2.06 D
T_b: 57.4°C (330.5 K)	P_c: 5.07 MPa	IP: 11.06 eV

TRANSITION PROPERTIES

$\Delta_{fus}H$ (T_m): 8.84 kJ/mol
$\Delta_{vap}H$ (T_b): 28.85 kJ/mol
$\Delta_{vap}H$ (25°C): 30.62 kJ/mol

Vapor pressure (0°C): 9.55 kPa
Vapor pressure (25°C): 30.5 kPa
Vapor pressure (100°C):

PROPERTIES AT 25°C AND 100 kPa

	Solid	Liquid	Gas		Liquid
$\Delta_f H°$/kJ mol^{-1}:		-158.4	-127.7	d:	1.1680 g/mL
$S°$/J mol^{-1}K^{-1}:		211.8	305.1	η:	0.464 mPa s
C_p/J mol^{-1}K^{-1}:		126.3	76.2	k:	

COMMENTS: TLV=100 ppm; highly toxic

Name: 1,2-Dichloroethane
Synonym: Ethylene dichloride

Mol. Form.: $C_2H_4Cl_2$

CAS RN: 107-06-2
Merck No.: 3754
Mol. Wt.: 98.959

PHYSICAL CONSTANTS

T_m: -35.5°C (237.6 K)
T_b: 83.5°C (356.6 K)

T_c: 288°C (561 K)
P_c: 5.4 MPa

μ:
IP: 11.04 eV

TRANSITION PROPERTIES

$\Delta_{fus}H$ (T_m): 8.83 kJ/mol
$\Delta_{vap}H$ (T_b): 31.98 kJ/mol
$\Delta_{vap}H$ (25°C): 35.16 kJ/mol

Vapor pressure (0°C): 2.84 kPa
Vapor pressure (25°C): 10.6 kPa
Vapor pressure (100°C): 162 kPa

PROPERTIES AT 25°C AND 100 kPa

	Solid	Liquid	Gas		Liquid
$\Delta_f H°$/kJ mol^{-1}:		-167.4	-126.9		d: 1.2457 g/mL
$S°$/J mol^{-1}K^{-1}:			308.4		η: 0.779 mPa s
C_p/J mol^{-1}K^{-1}:		128.4	78.7		k:

COMMENTS: TLV=10 ppm; carcinogen; highly toxic; flammable

Name: 1,1-Dichloroethylene
Synonym: Vinylidene chloride

Mol. Form.: $C_2H_2Cl_2$

CAS RN: 75-35-4
Merck No.: 9900
Mol. Wt.: 96.943

PHYSICAL CONSTANTS

T_m: -122.5°C (150.6 K)
T_b: 31.6°C (304.7 K)

T_c:
P_c:

μ: 1.34 D
IP: 9.79 eV

TRANSITION PROPERTIES

$\Delta_{fus}H$ (T_m): 6.51 kJ/mol
$\Delta_{vap}H$ (T_b): 26.14 kJ/mol
$\Delta_{vap}H$ (25°C): 26.48 kJ/mol

Vapor pressure (0°C): 28.9 kPa
Vapor pressure (25°C): 80.0 kPa
Vapor pressure (100°C):

PROPERTIES AT 25°C AND 100 kPa

	Solid	Liquid	Gas		Liquid
$\Delta_f H°$/kJ mol^{-1}:		-23.9	2.6		d: 1.18 g/mL
$S°$/J mol^{-1}K^{-1}:		201.5	289.0		η: 0.34 mPa s
C_p/J mol^{-1}K^{-1}:		111.3	67.1		k:

COMMENTS: TLV=5 ppm; carcinogen; highly toxic; very flammable

Name: *cis*-1,2-Dichloroethylene
Synonym: *cis*-1,2-Dichloroethene

Mol. Form.: $C_2H_2Cl_2$

CAS RN: 156-59-2
Merck No.: 86
Mol. Wt.: 96.943

PHYSICAL CONSTANTS

T_m: -80°C (193 K)
T_b: 60.19°C (333.34 K)

T_c: 271.1°C (544.2 K)
P_c:

μ: 1.90 D
IP: 9.66 eV

TRANSITION PROPERTIES

$\Delta_{fus}H$ (T_m): 7.20 kJ/mol
$\Delta_{vap}H$ (T_b): 30.23 kJ/mol
$\Delta_{vap}H$ (25°C): 31.57 kJ/mol

Vapor pressure (0°C):
Vapor pressure (25°C): 26.8 kPa
Vapor pressure (100°C):

PROPERTIES AT 25°C AND 100 kPa

	Solid	Liquid	Gas		Liquid
$\Delta_f H°$/kJ mol^{-1}:		-26.4	4.6		d: 1.2649 g/mL
$S°$/J mol^{-1}K^{-1}:		198.4	289.6		η: 0.445 mPa s
C_p/J mol^{-1}K^{-1}:		116.4	65.1		k:

COMMENTS: TLV=200 ppm; highly toxic

Name: *trans*-1,2-Dichloroethylene

Synonym: *trans*-1,2-Dichloroethene

Mol. Form.: $C_2H_2Cl_2$

CAS RN: 156-60-5

Merck No.: 86

Mol. Wt.: 96.943

PHYSICAL CONSTANTS

T_m: -49.8°C (223.3 K)

T_b: 48.73°C (321.88 K)

T_c: 243.4°C (516.5 K)

P_c: 5.51 MPa

μ: 0 D

IP: 9.65 eV

TRANSITION PROPERTIES

$\Delta_{fus}H$ (T_m): 11.98 kJ/mol

$\Delta_{vap}H$ (T_b): 28.89 kJ/mol

$\Delta_{vap}H$ (25°C): 30.04 kJ/mol

Vapor pressure (0°C): 14.7 kPa

Vapor pressure (25°C): 44.2 kPa

Vapor pressure (100°C):

PROPERTIES AT 25°C AND 100 kPa

	Solid	Liquid	Gas		Liquid
$\Delta_fH°$/kJ mol^{-1}:		-23.1	6.2		d: 1.2444 g/mL
$S°$/J mol^{-1}K^{-1}:		195.9	290.0		η: 0.317 mPa s
C_p/J mol^{-1}K^{-1}:		116.8	66.7		k:

COMMENTS: TLV=200 ppm; highly toxic

Name: Dichlorofluoromethane

Synonyms: Refrigerant 21

         CFC-21

Mol. Form.: $CHCl_2F$

CAS RN: 75-43-4

Merck No.:

Mol. Wt.: 102.923

PHYSICAL CONSTANTS

T_m: -135°C (138 K)

T_b: 8.95°C (282.10 K)

T_c: 178.43°C (451.58 K)

P_c: 5.18 MPa

μ: 1.29 D

IP: 11.50 eV

TRANSITION PROPERTIES

$\Delta_{fus}H$ (T_m):

$\Delta_{vap}H$ (T_b): 25.15 kJ/mol

$\Delta_{vap}H$ (25°C): 24.23 kJ/mol

Vapor pressure (0°C): 67.2 kPa

Vapor pressure (25°C): 187 kPa

Vapor pressure (100°C):

PROPERTIES AT 25°C AND 100 kPa

	Solid	Liquid	Gas		Gas
$\Delta_fH°$/kJ mol^{-1}:					d: 4.207 g/L
$S°$/J mol^{-1}K^{-1}:			293.1		η:
C_p/J mol^{-1}K^{-1}:			60.9		k:

COMMENTS: TLV=10 ppm

Name: Dichloromethane

Synonyms: Methylene chloride

         Refrigerant 30

Mol. Form.: CH_2Cl_2

CAS RN: 75-09-2

Merck No.: 5982

Mol. Wt.: 84.932

PHYSICAL CONSTANTS

T_m: -95.14°C (178.01 K)

T_b: 40°C (313 K)

T_c: 237°C (510 K)

P_c: 6.10 MPa

μ: 1.60 D

IP: 11.32 eV

TRANSITION PROPERTIES

$\Delta_{fus}H$ (T_m): 6.00 kJ/mol

$\Delta_{vap}H$ (T_b): 28.06 kJ/mol

$\Delta_{vap}H$ (25°C): 28.82 kJ/mol

Vapor pressure (0°C): 19.2 kPa

Vapor pressure (25°C): 58.2 kPa

Vapor pressure (100°C):

PROPERTIES AT 25°C AND 100 kPa

	Solid	Liquid	Gas		Liquid
$\Delta_fH°$/kJ mol^{-1}:		-124.1	-95.6		d: 1.3168 g/mL
$S°$/J mol^{-1}K^{-1}:		177.8	270.2		η: 0.413 mPa s
C_p/J mol^{-1}K^{-1}:		101.2	51.0		k:

COMMENTS: TLV=50 ppm; carcinogen

Name: 1,2-Dichloropropane
Synonym: Propylene dichloride

Mol. Form.: $C_3H_6Cl_2$

CAS RN: 78-87-5
Merck No.: 7867
Mol. Wt.: 112.986

PHYSICAL CONSTANTS

T_m: -100.5°C (172.6 K)
T_b: 96.37°C (369.52 K)

T_c:
P_c:

μ:
IP: 10.87 eV

TRANSITION PROPERTIES

$\Delta_{fus}H$ (T_m): 6.40 kJ/mol
$\Delta_{vap}H$ (T_b): 32.00 kJ/mol
$\Delta_{vap}H$ (25°C): 36.40 kJ/mol

Vapor pressure (0°C):
Vapor pressure (25°C): 7.08 kPa
Vapor pressure (100°C): 113 kPa

PROPERTIES AT 25°C AND 100 kPa

	Solid	Liquid	Gas		Liquid
$\Delta_f H°$/kJ mol^{-1}:		-198.8	-162.8		*d*: 1.1496 g/mL
$S°$/J mol^{-1}K^{-1}:					η:
C_p/J mol^{-1}K^{-1}:		149.1			*k*:

COMMENTS: TLV=75 ppm; carcinogen; highly toxic; flammable

Name: Dichlorosilane

Mol. Form.: Cl_2H_2Si

CAS RN: 4109-96-0
Merck No.:
Mol. Wt.: 101.007

PHYSICAL CONSTANTS

T_m: -122°C (151 K)
T_b: 8.3°C (281.4 K)

T_c:
P_c:

μ: 1.17 D
IP: 11.40 eV

TRANSITION PROPERTIES

$\Delta_{fus}H$ (T_m):
$\Delta_{vap}H$ (T_b): 25.00 kJ/mol
$\Delta_{vap}H$ (25°C): 24.20 kJ/mol

Vapor pressure (0°C):
Vapor pressure (25°C):
Vapor pressure (100°C):

PROPERTIES AT 25°C AND 100 kPa

	Solid	Liquid	Gas		Gas
$\Delta_f H°$/kJ mol^{-1}:					*d*: 4.129 g/L
$S°$/J mol^{-1}K^{-1}:			285.7		η:
C_p/J mol^{-1}K^{-1}:			60.5		*k*:

COMMENTS: Highly toxic

Name: 1,2-Dichlorotetrafluoroethane
Synonyms: Refrigerant 114
CFC-114
Mol. Form.: $C_2Cl_2F_4$

CAS RN: 76-14-2
Merck No.: 2608
Mol. Wt.: 170.921

PHYSICAL CONSTANTS

T_m: -94°C (179 K)
T_b: 3.8°C (276.9 K)

T_c: 145.63°C (418.78 K)
P_c: 3.252 MPa

μ: 0.5 D
IP: 12.20 eV

TRANSITION PROPERTIES

$\Delta_{fus}H$ (T_m): 6.32 kJ/mol
$\Delta_{vap}H$ (T_b): 23.25 kJ/mol
$\Delta_{vap}H$ (25°C): 23.4 kJ/mol

Vapor pressure (0°C): 88.3 kPa
Vapor pressure (25°C): 215 kPa
Vapor pressure (100°C): 1413 kPa

PROPERTIES AT 25°C AND 100 kPa

	Solid	Liquid	Gas		Gas
$\Delta_f H°$/kJ mol^{-1}:		-939.7	-916.3		*d*: 6.986 g/L
$S°$/J mol^{-1}K^{-1}:					η:
C_p/J mol^{-1}K^{-1}:		164.2			*k*: 0.01025 W/m K

COMMENTS: TLV=1000 ppm; highly toxic

Name: 2,4-Dichlorotoluene   CAS RN: 95-73-8
Synonym: 2,4-Dichloro-1-methylbenzene   Merck No.:
                                                                   Mol. Wt.: 161.030
Mol. Form.: $C_7H_6Cl_2$

PHYSICAL CONSTANTS

T_m: -13.5°C (259.6 K) T_c: μ:
T_b: 201°C (474 K) P_c: IP:

TRANSITION PROPERTIES

$\Delta_{fus}H$ (T_m):   Vapor pressure (0°C):
$\Delta_{vap}H$ (T_b):   Vapor pressure (25°C): 0.055 kPa
$\Delta_{vap}H$ (25°C):   Vapor pressure (100°C): 4.58 kPa

PROPERTIES AT 25°C AND 100 kPa

	Solid	Liquid	Gas	Liquid
$\Delta_f H°$/kJ mol^{-1}:				d: 1.241 g/mL
$S°$/J mol^{-1}K^{-1}:				η: 1.53 mPa s
C_p/J mol^{-1}K^{-1}:				k:
COMMENTS:				

Name: Diethanolamine   CAS RN: 111-42-2
Synonyms: Bis(2-hydroxyethyl)amine   Merck No.: 3097
          2,2′-Iminodiethanol   Mol. Wt.: 105.137
Mol. Form.: $C_4H_{11}NO_2$

PHYSICAL CONSTANTS

T_m: 28°C (301 K) T_c: μ:
T_b: 268.8°C (541.9 K) P_c: IP:

TRANSITION PROPERTIES

$\Delta_{fus}H$ (T_m): 25.10 kJ/mol   Vapor pressure (0°C):
$\Delta_{vap}H$ (T_b): 65.23 kJ/mol   Vapor pressure (25°C):
$\Delta_{vap}H$ (25°C):   Vapor pressure (100°C):

PROPERTIES AT 25°C AND 100 kPa

	Solid	Liquid	Gas	Solid
$\Delta_f H°$/kJ mol^{-1}:				d: 1.0899 g/mL
$S°$/J mol^{-1}K^{-1}:				η: N/A
C_p/J mol^{-1}K^{-1}:				k:
COMMENTS: TLV=3 ppm; highly toxic				

Name: 1,1-Diethoxyethane   CAS RN: 105-57-7
Synonyms: Acetal   Merck No.: 31
          Diethyl acetal   Mol. Wt.: 118.176
Mol. Form.: $C_6H_{14}O_2$

PHYSICAL CONSTANTS

T_m: -100°C (173 K) T_c: 254°C (527 K)   μ:
T_b: 102.25°C (375.40 K) P_c: IP: 9.78 eV

TRANSITION PROPERTIES

$\Delta_{fus}H$ (T_m):   Vapor pressure (0°C): 0.777 kPa
$\Delta_{vap}H$ (T_b): 36.28 kJ/mol   Vapor pressure (25°C): 3.68 kPa
$\Delta_{vap}H$ (25°C): 43.20 kJ/mol   Vapor pressure (100°C): 93.9 kPa

PROPERTIES AT 25°C AND 100 kPa

	Solid	Liquid	Gas	Liquid
$\Delta_f H°$/kJ mol^{-1}:		-491.4	-453.5	d: 0.822 g/mL
$S°$/J mol^{-1}K^{-1}:				η:
C_p/J mol^{-1}K^{-1}:		238.0		k:
COMMENTS:				

Name: Diethylamine
Synonym: *N*-Ethylethanamine

Mol. Form.: $C_4H_{11}N$

CAS RN: 109-89-7
Merck No.: 3100
Mol. Wt.: 73.138

PHYSICAL CONSTANTS

T_m: -49.8°C (223.3 K)
T_b: 55.5°C (328.6 K)

T_c: 226.84°C (499.99 K)
P_c: 3.758 MPa

μ: 0.92 D
IP: 8.01 eV

TRANSITION PROPERTIES

$\Delta_{fus}H$ (T_m):
$\Delta_{vap}H$ (T_b): 29.06 kJ/mol
$\Delta_{vap}H$ (25°C): 31.31 kJ/mol

Vapor pressure (0°C):
Vapor pressure (25°C): 30.1 kPa
Vapor pressure (100°C):

PROPERTIES AT 25°C AND 100 kPa

	Solid	Liquid	Gas		Liquid
$\Delta_f H°$/kJ mol^{-1}:		-103.7	-72.5		*d*: 0.7017 g/mL
$S°$/J mol^{-1}K^{-1}:					η: 0.319 mPa s
C_p/J mol^{-1}K^{-1}:		169.2			*k*:

COMMENTS: TLV=10 ppm; flammable

Name: *o*-Diethylbenzene
Synonym: 1,2-Diethylbenzene

Mol. Form.: $C_{10}H_{14}$

CAS RN: 135-01-3
Merck No.:
Mol. Wt.: 134.221

PHYSICAL CONSTANTS

T_m: -31.2°C (241.9 K)
T_b: 184°C (457 K)

T_c:
P_c:

μ:
IP: 8.51 eV

TRANSITION PROPERTIES

$\Delta_{fus}H$ (T_m):
$\Delta_{vap}H$ (T_b):
$\Delta_{vap}H$ (25°C):

Vapor pressure (0°C):
Vapor pressure (25°C):
Vapor pressure (100°C): 7.33 kPa

PROPERTIES AT 25°C AND 100 kPa

	Solid	Liquid	Gas		Liquid
$\Delta_f H°$/kJ mol^{-1}:		-68.5			*d*: 0.876 g/mL
$S°$/J mol^{-1}K^{-1}:					η:
C_p/J mol^{-1}K^{-1}:					*k*:

COMMENTS:

Name: *m*-Diethylbenzene
Synonym: 1,3-Diethylbenzene

Mol. Form.: $C_{10}H_{14}$

CAS RN: 141-93-5
Merck No.:
Mol. Wt.: 134.221

PHYSICAL CONSTANTS

T_m: -83.9°C (189.2 K)
T_b: 181.1°C (454.2 K)

T_c:
P_c:

μ:
IP: 8.49 eV

TRANSITION PROPERTIES

$\Delta_{fus}H$ (T_m):
$\Delta_{vap}H$ (T_b):
$\Delta_{vap}H$ (25°C):

Vapor pressure (0°C):
Vapor pressure (25°C):
Vapor pressure (100°C): 7.84 kPa

PROPERTIES AT 25°C AND 100 kPa

	Solid	Liquid	Gas		Liquid
$\Delta_f H°$/kJ mol^{-1}:		-73.5			*d*: 0.856 g/mL
$S°$/J mol^{-1}K^{-1}:					η:
C_p/J mol^{-1}K^{-1}:					*k*:

COMMENTS:

Name: *p*-Diethylbenzene
Synonym: 1,4-Diethylbenzene

Mol. Form.: $C_{10}H_{14}$

CAS RN: 105-05-5
Merck No.:
Mol. Wt.: 134.221

PHYSICAL CONSTANTS

T_m: -42.83°C (230.32 K) T_c: 384.73°C (657.88 K) μ:
T_b: 183.79°C (456.94 K) P_c: 2.803 MPa   IP: 8.40 eV

TRANSITION PROPERTIES

$\Delta_{fus}H$ (T_m):   Vapor pressure (0°C):
$\Delta_{vap}H$ (T_b):   Vapor pressure (25°C):
$\Delta_{vap}H$ (25°C):   Vapor pressure (100°C): 7.30 kPa

PROPERTIES AT 25°C AND 100 kPa

	Solid	Liquid	Gas		Liquid
$\Delta_f H°$/kJ mol^{-1}:		-72.8		*d*: 0.858 g/mL	
$S°$/J mol^{-1}K^{-1}:				η:	
C_p/J mol^{-1}K^{-1}:				*k*:	
COMMENTS:					

Name: Diethyl carbonate
Synonyms: Carbonic acid, diethyl ester
         Ethyl carbonate
Mol. Form.: $C_5H_{10}O_3$

CAS RN: 105-58-8
Merck No.: 3738
Mol. Wt.: 118.133

PHYSICAL CONSTANTS

T_m: -43°C (230 K) T_c: μ: 1.10 D
T_b: 126°C (399 K) P_c: IP:

TRANSITION PROPERTIES

$\Delta_{fus}H$ (T_m):   Vapor pressure (0°C):
$\Delta_{vap}H$ (T_b): 36.15 kJ/mol   Vapor pressure (25°C): 1.63 kPa
$\Delta_{vap}H$ (25°C): 43.60 kJ/mol   Vapor pressure (100°C): 44.0 kPa

PROPERTIES AT 25°C AND 100 kPa

	Solid	Liquid	Gas		Liquid
$\Delta_f H°$/kJ mol^{-1}:		-681.5	-637.9	*d*: 0.975 g/mL	
$S°$/J mol^{-1}K^{-1}:				η: 0.748 mPa s	
C_p/J mol^{-1}K^{-1}:		212.4		*k*:	
COMMENTS: Flammable					

Name: Diethylene glycol
Synonyms: Bis(2-hydroxyethyl) ether
         2,2′-Oxybisethanol
Mol. Form.: $C_4H_{10}O_3$

CAS RN: 111-46-6
Merck No.: 3109
Mol. Wt.: 106.122

PHYSICAL CONSTANTS

T_m: -10.45°C (262.70 K) T_c: μ:
T_b: 245.8°C (518.9 K) P_c: IP:

TRANSITION PROPERTIES

$\Delta_{fus}H$ (T_m):   Vapor pressure (0°C):
$\Delta_{vap}H$ (T_b): 52.26 kJ/mol   Vapor pressure (25°C): 0.001 kPa
$\Delta_{vap}H$ (25°C): 57.3 kJ/mol   Vapor pressure (100°C): 0.273 kPa

PROPERTIES AT 25°C AND 100 kPa

	Solid	Liquid	Gas		Liquid
$\Delta_f H°$/kJ mol^{-1}:		-628.5	-571.2	*d*: 1.1150 g/mL	
$S°$/J mol^{-1}K^{-1}:				η: 30.2 mPa s	
C_p/J mol^{-1}K^{-1}:		244.8		*k*:	
COMMENTS:					

Name: Diethylene glycol diethyl ether   CAS RN: 112-36-7
Synonym: Bis(2-ethoxyethyl) ether   Merck No.: 3108
                                                                      Mol. Wt.: 162.229
Mol. Form.: $C_8H_{18}O_3$

PHYSICAL CONSTANTS

T_m: -44.3°C (228.8 K) T_c: μ:
T_b: 188°C (461 K) P_c: IP:

TRANSITION PROPERTIES

$\Delta_{fus}H$ (T_m):   Vapor pressure (0°C):
$\Delta_{vap}H$ (T_b): 49.00 kJ/mol   Vapor pressure (25°C):
$\Delta_{vap}H$ (25°C): 58.40 kJ/mol   Vapor pressure (100°C): 5.05 kPa

PROPERTIES AT 25°C AND 100 kPa

	Solid	Liquid	Gas		Liquid
$\Delta_fH°$/kJ mol⁻¹:					d: 0.902 g/mL
$S°$/J mol⁻¹K⁻¹:					η:
C_p/J mol⁻¹K⁻¹:					k:
COMMENTS:					

Name: Diethylene glycol dimethyl ether   CAS RN: 111-96-6
Synonyms: Bis(2-methoxyethyl) ether   Merck No.: 3148
          Diglyme   Mol. Wt.: 134.175
Mol. Form.: $C_6H_{14}O_3$

PHYSICAL CONSTANTS

T_m: -64°C (209 K) T_c: μ:
T_b: 162°C (435 K) P_c: IP: 9.80 eV

TRANSITION PROPERTIES

$\Delta_{fus}H$ (T_m): 13.60 kJ/mol   Vapor pressure (0°C): 0.054 kPa
$\Delta_{vap}H$ (T_b): 36.17 kJ/mol   Vapor pressure (25°C): 0.315 kPa
$\Delta_{vap}H$ (25°C): 44.69 kJ/mol   Vapor pressure (100°C): 13.6 kPa

PROPERTIES AT 25°C AND 100 kPa

	Solid	Liquid	Gas		Liquid
$\Delta_fH°$/kJ mol⁻¹:					d: 0.939 g/mL
$S°$/J mol⁻¹K⁻¹:					η: 0.989 mPa s
C_p/J mol⁻¹K⁻¹:		274.1			k:
COMMENTS:					

Name: Diethyl ether   CAS RN: 60-29-7
Synonyms: 1,1′-Oxybisethane   Merck No.: 3762
          Ethyl ether   Mol. Wt.: 74.123
Mol. Form.: $C_4H_{10}O$

PHYSICAL CONSTANTS

T_m: -116.3°C (156.8 K) T_c: 193.59°C (466.74 K)   μ: 1.15 D
T_b: 34.5°C (307.6 K) P_c: 3.638 MPa   IP: 9.51 eV

TRANSITION PROPERTIES

$\Delta_{fus}H$ (T_m): 7.27 kJ/mol   Vapor pressure (0°C): 24.9 kPa
$\Delta_{vap}H$ (T_b): 26.52 kJ/mol   Vapor pressure (25°C): 71.7 kPa
$\Delta_{vap}H$ (25°C): 27.10 kJ/mol   Vapor pressure (100°C):

PROPERTIES AT 25°C AND 100 kPa

	Solid	Liquid	Gas		Liquid
$\Delta_fH°$/kJ mol⁻¹:		-279.3	-252.1		d: 0.7078 g/mL
$S°$/J mol⁻¹K⁻¹:		172.4	342.7		η: 0.224 mPa s
C_p/J mol⁻¹K⁻¹:		175.6	119.5		k: 0.130 W/m K
COMMENTS: TLV=400 ppm; very flammable					

Name: Diethyl malonate
Synonym: Diethyl propanedioate

Mol. Form.: $C_7H_{12}O_4$

CAS RN: 105-53-3
Merck No.: 3779
Mol. Wt.: 160.170

PHYSICAL CONSTANTS

T_m: -50°C (223 K) T_c: μ:
T_b: 200°C (473 K) P_c: IP:

TRANSITION PROPERTIES

$\Delta_{fus}H$ (T_m):   Vapor pressure (0°C):
$\Delta_{vap}H$ (T_b): 54.81 kJ/mol   Vapor pressure (25°C):
$\Delta_{vap}H$ (25°C):   Vapor pressure (100°C):

PROPERTIES AT 25°C AND 100 kPa

	Solid	Liquid	Gas	Liquid
$\Delta_f H°$/kJ mol^{-1}:				d: 1.0499 g/mL
$S°$/J mol^{-1}K^{-1}:				η: 1.94 mPa s
C_p/J mol^{-1}K^{-1}:		285.0		k:
COMMENTS:				

Name: Diethyl oxalate
Synonym: Diethyl ethanedioate

Mol. Form.: $C_6H_{10}O_4$

CAS RN: 95-92-1
Merck No.: 3115
Mol. Wt.: 146.143

PHYSICAL CONSTANTS

T_m: -40.6°C (232.5 K) T_c: μ:
T_b: 185.7°C (458.8 K) P_c: IP: 9.80 eV

TRANSITION PROPERTIES

$\Delta_{fus}H$ (T_m):   Vapor pressure (0°C):
$\Delta_{vap}H$ (T_b): 42.01 kJ/mol   Vapor pressure (25°C): 0.030 kPa
$\Delta_{vap}H$ (25°C): 63.5 kJ/mol   Vapor pressure (100°C): 3.10 kPa

PROPERTIES AT 25°C AND 100 kPa

	Solid	Liquid	Gas	Liquid
$\Delta_f H°$/kJ mol^{-1}:		-805.5	-742.0	d: 1.073 g/mL
$S°$/J mol^{-1}K^{-1}:				η: 1.8 mPa s
C_p/J mol^{-1}K^{-1}:				k:
COMMENTS:				

Name: 3,3-Diethylpentane
Synonym: Tetraethylmethane

Mol. Form.: C_9H_{20}

CAS RN: 1067-20-5
Merck No.:
Mol. Wt.: 128.258

PHYSICAL CONSTANTS

T_m: -33.1°C (240.0 K) T_c: μ:
T_b: 146.3°C (419.4 K) P_c: IP:

TRANSITION PROPERTIES

$\Delta_{fus}H$ (T_m): 10.09 kJ/mol   Vapor pressure (0°C):
$\Delta_{vap}H$ (T_b): 34.61 kJ/mol   Vapor pressure (25°C):
$\Delta_{vap}H$ (25°C): 42.00 kJ/mol   Vapor pressure (100°C): 26.2 kPa

PROPERTIES AT 25°C AND 100 kPa

	Solid	Liquid	Gas	Liquid
$\Delta_f H°$/kJ mol^{-1}:		-275.4	-232.3	d: 0.750 g/mL
$S°$/J mol^{-1}K^{-1}:				η:
C_p/J mol^{-1}K^{-1}:		278.2		k:
COMMENTS: Flammable				

Name: Diethyl phthalate
Synonym: Diethyl 1,2-benzenedicarboxylate

CAS RN: 84-66-2
Merck No.: 7345
Mol. Wt.: 222.241

Mol. Form.: $C_{12}H_{14}O_4$

PHYSICAL CONSTANTS

T_m: -40.5°C (232.6 K) T_c: μ:
T_b: 295°C (568 K) P_c: IP:

TRANSITION PROPERTIES

$\Delta_{fus}H$ (T_m):   Vapor pressure (0°C):
$\Delta_{vap}H$ (T_b):   Vapor pressure (25°C):
$\Delta_{vap}H$ (25°C): 88.2 kJ/mol   Vapor pressure (100°C):

PROPERTIES AT 25°C AND 100 kPa

	Solid	Liquid	Gas		Liquid
$\Delta_fH°$/kJ mol^{-1}:		-776.6	-688.4		d: 1.22 g/mL
$S°$/J mol^{-1}K^{-1}:		425.1			η:
C_p/J mol^{-1}K^{-1}:		366.1			k:
COMMENTS:					

Name: Diethyl sulfide
Synonyms: 3-Thiapentane
          1,1'-Thiobisethane
Mol. Form.: $C_4H_{10}S$

CAS RN: 352-93-2
Merck No.: 3809
Mol. Wt.: 90.189

PHYSICAL CONSTANTS

T_m: -103.95°C (169.20 K) T_c: 284°C (557 K)   μ: 1.54 D
T_b: 92.1°C (365.2 K) P_c: 3.96 MPa   IP: 8.43 eV

TRANSITION PROPERTIES

$\Delta_{fus}H$ (T_m): 11.90 kJ/mol   Vapor pressure (0°C):
$\Delta_{vap}H$ (T_b): 31.77 kJ/mol   Vapor pressure (25°C): 7.78 kPa
$\Delta_{vap}H$ (25°C): 35.80 kJ/mol   Vapor pressure (100°C): 128 kPa

PROPERTIES AT 25°C AND 100 kPa

	Solid	Liquid	Gas		Liquid
$\Delta_fH°$/kJ mol^{-1}:		-119.4	-83.6		d: 0.8312 g/mL
$S°$/J mol^{-1}K^{-1}:		269.3	368.1		η: 0.422 mPa s
C_p/J mol^{-1}K^{-1}:		171.4	117.0		k:
COMMENTS: Flammable					

Name: *p*-Difluorobenzene
Synonym: 1,4-Difluorobenzene

CAS RN: 540-36-3
Merck No.: 3132
Mol. Wt.: 114.095

Mol. Form.: $C_6H_4F_2$

PHYSICAL CONSTANTS

T_m: -13°C (260 K) T_c: 283°C (556 K)   μ: 0 D
T_b: 89°C (362 K) P_c: 4.40 MPa   IP: 9.14 eV

TRANSITION PROPERTIES

$\Delta_{fus}H$ (T_m):   Vapor pressure (0°C):
$\Delta_{vap}H$ (T_b): 31.77 kJ/mol   Vapor pressure (25°C):
$\Delta_{vap}H$ (25°C): 35.54 kJ/mol   Vapor pressure (100°C):

PROPERTIES AT 25°C AND 100 kPa

	Solid	Liquid	Gas		Liquid
$\Delta_fH°$/kJ mol^{-1}:		-342.3	-306.7		d: 1.163 g/mL
$S°$/J mol^{-1}K^{-1}:					η: 0.59 mPa s
C_p/J mol^{-1}K^{-1}:		157.5			k:
COMMENTS:					

Name: 1,1-Difluoroethane
Synonyms: Ethylidene difluoride
           Refrigerant 152a
Mol. Form.: $C_2H_4F_2$

CAS RN: 75-37-6
Merck No.:
Mol. Wt.: 66.051

PHYSICAL CONSTANTS

T_m: -117°C (156 K)
T_b: -24.95°C (248.20 K)

T_c: 113.6°C (386.7 K)
P_c: 4.50 MPa

μ: 2.27 D
IP: 11.87 eV

TRANSITION PROPERTIES

$\Delta_{fus}H$ (T_m):
$\Delta_{vap}H$ (T_b): 21.56 kJ/mol
$\Delta_{vap}H$ (25°C): 19.08 kJ/mol

Vapor pressure (0°C):
Vapor pressure (25°C): 601 kPa
Vapor pressure (100°C): 3518 kPa

PROPERTIES AT 25°C AND 100 kPa

	Solid	Liquid	Gas		Gas
$\Delta_f H°$/kJ mol^{-1}:			-497.0		d: 2.700 g/L
$S°$/J mol^{-1}K^{-1}:			282.5		η:
C_p/J mol^{-1}K^{-1}:			67.8		k:
COMMENTS:					

Name: 1,1-Difluoroethylene
Synonym: Vinylidene fluoride

Mol. Form.: $C_2H_2F_2$

CAS RN: 75-38-7
Merck No.:
Mol. Wt.: 64.035

PHYSICAL CONSTANTS

T_m: -144°C (129 K)
T_b: -85.7°C (187.4 K)

T_c: 29.8°C (302.9 K)
P_c: 4.46 MPa

μ: 1.368 D
IP: 10.29 eV

TRANSITION PROPERTIES

$\Delta_{fus}H$ (T_m):
$\Delta_{vap}H$ (T_b):
$\Delta_{vap}H$ (25°C):

Vapor pressure (0°C):
Vapor pressure (25°C):
Vapor pressure (100°C): N/A

PROPERTIES AT 25°C AND 100 kPa

	Solid	Liquid	Gas		Gas
$\Delta_f H°$/kJ mol^{-1}:			-335.0		d: 2.617 g/L
$S°$/J mol^{-1}K^{-1}:			266.2		η:
C_p/J mol^{-1}K^{-1}:			60.1		k:
COMMENTS: Very flammable					

Name: Difluoromethane
Synonyms: Methylene fluoride
          Refrigerant 32
Mol. Form.: CH_2F_2

CAS RN: 75-10-5
Merck No.:
Mol. Wt.: 52.024

PHYSICAL CONSTANTS

T_m: -136°C (137 K)
T_b: -51.69°C (221.46 K)

T_c: 78.5°C (351.6 K)
P_c: 5.830 MPa

μ: 1.978 D
IP: 12.71 eV

TRANSITION PROPERTIES

$\Delta_{fus}H$ (T_m):
$\Delta_{vap}H$ (T_b):
$\Delta_{vap}H$ (25°C):

Vapor pressure (0°C):
Vapor pressure (25°C):
Vapor pressure (100°C): N/A

PROPERTIES AT 25°C AND 100 kPa

	Solid	Liquid	Gas		Gas
$\Delta_f H°$/kJ mol^{-1}:			-452.2		d: 2.126 g/L
$S°$/J mol^{-1}K^{-1}:			246.7		η:
C_p/J mol^{-1}K^{-1}:			42.9		k:
COMMENTS:					

Name: Diiodomethane
Synonym: Methylene iodide

Mol. Form.: CH_2I_2

CAS RN: 75-11-6
Merck No.: 5985
Mol. Wt.: 267.836

PHYSICAL CONSTANTS

T_m: 6.1°C (279.2 K)　　T_c:　　　　μ:
T_b: 182°C (455 K)　　　P_c:　　　　IP: 9.46 eV

TRANSITION PROPERTIES

$\Delta_{fus}H$ (T_m): 44.80 kJ/mol
$\Delta_{vap}H$ (T_b): 42.49 kJ/mol
$\Delta_{vap}H$ (25°C): 49.38 kJ/mol

Vapor pressure (0°C):
Vapor pressure (25°C): 0.172 kPa
Vapor pressure (100°C): 7.93 kPa

PROPERTIES AT 25°C AND 100 kPa

	Solid	Liquid	Gas		Liquid
$\Delta_f H°$/kJ mol^{-1}:		66.9	113.0		d: 3.3079 g/mL
$S°$/J mol^{-1}K^{-1}:		174.1	309.7		η: 2.6 mPa s
C_p/J mol^{-1}K^{-1}:		134.0	57.7		k:
COMMENTS:					

Name: Diisopropylamine
Synonym: N-(1-methylethyl)-2-propanamine

Mol. Form.: $C_6H_{15}N$

CAS RN: 108-18-9
Merck No.: 3181
Mol. Wt.: 101.192

PHYSICAL CONSTANTS

T_m: -61°C (212 K)　　T_c: 250.0°C (523.1 K)　　μ:
T_b: 83.9°C (357.0 K)　P_c: 3.02 MPa　　　　IP: 7.73 eV

TRANSITION PROPERTIES

$\Delta_{fus}H$ (T_m):
$\Delta_{vap}H$ (T_b): 30.40 kJ/mol
$\Delta_{vap}H$ (25°C): 34.61 kJ/mol

Vapor pressure (0°C): 2.96 kPa
Vapor pressure (25°C): 10.7 kPa
Vapor pressure (100°C): 160 kPa

PROPERTIES AT 25°C AND 100 kPa

	Solid	Liquid	Gas		Liquid
$\Delta_f H°$/kJ mol^{-1}:		-178.5	-144.0		d: 0.7100 g/mL
$S°$/J mol^{-1}K^{-1}:					η: 0.393 mPa s
C_p/J mol^{-1}K^{-1}:					k:
COMMENTS: TLV=5 ppm; highly toxic; flammable					

Name: Diisopropyl ether
Synonyms: 2,2'-Oxybispropane
　　　　　Isopropyl ether
Mol. Form.: $C_6H_{14}O$

CAS RN: 108-20-3
Merck No.: 5100
Mol. Wt.: 102.177

PHYSICAL CONSTANTS

T_m: -86.8°C (186.3 K)　　T_c: 227.17°C (500.32 K)　μ: 1.13 D
T_b: 68.51°C (341.66 K)　P_c: 2.832 MPa　　　　IP: 9.20 eV

TRANSITION PROPERTIES

$\Delta_{fus}H$ (T_m): 11.03 kJ/mol
$\Delta_{vap}H$ (T_b): 29.10 kJ/mol
$\Delta_{vap}H$ (25°C): 32.12 kJ/mol

Vapor pressure (0°C):
Vapor pressure (25°C): 19.9 kPa
Vapor pressure (100°C):

PROPERTIES AT 25°C AND 100 kPa

	Solid	Liquid	Gas		Liquid
$\Delta_f H°$/kJ mol^{-1}:		-351.5	-319.2		d: 0.7207 g/mL
$S°$/J mol^{-1}K^{-1}:					η: 0.379 mPa s
C_p/J mol^{-1}K^{-1}:		216.8			k:
COMMENTS: Flammable					

Name: 1,2-Dimethoxyethane
Synonyms: Ethylene glycol dimethyl ether
           Dimethyl cellosolve
Mol. Form.: $C_4H_{10}O_2$

CAS RN: 110-71-4
Merck No.: 3213
Mol. Wt.: 90.122

PHYSICAL CONSTANTS

T_m: -58°C (215 K)	T_c: 263°C (536 K)	μ:
T_b: 85°C (358 K)	P_c: 3.87 MPa	IP: 9.30 eV

TRANSITION PROPERTIES

$\Delta_{fus}H$ (T_m): 12.60 kJ/mol
$\Delta_{vap}H$ (T_b): 32.42 kJ/mol
$\Delta_{vap}H$ (25°C): 36.39 kJ/mol

Vapor pressure (0°C): 2.62 kPa
Vapor pressure (25°C): 9.93 kPa
Vapor pressure (100°C): 155 kPa

PROPERTIES AT 25°C AND 100 kPa

	Solid	Liquid	Gas		Liquid
$\Delta_fH°$/kJ mol^{-1}:		-376.6			d: 0.859 g/mL
$S°$/J mol^{-1}K^{-1}:					η: 0.455 mPa s
C_p/J mol^{-1}K^{-1}:		193.3			k:
COMMENTS:					

Name: Dimethoxymethane
Synonyms: Methylal
           Formal
Mol. Form.: $C_3H_8O_2$

CAS RN: 109-87-5
Merck No.: 5936
Mol. Wt.: 76.095

PHYSICAL CONSTANTS

T_m: -104.8°C (168.3 K)	T_c:	μ:
T_b: 42°C (315 K)	P_c:	IP: 9.50 eV

TRANSITION PROPERTIES

$\Delta_{fus}H$ (T_m): 8.33 kJ/mol
$\Delta_{vap}H$ (T_b):
$\Delta_{vap}H$ (25°C): 28.89 kJ/mol

Vapor pressure (0°C): 16.7 kPa
Vapor pressure (25°C): 53.1 kPa
Vapor pressure (100°C):

PROPERTIES AT 25°C AND 100 kPa

	Solid	Liquid	Gas		Liquid
$\Delta_fH°$/kJ mol^{-1}:		-377.7	-348.4		d: 0.8538 g/mL
$S°$/J mol^{-1}K^{-1}:		244.0			η: 0.33 mPa s
C_p/J mol^{-1}K^{-1}:		162.0			k:
COMMENTS: TLV=1000 ppm; flammable					

Name: *N,N*-Dimethylacetamide
Synonym: *N,N*-Dimethylethanamide

Mol. Form.: C_4H_9NO

CAS RN: 127-19-5
Merck No.: 3216
Mol. Wt.: 87.122

PHYSICAL CONSTANTS

T_m: -20°C (253 K)	T_c:	μ:
T_b: 165°C (438 K)	P_c:	IP: 8.81 eV

TRANSITION PROPERTIES

$\Delta_{fus}H$ (T_m): 10.42 kJ/mol
$\Delta_{vap}H$ (T_b): 43.35 kJ/mol
$\Delta_{vap}H$ (25°C): 50.24 kJ/mol

Vapor pressure (0°C):
Vapor pressure (25°C): 0.075 kPa
Vapor pressure (100°C): 10.9 kPa

PROPERTIES AT 25°C AND 100 kPa

	Solid	Liquid	Gas		Liquid
$\Delta_fH°$/kJ mol^{-1}:		-278.3			d: 0.9365 g/mL
$S°$/J mol^{-1}K^{-1}:					η: 1.96 mPa s
C_p/J mol^{-1}K^{-1}:		175.6			k:
COMMENTS: TLV=10 ppm					

Name: Dimethylamine

Synonym: *N*-Methylmethanamine

Mol. Form.: C_2H_7N

CAS RN: 124-40-3

Merck No.: 3217

Mol. Wt.: 45.084

PHYSICAL CONSTANTS

T_m: -92.2°C (180.9 K)

T_b: 6.88°C (280.03 K)

T_c: 164.07°C (437.22 K)

P_c: 5.340 MPa

μ: 1.01 D

IP: 8.23 eV

TRANSITION PROPERTIES

$\Delta_{fus}H$ (T_m): 5.94 kJ/mol

$\Delta_{vap}H$ (T_b): 26.40 kJ/mol

$\Delta_{vap}H$ (25°C): 25.05 kJ/mol

Vapor pressure (0°C): 75.0 kPa

Vapor pressure (25°C): 203 kPa

Vapor pressure (100°C):

PROPERTIES AT 25°C AND 100 kPa

	Solid	Liquid	Gas		Gas
$\Delta_f H°$/kJ mol^{-1}:		-43.9	-18.5	d:	1.843 g/L
$S°$/J mol^{-1}K^{-1}:		182.3	273.1	η:	
C_p/J mol^{-1}K^{-1}:		137.7	70.7	k:	

COMMENTS: TLV=5 ppm; highly toxic; very flammable

Name: *N,N*-Dimethylaniline

Synonym: *N,N*-Dimethylbenzenamine

Mol. Form.: $C_8H_{11}N$

CAS RN: 121-69-7

Merck No.: 3223

Mol. Wt.: 121.182

PHYSICAL CONSTANTS

T_m: 2.45°C (275.60 K)

T_b: 194.15°C (467.30 K)

T_c: 414°C (687 K)

P_c: 3.63 MPa

μ: 1.68 D

IP: 7.12 eV

TRANSITION PROPERTIES

$\Delta_{fus}H$ (T_m):

$\Delta_{vap}H$ (T_b):

$\Delta_{vap}H$ (25°C): 52.83 kJ/mol

Vapor pressure (0°C):

Vapor pressure (25°C):

Vapor pressure (100°C): 4.90 kPa

PROPERTIES AT 25°C AND 100 kPa

	Solid	Liquid	Gas		Liquid
$\Delta_f H°$/kJ mol^{-1}:		47.7	100.5	d:	0.9523 g/mL
$S°$/J mol^{-1}K^{-1}:				η:	1.30 mPa s
C_p/J mol^{-1}K^{-1}:				k:	

COMMENTS: TLV=5 ppm; highly toxic

Name: 2,2-Dimethylbutane

Synonym: Neohexane

Mol. Form.: C_6H_{14}

CAS RN: 75-83-2

Merck No.:

Mol. Wt.: 86.177

PHYSICAL CONSTANTS

T_m: -99°C (174 K)

T_b: 49.73°C (322.88 K)

T_c: 215.7°C (488.8 K)

P_c: 3.090 MPa

μ:

IP: 10.06 eV

TRANSITION PROPERTIES

$\Delta_{fus}H$ (T_m): 0.58 kJ/mol

$\Delta_{vap}H$ (T_b): 26.31 kJ/mol

$\Delta_{vap}H$ (25°C): 27.68 kJ/mol

Vapor pressure (0°C):

Vapor pressure (25°C): 42.5 kPa

Vapor pressure (100°C):

PROPERTIES AT 25°C AND 100 kPa

	Solid	Liquid	Gas		Liquid
$\Delta_f H°$/kJ mol^{-1}:		-213.8	-186.1	d:	0.645 g/mL
$S°$/J mol^{-1}K^{-1}:		272.5		η:	
C_p/J mol^{-1}K^{-1}:		191.9		k:	

COMMENTS: Flammable

Name: 2,3-Dimethylbutane
Synonym: Diisopropyl

Mol. Form.: C_6H_{14}

CAS RN: 79-29-8
Merck No.:
Mol. Wt.: 86.177

PHYSICAL CONSTANTS

T_m: -128.8°C (144.3 K)
T_b: 57.98°C (331.13 K)

T_c: 226.9°C (500.0 K)
P_c: 3.131 MPa

μ:
IP: 10.02 eV

TRANSITION PROPERTIES

$\Delta_{fus}H$ (T_m): 0.80 kJ/mol
$\Delta_{vap}H$ (T_b): 27.38 kJ/mol
$\Delta_{vap}H$ (25°C): 29.12 kJ/mol

Vapor pressure (0°C):
Vapor pressure (25°C): 31.3 kPa
Vapor pressure (100°C):

PROPERTIES AT 25°C AND 100 kPa

	Solid	Liquid	Gas		Liquid
$\Delta_f H°$/kJ mol^{-1}:		-207.4	-178.3		d: 0.6570 g/mL
$S°$/J mol^{-1}K^{-1}:		287.8			η:
C_p/J mol^{-1}K^{-1}:		189.7			k:

COMMENTS: Flammable

Name: 3,3-Dimethyl-2-butanone
Synonyms: *tert*-Butyl methyl ketone
             Pinacolone
Mol. Form.: $C_6H_{12}O$

CAS RN: 75-97-8
Merck No.: 7409
Mol. Wt.: 100.161

PHYSICAL CONSTANTS

T_m: -52.5°C (220.6 K)
T_b: 106.1°C (379.2 K)

T_c:
P_c:

μ:
IP: 9.11 eV

TRANSITION PROPERTIES

$\Delta_{fus}H$ (T_m):
$\Delta_{vap}H$ (T_b): 33.39 kJ/mol
$\Delta_{vap}H$ (25°C): 37.91 kJ/mol

Vapor pressure (0°C):
Vapor pressure (25°C): 4.27 kPa
Vapor pressure (100°C): 84.5 kPa

PROPERTIES AT 25°C AND 100 kPa

	Solid	Liquid	Gas		Liquid
$\Delta_f H°$/kJ mol^{-1}:		-328.6	-290.7		d: 0.7229 g/mL
$S°$/J mol^{-1}K^{-1}:					η:
C_p/J mol^{-1}K^{-1}:					k:

COMMENTS:

Name: 2,3-Dimethyl-1-butene

Mol. Form.: C_6H_{12}

CAS RN: 563-78-0
Merck No.:
Mol. Wt.: 84.161

PHYSICAL CONSTANTS

T_m: -157°C (116 K)
T_b: 55.6°C (328.7 K)

T_c:
P_c:

μ:
IP: 9.07 eV

TRANSITION PROPERTIES

$\Delta_{fus}H$ (T_m):
$\Delta_{vap}H$ (T_b):
$\Delta_{vap}H$ (25°C): 29.18 kJ/mol

Vapor pressure (0°C): 11.0 kPa
Vapor pressure (25°C): 33.6 kPa
Vapor pressure (100°C):

PROPERTIES AT 25°C AND 100 kPa

	Solid	Liquid	Gas		Liquid
$\Delta_f H°$/kJ mol^{-1}:		-93.3	-62.6		d: 0.6732 g/mL
$S°$/J mol^{-1}K^{-1}:					η:
C_p/J mol^{-1}K^{-1}:					k:

COMMENTS: Flammable

Name: 2,3-Dimethyl-2-butene
Synonym: Tetramethylethylene

Mol. Form.: C_6H_{12}

CAS RN: 563-79-1
Merck No.:
Mol. Wt.: 84.161

PHYSICAL CONSTANTS

T_m: -74.6°C (198.5 K)
T_b: 73.3°C (346.4 K)

T_c:
P_c:

μ:
IP: 8.27 eV

TRANSITION PROPERTIES

$\Delta_{fus}H$ (T_m): 5.46 kJ/mol
$\Delta_{vap}H$ (T_b): 29.64 kJ/mol
$\Delta_{vap}H$ (25°C): 32.53 kJ/mol

Vapor pressure (0°C):
Vapor pressure (25°C): 16.7 kPa
Vapor pressure (100°C):

PROPERTIES AT 25°C AND 100 kPa

	Solid	Liquid	Gas	Liquid
$\Delta_f H°$/kJ mol^{-1}:		-101.5	-68.2	d: 0.7037 g/mL
$S°$/J mol^{-1}K^{-1}:		270.2		η:
C_p/J mol^{-1}K^{-1}:		174.7		k:
COMMENTS: Flammable				

Name: 1,1-Dimethylcyclohexane

Mol. Form.: C_8H_{16}

CAS RN: 590-66-9
Merck No.:
Mol. Wt.: 112.215

PHYSICAL CONSTANTS

T_m: -33.3°C (239.8 K)
T_b: 119.6°C (392.7 K)

T_c:
P_c:

μ:
IP: 9.42 eV

TRANSITION PROPERTIES

$\Delta_{fus}H$ (T_m): 2.06 kJ/mol
$\Delta_{vap}H$ (T_b): 32.51 kJ/mol
$\Delta_{vap}H$ (25°C): 37.92 kJ/mol

Vapor pressure (0°C):
Vapor pressure (25°C): 3.02 kPa
Vapor pressure (100°C): 58.2 kPa

PROPERTIES AT 25°C AND 100 kPa

	Solid	Liquid	Gas	Liquid
$\Delta_f H°$/kJ mol^{-1}:		-218.7	-180.9	d: 0.777 g/mL
$S°$/J mol^{-1}K^{-1}:		267.2		η:
C_p/J mol^{-1}K^{-1}:		209.2		k:
COMMENTS:				

Name: *cis*-1,2-Dimethylcyclohexane

Mol. Form.: C_8H_{16}

CAS RN: 2207-01-4
Merck No.:
Mol. Wt.: 112.215

PHYSICAL CONSTANTS

T_m: -49.9°C (223.2 K)
T_b: 129.8°C (402.9 K)

T_c:
P_c:

μ:
IP: 9.78 eV

TRANSITION PROPERTIES

$\Delta_{fus}H$ (T_m): 1.64 kJ/mol
$\Delta_{vap}H$ (T_b): 33.47 kJ/mol
$\Delta_{vap}H$ (25°C): 39.70 kJ/mol

Vapor pressure (0°C):
Vapor pressure (25°C): 1.93 kPa
Vapor pressure (100°C): 42.8 kPa

PROPERTIES AT 25°C AND 100 kPa

	Solid	Liquid	Gas	Liquid
$\Delta_f H°$/kJ mol^{-1}:		-211.8	-172.1	d: 0.792 g/mL
$S°$/J mol^{-1}K^{-1}:		274.1		η: 1.0 mPa s
C_p/J mol^{-1}K^{-1}:		210.2		k:
COMMENTS:				

Name: *trans*-1,2-Dimethylcyclohexane

CAS RN: 6876-23-9
Merck No.:
Mol. Wt.: 112.215

Mol. Form.: C_8H_{16}

PHYSICAL CONSTANTS

T_m: -90°C (183 K)	T_c:	μ:
T_b: 123.5°C (396.6 K)	P_c:	IP: 9.41 eV

TRANSITION PROPERTIES

$\Delta_{fus}H$ (T_m): 10.49 kJ/mol
$\Delta_{vap}H$ (T_b): 32.96 kJ/mol
$\Delta_{vap}H$ (25°C): 38.36 kJ/mol

Vapor pressure (0°C):
Vapor pressure (25°C): 2.59 kPa
Vapor pressure (100°C): 51.7 kPa

PROPERTIES AT 25°C AND 100 kPa

	Solid	Liquid	Gas		Liquid
$\Delta_f H°$/kJ mol^{-1}:		-218.2	-179.9		d: 0.772 g/mL
$S°$/J mol^{-1}K^{-1}:		273.2			η: 0.8 mPa s
C_p/J mol^{-1}K^{-1}:		209.4			k:
COMMENTS:					

Name: *cis*-1,3-Dimethylcyclohexane

CAS RN: 638-04-0
Merck No.:
Mol. Wt.: 112.215

Mol. Form.: C_8H_{16}

PHYSICAL CONSTANTS

T_m: -75.6°C (197.5 K)	T_c:	μ:
T_b: 120.1°C (393.2 K)	P_c:	IP: 9.98 eV

TRANSITION PROPERTIES

$\Delta_{fus}H$ (T_m): 10.82 kJ/mol
$\Delta_{vap}H$ (T_b): 32.91 kJ/mol
$\Delta_{vap}H$ (25°C): 38.26 kJ/mol

Vapor pressure (0°C):
Vapor pressure (25°C): 2.87 kPa
Vapor pressure (100°C): 57.0 kPa

PROPERTIES AT 25°C AND 100 kPa

	Solid	Liquid	Gas		Liquid
$\Delta_f H°$/kJ mol^{-1}:		-222.9	-184.6		d: 0.762 g/mL
$S°$/J mol^{-1}K^{-1}:		272.6			η:
C_p/J mol^{-1}K^{-1}:		209.4			k:
COMMENTS:					

Name: *trans*-1,3-Dimethylcyclohexane

CAS RN: 2207-03-6
Merck No.:
Mol. Wt.: 112.215

Mol. Form.: C_8H_{16}

PHYSICAL CONSTANTS

T_m: -90.1°C (183.0 K)	T_c:	μ:
T_b: 124.5°C (397.6 K)	P_c:	IP: 9.53 eV

TRANSITION PROPERTIES

$\Delta_{fus}H$ (T_m): 9.86 kJ/mol
$\Delta_{vap}H$ (T_b): 33.39 kJ/mol
$\Delta_{vap}H$ (25°C): 39.16 kJ/mol

Vapor pressure (0°C):
Vapor pressure (25°C): 2.34 kPa
Vapor pressure (100°C): 50.0 kPa

PROPERTIES AT 25°C AND 100 kPa

	Solid	Liquid	Gas		Liquid
$\Delta_f H°$/kJ mol^{-1}:		-215.7	-176.5		d:
$S°$/J mol^{-1}K^{-1}:		276.3			η:
C_p/J mol^{-1}K^{-1}:		212.8			k:
COMMENTS:					

Name: *cis*-1,4-Dimethylcyclohexane

CAS RN: 624-29-3
Merck No.:
Mol. Wt.: 112.215

Mol. Form.: C_8H_{16}

PHYSICAL CONSTANTS

T_m: -87.4°C (185.7 K) T_c: μ:
T_b: 124.4°C (397.5 K) P_c: IP: 9.93 eV

TRANSITION PROPERTIES

$\Delta_{fus}H$ (T_m): 9.31 kJ/mol   Vapor pressure (0°C):
$\Delta_{vap}H$ (T_b): 33.28 kJ/mol   Vapor pressure (25°C): 2.39 kPa
$\Delta_{vap}H$ (25°C): 39.02 kJ/mol   Vapor pressure (100°C): 50.3 kPa

PROPERTIES AT 25°C AND 100 kPa

	Solid	Liquid	Gas	Liquid
$\Delta_f H°$/kJ mol^{-1}:		-215.6	-176.6	d: 0.779 g/mL
$S°$/J mol^{-1}K^{-1}:		271.1		η:
C_p/J mol^{-1}K^{-1}:		212.1		k:
COMMENTS:				

Name: *trans*-1,4-Dimethylcyclohexane

CAS RN: 2207-04-7
Merck No.:
Mol. Wt.: 112.215

Mol. Form.: C_8H_{16}

PHYSICAL CONSTANTS

T_m: -36.9°C (236.2 K) T_c: 314.6°C (587.7 K) μ:
T_b: 119.4°C (392.5 K) P_c: IP: 9.56 eV

TRANSITION PROPERTIES

$\Delta_{fus}H$ (T_m): 12.33 kJ/mol   Vapor pressure (0°C):
$\Delta_{vap}H$ (T_b): 32.56 kJ/mol   Vapor pressure (25°C): 3.02 kPa
$\Delta_{vap}H$ (25°C): 37.90 kJ/mol   Vapor pressure (100°C): 58.4 kPa

PROPERTIES AT 25°C AND 100 kPa

	Solid	Liquid	Gas	Liquid
$\Delta_f H°$/kJ mol^{-1}:		-222.4	-184.5	d:
$S°$/J mol^{-1}K^{-1}:		268.0		η:
C_p/J mol^{-1}K^{-1}:		210.2		k:
COMMENTS:				

Name: *cis*-1,2-Dimethylcyclopentane

CAS RN: 1192-18-3
Merck No.:
Mol. Wt.: 98.188

Mol. Form.: C_7H_{14}

PHYSICAL CONSTANTS

T_m: -54°C (219 K) T_c: μ:
T_b: 99.5°C (372.6 K) P_c: IP: 9.92 eV

TRANSITION PROPERTIES

$\Delta_{fus}H$ (T_m):   Vapor pressure (0°C): 1.63 kPa
$\Delta_{vap}H$ (T_b):   Vapor pressure (25°C): 6.30 kPa
$\Delta_{vap}H$ (25°C): 35.8 kJ/mol   Vapor pressure (100°C): 103 kPa

PROPERTIES AT 25°C AND 100 kPa

	Solid	Liquid	Gas	Liquid
$\Delta_f H°$/kJ mol^{-1}:		-165.3	-129.5	d: 0.768 g/mL
$S°$/J mol^{-1}K^{-1}:		269.2		η:
C_p/J mol^{-1}K^{-1}:				k:
COMMENTS: Flammable				

Name: *trans*-1,2-Dimethylcyclopentane

Mol. Form.: C_7H_{14}

CAS RN: 822-50-4
Merck No.:
Mol. Wt.: 98.188

PHYSICAL CONSTANTS

T_m: -117.5°C (155.6 K)	T_c:	μ:
T_b: 91.9°C (365.0 K)	P_c:	IP: 9.95 eV

TRANSITION PROPERTIES

$\Delta_{fus}H$ (T_m):
$\Delta_{vap}H$ (T_b):
$\Delta_{vap}H$ (25°C): 34.6 kJ/mol

Vapor pressure (0°C): 2.31 kPa
Vapor pressure (25°C): 8.54 kPa
Vapor pressure (100°C): 128 kPa

PROPERTIES AT 25°C AND 100 kPa

	Solid	Liquid	Gas		Liquid
$\Delta_f H°$/kJ mol^{-1}:		-171.2	-136.6		d: 0.747 g/mL
$S°$/J mol^{-1}K^{-1}:					η:
C_p/J mol^{-1}K^{-1}:					k:

COMMENTS: Flammable

Name: Dimethyl disulfide
Synonyms: 2,3-Dithiabutane
            Methyl disulfide
Mol. Form.: $C_2H_6S_2$

CAS RN: 624-92-0
Merck No.:
Mol. Wt.: 94.202

PHYSICAL CONSTANTS

T_m: -85°C (188 K)	T_c:	μ:
T_b: 109.8°C (382.9 K)	P_c:	IP: 7.40 eV

TRANSITION PROPERTIES

$\Delta_{fus}H$ (T_m): 9.19 kJ/mol
$\Delta_{vap}H$ (T_b): 33.78 kJ/mol
$\Delta_{vap}H$ (25°C): 37.86 kJ/mol

Vapor pressure (0°C): 0.893 kPa
Vapor pressure (25°C): 3.82 kPa
Vapor pressure (100°C): 76.0 kPa

PROPERTIES AT 25°C AND 100 kPa

	Solid	Liquid	Gas		Liquid
$\Delta_f H°$/kJ mol^{-1}:		-62.6	-24.2		d: 1.057 g/mL
$S°$/J mol^{-1}K^{-1}:		235.4			η: 0.585 mPa s
C_p/J mol^{-1}K^{-1}:		146.1			k:

COMMENTS: Very flammable

Name: Dimethyl ether
Synonyms: Oxybismethane
            Methyl ether
Mol. Form.: C_2H_6O

CAS RN: 115-10-6
Merck No.: 5990
Mol. Wt.: 46.069

PHYSICAL CONSTANTS

T_m: -141.5°C (131.6 K)	T_c: 126.9°C (400.0 K)	μ: 1.30 D
T_b: -24.8°C (248.3 K)	P_c: 5.37 MPa	IP: 10.03 eV

TRANSITION PROPERTIES

$\Delta_{fus}H$ (T_m): 4.94 kJ/mol
$\Delta_{vap}H$ (T_b): 21.51 kJ/mol
$\Delta_{vap}H$ (25°C): 18.51 kJ/mol

Vapor pressure (0°C): 273 kPa
Vapor pressure (25°C):
Vapor pressure (100°C):

PROPERTIES AT 25°C AND 100 kPa

	Solid	Liquid	Gas		Gas
$\Delta_f H°$/kJ mol^{-1}:			-184.1		d: 1.883 g/L
$S°$/J mol^{-1}K^{-1}:			266.4		η:
C_p/J mol^{-1}K^{-1}:			64.4		k:

COMMENTS: Very flammable

Name: *N,N*-Dimethylformamide
Synonym: *N,N*-Dimethylmethanamide

Mol. Form.: C_3H_7NO

CAS RN: 68-12-2
Merck No.: 3232
Mol. Wt.: 73.095

PHYSICAL CONSTANTS

T_m: -60.43°C (212.72 K)
T_b: 153°C (426 K)

T_c: 376.5°C (649.6 K)
P_c:

μ: 3.82 D
IP: 9.13 eV

TRANSITION PROPERTIES

$\Delta_{fus}H$ (T_m): 16.15 kJ/mol
$\Delta_{vap}H$ (T_b): 38.44 kJ/mol
$\Delta_{vap}H$ (25°C): 46.89 kJ/mol

Vapor pressure (0°C):
Vapor pressure (25°C): 0.439 kPa
Vapor pressure (100°C):

PROPERTIES AT 25°C AND 100 kPa

	Solid	Liquid	Gas		Liquid
$\Delta_f H°$/kJ mol^{-1}:		-239.3	-191.7		d: 0.9447 g/mL
$S°$/J mol^{-1}K^{-1}:					η: 0.794 mPa s
C_p/J mol^{-1}K^{-1}:		150.6			k: 0.184 W/m K

COMMENTS: TLV=10 ppm

Name: 2,6-Dimethyl-4-heptanone
Synonyms: Diisobytyl ketone
　　　　　 Isovalerone
Mol. Form.: $C_9H_{18}O$

CAS RN: 108-83-8
Merck No.:
Mol. Wt.: 142.241

PHYSICAL CONSTANTS

T_m: -41.5°C (231.6 K)
T_b: 169.4°C (442.5 K)

T_c:
P_c:

μ:
IP: 9.04 eV

TRANSITION PROPERTIES

$\Delta_{fus}H$ (T_m):
$\Delta_{vap}H$ (T_b): 39.92 kJ/mol
$\Delta_{vap}H$ (25°C): 50.92 kJ/mol

Vapor pressure (0°C):
Vapor pressure (25°C): 0.230 kPa
Vapor pressure (100°C): 11.6 kPa

PROPERTIES AT 25°C AND 100 kPa

	Solid	Liquid	Gas		Liquid
$\Delta_f H°$/kJ mol^{-1}:		-408.5	-357.6		d: 0.802 g/mL
$S°$/J mol^{-1}K^{-1}:					η: 0.83 mPa s
C_p/J mol^{-1}K^{-1}:		297.3			k:

COMMENTS: TLV=25 ppm

Name: 2,2-Dimethylhexane

Mol. Form.: C_8H_{18}

CAS RN: 590-73-8
Merck No.:
Mol. Wt.: 114.231

PHYSICAL CONSTANTS

T_m: -121.18°C (151.97 K)
T_b: 106.86°C (380.01 K)

T_c: 276.8°C (549.9 K)
P_c: 2.529 MPa

μ:
IP:

TRANSITION PROPERTIES

$\Delta_{fus}H$ (T_m):
$\Delta_{vap}H$ (T_b): 32.07 kJ/mol
$\Delta_{vap}H$ (25°C): 37.28 kJ/mol

Vapor pressure (0°C): 1.10 kPa
Vapor pressure (25°C): 4.54 kPa
Vapor pressure (100°C): 83.1 kPa

PROPERTIES AT 25°C AND 100 kPa

	Solid	Liquid	Gas		Liquid
$\Delta_f H°$/kJ mol^{-1}:		-261.9	-224.6		d: 0.6911 g/mL
$S°$/J mol^{-1}K^{-1}:					η:
C_p/J mol^{-1}K^{-1}:					k:

COMMENTS: Flammable

Name: 2,3-Dimethylhexane

Mol. Form.: C_8H_{18}

CAS RN: 584-94-1
Merck No.:
Mol. Wt.: 114.231

PHYSICAL CONSTANTS

T_m:
T_b: 115.62°C (388.77 K)

T_c: 290.4°C (563.5 K)
P_c: 2.628 MPa

μ:
IP:

TRANSITION PROPERTIES

$\Delta_{fus}H$ (T_m):
$\Delta_{vap}H$ (T_b): 33.17 kJ/mol
$\Delta_{vap}H$ (25°C): 38.78 kJ/mol

Vapor pressure (0°C):
Vapor pressure (25°C): 3.13 kPa
Vapor pressure (100°C): 64.3 kPa

PROPERTIES AT 25°C AND 100 kPa

	Solid	Liquid	Gas		Liquid
$\Delta_f H°$/kJ mol^{-1}:		-252.6	-213.8	d:	0.7081 g/mL
$S°$/J mol^{-1}K^{-1}:				η:	
C_p/J mol^{-1}K^{-1}:				k:	

COMMENTS: Flammable

Name: 2,4-Dimethylhexane

Mol. Form.: C_8H_{18}

CAS RN: 589-43-5
Merck No.:
Mol. Wt.: 114.231

PHYSICAL CONSTANTS

T_m:
T_b: 109.5°C (382.6 K)

T_c: 280.5°C (553.6 K)
P_c: 2.556 MPa

μ:
IP:

TRANSITION PROPERTIES

$\Delta_{fus}H$ (T_m):
$\Delta_{vap}H$ (T_b): 32.51 kJ/mol
$\Delta_{vap}H$ (25°C): 37.76 kJ/mol

Vapor pressure (0°C):
Vapor pressure (25°C): 4.05 kPa
Vapor pressure (100°C): 77.0 kPa

PROPERTIES AT 25°C AND 100 kPa

	Solid	Liquid	Gas		Liquid
$\Delta_f H°$/kJ mol^{-1}:		-257.0	-219.2	d:	0.6962 g/mL
$S°$/J mol^{-1}K^{-1}:				η:	
C_p/J mol^{-1}K^{-1}:				k:	

COMMENTS: Flammable

Name: 2,5-Dimethylhexane
Synonym: Biisobutyl

Mol. Form.: C_8H_{18}

CAS RN: 592-13-2
Merck No.:
Mol. Wt.: 114.231

PHYSICAL CONSTANTS

T_m: -91°C (182 K)
T_b: 109.12°C (382.27 K)

T_c: 277.0°C (550.1 K)
P_c: 2.487 MPa

μ:
IP:

TRANSITION PROPERTIES

$\Delta_{fus}H$ (T_m):
$\Delta_{vap}H$ (T_b): 32.54 kJ/mol
$\Delta_{vap}H$ (25°C): 37.85 kJ/mol

Vapor pressure (0°C):
Vapor pressure (25°C): 4.06 kPa
Vapor pressure (100°C): 77.7 kPa

PROPERTIES AT 25°C AND 100 kPa

	Solid	Liquid	Gas		Liquid
$\Delta_f H°$/kJ mol^{-1}:		-260.4	-222.5	d:	0.6893 g/mL
$S°$/J mol^{-1}K^{-1}:				η:	
C_p/J mol^{-1}K^{-1}:		249.2		k:	

COMMENTS: Flammable

Name: 3,3-Dimethylhexane

CAS RN: 563-16-6
Merck No.:
Mol. Wt.: 114.231

Mol. Form.: C_8H_{18}

PHYSICAL CONSTANTS

T_m: -126.1°C (147.0 K)	T_c: 289.0°C (562.1 K)	μ:
T_b: 111.97°C (385.12 K)	P_c: 2.654 MPa	IP:

TRANSITION PROPERTIES

$\Delta_{fus}H$ (T_m):   Vapor pressure (0°C):
$\Delta_{vap}H$ (T_b): 32.31 kJ/mol   Vapor pressure (25°C): 3.82 kPa
$\Delta_{vap}H$ (25°C): 37.53 kJ/mol   Vapor pressure (100°C): 71.8 kPa

PROPERTIES AT 25°C AND 100 kPa

	Solid	Liquid	Gas		Liquid
$\Delta_f H°$/kJ mol^{-1}:		-257.5	-220.0		d: 0.7060 g/mL
$S°$/J mol^{-1}K^{-1}:					η:
C_p/J mol^{-1}K^{-1}:		246.6			k:
COMMENTS: Flammable					

Name: 3,4-Dimethylhexane

CAS RN: 583-48-2
Merck No.:
Mol. Wt.: 114.231

Mol. Form.: C_8H_{18}

PHYSICAL CONSTANTS

T_m:	T_c: 295.8°C (568.9 K)	μ:
T_b: 117.73°C (390.88 K)	P_c: 2.692 MPa	IP:

TRANSITION PROPERTIES

$\Delta_{fus}H$ (T_m):   Vapor pressure (0°C):
$\Delta_{vap}H$ (T_b): 33.24 kJ/mol   Vapor pressure (25°C): 2.89 kPa
$\Delta_{vap}H$ (25°C): 38.97 kJ/mol   Vapor pressure (100°C): 60.3 kPa

PROPERTIES AT 25°C AND 100 kPa

	Solid	Liquid	Gas		Liquid
$\Delta_f H°$/kJ mol^{-1}:		-251.8	-212.8		d: 0.715 g/mL
$S°$/J mol^{-1}K^{-1}:					η:
C_p/J mol^{-1}K^{-1}:					k:
COMMENTS: Flammable					

Name: 1,1-Dimethylhydrazine
Synonym: Dimazine

CAS RN: 57-14-7
Merck No.: 3236
Mol. Wt.: 60.099

Mol. Form.: $C_2H_8N_2$

PHYSICAL CONSTANTS

T_m: -58°C (215 K)	T_c:	μ:
T_b: 63.9°C (337.0 K)	P_c:	IP: 7.28 eV

TRANSITION PROPERTIES

$\Delta_{fus}H$ (T_m):   Vapor pressure (0°C): 5.53 kPa
$\Delta_{vap}H$ (T_b): 32.55 kJ/mol   Vapor pressure (25°C):
$\Delta_{vap}H$ (25°C): 35.00 kJ/mol   Vapor pressure (100°C):

PROPERTIES AT 25°C AND 100 kPa

	Solid	Liquid	Gas		Liquid
$\Delta_f H°$/kJ mol^{-1}:		48.9	83.9		d: 0.785 g/mL
$S°$/J mol^{-1}K^{-1}:		198.0			η:
C_p/J mol^{-1}K^{-1}:		164.1			k:
COMMENTS: TLV=0.01 ppm; carcinogen; highly toxic; flammable					

Name: Dimethyl mercury
Synonym: Methyl mercury

Mol. Form.: C_2H_6Hg

CAS RN: 593-74-8
Merck No.: 3238
Mol. Wt.: 230.660

PHYSICAL CONSTANTS

T_m:	T_c:	μ: 0 D
T_b: 93°C (366 K)	P_c:	IP:

TRANSITION PROPERTIES

$\Delta_{fus}H$ (T_m):
$\Delta_{vap}H$ (T_b):
$\Delta_{vap}H$ (25°C): 34.6 kJ/mol

Vapor pressure (0°C):
Vapor pressure (25°C): 2.27 kPa
Vapor pressure (100°C): 8.30 kPa

PROPERTIES AT 25°C AND 100 kPa

	Solid	Liquid	Gas		Liquid
$\Delta_f H°$/kJ mol^{-1}:		59.8	94.4		d: 3.17 g/mL
$S°$/J mol^{-1}K^{-1}:		209.0	306.0		η:
C_p/J mol^{-1}K^{-1}:			83.3		k:
COMMENTS: Highly toxic					

Name: 2,2-Dimethylpentane

Mol. Form.: C_7H_{16}

CAS RN: 590-35-2
Merck No.:
Mol. Wt.: 100.204

PHYSICAL CONSTANTS

T_m: -123.8°C (149.3 K)	T_c: 247.4°C (520.5 K)	μ:
T_b: 79.2°C (352.3 K)	P_c: 2.773 MPa	IP:

TRANSITION PROPERTIES

$\Delta_{fus}H$ (T_m): 5.86 kJ/mol
$\Delta_{vap}H$ (T_b): 29.23 kJ/mol
$\Delta_{vap}H$ (25°C): 32.42 kJ/mol

Vapor pressure (0°C): 4.08 kPa
Vapor pressure (25°C): 14.0 kPa
Vapor pressure (100°C): 182 kPa

PROPERTIES AT 25°C AND 100 kPa

	Solid	Liquid	Gas		Liquid
$\Delta_f H°$/kJ mol^{-1}:		-238.3	-205.9		d: 0.6695 g/mL
$S°$/J mol^{-1}K^{-1}:		300.3			η:
C_p/J mol^{-1}K^{-1}:		221.1			k:
COMMENTS: Flammable					

Name: 2,3-Dimethylpentane

Mol. Form.: C_7H_{16}

CAS RN: 565-59-3
Merck No.:
Mol. Wt.: 100.204

PHYSICAL CONSTANTS

T_m:	T_c: 264.3°C (537.4 K)	μ:
T_b: 89.78°C (362.93 K)	P_c: 2.908 MPa	IP:

TRANSITION PROPERTIES

$\Delta_{fus}H$ (T_m):
$\Delta_{vap}H$ (T_b): 30.46 kJ/mol
$\Delta_{vap}H$ (25°C): 34.26 kJ/mol

Vapor pressure (0°C): 2.48 kPa
Vapor pressure (25°C): 9.18 kPa
Vapor pressure (100°C): 135 kPa

PROPERTIES AT 25°C AND 100 kPa

	Solid	Liquid	Gas		Liquid
$\Delta_f H°$/kJ mol^{-1}:		-233.1	-198.9		d: 0.6909 g/mL
$S°$/J mol^{-1}K^{-1}:					η: 0.42 mPa s
C_p/J mol^{-1}K^{-1}:		218.3			k:
COMMENTS: Flammable					

Name: 2,4-Dimethylpentane

CAS RN: 108-08-7
Merck No.:
Mol. Wt.: 100.204

Mol. Form.: C_7H_{16}

PHYSICAL CONSTANTS

T_m: -119.9°C (153.2 K) T_c: 246.7°C (519.8 K)   μ:
T_b: 80.49°C (353.64 K) P_c: 2.737 MPa   IP:

TRANSITION PROPERTIES

$\Delta_{fus}H$ (T_m): 6.69 kJ/mol   Vapor pressure (0°C): 3.71 kPa
$\Delta_{vap}H$ (T_b): 29.55 kJ/mol   Vapor pressure (25°C): 13.1 kPa
$\Delta_{vap}H$ (25°C): 32.88 kJ/mol   Vapor pressure (100°C): 175 kPa

PROPERTIES AT 25°C AND 100 kPa

	Solid	Liquid	Gas		Liquid
$\Delta_f H°$/kJ mol^{-1}:		-234.6	-201.7		d: 0.6683 g/mL
$S°$/J mol^{-1}K^{-1}:		303.2			η: 0.35 mPa s
C_p/J mol^{-1}K^{-1}:		224.2			k:

COMMENTS: Flammable

Name: 3,3-Dimethylpentane

CAS RN: 562-49-2
Merck No.:
Mol. Wt.: 100.204

Mol. Form.: C_7H_{16}

PHYSICAL CONSTANTS

T_m: -134.9°C (138.2 K) T_c: 263.3°C (536.4 K)   μ:
T_b: 86.06°C (359.21 K) P_c: 2.946 MPa   IP:

TRANSITION PROPERTIES

$\Delta_{fus}H$ (T_m): 7.07 kJ/mol   Vapor pressure (0°C): 3.13 kPa
$\Delta_{vap}H$ (T_b): 29.62 kJ/mol   Vapor pressure (25°C): 11.0 kPa
$\Delta_{vap}H$ (25°C): 33.03 kJ/mol   Vapor pressure (100°C): 149 kPa

PROPERTIES AT 25°C AND 100 kPa

	Solid	Liquid	Gas		Liquid
$\Delta_f H°$/kJ mol^{-1}:		-234.2	-201.2		d: 0.6891 g/mL
$S°$/J mol^{-1}K^{-1}:					η:
C_p/J mol^{-1}K^{-1}:					k:

COMMENTS: Flammable

Name: Dimethyl phthalate
Synonym: Dimethyl 1,2-benzenedicarboxylate

CAS RN: 131-11-3
Merck No.: 3243
Mol. Wt.: 194.187

Mol. Form.: $C_{10}H_{10}O_4$

PHYSICAL CONSTANTS

T_m: 5.5°C (278.6 K) T_c: μ:
T_b: 283.7°C (556.8 K) P_c: IP: 9.64 eV

TRANSITION PROPERTIES

$\Delta_{fus}H$ (T_m):   Vapor pressure (0°C):
$\Delta_{vap}H$ (T_b):   Vapor pressure (25°C):
$\Delta_{vap}H$ (25°C):   Vapor pressure (100°C):

PROPERTIES AT 25°C AND 100 kPa

	Solid	Liquid	Gas		Liquid
$\Delta_f H°$/kJ mol^{-1}:					d: 1.193 g/mL
$S°$/J mol^{-1}K^{-1}:					η: 14.4 mPa s
C_p/J mol^{-1}K^{-1}:					k: 0.1473 W/m K

COMMENTS:

Name: 2,2-Dimethyl-1-propanol
Synonyms: Neopentyl alcohol
      tert-Butylcarbinol
Mol. Form.: $C_5H_{12}O$

CAS RN: 75-84-3
Merck No.: 6373
Mol. Wt.: 88.150

PHYSICAL CONSTANTS

T_m: 52.5°C (325.6 K)	T_c:	μ:
T_b: 113.5°C (386.6 K)	P_c:	IP:

TRANSITION PROPERTIES

$\Delta_{fus}H$ (T_m):
$\Delta_{vap}H$ (T_b):
$\Delta_{vap}H$ (25°C):

Vapor pressure (0°C):
Vapor pressure (25°C):
Vapor pressure (100°C): 59.8 kPa

PROPERTIES AT 25°C AND 100 kPa

	Solid	Liquid	Gas		Solid
$\Delta_fH°$/kJ mol^{-1}:		-399.4			*d*: 0.808 g/mL
$S°$/J mol^{-1}K^{-1}:					η: N/A
C_p/J mol^{-1}K^{-1}:					*k*:
COMMENTS:					

Name: 2,6-Dimethylpyridine
Synonym: 2,6-Lutidine

Mol. Form.: C_7H_9N

CAS RN: 108-48-5
Merck No.: 5485
Mol. Wt.: 107.155

PHYSICAL CONSTANTS

T_m: -6.1°C (267.0 K)	T_c: 350.7°C (623.8 K)	μ:
T_b: 144.1°C (417.2 K)	P_c:	IP: 8.86 eV

TRANSITION PROPERTIES

$\Delta_{fus}H$ (T_m): 10.04 kJ/mol
$\Delta_{vap}H$ (T_b): 37.46 kJ/mol
$\Delta_{vap}H$ (25°C): 45.36 kJ/mol

Vapor pressure (0°C):
Vapor pressure (25°C): 0.746 kPa
Vapor pressure (100°C): 25.6 kPa

PROPERTIES AT 25°C AND 100 kPa

	Solid	Liquid	Gas		Liquid
$\Delta_fH°$/kJ mol^{-1}:		12.7	58.7		*d*: 0.9181 g/mL
$S°$/J mol^{-1}K^{-1}:		244.2			η: 0.82 mPa s
C_p/J mol^{-1}K^{-1}:		185.2			*k*:
COMMENTS:					

Name: Dimethyl sulfide
Synonym: 2-Thiapropane

Mol. Form.: C_2H_6S

CAS RN: 75-18-3
Merck No.: 6042
Mol. Wt.: 62.136

PHYSICAL CONSTANTS

T_m: -98.3°C (174.8 K)	T_c: 229.9°C (503.0 K)	μ: 1.554 D
T_b: 37.33°C (310.48 K)	P_c: 5.53 MPa	IP: 8.69 eV

TRANSITION PROPERTIES

$\Delta_{fus}H$ (T_m): 7.99 kJ/mol
$\Delta_{vap}H$ (T_b): 27.00 kJ/mol
$\Delta_{vap}H$ (25°C): 27.65 kJ/mol

Vapor pressure (0°C): 22.3 kPa
Vapor pressure (25°C): 64.4 kPa
Vapor pressure (100°C):

PROPERTIES AT 25°C AND 100 kPa

	Solid	Liquid	Gas		Liquid
$\Delta_fH°$/kJ mol^{-1}:		-65.4	-37.5		*d*: 0.8423 g/mL
$S°$/J mol^{-1}K^{-1}:		196.4	286.0		η: 0.284 mPa s
C_p/J mol^{-1}K^{-1}:		118.1	74.1		*k*:
COMMENTS:					

Name: Dimethyl sulfone
Synonym: Methyl sulfone

CAS RN: 67-71-0
Merck No.: 3246
Mol. Wt.: 94.134

Mol. Form.: $C_2H_6O_2S$

PHYSICAL CONSTANTS

T_m: 109°C (382 K) T_c: μ:
T_b: 238°C (511 K) P_c: IP:

TRANSITION PROPERTIES

$\Delta_{fus}H$ (T_m):   Vapor pressure (0°C):
$\Delta_{vap}H$ (T_b):   Vapor pressure (25°C):
$\Delta_{vap}H$ (25°C):   Vapor pressure (100°C):

PROPERTIES AT 25°C AND 100 kPa

	Solid	Liquid	Gas		Solid
$\Delta_fH°$/kJ mol^{-1}:	-451.0		-371.1	d:	
$S°$/J mol^{-1}K^{-1}:	142.0		310.6	η: N/A	
C_p/J mol^{-1}K^{-1}:			100.0	k:	
COMMENTS:					

Name: Dimethyl sulfoxide
Synonyms: Sulfinylbismethane
         DMSO

CAS RN: 67-68-5
Merck No.: 3247
Mol. Wt.: 78.135

Mol. Form.: C_2H_6OS

PHYSICAL CONSTANTS

T_m: 18.52°C (291.67 K) T_c: μ: 3.96 D
T_b: 189°C (462 K) P_c: IP: 9.01 eV

TRANSITION PROPERTIES

$\Delta_{fus}H$ (T_m): 14.37 kJ/mol   Vapor pressure (0°C):
$\Delta_{vap}H$ (T_b): 43.14 kJ/mol   Vapor pressure (25°C): 0.084 kPa
$\Delta_{vap}H$ (25°C): 52.88 kJ/mol   Vapor pressure (100°C): 5.27 kPa

PROPERTIES AT 25°C AND 100 kPa

	Solid	Liquid	Gas		Liquid
$\Delta_fH°$/kJ mol^{-1}:		-204.2	-151.3	d:	1.0955 g/mL
$S°$/J mol^{-1}K^{-1}:		188.3		η:	1.99 mPa s
C_p/J mol^{-1}K^{-1}:		153.0		k:	
COMMENTS:					

Name: Dimethyl terephthalate
Synonym: Dimethyl 1,4-benzenedicarboxylate

CAS RN: 120-61-6
Merck No.:
Mol. Wt.: 194.187

Mol. Form.: $C_{10}H_{10}O_4$

PHYSICAL CONSTANTS

T_m: 141°C (414 K) T_c: μ:
T_b: 288°C (561 K) P_c: IP:

TRANSITION PROPERTIES

$\Delta_{fus}H$ (T_m):   Vapor pressure (0°C):
$\Delta_{vap}H$ (T_b):   Vapor pressure (25°C):
$\Delta_{vap}H$ (25°C):   Vapor pressure (100°C):

PROPERTIES AT 25°C AND 100 kPa

	Solid	Liquid	Gas		Solid
$\Delta_fH°$/kJ mol^{-1}:	-732.6			d:	
$S°$/J mol^{-1}K^{-1}:				η: N/A	
C_p/J mol^{-1}K^{-1}:	261.1			k:	
COMMENTS:					

Name: 1,3-Dioxane
Synonym: 1,3-Dioxacyclohexane

Mol. Form.: $C_4H_8O_2$

CAS RN: 505-22-6
Merck No.:
Mol. Wt.: 88.106

PHYSICAL CONSTANTS

T_m: -45°C (228 K)
T_b: 106.1°C (379.2 K)

T_c:
P_c:

μ: 2.06 D
IP: 9.80 eV

TRANSITION PROPERTIES

$\Delta_{fus}H$ (T_m):
$\Delta_{vap}H$ (T_b): 34.37 kJ/mol
$\Delta_{vap}H$ (25°C): 39.09 kJ/mol

Vapor pressure (0°C): 1.16 kPa
Vapor pressure (25°C):
Vapor pressure (100°C):

PROPERTIES AT 25°C AND 100 kPa

	Solid	Liquid	Gas		Liquid
$\Delta_f H°$/kJ mol^{-1}:		-379.7	-342.3		d: 1.029 g/mL
$S°$/J mol^{-1}K^{-1}:					η:
C_p/J mol^{-1}K^{-1}:		143.9			k:

COMMENTS:

Name: 1,4-Dioxane
Synonym: 1,4-Dioxacyclohexane

Mol. Form.: $C_4H_8O_2$

CAS RN: 123-91-1
Merck No.: 3294
Mol. Wt.: 88.106

PHYSICAL CONSTANTS

T_m: 11.8°C (284.9 K)
T_b: 101.5°C (374.6 K)

T_c: 314°C (587 K)
P_c: 5.21 MPa

μ: 0 D
IP: 9.19 eV

TRANSITION PROPERTIES

$\Delta_{fus}H$ (T_m): 12.85 kJ/mol
$\Delta_{vap}H$ (T_b): 34.16 kJ/mol
$\Delta_{vap}H$ (25°C): 38.60 kJ/mol

Vapor pressure (0°C):
Vapor pressure (25°C): 4.95 kPa
Vapor pressure (100°C): 97.2 kPa

PROPERTIES AT 25°C AND 100 kPa

	Solid	Liquid	Gas		Liquid
$\Delta_f H°$/kJ mol^{-1}:		-353.9	-315.8		d: 1.0286 g/mL
$S°$/J mol^{-1}K^{-1}:		270.2			η: 1.18 mPa s
C_p/J mol^{-1}K^{-1}:		152.1			k: 0.159 W/m K

COMMENTS: TLV=25 ppm; carcinogen; flammable

Name: 1,3-Dioxolane
Synonym: 1,3-Dioxacyclopentane

Mol. Form.: $C_3H_6O_2$

CAS RN: 646-06-0
Merck No.:
Mol. Wt.: 74.079

PHYSICAL CONSTANTS

T_m: -95°C (178 K)
T_b: 78°C (351 K)

T_c:
P_c:

μ: 1.19 D
IP: 9.90 eV

TRANSITION PROPERTIES

$\Delta_{fus}H$ (T_m): 27.48 kJ/mol
$\Delta_{vap}H$ (T_b):
$\Delta_{vap}H$ (25°C): 35.60 kJ/mol

Vapor pressure (0°C): 3.51 kPa
Vapor pressure (25°C): 13.5 kPa
Vapor pressure (100°C):

PROPERTIES AT 25°C AND 100 kPa

	Solid	Liquid	Gas		Liquid
$\Delta_f H°$/kJ mol^{-1}:		-333.5	-298.0		d: 1.055 g/mL
$S°$/J mol^{-1}K^{-1}:					η:
C_p/J mol^{-1}K^{-1}:		118.0			k:

COMMENTS: Flammable

Name: Diphenylacetylene

Synonym: Diphenylethyne

CAS RN: 501-65-5
Merck No.: 9428
Mol. Wt.: 178.233

Mol. Form.: $C_{14}H_{10}$

PHYSICAL CONSTANTS

T_m: 62.5°C (335.6 K) T_c: μ:

T_b: 300°C (573 K) P_c: IP: 7.90 eV

TRANSITION PROPERTIES

$\Delta_{fus}H$ (T_m): Vapor pressure (0°C):

$\Delta_{vap}H$ (T_b): Vapor pressure (25°C):

$\Delta_{vap}H$ (25°C): Vapor pressure (100°C):

PROPERTIES AT 25°C AND 100 kPa

	Solid	Liquid	Gas		Solid
$\Delta_f H°$/kJ mol^{-1}:	312.4			d:	
$S°$/J mol^{-1}K^{-1}:				η: N/A	
C_p/J mol^{-1}K^{-1}:	225.9			k:	

COMMENTS:

Name: Diphenylamine

Synonym: *N*-Phenylaniline

CAS RN: 122-39-4
Merck No.: 3317
Mol. Wt.: 169.226

Mol. Form.: $C_{12}H_{11}N$

PHYSICAL CONSTANTS

T_m: 52.98°C (326.13 K) T_c: μ:

T_b: 302°C (575 K) P_c: IP: 7.16 eV

TRANSITION PROPERTIES

$\Delta_{fus}H$ (T_m): 17.86 kJ/mol Vapor pressure (0°C):

$\Delta_{vap}H$ (T_b): Vapor pressure (25°C):

$\Delta_{vap}H$ (25°C): Vapor pressure (100°C):

PROPERTIES AT 25°C AND 100 kPa

	Solid	Liquid	Gas		Solid
$\Delta_f H°$/kJ mol^{-1}:	130.2		219.3	d:	1.16 g/mL
$S°$/J mol^{-1}K^{-1}:				η: N/A	
C_p/J mol^{-1}K^{-1}:				k:	

COMMENTS: Highly toxic

Name: 1,1-Diphenylethane

Synonym: 1,1'-Ethylidenebisbenzene

CAS RN: 612-00-0
Merck No.:
Mol. Wt.: 182.265

Mol. Form.: $C_{14}H_{14}$

PHYSICAL CONSTANTS

T_m: -17.9°C (255.2 K) T_c: μ:

T_b: 272.6°C (545.7 K) P_c: IP:

TRANSITION PROPERTIES

$\Delta_{fus}H$ (T_m): Vapor pressure (0°C):

$\Delta_{vap}H$ (T_b): Vapor pressure (25°C):

$\Delta_{vap}H$ (25°C): Vapor pressure (100°C):

PROPERTIES AT 25°C AND 100 kPa

	Solid	Liquid	Gas		Liquid
$\Delta_f H°$/kJ mol^{-1}:		48.7		d:	0.995 g/mL
$S°$/J mol^{-1}K^{-1}:				η:	
C_p/J mol^{-1}K^{-1}:				k:	

COMMENTS:

Name: 1,2-Diphenylethane
Synonyms: Dibenzyl
         Dihydrostilbene
Mol. Form.: $C_{14}H_{14}$

CAS RN: 103-29-7
Merck No.: 1219
Mol. Wt.: 182.265

PHYSICAL CONSTANTS

T_m: 52.5°C (325.6 K)	T_c:	μ:
T_b: 284°C (557 K)	P_c:	IP: 8.70 eV

TRANSITION PROPERTIES

$\Delta_{fus}H$ (T_m):
$\Delta_{vap}H$ (T_b):
$\Delta_{vap}H$ (25°C):

Vapor pressure (0°C):
Vapor pressure (25°C):
Vapor pressure (100°C):

PROPERTIES AT 25°C AND 100 kPa

	Solid	Liquid	Gas		Solid
$\Delta_f H°$/kJ mol^{-1}:	51.5		142.9		d: 0.9780 g/mL
$S°$/J mol^{-1}K^{-1}:					η: N/A
C_p/J mol^{-1}K^{-1}:					k:
COMMENTS:					

Name: Diphenyl ether
Synonyms: Oxybisbenzene
         Phenyl ether
Mol. Form.: $C_{12}H_{10}O$

CAS RN: 101-84-8
Merck No.: 7259
Mol. Wt.: 170.211

PHYSICAL CONSTANTS

T_m: 26.87°C (300.02 K)	T_c: 493.7°C (766.8 K)	μ: 1.3 D
T_b: 258.05°C (531.20 K)	P_c:	IP: 8.09 eV

TRANSITION PROPERTIES

$\Delta_{fus}H$ (T_m): 17.22 kJ/mol
$\Delta_{vap}H$ (T_b): 48.20 kJ/mol
$\Delta_{vap}H$ (25°C): 66.90 kJ/mol

Vapor pressure (0°C):
Vapor pressure (25°C): 0.003 kPa
Vapor pressure (100°C):

PROPERTIES AT 25°C AND 100 kPa

	Solid	Liquid	Gas		Solid
$\Delta_f H°$/kJ mol^{-1}:	-32.1		52.0		d:
$S°$/J mol^{-1}K^{-1}:	233.9				η: N/A
C_p/J mol^{-1}K^{-1}:	216.6				k:
COMMENTS:					

Name: Diphenylmethane
Synonym: Benzylbenzene

Mol. Form.: $C_{13}H_{12}$

CAS RN: 101-81-5
Merck No.: 3329
Mol. Wt.: 168.238

PHYSICAL CONSTANTS

T_m: 25.24°C (298.39 K)	T_c: 497°C (770 K)	μ:
T_b: 265.05°C (538.20 K)	P_c: 2.86 MPa	IP: 8.55 eV

TRANSITION PROPERTIES

$\Delta_{fus}H$ (T_m): 18.2 kJ/mol
$\Delta_{vap}H$ (T_b):
$\Delta_{vap}H$ (25°C): 49.3 kJ/mol

Vapor pressure (0°C):
Vapor pressure (25°C):
Vapor pressure (100°C):

PROPERTIES AT 25°C AND 100 kPa

	Solid	Liquid	Gas		Liquid
$\Delta_f H°$/kJ mol^{-1}:	71.5	89.7	139.0		d: 1.001 g/mL
$S°$/J mol^{-1}K^{-1}:	239.3				η:
C_p/J mol^{-1}K^{-1}:					k:
COMMENTS:					

Name: Dipropylamine
Synonym: *N*-Propyl-1-propanamine

Mol. Form.: $C_6H_{15}N$

CAS RN: 142-84-7
Merck No.: 3350
Mol. Wt.: 101.192

PHYSICAL CONSTANTS

T_m: -63°C (210 K)
T_b: 109.3°C (382.4 K)

T_c: 282.7°C (555.8 K)
P_c: 3.63 MPa

μ:
IP: 7.84 eV

TRANSITION PROPERTIES

$\Delta_{fus}H$ (T_m):
$\Delta_{vap}H$ (T_b): 33.47 kJ/mol
$\Delta_{vap}H$ (25°C): 40.04 kJ/mol

Vapor pressure (0°C):
Vapor pressure (25°C): 3.21 kPa
Vapor pressure (100°C): 75.6 kPa

PROPERTIES AT 25°C AND 100 kPa

	Solid	Liquid	Gas		Liquid
$\Delta_f H°$/kJ mol^{-1}:		-156.1	-116.1		d: 0.7329 g/mL
$S°$/J mol^{-1}K^{-1}:					η: 0.517 mPa s
C_p/J mol^{-1}K^{-1}:					k:
COMMENTS: Flammable					

Name: Dipropyl ether
Synonyms: 1,1'-Oxybispropane
Propyl ether
Mol. Form.: $C_6H_{14}O$

CAS RN: 111-43-3
Merck No.: 7870
Mol. Wt.: 102.177

PHYSICAL CONSTANTS

T_m: -126.1°C (147.0 K)
T_b: 90.08°C (363.23 K)

T_c: 257.5°C (530.6 K)
P_c: 3.028 MPa

μ: 1.21 D
IP: 9.27 eV

TRANSITION PROPERTIES

$\Delta_{fus}H$ (T_m): 8.83 kJ/mol
$\Delta_{vap}H$ (T_b): 31.31 kJ/mol
$\Delta_{vap}H$ (25°C): 35.69 kJ/mol

Vapor pressure (0°C):
Vapor pressure (25°C): 8.35 kPa
Vapor pressure (100°C): 135 kPa

PROPERTIES AT 25°C AND 100 kPa

	Solid	Liquid	Gas		Liquid
$\Delta_f H°$/kJ mol^{-1}:		-328.8	-292.9		d: 0.7419 g/mL
$S°$/J mol^{-1}K^{-1}:		323.9			η: 0.396 mPa s
C_p/J mol^{-1}K^{-1}:		221.6			k:
COMMENTS: Flammable					

Name: Disilane

Mol. Form.: H_6Si_2

CAS RN: 1590-87-0
Merck No.: 3362
Mol. Wt.: 62.219

PHYSICAL CONSTANTS

T_m: -132.5°C (140.8 K)
T_b: -14.3°C (258.8 K)

T_c:
P_c:

μ: 0 D
IP: 9.70 eV

TRANSITION PROPERTIES

$\Delta_{fus}H$ (T_m):
$\Delta_{vap}H$ (T_b): 21.20 kJ/mol
$\Delta_{vap}H$ (25°C):

Vapor pressure (0°C):
Vapor pressure (25°C):
Vapor pressure (100°C):

PROPERTIES AT 25°C AND 100 kPa

	Solid	Liquid	Gas		Gas
$\Delta_f H°$/kJ mol^{-1}:			80.3		d: 2.543 g/L
$S°$/J mol^{-1}K^{-1}:			272.7		η:
C_p/J mol^{-1}K^{-1}:			80.8		k:
COMMENTS: Flammable					

Name: Divinyl ether
Synonyms: 1,1′-Oxybisethene
        Vinyl ether
Mol. Form.: C_4H_6O

CAS RN: 109-93-3
Merck No.: 9899
Mol. Wt.: 70.091

PHYSICAL CONSTANTS

T_m: -101°C (172 K)	T_c:	μ: 0.78 D
T_b: 28.3°C (301.4 K)	P_c:	IP: 8.70 eV

TRANSITION PROPERTIES

$\Delta_{fus}H$ (T_m):	Vapor pressure (0°C): 31.2 kPa
$\Delta_{vap}H$ (T_b):	Vapor pressure (25°C): 89.3 kPa
$\Delta_{vap}H$ (25°C): 26.2 kJ/mol	Vapor pressure (100°C):

PROPERTIES AT 25°C AND 100 kPa

	Solid	Liquid	Gas		Liquid
$\Delta_f H°$/kJ mol^{-1}:		-39.8	-13.6	d:	0.767 g/mL
$S°$/J mol^{-1}K^{-1}:				η:	
C_p/J mol^{-1}K^{-1}:				k:	
COMMENTS: Flammable					

Name: Dodecane

CAS RN: 112-40-3
Merck No.:
Mol. Wt.: 170.338

Mol. Form.: $C_{12}H_{26}$

PHYSICAL CONSTANTS

T_m: -9.6°C (263.5 K)	T_c: 385.50°C (658.65 K)	μ:
T_b: 216.32°C (489.47 K)	P_c: 1.830 MPa	IP:

TRANSITION PROPERTIES

$\Delta_{fus}H$ (T_m): 36.58 kJ/mol	Vapor pressure (0°C):
$\Delta_{vap}H$ (T_b): 44.5 kJ/mol	Vapor pressure (25°C): 0.016 kPa
$\Delta_{vap}H$ (25°C): 61.51 kJ/mol	Vapor pressure (100°C): 2.01 kPa

PROPERTIES AT 25°C AND 100 kPa

	Solid	Liquid	Gas		Liquid
$\Delta_f H°$/kJ mol^{-1}:		-350.9	-289.7	d:	0.7452 g/mL
$S°$/J mol^{-1}K^{-1}:				η:	1.38 mPa s
C_p/J mol^{-1}K^{-1}:		375.8		k:	0.152 W/m K
COMMENTS:					

Name: Dodecanoic acid
Synonyms: Lauric acid
        Dodecylic acid
Mol. Form.: $C_{12}H_{24}O_2$

CAS RN: 143-07-7
Merck No.: 5254
Mol. Wt.: 200.321

PHYSICAL CONSTANTS

T_m: 43.22°C (316.37 K)	T_c:	μ:
T_b: 91.4°C (364.5 K)	P_c:	IP:

TRANSITION PROPERTIES

$\Delta_{fus}H$ (T_m): 36.64 kJ/mol	Vapor pressure (0°C):
$\Delta_{vap}H$ (T_b):	Vapor pressure (25°C):
$\Delta_{vap}H$ (25°C): 95.9 kJ/mol	Vapor pressure (100°C):

PROPERTIES AT 25°C AND 100 kPa

	Solid	Liquid	Gas		Solid
$\Delta_f H°$/kJ mol^{-1}:	-774.6	-737.9	-642.0	d:	1.032 g/mL
$S°$/J mol^{-1}K^{-1}:				η:	N/A
C_p/J mol^{-1}K^{-1}:	404.3			k:	
COMMENTS:					

Name: 1-Dodecanol
Synonyms: Lauryl alcohol
          Dodecyl alcohol
Mol. Form.: $C_{12}H_{26}O$

CAS RN: 112-53-8
Merck No.: 3402
Mol. Wt.: 186.338

PHYSICAL CONSTANTS

T_m: 24°C (297 K) T_c: 447°C (720 K) μ:
T_b: 259°C (532 K) P_c: 2.08 MPa   IP:

TRANSITION PROPERTIES

$\Delta_{fus}H$ (T_m):   Vapor pressure (0°C):
$\Delta_{vap}H$ (T_b):   Vapor pressure (25°C):
$\Delta_{vap}H$ (25°C): 91.96 kJ/mol   Vapor pressure (100°C): 0.129 kPa

PROPERTIES AT 25°C AND 100 kPa

	Solid	Liquid	Gas		Liquid
$\Delta_f H°$/kJ mol^{-1}:		-528.5	-436.6		d: 0.8308 g/mL
$S°$/J mol^{-1}K^{-1}:					η:
C_p/J mol^{-1}K^{-1}:		438.1			k: 0.146 W/m K
COMMENTS:					

Name: 1-Dodecene

Mol. Form.: $C_{12}H_{24}$

CAS RN: 112-41-4
Merck No.:
Mol. Wt.: 168.323

PHYSICAL CONSTANTS

T_m: -35.2°C (237.9 K) T_c: 384.5°C (657.6 K) μ:
T_b: 213.8°C (486.9 K) P_c: 1.930 MPa   IP:

TRANSITION PROPERTIES

$\Delta_{fus}H$ (T_m): 17.42 kJ/mol   Vapor pressure (0°C):
$\Delta_{vap}H$ (T_b): 43.97 kJ/mol   Vapor pressure (25°C): 0.019 kPa
$\Delta_{vap}H$ (25°C): 60.78 kJ/mol   Vapor pressure (100°C): 2.28 kPa

PROPERTIES AT 25°C AND 100 kPa

	Solid	Liquid	Gas		Liquid
$\Delta_f H°$/kJ mol^{-1}:		-226.2	-165.4		d: 0.755 g/mL
$S°$/J mol^{-1}K^{-1}:		484.8			η: 1.20 mPa s
C_p/J mol^{-1}K^{-1}:		360.7			k:
COMMENTS:					

Name: Eicosane
Synonym: Icosane

Mol. Form.: $C_{20}H_{42}$

CAS RN: 112-95-8
Merck No.:
Mol. Wt.: 282.554

PHYSICAL CONSTANTS

T_m: 36.8°C (309.9 K) T_c: 496°C (769 K) μ:
T_b: 343°C (616 K) P_c: 1.16 MPa   IP:

TRANSITION PROPERTIES

$\Delta_{fus}H$ (T_m): 69.88 kJ/mol   Vapor pressure (0°C):
$\Delta_{vap}H$ (T_b): 59 kJ/mol   Vapor pressure (25°C):
$\Delta_{vap}H$ (25°C):   Vapor pressure (100°C):

PROPERTIES AT 25°C AND 100 kPa

	Solid	Liquid	Gas		Solid
$\Delta_f H°$/kJ mol^{-1}:					d: 0.7823 g/mL
$S°$/J mol^{-1}K^{-1}:					η: N/A
C_p/J mol^{-1}K^{-1}:					k:
COMMENTS:					

Name: 1-Eicosanol   CAS RN: 629-96-9
Synonyms: Arachic alcohol   Merck No.:
          1-Icosanol   Mol. Wt.: 298.553
Mol. Form.: $C_{20}H_{42}O$

PHYSICAL CONSTANTS

T_m: 66.1°C (339.2 K) T_c: 536°C (809 K)   μ:
T_b: P_c: 1.30 MPa   IP:

TRANSITION PROPERTIES

$\Delta_{fus}H$ (T_m):   Vapor pressure (0°C):
$\Delta_{vap}H$ (T_b):   Vapor pressure (25°C):
$\Delta_{vap}H$ (25°C):   Vapor pressure (100°C):

PROPERTIES AT 25°C AND 100 kPa

	Solid	Liquid	Gas		Solid
$\Delta_f H°$/kJ mol^{-1}:				d:	
$S°$/J mol^{-1}K^{-1}:				η: N/A	
C_p/J mol^{-1}K^{-1}:				k:	
COMMENTS:					

Name: Epichlorohydrin   CAS RN: 106-89-8
Synonyms: (Chloromethyl)oxirane   Merck No.: 3563
        Chloropropylene oxide   Mol. Wt.: 92.525
Mol. Form.: C_3H_5ClO

PHYSICAL CONSTANTS

T_m: -57.2°C (215.9 K) T_c: μ:
T_b: 116.11°C (389.26 K) P_c: IP: 10.20 eV

TRANSITION PROPERTIES

$\Delta_{fus}H$ (T_m):   Vapor pressure (0°C):
$\Delta_{vap}H$ (T_b): 37.91 kJ/mol   Vapor pressure (25°C): 2.20 kPa
$\Delta_{vap}H$ (25°C): 40.6 kJ/mol   Vapor pressure (100°C): 60.8 kPa

PROPERTIES AT 25°C AND 100 kPa

	Solid	Liquid	Gas		Liquid
$\Delta_f H°$/kJ mol^{-1}:		-148.4	-107.8	d: 1.1746 g/mL	
$S°$/J mol^{-1}K^{-1}:				η: 1.07 mPa s	
C_p/J mol^{-1}K^{-1}:		131.6		k: 0.131 W/m K	
COMMENTS: TLV=0.1 ppm; carcinogen; highly toxic					

Name: 1,2-Epoxybutane   CAS RN: 106-88-7
Synonyms: Ethyloxirane   Merck No.:
        1,2-Butylene oxide   Mol. Wt.: 72.107
Mol. Form.: C_4H_8O

PHYSICAL CONSTANTS

T_m: -150°C (123 K) T_c: μ:
T_b: 63.3°C (336.4 K) P_c: IP: 10.15 eV

TRANSITION PROPERTIES

$\Delta_{fus}H$ (T_m):   Vapor pressure (0°C): 12.5 kPa
$\Delta_{vap}H$ (T_b): 30.30 kJ/mol   Vapor pressure (25°C): 31.7 kPa
$\Delta_{vap}H$ (25°C):   Vapor pressure (100°C):

PROPERTIES AT 25°C AND 100 kPa

	Solid	Liquid	Gas		Liquid
$\Delta_f H°$/kJ mol^{-1}:		-168.9		d: 0.824 g/mL	
$S°$/J mol^{-1}K^{-1}:		230.9		η: 0.40 mPa s	
C_p/J mol^{-1}K^{-1}:		147.0		k:	
COMMENTS: Highly toxic; flammable					

Name: Ethane

CAS RN: 74-84-0
Merck No.: 3676
Mol. Wt.: 30.070

Mol. Form.: C_2H_6

PHYSICAL CONSTANTS

T_m: -182.8°C (90.3 K)	T_c: 32.3°C (305.4 K)	μ: 0 D
T_b: -88.6°C (184.5 K)	P_c: 4.884 MPa	IP: 11.52 eV

TRANSITION PROPERTIES

$\Delta_{fus}H$ (T_m): 2.86 kJ/mol

Vapor pressure (0°C): 2390 kPa

$\Delta_{vap}H$ (T_b): 14.69 kJ/mol

Vapor pressure (25°C): 4163 kPa

$\Delta_{vap}H$ (25°C): 5.16 kJ/mol

Vapor pressure (100°C): N/A

PROPERTIES AT 25°C AND 100 kPa

	Solid	Liquid	Gas		Gas
$\Delta_fH°$/kJ mol^{-1}:			-83.8	d:	1.229 g/L
$S°$/J mol^{-1}K^{-1}:			229.6	η:	9.4 µPa s
C_p/J mol^{-1}K^{-1}:			52.6	k:	0.0213 W/m K

COMMENTS: Very flammable

Name: 1,2-Ethanediamine
Synonym: Ethylenediamine

CAS RN: 107-15-3
Merck No.: 3752
Mol. Wt.: 60.099

Mol. Form.: $C_2H_8N_2$

PHYSICAL CONSTANTS

T_m: 11.1°C (284.2 K)	T_c:	μ: 1.99 D
T_b: 117°C (390 K)	P_c:	IP: 8.60 eV

TRANSITION PROPERTIES

$\Delta_{fus}H$ (T_m): 22.58 kJ/mol

Vapor pressure (0°C):

$\Delta_{vap}H$ (T_b): 37.98 kJ/mol

Vapor pressure (25°C): 1.62 kPa

$\Delta_{vap}H$ (25°C): 44.98 kJ/mol

Vapor pressure (100°C): 57.4 kPa

PROPERTIES AT 25°C AND 100 kPa

	Solid	Liquid	Gas		Liquid
$\Delta_fH°$/kJ mol^{-1}:		-63.0	-17.6	d:	0.8931 g/mL
$S°$/J mol^{-1}K^{-1}:				η:	1.54 mPa s
C_p/J mol^{-1}K^{-1}:		172.6		k:	

COMMENTS: TLV=10 ppm; highly toxic

Name: Ethanethiol
Synonyms: Ethyl mercaptan
         Thioethanol
Mol. Form.: C_2H_6S

CAS RN: 75-08-1
Merck No.: 3680
Mol. Wt.: 62.136

PHYSICAL CONSTANTS

T_m: -147.89°C (125.26 K)	T_c: 226°C (499 K)	μ:
T_b: 35.1°C (308.2 K)	P_c: 5.49 MPa	IP: 9.29 eV

TRANSITION PROPERTIES

$\Delta_{fus}H$ (T_m): 4.98 kJ/mol

Vapor pressure (0°C): 24.5 kPa

$\Delta_{vap}H$ (T_b): 26.79 kJ/mol

Vapor pressure (25°C): 70.3 kPa

$\Delta_{vap}H$ (25°C): 27.30 kJ/mol

Vapor pressure (100°C):

PROPERTIES AT 25°C AND 100 kPa

	Solid	Liquid	Gas		Liquid
$\Delta_fH°$/kJ mol^{-1}:		-73.6	-45.3	d:	0.8315 g/mL
$S°$/J mol^{-1}K^{-1}:		207.0	296.2	η:	0.287 mPa s
C_p/J mol^{-1}K^{-1}:		117.9	72.7	k:	

COMMENTS: TLV=0.5 ppm; very flammable

Name: Ethanol CAS RN: 64-17-5
Synonym: Ethyl alcohol Merck No.: 3716
  Mol. Wt.: 46.069

Mol. Form.: C_2H_6O

PHYSICAL CONSTANTS

T_m: -114.1°C (159.0 K) T_c: 240.77°C (513.92 K) μ: 1.69 D
T_b: 78.29°C (351.44 K) P_c: 6.132 MPa IP: 10.47 eV

TRANSITION PROPERTIES

$\Delta_{fus}H$ (T_m): 5.02 kJ/mol Vapor pressure (0°C): 1.50 kPa
$\Delta_{vap}H$ (T_b): 38.56 kJ/mol Vapor pressure (25°C): 7.87 kPa
$\Delta_{vap}H$ (25°C): 42.32 kJ/mol Vapor pressure (100°C): 224 kPa

PROPERTIES AT 25°C AND 100 kPa

	Solid	Liquid	Gas	Liquid
$\Delta_fH°$/kJ mol^{-1}:		-277.7	-235.1	d: 0.7849 g/mL
$S°$/J mol^{-1}K^{-1}:		160.7	282.7	η: 1.07 mPa s
C_p/J mol^{-1}K^{-1}:		112.3	65.4	k: 0.169 W/m K

COMMENTS: TLV=1000 ppm; flammable

Name: Ethanolamine CAS RN: 141-43-5
Synonyms: 2-Aminoethanol Merck No.: 3681
  Glycinol Mol. Wt.: 61.084
Mol. Form.: C_2H_7NO

PHYSICAL CONSTANTS

T_m: 10.5°C (283.6 K) T_c: μ:
T_b: 171°C (444 K) P_c: IP: 8.96 eV

TRANSITION PROPERTIES

$\Delta_{fus}H$ (T_m): 20.50 kJ/mol Vapor pressure (0°C):
$\Delta_{vap}H$ (T_b): 49.83 kJ/mol Vapor pressure (25°C): 0.050 kPa
$\Delta_{vap}H$ (25°C): Vapor pressure (100°C): 6.56 kPa

PROPERTIES AT 25°C AND 100 kPa

	Solid	Liquid	Gas	Liquid
$\Delta_fH°$/kJ mol^{-1}:				d: 1.0136 g/mL
$S°$/J mol^{-1}K^{-1}:				η: 21.1 mPa s
C_p/J mol^{-1}K^{-1}:		195.5		k: 0.299 W/m K

COMMENTS: TLV=3 ppm

Name: 2-Ethoxyethanol CAS RN: 110-80-5
Synonym: Ethylene glycol monoethyl ether Merck No.: 3707
  Mol. Wt.: 90.122

Mol. Form.: $C_4H_{10}O_2$

PHYSICAL CONSTANTS

T_m: -70°C (203 K) T_c: μ:
T_b: 135°C (408 K) P_c: IP: 9.60 eV

TRANSITION PROPERTIES

$\Delta_{fus}H$ (T_m): Vapor pressure (0°C):
$\Delta_{vap}H$ (T_b): 39.22 kJ/mol Vapor pressure (25°C): 0.710 kPa
$\Delta_{vap}H$ (25°C): 48.21 kJ/mol Vapor pressure (100°C): 29.8 kPa

PROPERTIES AT 25°C AND 100 kPa

	Solid	Liquid	Gas	Liquid
$\Delta_fH°$/kJ mol^{-1}:				d: 0.9247 g/mL
$S°$/J mol^{-1}K^{-1}:				η: 1.85 mPa s
C_p/J mol^{-1}K^{-1}:		210.8		k:

COMMENTS: TLV=5 ppm; highly toxic

Name: 2-(2-Ethoxyethoxy)ethanol
Synonyms: Diethylene glycol monoethyl ether
        Carbitol
Mol. Form.: $C_6H_{14}O_3$

CAS RN: 111-90-0
Merck No.: 1806
Mol. Wt.: 134.175

PHYSICAL CONSTANTS

T_m: -76°C (197 K)	T_c:	μ:
T_b: 196°C (469 K)	P_c:	IP:

TRANSITION PROPERTIES

$\Delta_{fus}H\ (T_m)$:
$\Delta_{vap}H\ (T_b)$: 47.45 kJ/mol
$\Delta_{vap}H$ (25°C):

Vapor pressure (0°C):
Vapor pressure (25°C): 0.017 kPa
Vapor pressure (100°C): 2.55 kPa

PROPERTIES AT 25°C AND 100 kPa

	Solid	Liquid	Gas	Liquid
$\Delta_f H°$/kJ mol^{-1}:				d: 0.9841 g/mL
$S°$/J mol^{-1}K^{-1}:				η: 3.85 mPa s
C_p/J mol^{-1}K^{-1}:		301.0		k:
COMMENTS:				

Name: 2-Ethoxyethyl acetate
Synonyms: Ethylene glycol ethyl ether acetate
        3-Oxapentyl ethanoate
Mol. Form.: $C_6H_{12}O_3$

CAS RN: 111-15-9
Merck No.: 3708
Mol. Wt.: 132.160

PHYSICAL CONSTANTS

T_m: -61.7°C (211.4 K)	T_c: 334.2°C (607.3 K)	μ:
T_b: 156.4°C (429.5 K)	P_c: 3.166 MPa	IP:

TRANSITION PROPERTIES

$\Delta_{fus}H\ (T_m)$:
$\Delta_{vap}H\ (T_b)$:
$\Delta_{vap}H$ (25°C): 52.69 kJ/mol

Vapor pressure (0°C):
Vapor pressure (25°C): 0.240 kPa
Vapor pressure (100°C): 16.4 kPa

PROPERTIES AT 25°C AND 100 kPa

	Solid	Liquid	Gas	Liquid
$\Delta_f H°$/kJ mol^{-1}:				d: 0.9730 g/mL
$S°$/J mol^{-1}K^{-1}:				η: 1.025 mPa s
C_p/J mol^{-1}K^{-1}:		376.0		k:
COMMENTS:				

Name: Ethyl acetate
Synonym: Ethyl ethanoate

Mol. Form.: $C_4H_8O_2$

CAS RN: 141-78-6
Merck No.: 3713
Mol. Wt.: 88.106

PHYSICAL CONSTANTS

T_m: -83.6°C (189.5 K)	T_c: 250.2°C (523.3 K)	μ: 1.78 D
T_b: 77.11°C (350.26 K)	P_c: 3.882 MPa	IP: 10.01 eV

TRANSITION PROPERTIES

$\Delta_{fus}H\ (T_m)$: 10.48 kJ/mol
$\Delta_{vap}H\ (T_b)$: 31.94 kJ/mol
$\Delta_{vap}H$ (25°C): 35.60 kJ/mol

Vapor pressure (0°C): 3.25 kPa
Vapor pressure (25°C): 12.6 kPa
Vapor pressure (100°C): 204 kPa

PROPERTIES AT 25°C AND 100 kPa

	Solid	Liquid	Gas	Liquid
$\Delta_f H°$/kJ mol^{-1}:		-479.3	-444.1	d: 0.8945 g/mL
$S°$/J mol^{-1}K^{-1}:		257.7		η: 0.423 mPa s
C_p/J mol^{-1}K^{-1}:		170.7		k: 0.144 W/m K
COMMENTS: TLV=400 ppm; flammable				

Name: Ethyl acetoacetate
Synonyms: Ethyl 3-oxobutanoate
        1-Ethoxybutane-1,3-dione
Mol. Form.: $C_6H_{10}O_3$

CAS RN: 141-97-9
Merck No.: 3714
Mol. Wt.: 130.144

PHYSICAL CONSTANTS

T_m: -45°C (228 K) T_c: μ:
T_b: 180.8°C (453.9 K) P_c: IP:

TRANSITION PROPERTIES

$\Delta_{fus}H$ (T_m):   Vapor pressure (0°C):
$\Delta_{vap}H$ (T_b):   Vapor pressure (25°C): 0.095 kPa
$\Delta_{vap}H$ (25°C):   Vapor pressure (100°C): 6.26 kPa

PROPERTIES AT 25°C AND 100 kPa

	Solid	Liquid	Gas	Liquid
$\Delta_f H°$/kJ mol^{-1}:				d: 1.022 g/mL
$S°$/J mol^{-1}K^{-1}:				η: 1.508 mPa s
C_p/J mol^{-1}K^{-1}:		248.0		k:

COMMENTS:

Name: Ethyl acrylate
Synonym: Ethyl propenoate

Mol. Form.: $C_5H_8O_2$

CAS RN: 140-88-5
Merck No.: 3715
Mol. Wt.: 100.117

PHYSICAL CONSTANTS

T_m: -71.2°C (201.9 K) T_c: μ:
T_b: 99.4°C (372.5 K) P_c: IP: 10.30 eV

TRANSITION PROPERTIES

$\Delta_{fus}H$ (T_m):   Vapor pressure (0°C): 1.20 kPa
$\Delta_{vap}H$ (T_b): 34.70 kJ/mol   Vapor pressure (25°C): 5.14 kPa
$\Delta_{vap}H$ (25°C):   Vapor pressure (100°C): 103 kPa

PROPERTIES AT 25°C AND 100 kPa

	Solid	Liquid	Gas	Liquid
$\Delta_f H°$/kJ mol^{-1}:				d: 0.917 g/mL
$S°$/J mol^{-1}K^{-1}:				η:
C_p/J mol^{-1}K^{-1}:				k:

COMMENTS: TLV=5 ppm; carcinogen; highly toxic; flammable

Name: Ethylamine
Synonyms: Ethanamine
        Aminoethane
Mol. Form.: C_2H_7N

CAS RN: 75-04-7
Merck No.: 3718
Mol. Wt.: 45.084

PHYSICAL CONSTANTS

T_m: -80.53°C (192.62 K) T_c: 183°C (456 K) μ: 1.22 D
T_b: 16.55°C (289.70 K) P_c: 5.62 MPa   IP: 8.86 eV

TRANSITION PROPERTIES

$\Delta_{fus}H$ (T_m):   Vapor pressure (0°C): 48.9 kPa
$\Delta_{vap}H$ (T_b):   Vapor pressure (25°C): 142 kPa
$\Delta_{vap}H$ (25°C): 26.6 kJ/mol   Vapor pressure (100°C):

PROPERTIES AT 25°C AND 100 kPa

	Solid	Liquid	Gas	Gas
$\Delta_f H°$/kJ mol^{-1}:		-74.1	-47.5	d: 1.843 g/L
$S°$/J mol^{-1}K^{-1}:			283.8	η:
C_p/J mol^{-1}K^{-1}:		130.0	71.5	k:

COMMENTS: TLV=10 ppm; highly toxic; very flammable

Name: *N*-Ethylaniline
Synonym: Ethylphenylamine

Mol. Form.: $C_8H_{11}N$

CAS RN: 103-69-5
Merck No.: 3722
Mol. Wt.: 121.182

PHYSICAL CONSTANTS

T_m: -63.5°C (209.6 K)
T_b: 203.05°C (476.20 K)

T_c: 425°C (698 K)
P_c:

μ:
IP: 7.67 eV

TRANSITION PROPERTIES

$\Delta_{fus}H$ (T_m):
$\Delta_{vap}H$ (T_b):
$\Delta_{vap}H$ (25°C): 52.3 kJ/mol

Vapor pressure (0°C):
Vapor pressure (25°C):
Vapor pressure (100°C): 3.11 kPa

PROPERTIES AT 25°C AND 100 kPa

	Solid	Liquid	Gas	Liquid
$\Delta_f H°$/kJ mol^{-1}:		4.0	56.3	d: 0.961 g/mL
$S°$/J mol^{-1}K^{-1}:				η: 2.05 mPa s
C_p/J mol^{-1}K^{-1}:				k:

COMMENTS: Highly toxic

Name: Ethylbenzene
Synonym: Phenylethane

Mol. Form.: C_8H_{10}

CAS RN: 100-41-4
Merck No.: 3723
Mol. Wt.: 106.167

PHYSICAL CONSTANTS

T_m: -94.95°C (178.20 K)
T_b: 136.19°C (409.34 K)

T_c: 344.1°C (617.2 K)
P_c: 3.600 MPa

μ: 0.59 D
IP: 8.77 eV

TRANSITION PROPERTIES

$\Delta_{fus}H$ (T_m): 9.18 kJ/mol
$\Delta_{vap}H$ (T_b): 35.57 kJ/mol
$\Delta_{vap}H$ (25°C): 42.24 kJ/mol

Vapor pressure (0°C):
Vapor pressure (25°C): 1.28 kPa
Vapor pressure (100°C): 34.2 kPa

PROPERTIES AT 25°C AND 100 kPa

	Solid	Liquid	Gas	Liquid
$\Delta_f H°$/kJ mol^{-1}:		-12.3	29.9	d: 0.8625 g/mL
$S°$/J mol^{-1}K^{-1}:				η: 0.647 mPa s
C_p/J mol^{-1}K^{-1}:		183.2		k: 0.130 W/m K

COMMENTS: TLV=100 ppm; highly toxic; flammable

Name: Ethyl benzoate
Synonym: Benzoyl ethyl ether

Mol. Form.: $C_9H_{10}O_2$

CAS RN: 93-89-0
Merck No.: 3725
Mol. Wt.: 150.177

PHYSICAL CONSTANTS

T_m: -34°C (239 K)
T_b: 212°C (485 K)

T_c:
P_c:

μ: 2.00 D
IP: 8.90 eV

TRANSITION PROPERTIES

$\Delta_{fus}H$ (T_m):
$\Delta_{vap}H$ (T_b):
$\Delta_{vap}H$ (25°C):

Vapor pressure (0°C):
Vapor pressure (25°C):
Vapor pressure (100°C): 2.51 kPa

PROPERTIES AT 25°C AND 100 kPa

	Solid	Liquid	Gas	Liquid
$\Delta_f H°$/kJ mol^{-1}:				d: 1.042 g/mL
$S°$/J mol^{-1}K^{-1}:				η: 2.0 mPa s
C_p/J mol^{-1}K^{-1}:		246.0		k:

COMMENTS:

Name: Ethyl butanoate   CAS RN: 105-54-4
Synonym: Ethyl butyrate   Merck No.: 3733
                                                                                Mol. Wt.: 116.160
Mol. Form.: $C_6H_{12}O_2$

PHYSICAL CONSTANTS
T_m: -98°C (175 K) T_c: 293°C (566 K)   μ:
T_b: 121.5°C (394.6 K) P_c: 3.06 MPa   IP:

TRANSITION PROPERTIES
$\Delta_{fus}H$ (T_m):   Vapor pressure (0°C):
$\Delta_{vap}H$ (T_b): 35.47 kJ/mol   Vapor pressure (25°C): 1.99 kPa
$\Delta_{vap}H$ (25°C): 42.68 kJ/mol   Vapor pressure (100°C): 54.1 kPa

PROPERTIES AT 25°C AND 100 kPa

	Solid	Liquid	Gas	Liquid
$\Delta_f H°$/kJ mol^{-1}:				d: 0.874 g/mL
$S°$/J mol^{-1}K^{-1}:				η: 0.639 mPa s
C_p/J mol^{-1}K^{-1}:		228.0		k:

COMMENTS: Flammable

Name: 2-Ethyl-1-butanol   CAS RN: 97-95-0
Synonym: 2-Ethylbutyl alcohol   Merck No.:
                                                                                Mol. Wt.: 102.177
Mol. Form.: $C_6H_{14}O$

PHYSICAL CONSTANTS
T_m: -114.4°C (158.7 K) T_c: μ:
T_b: 146.5°C (419.6 K) P_c: IP:

TRANSITION PROPERTIES
$\Delta_{fus}H$ (T_m):   Vapor pressure (0°C):
$\Delta_{vap}H$ (T_b): 43.20 kJ/mol   Vapor pressure (25°C): 0.206 kPa
$\Delta_{vap}H$ (25°C):   Vapor pressure (100°C): 19.1 kPa

PROPERTIES AT 25°C AND 100 kPa

	Solid	Liquid	Gas	Liquid
$\Delta_f H°$/kJ mol^{-1}:				d: 0.8295 g/mL
$S°$/J mol^{-1}K^{-1}:				η: 5.89 mPa s
C_p/J mol^{-1}K^{-1}:				k:

COMMENTS:

Name: 2-Ethyl-1-butene   CAS RN: 760-21-4
Synonym: 3-Methylenepentane   Merck No.:
                                                                                Mol. Wt.: 84.161
Mol. Form.: C_6H_{12}

PHYSICAL CONSTANTS
T_m: -131.5°C (141.6 K) T_c: μ:
T_b: 64.7°C (337.8 K) P_c: IP: 9.06 eV

TRANSITION PROPERTIES
$\Delta_{fus}H$ (T_m):   Vapor pressure (0°C): 7.16 kPa
$\Delta_{vap}H$ (T_b):   Vapor pressure (25°C): 23.3 kPa
$\Delta_{vap}H$ (25°C): 31.13 kJ/mol   Vapor pressure (100°C):

PROPERTIES AT 25°C AND 100 kPa

	Solid	Liquid	Gas	Liquid
$\Delta_f H°$/kJ mol^{-1}:		-87.1	-56.0	d: 0.6848 g/mL
$S°$/J mol^{-1}K^{-1}:				η:
C_p/J mol^{-1}K^{-1}:				k:

COMMENTS: Flammable

Name: Ethyl chloroacetate
Synonym: Ethyl chloroethanoate

CAS RN: 105-39-5
Merck No.: 3741
Mol. Wt.: 122.551

Mol. Form.: $C_4H_7ClO_2$

PHYSICAL CONSTANTS

T_m: -21°C (252 K) T_c: μ:
T_b: 144.3°C (417.4 K) P_c: IP:

TRANSITION PROPERTIES

$\Delta_{fus}H$ (T_m):   Vapor pressure (0°C): 0.121 kPa
$\Delta_{vap}H$ (T_b): 40.43 kJ/mol   Vapor pressure (25°C): 0.640 kPa
$\Delta_{vap}H$ (25°C): 49.47 kJ/mol   Vapor pressure (100°C): 23.1 kPa

PROPERTIES AT 25°C AND 100 kPa

	Solid	Liquid	Gas		Liquid
$\Delta_fH°$/kJ mol^{-1}:				d: 1.151 g/mL
$S°$/J mol^{-1}K^{-1}:				η:
C_p/J mol^{-1}K^{-1}:				k:
COMMENTS: Flammable

Name: Ethylcyclohexane

CAS RN: 1678-91-7
Merck No.:
Mol. Wt.: 112.215

Mol. Form.: C_8H_{16}

PHYSICAL CONSTANTS

T_m: -111.3°C (161.8 K) T_c: μ:
T_b: 131.9°C (405.0 K) P_c: IP: 9.54 eV

TRANSITION PROPERTIES

$\Delta_{fus}H$ (T_m): 8.33 kJ/mol   Vapor pressure (0°C):
$\Delta_{vap}H$ (T_b): 34.04 kJ/mol   Vapor pressure (25°C): 1.70 kPa
$\Delta_{vap}H$ (25°C): 40.56 kJ/mol   Vapor pressure (100°C): 40.1 kPa

PROPERTIES AT 25°C AND 100 kPa

	Solid	Liquid	Gas		Liquid
$\Delta_fH°$/kJ mol^{-1}:		-211.9	-171.7	d: 0.784 g/mL
$S°$/J mol^{-1}K^{-1}:		280.9		η: 0.784 mPa s
C_p/J mol^{-1}K^{-1}:		211.8		k:
COMMENTS:

Name: Ethylcyclopentane

CAS RN: 1640-89-7
Merck No.:
Mol. Wt.: 98.188

Mol. Form.: C_7H_{14}

PHYSICAL CONSTANTS

T_m: -138.44°C (134.71 K) T_c: 296.4°C (569.5 K) μ:
T_b: 103.5°C (376.6 K) P_c: 3.397 MPa   IP: 10.12 eV

TRANSITION PROPERTIES

$\Delta_{fus}H$ (T_m):   Vapor pressure (0°C): 1.34 kPa
$\Delta_{vap}H$ (T_b): 31.96 kJ/mol   Vapor pressure (25°C): 5.32 kPa
$\Delta_{vap}H$ (25°C): 36.40 kJ/mol   Vapor pressure (100°C): 91.7 kPa

PROPERTIES AT 25°C AND 100 kPa

	Solid	Liquid	Gas		Liquid
$\Delta_fH°$/kJ mol^{-1}:		-163.4	-126.9	d: 0.763 g/mL
$S°$/J mol^{-1}K^{-1}:		279.9		η:
C_p/J mol^{-1}K^{-1}:				k:
COMMENTS: Flammable

Name: Ethylene
Synonym: Ethene

Mol. Form.: C_2H_4

CAS RN: 74-85-1
Merck No.: 3748
Mol. Wt.: 28.054

PHYSICAL CONSTANTS

T_m: -169°C (104 K) T_c: 9.19°C (282.34 K)   μ: 0 D
T_b: -103.77°C (169.38 K) P_c: 5.041 MPa   IP: 10.51 eV

TRANSITION PROPERTIES

$\Delta_{fus}H$ (T_m):   Vapor pressure (0°C): 4100 kPa
$\Delta_{vap}H$ (T_b): 13.53 kJ/mol   Vapor pressure (25°C): N/A
$\Delta_{vap}H$ (25°C):   Vapor pressure (100°C): N/A

PROPERTIES AT 25°C AND 100 kPa

	Solid	Liquid	Gas	Gas
$\Delta_f H°$/kJ mol^{-1}:			52.5	d: 1.147 g/L
$S°$/J mol^{-1}K^{-1}:			219.6	η: 10.3 μPa s
C_p/J mol^{-1}K^{-1}:			43.6	k: 0.0205 W/m K
COMMENTS: Very flammable				

Name: Ethylene glycol
Synonyms: 1,2-Ethanediol
        2-Hydroxyethanol
Mol. Form.: $C_2H_6O_2$

CAS RN: 107-21-1
Merck No.: 3755
Mol. Wt.: 62.068

PHYSICAL CONSTANTS

T_m: -13°C (260 K) T_c: 445°C (718 K)   μ: 2.28 D
T_b: 197.34°C (470.49 K) P_c: IP: 10.16 eV

TRANSITION PROPERTIES

$\Delta_{fus}H$ (T_m): 11.23 kJ/mol   Vapor pressure (0°C):
$\Delta_{vap}H$ (T_b): 50.46 kJ/mol   Vapor pressure (25°C): 0.010 kPa
$\Delta_{vap}H$ (25°C): 67.80 kJ/mol   Vapor pressure (100°C): 2.14 kPa

PROPERTIES AT 25°C AND 100 kPa

	Solid	Liquid	Gas	Liquid
$\Delta_f H°$/kJ mol^{-1}:		-455.3	-387.5	d: 1.1101 g/mL
$S°$/J mol^{-1}K^{-1}:		163.2	303.8	η: 16.1 mPa s
C_p/J mol^{-1}K^{-1}:		148.6	82.7	k: 0.256 W/m K
COMMENTS: TLV=50 ppm; highly toxic				

Name: Ethylene glycol diacetate
Synonyms: 1,2-Ethanediol diacetate
        1,2-Diacetoxyethane
Mol. Form.: $C_6H_{10}O_4$

CAS RN: 111-55-7
Merck No.: 3756
Mol. Wt.: 146.143

PHYSICAL CONSTANTS

T_m: -31°C (242 K) T_c: μ:
T_b: 190°C (463 K) P_c: IP:

TRANSITION PROPERTIES

$\Delta_{fus}H$ (T_m):   Vapor pressure (0°C):
$\Delta_{vap}H$ (T_b): 45.52 kJ/mol   Vapor pressure (25°C): 0.030 kPa
$\Delta_{vap}H$ (25°C): 61.44 kJ/mol   Vapor pressure (100°C): 4.07 kPa

PROPERTIES AT 25°C AND 100 kPa

	Solid	Liquid	Gas	Liquid
$\Delta_f H°$/kJ mol^{-1}:				d: 1.0991 g/mL
$S°$/J mol^{-1}K^{-1}:				η:
C_p/J mol^{-1}K^{-1}:				k:
COMMENTS:				

Name: Ethyleneimine
Synonyms: Aziridine
         Azacyclopropane
Mol. Form.: C_2H_5N

CAS RN: 151-56-4
Merck No.: 3760
Mol. Wt.: 43.068

PHYSICAL CONSTANTS

T_m: -77.95°C (195.20 K)	T_c:	μ: 1.90 D
T_b: 56°C (329 K)	P_c:	IP: 9.20 eV

TRANSITION PROPERTIES

$\Delta_{fus}H$ (T_m):	Vapor pressure (0°C): 7.90 kPa
$\Delta_{vap}H$ (T_b):	Vapor pressure (25°C): 28.9 kPa
$\Delta_{vap}H$ (25°C): 34.6 kJ/mol	Vapor pressure (100°C):

PROPERTIES AT 25°C AND 100 kPa

	Solid	Liquid	Gas		Liquid
$\Delta_f H°$/kJ mol^{-1}:		91.9	126.5		d: 0.832 g/mL
$S°$/J mol^{-1}K^{-1}:					η: 0.418 mPa s
C_p/J mol^{-1}K^{-1}:					k:

COMMENTS: TLV=0.05 ppm; carcinogen; highly toxic; flammable

Name: Ethylene oxide
Synonyms: Oxirane
         Epoxyethane
Mol. Form.: C_2H_4O

CAS RN: 75-21-8
Merck No.: 3758
Mol. Wt.: 44.053

PHYSICAL CONSTANTS

T_m: -111.7°C (161.4 K)	T_c: 196°C (469 K)	μ: 1.89 D
T_b: 10.6°C (283.7 K)	P_c: 7.19 MPa	IP: 10.57 eV

TRANSITION PROPERTIES

$\Delta_{fus}H$ (T_m):	Vapor pressure (0°C): 65.8 kPa
$\Delta_{vap}H$ (T_b): 25.54 kJ/mol	Vapor pressure (25°C):
$\Delta_{vap}H$ (25°C): 24.75 kJ/mol	Vapor pressure (100°C):

PROPERTIES AT 25°C AND 100 kPa

	Solid	Liquid	Gas		Gas
$\Delta_f H°$/kJ mol^{-1}:		-77.8	-52.6		d: 1.801 g/L
$S°$/J mol^{-1}K^{-1}:		153.9	242.5		η:
C_p/J mol^{-1}K^{-1}:		88.0	47.9		k:

COMMENTS: TLV=1 ppm; carcinogen; highly toxic

Name: Ethyl formate
Synonym: Ethyl methanoate

Mol. Form.: $C_3H_6O_2$

CAS RN: 109-94-4
Merck No.: 3763
Mol. Wt.: 74.079

PHYSICAL CONSTANTS

T_m: -79.6°C (193.5 K)	T_c: 235.4°C (508.5 K)	μ:
T_b: 54.4°C (327.5 K)	P_c: 4.74 MPa	IP: 10.61 eV

TRANSITION PROPERTIES

$\Delta_{fus}H$ (T_m): 9.20 kJ/mol	Vapor pressure (0°C): 9.58 kPa
$\Delta_{vap}H$ (T_b): 29.91 kJ/mol	Vapor pressure (25°C): 32.3 kPa
$\Delta_{vap}H$ (25°C): 31.96 kJ/mol	Vapor pressure (100°C): 393 kPa

PROPERTIES AT 25°C AND 100 kPa

	Solid	Liquid	Gas		Liquid
$\Delta_f H°$/kJ mol^{-1}:					d: 0.9153 g/mL
$S°$/J mol^{-1}K^{-1}:					η: 0.380 mPa s
C_p/J mol^{-1}K^{-1}:		149.3			k:

COMMENTS: TLV=100 ppm; flammable

Name: 3-Ethylhexane

Mol. Form.: C_8H_{18}

CAS RN: 619-99-8
Merck No.:
Mol. Wt.: 114.231

PHYSICAL CONSTANTS

T_m:
T_b: 118.6°C (391.7 K)

T_c: 292.4°C (565.5 K)
P_c: 2.608 MPa

μ:
IP:

TRANSITION PROPERTIES

$\Delta_{fus}H$ (T_m):
$\Delta_{vap}H$ (T_b): 33.59 kJ/mol
$\Delta_{vap}H$ (25°C): 39.64 kJ/mol

Vapor pressure (0°C):
Vapor pressure (25°C): 2.68 kPa
Vapor pressure (100°C): 58.6 kPa

PROPERTIES AT 25°C AND 100 kPa

	Solid	Liquid	Gas		Liquid
$\Delta_f H°$/kJ mol^{-1}:		-250.4	-210.7		d: 0.7095 g/mL
$S°$/J mol^{-1}K^{-1}:					η:
C_p/J mol^{-1}K^{-1}:					k:
COMMENTS: Flammable					

Name: 2-Ethyl-1-hexanol

Mol. Form.: $C_8H_{18}O$

CAS RN: 104-76-7
Merck No.: 3764
Mol. Wt.: 130.230

PHYSICAL CONSTANTS

T_m: -70°C (203 K)
T_b: 184.62°C (457.77 K)

T_c: 368°C (641 K)
P_c: 2.8 MPa

μ:
IP:

TRANSITION PROPERTIES

$\Delta_{fus}H$ (T_m):
$\Delta_{vap}H$ (T_b): 54.20 kJ/mol
$\Delta_{vap}H$ (25°C): 67.5 kJ/mol

Vapor pressure (0°C):
Vapor pressure (25°C): 0.019 kPa
Vapor pressure (100°C): 4.19 kPa

PROPERTIES AT 25°C AND 100 kPa

	Solid	Liquid	Gas		Liquid
$\Delta_f H°$/kJ mol^{-1}:		-432.8	-365.3		d: 0.8290 g/mL
$S°$/J mol^{-1}K^{-1}:		347.0			η: 6.27 mPa s
C_p/J mol^{-1}K^{-1}:		317.5			k:
COMMENTS:					

Name: 2-Ethylhexyl acetate
Synonym: 2-Ethylhexyl ethanoate

Mol. Form.: $C_{10}H_{20}O_2$

CAS RN: 103-09-3
Merck No.: 6683
Mol. Wt.: 172.268

PHYSICAL CONSTANTS

T_m: -80°C (193 K)
T_b: 199°C (472 K)

T_c:
P_c:

μ:
IP:

TRANSITION PROPERTIES

$\Delta_{fus}H$ (T_m):
$\Delta_{vap}H$ (T_b): 43.50 kJ/mol
$\Delta_{vap}H$ (25°C): 48.10 kJ/mol

Vapor pressure (0°C):
Vapor pressure (25°C):
Vapor pressure (100°C): 7.68 kPa

PROPERTIES AT 25°C AND 100 kPa

	Solid	Liquid	Gas		Liquid
$\Delta_f H°$/kJ mol^{-1}:					d: 0.867 g/mL
$S°$/J mol^{-1}K^{-1}:					η:
C_p/J mol^{-1}K^{-1}:					k:
COMMENTS:					

Name: Ethyl methacrylate
Synonym: Ethyl 2-methylpropenoate

Mol. Form.: $C_6H_{10}O_2$

CAS RN: 97-63-2
Merck No.:
Mol. Wt.: 114.144

PHYSICAL CONSTANTS

T_m: <-75°C (<198 K) T_c: μ:
T_b: 117°C (390 K) P_c: IP:

TRANSITION PROPERTIES

$\Delta_{fus}H$ (T_m):   Vapor pressure (0°C): 0.640 kPa
$\Delta_{vap}H$ (T_b):   Vapor pressure (25°C): 2.62 kPa
$\Delta_{vap}H$ (25°C):   Vapor pressure (100°C): 58.8 kPa

PROPERTIES AT 25°C AND 100 kPa

	Solid	Liquid	Gas	Liquid
$\Delta_fH°$/kJ mol^{-1}:				d: 0.909 g/mL
$S°$/J mol^{-1}K^{-1}:				η:
C_p/J mol^{-1}K^{-1}:				k:

COMMENTS: Flammable

Name: Ethyl 3-methylbutanoate
Synonyms: Ethyl isovalerate
          Ethyl isopentanoate
Mol. Form.: $C_7H_{14}O_2$

CAS RN: 108-64-5
Merck No.: 3772
Mol. Wt.: 130.187

PHYSICAL CONSTANTS

T_m: -99.3°C (173.8 K) T_c: 315°C (588 K)   μ:
T_b: 135.05°C (408.20 K) P_c: IP:

TRANSITION PROPERTIES

$\Delta_{fus}H$ (T_m):   Vapor pressure (0°C):
$\Delta_{vap}H$ (T_b): 36.95 kJ/mol   Vapor pressure (25°C): 1.07 kPa
$\Delta_{vap}H$ (25°C): 47.30 kJ/mol   Vapor pressure (100°C): 33.0 kPa

PROPERTIES AT 25°C AND 100 kPa

	Solid	Liquid	Gas	Liquid
$\Delta_fH°$/kJ mol^{-1}:		-570.9	-527.0	d: 0.861 g/mL
$S°$/J mol^{-1}K^{-1}:				η:
C_p/J mol^{-1}K^{-1}:				k:

COMMENTS:

Name: Ethyl methyl ether
Synonyms: Methoxyethane
          Methyl ethyl ether
Mol. Form.: C_3H_8O

CAS RN: 540-67-0
Merck No.: 3783
Mol. Wt.: 60.096

PHYSICAL CONSTANTS

T_m: -113°C (160 K) T_c: 164.7°C (437.8 K)   μ: 1.17 D
T_b: 7.45°C (280.60 K) P_c: 4.40 MPa   IP: 9.72 eV

TRANSITION PROPERTIES

$\Delta_{fus}H$ (T_m):   Vapor pressure (0°C): 73.9 kPa
$\Delta_{vap}H$ (T_b):   Vapor pressure (25°C): 195 kPa
$\Delta_{vap}H$ (25°C):   Vapor pressure (100°C):

PROPERTIES AT 25°C AND 100 kPa

	Solid	Liquid	Gas	Gas
$\Delta_fH°$/kJ mol^{-1}:			-216.4	d: 2.456 g/L
$S°$/J mol^{-1}K^{-1}:			309.2	η:
C_p/J mol^{-1}K^{-1}:			93.3	k:

COMMENTS: Very flammable

Name: 3-Ethyl-2-methylpentane

Mol. Form.: C_8H_{18}

CAS RN: 609-26-7
Merck No.:
Mol. Wt.: 114.231

PHYSICAL CONSTANTS

T_m: -114.95°C (158.20 K)
T_b: 115.66°C (388.81 K)

T_c: 294.0°C (567.1 K)
P_c: 2.700 MPa

μ:
IP:

TRANSITION PROPERTIES

$\Delta_{fus}H\ (T_m)$:
$\Delta_{vap}H\ (T_b)$: 32.93 kJ/mol
$\Delta_{vap}H\ (25°C)$: 38.52 kJ/mol

Vapor pressure (0°C):
Vapor pressure (25°C): 3.19 kPa
Vapor pressure (100°C): 64.3 kPa

PROPERTIES AT 25°C AND 100 kPa

	Solid	Liquid	Gas		Liquid
$\Delta_f H°$/kJ mol^{-1}:		-249.6	-211.0	d: 0.7152 g/mL
$S°$/J mol^{-1}K^{-1}:				η:
C_p/J mol^{-1}K^{-1}:				k:

COMMENTS: Flammable

Name: Ethyl 2-methylpropanoate
Synonyms: Ethyl 2-methylpropionate
          Ethyl isobutyrate
Mol. Form.: $C_6H_{12}O_2$

CAS RN: 97-62-1
Merck No.: 3770
Mol. Wt.: 116.160

PHYSICAL CONSTANTS

T_m: -88.2°C (184.9 K)
T_b: 110.1°C (383.2 K)

T_c: 280°C (553 K)
P_c: 3.07 MPa

μ:
IP:

TRANSITION PROPERTIES

$\Delta_{fus}H\ (T_m)$:
$\Delta_{vap}H\ (T_b)$: 33.67 kJ/mol
$\Delta_{vap}H\ (25°C)$: 39.83 kJ/mol

Vapor pressure (0°C): 0.693 kPa
Vapor pressure (25°C): 3.25 kPa
Vapor pressure (100°C): 74.2 kPa

PROPERTIES AT 25°C AND 100 kPa

	Solid	Liquid	Gas		Liquid
$\Delta_f H°$/kJ mol^{-1}:				d: 0.864 g/mL
$S°$/J mol^{-1}K^{-1}:				η:
C_p/J mol^{-1}K^{-1}:				k:

COMMENTS: Flammable

Name: Ethyl methyl sulfide
Synonym: 2-Thiabutane

Mol. Form.: C_3H_8S

CAS RN: 624-89-5
Merck No.:
Mol. Wt.: 76.162

PHYSICAL CONSTANTS

T_m: -105.9°C (167.2 K)
T_b: 66.7°C (339.8 K)

T_c: 260°C (533 K)
P_c: 4.26 MPa

μ: 1.56 D
IP: 8.54 eV

TRANSITION PROPERTIES

$\Delta_{fus}H\ (T_m)$:
$\Delta_{vap}H\ (T_b)$: 29.53 kJ/mol
$\Delta_{vap}H\ (25°C)$: 31.85 kJ/mol

Vapor pressure (0°C):
Vapor pressure (25°C): 21.3 kPa
Vapor pressure (100°C): 265 kPa

PROPERTIES AT 25°C AND 100 kPa

	Solid	Liquid	Gas		Liquid
$\Delta_f H°$/kJ mol^{-1}:		-91.6	-59.6	d: 0.838 g/mL
$S°$/J mol^{-1}K^{-1}:		239.1		η:
C_p/J mol^{-1}K^{-1}:		144.6		k:

COMMENTS:

Name: Ethyl nonanoate
Synonym: Ethyl pelargonate

Mol. Form.: $C_{11}H_{22}O_2$

CAS RN: 123-29-5
Merck No.: 3793
Mol. Wt.: 186.295

PHYSICAL CONSTANTS

T_m: -36.7°C (236.4 K)
T_b: 227.05°C (500.20 K)

T_c: 401°C (674 K)
P_c:

μ:
IP:

TRANSITION PROPERTIES

$\Delta_{fus}H$ (T_m):
$\Delta_{vap}H$ (T_b):
$\Delta_{vap}H$ (25°C):

Vapor pressure (0°C):
Vapor pressure (25°C):
Vapor pressure (100°C):

PROPERTIES AT 25°C AND 100 kPa

	Solid	Liquid	Gas		Liquid
$\Delta_f H°$/kJ mol^{-1}:					*d*: 0.861 g/mL
$S°$/J mol^{-1}K^{-1}:					η:
C_p/J mol^{-1}K^{-1}:					*k*:
COMMENTS:					

Name: Ethyl octanoate
Synonym: Ethyl caprylate

Mol. Form.: $C_{10}H_{20}O_2$

CAS RN: 106-32-1
Merck No.: 3736
Mol. Wt.: 172.268

PHYSICAL CONSTANTS

T_m: -43.1°C (230.0 K)
T_b: 208.55°C (481.70 K)

T_c: 386°C (659 K)
P_c:

μ:
IP:

TRANSITION PROPERTIES

$\Delta_{fus}H$ (T_m):
$\Delta_{vap}H$ (T_b):
$\Delta_{vap}H$ (25°C):

Vapor pressure (0°C):
Vapor pressure (25°C):
Vapor pressure (100°C): 2.46 kPa

PROPERTIES AT 25°C AND 100 kPa

	Solid	Liquid	Gas		Liquid
$\Delta_f H°$/kJ mol^{-1}:					*d*: 0.860 g/mL
$S°$/J mol^{-1}K^{-1}:					η:
C_p/J mol^{-1}K^{-1}:					*k*:
COMMENTS:					

Name: 3-Ethylpentane

Mol. Form.: C_7H_{16}

CAS RN: 617-78-7
Merck No.:
Mol. Wt.: 100.204

PHYSICAL CONSTANTS

T_m: -118.6°C (154.5 K)
T_b: 93.5°C (366.6 K)

T_c: 267.6°C (540.7 K)
P_c: 2.891 MPa

μ:
IP:

TRANSITION PROPERTIES

$\Delta_{fus}H$ (T_m): 9.55 kJ/mol
$\Delta_{vap}H$ (T_b): 31.12 kJ/mol
$\Delta_{vap}H$ (25°C): 35.22 kJ/mol

Vapor pressure (0°C): 2.03 kPa
Vapor pressure (25°C): 7.74 kPa
Vapor pressure (100°C): 122 kPa

PROPERTIES AT 25°C AND 100 kPa

	Solid	Liquid	Gas		Liquid
$\Delta_f H°$/kJ mol^{-1}:		-224.8	-189.6		*d*: 0.6940 g/mL
$S°$/J mol^{-1}K^{-1}:		314.5			η:
C_p/J mol^{-1}K^{-1}:		219.6			*k*:
COMMENTS: Flammable					

Name: *o*-Ethylphenol
Synonym: 2-Ethylphenol

Mol. Form.: $C_8H_{10}O$

CAS RN: 90-00-6
Merck No.: 7302
Mol. Wt.: 122.167

PHYSICAL CONSTANTS

T_m: 18°C (291 K) T_c: 429.9°C (703.0 K) μ:
T_b: 204.52°C (477.67 K) P_c: IP:

TRANSITION PROPERTIES

$\Delta_{fus}H$ (T_m):   Vapor pressure (0°C):
$\Delta_{vap}H$ (T_b):   Vapor pressure (25°C):
$\Delta_{vap}H$ (25°C): 63.6 kJ/mol   Vapor pressure (100°C): 2.41 kPa

PROPERTIES AT 25°C AND 100 kPa

	Solid	Liquid	Gas		Liquid
$\Delta_f H°$/kJ mol^{-1}:		-208.8	-145.2		*d*: 1.0146 g/mL
$S°$/J mol^{-1}K^{-1}:					η:
C_p/J mol^{-1}K^{-1}:					*k*:
COMMENTS:					

Name: *p*-Ethylphenol
Synonym: 4-Ethylphenol

Mol. Form.: $C_8H_{10}O$

CAS RN: 123-07-9
Merck No.:
Mol. Wt.: 122.167

PHYSICAL CONSTANTS

T_m: 45.08°C (318.23 K) T_c: 443.3°C (716.4 K) μ:
T_b: 217.98°C (491.13 K) P_c: IP: 7.84 eV

TRANSITION PROPERTIES

$\Delta_{fus}H$ (T_m):   Vapor pressure (0°C):
$\Delta_{vap}H$ (T_b):   Vapor pressure (25°C):
$\Delta_{vap}H$ (25°C):   Vapor pressure (100°C): 1.12 kPa

PROPERTIES AT 25°C AND 100 kPa

	Solid	Liquid	Gas		Solid
$\Delta_f H°$/kJ mol^{-1}:	-224.4		-144.1		*d*: 0.991 g/mL
$S°$/J mol^{-1}K^{-1}:					η: N/A
C_p/J mol^{-1}K^{-1}:	206.9				*k*:
COMMENTS:					

Name: Ethyl propanoate
Synonym: Ethyl propionate

Mol. Form.: $C_5H_{10}O_2$

CAS RN: 105-37-3
Merck No.: 3801
Mol. Wt.: 102.133

PHYSICAL CONSTANTS

T_m: -73.9°C (199.2 K) T_c: 272.9°C (546.0 K) μ:
T_b: 99.1°C (372.2 K) P_c: 3.362 MPa   IP: 10.00 eV

TRANSITION PROPERTIES

$\Delta_{fus}H$ (T_m):   Vapor pressure (0°C):
$\Delta_{vap}H$ (T_b): 33.88 kJ/mol   Vapor pressure (25°C): 4.97 kPa
$\Delta_{vap}H$ (25°C): 39.21 kJ/mol   Vapor pressure (100°C): 104 kPa

PROPERTIES AT 25°C AND 100 kPa

	Solid	Liquid	Gas		Liquid
$\Delta_f H°$/kJ mol^{-1}:		-502.7	-463.6		*d*: 0.8840 g/mL
$S°$/J mol^{-1}K^{-1}:					η: 0.501 mPa s
C_p/J mol^{-1}K^{-1}:		196.1			*k*:
COMMENTS: Flammable					

Name: *o*-Ethyltoluene

Synonym: 1-Ethyl-2-methylbenzene

Mol. Form.: C_9H_{12}

CAS RN: 611-14-3

Merck No.:

Mol. Wt.: 120.194

PHYSICAL CONSTANTS

T_m: -80.8°C (192.3 K) T_c: μ:

T_b: 165.2°C (438.3 K) P_c: IP:

TRANSITION PROPERTIES

$\Delta_{fus}H$ (T_m):

$\Delta_{vap}H$ (T_b):

$\Delta_{vap}H$ (25°C): 47.7 kJ/mol

Vapor pressure (0°C):

Vapor pressure (25°C):

Vapor pressure (100°C): 13.6 kPa

PROPERTIES AT 25°C AND 100 kPa

	Solid	Liquid	Gas		Liquid
$\Delta_f H°$/kJ mol^{-1}:		-46.4	1.3		*d*: 0.876 g/mL
$S°$/J mol^{-1}K^{-1}:					η:
C_p/J mol^{-1}K^{-1}:					*k*:
COMMENTS:					

Name: *m*-Ethyltoluene

Synonym: 1-Ethyl-3-methylbenzene

Mol. Form.: C_9H_{12}

CAS RN: 620-14-4

Merck No.:

Mol. Wt.: 120.194

PHYSICAL CONSTANTS

T_m: -95.5°C (177.6 K) T_c: μ:

T_b: 161.3°C (434.4 K) P_c: IP:

TRANSITION PROPERTIES

$\Delta_{fus}H$ (T_m):

$\Delta_{vap}H$ (T_b):

$\Delta_{vap}H$ (25°C): 46.9 kJ/mol

Vapor pressure (0°C):

Vapor pressure (25°C):

Vapor pressure (100°C): 15.3 kPa

PROPERTIES AT 25°C AND 100 kPa

	Solid	Liquid	Gas		Liquid
$\Delta_f H°$/kJ mol^{-1}:		-48.7	-1.8		*d*: 0.860 g/mL
$S°$/J mol^{-1}K^{-1}:					η:
C_p/J mol^{-1}K^{-1}:					*k*:
COMMENTS:					

Name: *p*-Ethyltoluene

Synonym: 1-Ethyl-4-methylbenzene

Mol. Form.: C_9H_{12}

CAS RN: 622-96-8

Merck No.:

Mol. Wt.: 120.194

PHYSICAL CONSTANTS

T_m: -62.3°C (210.8 K) T_c: μ:

T_b: 162°C (435 K) P_c: IP:

TRANSITION PROPERTIES

$\Delta_{fus}H$ (T_m):

$\Delta_{vap}H$ (T_b):

$\Delta_{vap}H$ (25°C): 46.6 kJ/mol

Vapor pressure (0°C):

Vapor pressure (25°C):

Vapor pressure (100°C): 15.1 kPa

PROPERTIES AT 25°C AND 100 kPa

	Solid	Liquid	Gas		Liquid
$\Delta_f H°$/kJ mol^{-1}:		-49.8	-3.2		*d*: 0.857 g/mL
$S°$/J mol^{-1}K^{-1}:					η:
C_p/J mol^{-1}K^{-1}:					*k*:
COMMENTS:					

Name: Ethyl vinyl ether   CAS RN: 109-92-2
Synonyms: Ethoxyethene   Merck No.:
           Vinyl ethyl ether   Mol. Wt.: 72.107
Mol. Form.: C_4H_8O

PHYSICAL CONSTANTS

T_m: -115.8°C (157.3 K) T_c: 202°C (475 K)   μ:
T_b: 35.55°C (308.70 K) P_c: 4.07 MPa   IP: 8.80 eV

TRANSITION PROPERTIES

$\Delta_{fus}H$ (T_m):   Vapor pressure (0°C): 23.1 kPa
$\Delta_{vap}H$ (T_b): 26.20 kJ/mol   Vapor pressure (25°C): 68.7 kPa
$\Delta_{vap}H$ (25°C): 27.50 kJ/mol   Vapor pressure (100°C):

PROPERTIES AT 25°C AND 100 kPa

	Solid	Liquid	Gas		Liquid
$\Delta_f H°$/kJ mol^{-1}:		-167.4	-140.8		d: 0.755 g/mL
$S°$/J mol^{-1}K^{-1}:					η:
C_p/J mol^{-1}K^{-1}:					k:

COMMENTS: Very flammable

Name: Fluoranthene   CAS RN: 206-44-0
Synonym: 1,2-(1,8-Naphthylene)benzene   Merck No.:
                                                                                Mol. Wt.: 202.255
Mol. Form.: $C_{16}H_{10}$

PHYSICAL CONSTANTS

T_m: 107.8°C (380.9 K) T_c: μ:
T_b: 384°C (657 K) P_c: IP: 7.95 eV

TRANSITION PROPERTIES

$\Delta_{fus}H$ (T_m): 18.87 kJ/mol   Vapor pressure (0°C):
$\Delta_{vap}H$ (T_b):   Vapor pressure (25°C):
$\Delta_{vap}H$ (25°C):   Vapor pressure (100°C):

PROPERTIES AT 25°C AND 100 kPa

	Solid	Liquid	Gas		Solid
$\Delta_f H°$/kJ mol^{-1}:	189.9		289.0		d:
$S°$/J mol^{-1}K^{-1}:	230.6				η: N/A
C_p/J mol^{-1}K^{-1}:	230.2				k:

COMMENTS: Highly toxic

Name: Fluorene   CAS RN: 86-73-7
Synonym: Diphenylenemethane   Merck No.: 4081
                                                                                Mol. Wt.: 166.222
Mol. Form.: $C_{13}H_{10}$

PHYSICAL CONSTANTS

T_m: 114.8°C (387.9 K) T_c: μ:
T_b: 295°C (568 K) P_c: IP: 7.89 eV

TRANSITION PROPERTIES

$\Delta_{fus}H$ (T_m): 19.58 kJ/mol   Vapor pressure (0°C):
$\Delta_{vap}H$ (T_b):   Vapor pressure (25°C):
$\Delta_{vap}H$ (25°C):   Vapor pressure (100°C):

PROPERTIES AT 25°C AND 100 kPa

	Solid	Liquid	Gas		Solid
$\Delta_f H°$/kJ mol^{-1}:					d: 1.202 g/mL
$S°$/J mol^{-1}K^{-1}:					η: N/A
C_p/J mol^{-1}K^{-1}:					k:

COMMENTS: Highly toxic

Name: Fluorine (F$_2$)
Synonym: Difluorine

Mol. Form.: F$_2$

CAS RN: 7782-41-4
Merck No.: 4090
Mol. Wt.: 37.997

PHYSICAL CONSTANTS

T_m: -219.66°C (53.49 K)
T_b: -188.12°C (85.03 K)

T_c: -129.02°C (144.13 K)
P_c: 5.172 MPa

μ: 0 D
IP: 15.70 eV

TRANSITION PROPERTIES

$\Delta_{fus}H$ (T_m): 0.51 kJ/mol
$\Delta_{vap}H$ (T_b): 6.62 kJ/mol
$\Delta_{vap}H$ (25°C):

Vapor pressure (0°C): N/A
Vapor pressure (25°C): N/A
Vapor pressure (100°C): N/A

PROPERTIES AT 25°C AND 100 kPa

	Solid	Liquid	Gas		Gas
$\Delta_f H°$/kJ mol^{-1}:			0.0	d:	1.553 g/L
$S°$/J mol^{-1}K^{-1}:			202.8	η:	23.7 μPa s
C_p/J mol^{-1}K^{-1}:			31.3	k:	0.0279 W/m K
COMMENTS: TLV=1 ppm; highly toxic					

Name: Fluorobenzene
Synonym: Phenyl fluoride

Mol. Form.: C$_6$H$_5$F

CAS RN: 462-06-6
Merck No.: 4099
Mol. Wt.: 96.104

PHYSICAL CONSTANTS

T_m: -42.21°C (230.94 K)
T_b: 84.73°C (357.88 K)

T_c: 286.94°C (560.09 K)
P_c: 4.551 MPa

μ: 1.60 D
IP: 9.20 eV

TRANSITION PROPERTIES

$\Delta_{fus}H$ (T_m): 11.31 kJ/mol
$\Delta_{vap}H$ (T_b): 31.19 kJ/mol
$\Delta_{vap}H$ (25°C): 34.58 kJ/mol

Vapor pressure (0°C): 2.84 kPa
Vapor pressure (25°C): 10.4 kPa
Vapor pressure (100°C): 156 kPa

PROPERTIES AT 25°C AND 100 kPa

	Solid	Liquid	Gas		Liquid
$\Delta_f H°$/kJ mol^{-1}:		-150.6	-116.0	d:	1.019 g/mL
$S°$/J mol^{-1}K^{-1}:		205.9		η:	0.550 mPa s
C_p/J mol^{-1}K^{-1}:		146.4		k:	
COMMENTS: Flammable					

Name: Fluoroethane
Synonyms: Ethyl fluoride
　　　　　 Refrigerant 161
Mol. Form.: C$_2$H$_5$F

CAS RN: 353-36-6
Merck No.:
Mol. Wt.: 48.060

PHYSICAL CONSTANTS

T_m: -143.2°C (129.9 K)
T_b: -37.65°C (235.50 K)

T_c: 102.16°C (375.31 K)
P_c: 5.028 MPa

μ: 1.94 D
IP: 11.60 eV

TRANSITION PROPERTIES

$\Delta_{fus}H$ (T_m):
$\Delta_{vap}H$ (T_b): 21.1 kJ/mol
$\Delta_{vap}H$ (25°C):

Vapor pressure (0°C):
Vapor pressure (25°C):
Vapor pressure (100°C):

PROPERTIES AT 25°C AND 100 kPa

	Solid	Liquid	Gas		Gas
$\Delta_f H°$/kJ mol^{-1}:				d:	1.964 g/L
$S°$/J mol^{-1}K^{-1}:			264.5	η:	
C_p/J mol^{-1}K^{-1}:			58.6	k:	
COMMENTS: Very flammable					

Name: Fluoroethylene
Synonym: Vinyl fluoride

Mol. Form.: C_2H_3F

CAS RN: 75-02-5
Merck No.:
Mol. Wt.: 46.044

PHYSICAL CONSTANTS

T_m: -160.5°C (112.6 K) T_c: 54.8°C (327.9 K) μ: 1.468 D
T_b: -72°C (201 K) P_c: 5.24 MPa   IP: 10.36 eV

TRANSITION PROPERTIES

$\Delta_{fus}H$ (T_m):   Vapor pressure (0°C):
$\Delta_{vap}H$ (T_b): 17.1 kJ/mol   Vapor pressure (25°C):
$\Delta_{vap}H$ (25°C):   Vapor pressure (100°C): N/A

PROPERTIES AT 25°C AND 100 kPa

	Solid	Liquid	Gas		Gas
$\Delta_f H°$/kJ mol^{-1}:			-138.8	d:	1.882 g/L
$S°$/J mol^{-1}K^{-1}:				η:	
C_p/J mol^{-1}K^{-1}:				k:	

COMMENTS: Very flammable

Name: Fluoromethane
Synonyms: Methyl fluoride
        Refrigerant 41
Mol. Form.: CH_3F

CAS RN: 593-53-3
Merck No.: 4103
Mol. Wt.: 34.033

PHYSICAL CONSTANTS

T_m: -141.8°C (131.3 K) T_c: 44.7°C (317.8 K) μ: 1.858 D
T_b: -78.41°C (194.74 K) P_c: 5.88 MPa   IP: 12.47 eV

TRANSITION PROPERTIES

$\Delta_{fus}H$ (T_m):   Vapor pressure (0°C):
$\Delta_{vap}H$ (T_b):   Vapor pressure (25°C):
$\Delta_{vap}H$ (25°C):   Vapor pressure (100°C): N/A

PROPERTIES AT 25°C AND 100 kPa

	Solid	Liquid	Gas		Gas
$\Delta_f H°$/kJ mol^{-1}:				d:	1.391 g/L
$S°$/J mol^{-1}K^{-1}:			222.9	η:	
C_p/J mol^{-1}K^{-1}:			37.5	k:	

COMMENTS:

Name: *o*-Fluorotoluene
Synonym: *o*-Tolyl fluoride

Mol. Form.: C_7H_7F

CAS RN: 95-52-3
Merck No.: 4108
Mol. Wt.: 110.131

PHYSICAL CONSTANTS

T_m: -62°C (211 K) T_c: μ: 1.37 D
T_b: 115°C (388 K) P_c: IP: 8.91 eV

TRANSITION PROPERTIES

$\Delta_{fus}H$ (T_m):   Vapor pressure (0°C):
$\Delta_{vap}H$ (T_b): 35.40 kJ/mol   Vapor pressure (25°C): 3.13 kPa
$\Delta_{vap}H$ (25°C):   Vapor pressure (100°C): 66.1 kPa

PROPERTIES AT 25°C AND 100 kPa

	Solid	Liquid	Gas		Liquid
$\Delta_f H°$/kJ mol^{-1}:				d:	0.992 g/mL
$S°$/J mol^{-1}K^{-1}:				η:	0.64 mPa s
C_p/J mol^{-1}K^{-1}:				k:	

COMMENTS:

Name: *m*-Fluorotoluene
Synonym: *m*-Tolyl fluoride

Mol. Form.: C$_7$H$_7$F

CAS RN: 352-70-5
Merck No.: 4108
Mol. Wt.: 110.131

PHYSICAL CONSTANTS

T_m: -87°C (186 K)	T_c:	μ: 1.82 D
T_b: 115°C (388 K)	P_c:	IP: 8.91 eV

TRANSITION PROPERTIES

$\Delta_{fus}H$ (T_m):
$\Delta_{vap}H$ (T_b):
$\Delta_{vap}H$ (25°C):

Vapor pressure (0°C):
Vapor pressure (25°C): 2.83 kPa
Vapor pressure (100°C): 61.8 kPa

PROPERTIES AT 25°C AND 100 kPa

	Solid	Liquid	Gas		Liquid
$\Delta_f H°$/kJ mol^{-1}:					d: 0.992 g/mL
$S°$/J mol^{-1}K^{-1}:					η: 0.57 mPa s
C_p/J mol^{-1}K^{-1}:					k:
COMMENTS:					

Name: *p*-Fluorotoluene
Synonym: *p*-Tolyl fluoride

Mol. Form.: C$_7$H$_7$F

CAS RN: 352-32-9
Merck No.: 4108
Mol. Wt.: 110.131

PHYSICAL CONSTANTS

T_m: -56°C (217 K)	T_c:	μ: 2.00 D
T_b: 116.6°C (389.7 K)	P_c:	IP: 8.79 eV

TRANSITION PROPERTIES

$\Delta_{fus}H$ (T_m):
$\Delta_{vap}H$ (T_b): 34.08 kJ/mol
$\Delta_{vap}H$ (25°C): 39.42 kJ/mol

Vapor pressure (0°C):
Vapor pressure (25°C): 3.00 kPa
Vapor pressure (100°C): 61.7 kPa

PROPERTIES AT 25°C AND 100 kPa

	Solid	Liquid	Gas		Liquid
$\Delta_f H°$/kJ mol^{-1}:		-186.9	-147.5		d: 0.992 g/mL
$S°$/J mol^{-1}K^{-1}:					η: 0.57 mPa s
C_p/J mol^{-1}K^{-1}:					k:
COMMENTS:					

Name: Formaldehyde
Synonyms: Methanal
            Formalin
Mol. Form.: CH$_2$O

CAS RN: 50-00-0
Merck No.: 4148
Mol. Wt.: 30.026

PHYSICAL CONSTANTS

T_m: -92°C (181 K)	T_c:	μ: 2.332 D
T_b: -19.1°C (254.0 K)	P_c:	IP: 10.87 eV

TRANSITION PROPERTIES

$\Delta_{fus}H$ (T_m):
$\Delta_{vap}H$ (T_b): 23.3 kJ/mol
$\Delta_{vap}H$ (25°C):

Vapor pressure (0°C): 220 kPa
Vapor pressure (25°C):
Vapor pressure (100°C):

PROPERTIES AT 25°C AND 100 kPa

	Solid	Liquid	Gas		Gas
$\Delta_f H°$/kJ mol^{-1}:			-108.6		d: 1.227 g/L
$S°$/J mol^{-1}K^{-1}:			218.8		η:
C_p/J mol^{-1}K^{-1}:			35.4		k:
COMMENTS: TLV=0.3 ppm; carcinogen; highly toxic; very flammable					

Name: Formamide
Synonyms: Methanamide
 Carbamaldehyde
Mol. Form.: CH_3NO

CAS RN: 75-12-7
Merck No.: 4151
Mol. Wt.: 45.041

PHYSICAL CONSTANTS

T_m: 2.55°C (275.70 K)	T_c:	μ: 3.73 D
T_b: 220°C (493 K)	P_c:	IP: 10.16 eV

TRANSITION PROPERTIES

$\Delta_{fus}H$ (T_m): 6.69 kJ/mol   Vapor pressure (0°C):
$\Delta_{vap}H$ (T_b):   Vapor pressure (25°C):
$\Delta_{vap}H$ (25°C): 60.15 kJ/mol   Vapor pressure (100°C):

PROPERTIES AT 25°C AND 100 kPa

	Solid	Liquid	Gas		Liquid
$\Delta_f H°$/kJ mol^{-1}:		-254.0		d:	1.1291 g/mL
$S°$/J mol^{-1}K^{-1}:				η:	3.34 mPa s
C_p/J mol^{-1}K^{-1}:		107.6		k:	
COMMENTS: TLV=10 ppm					

Name: Formanilide
Synonyms: *N*-Phenylformamide
 Formylaniline
Mol. Form.: C_7H_7NO

CAS RN: 103-70-8
Merck No.: 4152
Mol. Wt.: 121.139

PHYSICAL CONSTANTS

T_m: 46°C (319 K)	T_c:	μ:
T_b: 271°C (544 K)	P_c:	IP:

TRANSITION PROPERTIES

$\Delta_{fus}H$ (T_m):   Vapor pressure (0°C):
$\Delta_{vap}H$ (T_b):   Vapor pressure (25°C):
$\Delta_{vap}H$ (25°C):   Vapor pressure (100°C):

PROPERTIES AT 25°C AND 100 kPa

	Solid	Liquid	Gas		Solid
$\Delta_f H°$/kJ mol^{-1}:				d:	
$S°$/J mol^{-1}K^{-1}:				η:	N/A
C_p/J mol^{-1}K^{-1}:				k:	
COMMENTS:					

Name: Formic acid
Synonym: Methanoic acid

Mol. Form.: CH_2O_2

CAS RN: 64-18-6
Merck No.: 4153
Mol. Wt.: 46.026

PHYSICAL CONSTANTS

T_m: 8.3°C (281.4 K)	T_c: 315°C (588 K)	μ: 1.41 D
T_b: 101°C (374 K)	P_c:	IP: 11.33 eV

TRANSITION PROPERTIES

$\Delta_{fus}H$ (T_m): 12.72 kJ/mol   Vapor pressure (0°C):
$\Delta_{vap}H$ (T_b): 22.69 kJ/mol   Vapor pressure (25°C): 5.75 kPa
$\Delta_{vap}H$ (25°C): 20.10 kJ/mol   Vapor pressure (100°C): 99.6 kPa

PROPERTIES AT 25°C AND 100 kPa

	Solid	Liquid	Gas		Liquid
$\Delta_f H°$/kJ mol^{-1}:		-424.7	-378.6	d:	1.214 g/mL
$S°$/J mol^{-1}K^{-1}:		129.0		η:	1.61 mPa s
C_p/J mol^{-1}K^{-1}:		99.0		k:	
COMMENTS: TLV=5 ppm; highly toxic					

Name: Fumaric acid
Synonym: *trans*-2-Butenedioic acid

Mol. Form.: $C_4H_4O_4$

CAS RN: 110-17-8
Merck No.: 4200
Mol. Wt.: 116.073

PHYSICAL CONSTANTS

T_m: 287°C (560 K)	T_c:	μ:
T_b:	P_c:	IP: 10.70 eV

TRANSITION PROPERTIES

$\Delta_{fus}H$ (T_m):   Vapor pressure (0°C): N/A
$\Delta_{vap}H$ (T_b):   Vapor pressure (25°C): N/A
$\Delta_{vap}H$ (25°C):   Vapor pressure (100°C): N/A

PROPERTIES AT 25°C AND 100 kPa

	Solid	Liquid	Gas		Liquid
$\Delta_f H°$/kJ mol^{-1}:	-811.7		-675.8	d:	
$S°$/J mol^{-1}K^{-1}:	168.0			η:	
C_p/J mol^{-1}K^{-1}:	142.0			k:	

COMMENTS:

Name: Furan
Synonym: Oxacyclopentadiene

Mol. Form.: C_4H_4O

CAS RN: 110-00-9
Merck No.: 4206
Mol. Wt.: 68.075

PHYSICAL CONSTANTS

T_m: -85.6°C (187.5 K)	T_c: 217.1°C (490.2 K)	μ: 0.66 D
T_b: 31.5°C (304.6 K)	P_c: 5.50 MPa	IP: 8.88 eV

TRANSITION PROPERTIES

$\Delta_{fus}H$ (T_m): 3.80 kJ/mol   Vapor pressure (0°C): 27.7 kPa
$\Delta_{vap}H$ (T_b): 27.10 kJ/mol   Vapor pressure (25°C): 80.0 kPa
$\Delta_{vap}H$ (25°C): 27.45 kJ/mol   Vapor pressure (100°C): 730 kPa

PROPERTIES AT 25°C AND 100 kPa

	Solid	Liquid	Gas		Liquid
$\Delta_f H°$/kJ mol^{-1}:		-62.3	-34.9	d:	0.9348 g/mL
$S°$/J mol^{-1}K^{-1}:		177.0	267.2	η:	0.361 mPa s
C_p/J mol^{-1}K^{-1}:		115.3	65.4	k:	0.126 W/m K

COMMENTS: Highly toxic; very flammable

Name: Furfural
Synonyms: 2-Furaldehyde
            2-Furancarboxaldehyde
Mol. Form.: $C_5H_4O_2$

CAS RN: 98-01-1
Merck No.: 4214
Mol. Wt.: 96.086

PHYSICAL CONSTANTS

T_m: -36.5°C (236.6 K)	T_c: 397°C (670 K)	μ:
T_b: 161.75°C (434.90 K)	P_c: 5.89 MPa	IP: 9.21 eV

TRANSITION PROPERTIES

$\Delta_{fus}H$ (T_m): 14.35 kJ/mol   Vapor pressure (0°C):
$\Delta_{vap}H$ (T_b): 43.22 kJ/mol   Vapor pressure (25°C): 0.290 kPa
$\Delta_{vap}H$ (25°C): 50.6 kJ/mol   Vapor pressure (100°C): 13.7 kPa

PROPERTIES AT 25°C AND 100 kPa

	Solid	Liquid	Gas		Liquid
$\Delta_f H°$/kJ mol^{-1}:		-201.6	-151.0	d:	1.1554 g/mL
$S°$/J mol^{-1}K^{-1}:				η:	1.49 mPa s
C_p/J mol^{-1}K^{-1}:		163.2		k:	

COMMENTS: TLV=2 ppm

Name: Furfuryl alcohol
Synonyms: 2-Furanmethanol
         2-(Hydroxymethyl)furan
Mol. Form.: $C_5H_6O_2$

CAS RN: 98-00-0
Merck No.: 4215
Mol. Wt.: 98.101

PHYSICAL CONSTANTS

T_m: -31°C (242 K)	T_c:	μ:
T_b: 171°C (444 K)	P_c:	IP:

TRANSITION PROPERTIES

$\Delta_{fus}H$ (T_m): 13.13 kJ/mol
$\Delta_{vap}H$ (T_b): 53.64 kJ/mol
$\Delta_{vap}H$ (25°C): 64.43 kJ/mol

Vapor pressure (0°C):
Vapor pressure (25°C): 0.097 kPa
Vapor pressure (100°C): 6.60 kPa

PROPERTIES AT 25°C AND 100 kPa

	Solid	Liquid	Gas		Liquid
$\Delta_f H°$/kJ mol^{-1}:		-276.2	-211.8	d:	1.1308 g/mL
$S°$/J mol^{-1}K^{-1}:				η:	1.59 mPa s
C_p/J mol^{-1}K^{-1}:		204.0		k:	
COMMENTS: TLV=10 ppm					

Name: Gallium

CAS RN: 7440-55-3
Merck No.: 4252
Mol. Wt.: 69.723

Mol. Form.: Ga

PHYSICAL CONSTANTS

T_m: 29.76°C (302.91 K)	T_c:	μ:
T_b: 2204°C (2477 K)	P_c:	IP: 6.00 eV

TRANSITION PROPERTIES

$\Delta_{fus}H$ (T_m): 5.59 kJ/mol
$\Delta_{vap}H$ (T_b): 254.00 kJ/mol
$\Delta_{vap}H$ (25°C): 271.4 kJ/mol

Vapor pressure (0°C):
Vapor pressure (25°C):
Vapor pressure (100°C):

PROPERTIES AT 25°C AND 100 kPa

	Solid	Liquid	Gas		Solid
$\Delta_f H°$/kJ mol^{-1}:	0.0	5.6	277.0	d:	5.91 g/mL
$S°$/J mol^{-1}K^{-1}:	40.9		169.1	η:	N/A
C_p/J mol^{-1}K^{-1}:	25.9		25.4	k:	40.8 W/m K
COMMENTS:					

Name: Germane
Synonym: Germanium hydride

CAS RN: 7782-65-2
Merck No.: 4301
Mol. Wt.: 76.642

Mol. Form.: GeH_4

PHYSICAL CONSTANTS

T_m: -165°C (108 K)	T_c: 39.1°C (312.2 K)	μ: 0 D
T_b: -90°C (183 K)	P_c: 4.95 MPa	IP: 11.33 eV

TRANSITION PROPERTIES

$\Delta_{fus}H$ (T_m):
$\Delta_{vap}H$ (T_b): 14.06 kJ/mol
$\Delta_{vap}H$ (25°C):

Vapor pressure (0°C):
Vapor pressure (25°C):
Vapor pressure (100°C): N/A

PROPERTIES AT 25°C AND 100 kPa

	Solid	Liquid	Gas		Gas
$\Delta_f H°$/kJ mol^{-1}:			90.8	d:	3.133 g/L
$S°$/J mol^{-1}K^{-1}:			217.1	η:	
C_p/J mol^{-1}K^{-1}:			45.0	k:	
COMMENTS:					

Name: Germanium

CAS RN: 7440-56-4
Merck No.: 4302
Mol. Wt.: 72.610

Mol. Form.: Ge

PHYSICAL CONSTANTS

T_m: 938.25°C (1211.40 K)	T_c:	μ:
T_b: 2833°C (3106 K)	P_c:	IP: 7.90 eV

TRANSITION PROPERTIES

$\Delta_{fus}H$ (T_m): 36.94 kJ/mol   Vapor pressure (0°C): N/A
$\Delta_{vap}H$ (T_b): 334.00 kJ/mol   Vapor pressure (25°C): N/A
$\Delta_{vap}H$ (25°C):   Vapor pressure (100°C): N/A

PROPERTIES AT 25°C AND 100 kPa

	Solid	Liquid	Gas		Solid
$\Delta_f H°$/kJ mol^{-1}:	0.0		372.0	d:	5.3234 g/mL
$S°$/J mol^{-1}K^{-1}:	31.1		167.9	η:	N/A
C_p/J mol^{-1}K^{-1}:	23.3		30.7	k:	60.2 W/m K
COMMENTS:					

Name: Germanium tetrachloride
Synonym: Tetrachlorogermane

CAS RN: 10038-98-9
Merck No.: 4305
Mol. Wt.: 214.421

Mol. Form.: Cl$_4$Ge

PHYSICAL CONSTANTS

T_m: -49.5°C (223.6 K)	T_c: 280.1°C (553.2 K)	μ: 0 D
T_b: 86.55°C (359.70 K)	P_c: 3.861 MPa	IP: 11.68 eV

TRANSITION PROPERTIES

$\Delta_{fus}H$ (T_m):   Vapor pressure (0°C):
$\Delta_{vap}H$ (T_b): 27.90 kJ/mol   Vapor pressure (25°C):
$\Delta_{vap}H$ (25°C): 36.0 kJ/mol   Vapor pressure (100°C):

PROPERTIES AT 25°C AND 100 kPa

	Solid	Liquid	Gas		Liquid
$\Delta_f H°$/kJ mol^{-1}:		-531.8	-495.8	d:	1.88 g/mL
$S°$/J mol^{-1}K^{-1}:		245.6	347.7	η:	
C_p/J mol^{-1}K^{-1}:			96.1	k:	
COMMENTS:					

Name: Glutaric acid
Synonym: Pentanedioic acid

CAS RN: 110-94-1
Merck No.: 4367
Mol. Wt.: 132.116

Mol. Form.: C$_5$H$_8$O$_4$

PHYSICAL CONSTANTS

T_m: 97.8°C (370.9 K)	T_c:	μ:
T_b: 303°C (576 K)	P_c:	IP:

TRANSITION PROPERTIES

$\Delta_{fus}H$ (T_m): 20.90 kJ/mol   Vapor pressure (0°C):
$\Delta_{vap}H$ (T_b):   Vapor pressure (25°C):
$\Delta_{vap}H$ (25°C):   Vapor pressure (100°C):

PROPERTIES AT 25°C AND 100 kPa

	Solid	Liquid	Gas		Solid
$\Delta_f H°$/kJ mol^{-1}:	-960.0			d:	1.415 g/mL
$S°$/J mol^{-1}K^{-1}:				η:	N/A
C_p/J mol^{-1}K^{-1}:				k:	
COMMENTS:					

Name: Glycerol
Synonyms: 1,2,3-Propanetriol
    1,2,3-Trihydroxypropane
Mol. Form.: $C_3H_8O_3$

CAS RN: 56-81-5
Merck No.: 4379
Mol. Wt.: 92.095

PHYSICAL CONSTANTS

T_m: 18.2°C (291.3 K)	T_c:	μ:
T_b: 290°C (563 K)	P_c:	IP:

TRANSITION PROPERTIES

$\Delta_{fus}H$ (T_m): 18.28 kJ/mol   Vapor pressure (0°C):
$\Delta_{vap}H$ (T_b): 61.04 kJ/mol   Vapor pressure (25°C):
$\Delta_{vap}H$ (25°C): 85.8 kJ/mol   Vapor pressure (100°C):

PROPERTIES AT 25°C AND 100 kPa

	Solid	Liquid	Gas		Liquid
$\Delta_f H°$/kJ mol^{-1}:		-668.5	-582.7	d:	1.2567 g/mL
$S°$/J mol^{-1}K^{-1}:		206.3		η:	923 mPa s
C_p/J mol^{-1}K^{-1}:		218.9		k:	0.292 W/m K
COMMENTS:					

Name: Glycolic acid
Synonym: Hydroxyacetic acid

Mol. Form.: $C_2H_4O_3$

CAS RN: 79-14-1
Merck No.: 4394
Mol. Wt.: 76.052

PHYSICAL CONSTANTS

T_m: 79.5°C (352.6 K)	T_c:	μ:
T_b: 100°C (373 K)	P_c:	IP:

TRANSITION PROPERTIES

$\Delta_{fus}H$ (T_m):   Vapor pressure (0°C):
$\Delta_{vap}H$ (T_b):   Vapor pressure (25°C):
$\Delta_{vap}H$ (25°C):   Vapor pressure (100°C): 101 kPa

PROPERTIES AT 25°C AND 100 kPa

	Solid	Liquid	Gas		Solid
$\Delta_f H°$/kJ mol^{-1}:				d:	
$S°$/J mol^{-1}K^{-1}:				η:	N/A
C_p/J mol^{-1}K^{-1}:				k:	
COMMENTS:					

Name: Glyoxal
Synonyms: Ethanedial
    Biformyl
Mol. Form.: $C_2H_2O_2$

CAS RN: 107-22-2
Merck No.: 4405
Mol. Wt.: 58.037

PHYSICAL CONSTANTS

T_m: 15°C (288 K)	T_c:	μ:
T_b: 50.4°C (323.5 K)	P_c:	IP: 10.10 eV

TRANSITION PROPERTIES

$\Delta_{fus}H$ (T_m):   Vapor pressure (0°C):
$\Delta_{vap}H$ (T_b): 38 kJ/mol   Vapor pressure (25°C):
$\Delta_{vap}H$ (25°C):   Vapor pressure (100°C):

PROPERTIES AT 25°C AND 100 kPa

	Solid	Liquid	Gas		Liquid
$\Delta_f H°$/kJ mol^{-1}:			-212.0	d:	1.134 g/mL
$S°$/J mol^{-1}K^{-1}:				η:	
C_p/J mol^{-1}K^{-1}:				k:	
COMMENTS:					

Name: Gold

CAS RN: 7440-57-5
Merck No.: 4412
Mol. Wt.: 196.967

Mol. Form.: Au

PHYSICAL CONSTANTS

T_m: 1064.18°C (1337.33 K) T_c: μ:
T_b: 2856°C (3129 K) P_c: IP: 9.23 eV

TRANSITION PROPERTIES

$\Delta_{fus}H$ (T_m): 12.55 kJ/mol   Vapor pressure (0°C): N/A
$\Delta_{vap}H$ (T_b): 324.00 kJ/mol   Vapor pressure (25°C): N/A
$\Delta_{vap}H$ (25°C):   Vapor pressure (100°C): N/A

PROPERTIES AT 25°C AND 100 kPa

	Solid	Liquid	Gas		Solid
$\Delta_f H°$/kJ mol^{-1}:	0.0		366.1	d:	19.3 g/mL
$S°$/J mol^{-1}K^{-1}:	47.4		180.5	η:	N/A
C_p/J mol^{-1}K^{-1}:	25.4		20.8	k:	318 W/m K

COMMENTS:

Name: Hafnium

CAS RN: 7440-58-6
Merck No.: 4501
Mol. Wt.: 178.490

Mol. Form.: Hf

PHYSICAL CONSTANTS

T_m: 2233°C (2506 K) T_c: μ:
T_b: 4603°C (4876 K) P_c: IP: 6.83 eV

TRANSITION PROPERTIES

$\Delta_{fus}H$ (T_m): 27.20 kJ/mol   Vapor pressure (0°C): N/A
$\Delta_{vap}H$ (T_b):   Vapor pressure (25°C): N/A
$\Delta_{vap}H$ (25°C):   Vapor pressure (100°C): N/A

PROPERTIES AT 25°C AND 100 kPa

	Solid	Liquid	Gas		Solid
$\Delta_f H°$/kJ mol^{-1}:	0.0		619.2	d:	13.3 g/mL
$S°$/J mol^{-1}K^{-1}:	43.6		186.9	η:	N/A
C_p/J mol^{-1}K^{-1}:	25.7		20.8	k:	23.0 W/m K

COMMENTS:

Name: Helium

CAS RN: 7440-59-7
Merck No.: 4546
Mol. Wt.: 4.003

Mol. Form.: He

PHYSICAL CONSTANTS

T_m: T_c: -267.96°C (5.19 K)   μ: 0 D
T_b: -268.93°C (4.22 K) P_c: 0.227 MPa   IP: 24.59 eV

TRANSITION PROPERTIES

$\Delta_{fus}H$ (T_m):   Vapor pressure (0°C): N/A
$\Delta_{vap}H$ (T_b): 0.08 kJ/mol   Vapor pressure (25°C): N/A
$\Delta_{vap}H$ (25°C):   Vapor pressure (100°C): N/A

PROPERTIES AT 25°C AND 100 kPa

	Solid	Liquid	Gas		Gas
$\Delta_f H°$/kJ mol^{-1}:			0.0	d:	0.164 g/L
$S°$/J mol^{-1}K^{-1}:			126.2	η:	19.9 μPa s
C_p/J mol^{-1}K^{-1}:			20.8	k:	0.1567 W/m K

COMMENTS:

Name: Heptadecane

CAS RN: 629-78-7
Merck No.:
Mol. Wt.: 240.473

Mol. Form.: $C_{17}H_{36}$

PHYSICAL CONSTANTS

T_m: 22°C (295 K) T_c: 462°C (735 K) μ:
T_b: 302.0°C (575.1 K) P_c: 1.37 MPa IP:

TRANSITION PROPERTIES

$\Delta_{fus}H\ (T_m)$: Vapor pressure (0°C):
$\Delta_{vap}H\ (T_b)$: Vapor pressure (25°C):
$\Delta_{vap}H\ (25°C)$: 86.02 kJ/mol Vapor pressure (100°C):

PROPERTIES AT 25°C AND 100 kPa

	Solid	Liquid	Gas		Liquid
$\Delta_f H°$/kJ mol^{-1}:					d: 0.7746 g/mL
$S°$/J mol^{-1}K^{-1}:					η: 3.5 mPa s
C_p/J mol^{-1}K^{-1}:					k:
COMMENTS:					

Name: 1-Heptadecanol
Synonyms: Margaryl alcohol
        Heptadecyl alcohol
Mol. Form.: $C_{17}H_{36}O$

CAS RN: 1454-85-9
Merck No.:
Mol. Wt.: 256.472

PHYSICAL CONSTANTS

T_m: 53.8°C (326.9 K) T_c: 507°C (780 K) μ:
T_b: 333°C (606 K) P_c: 1.50 MPa IP:

TRANSITION PROPERTIES

$\Delta_{fus}H\ (T_m)$: Vapor pressure (0°C):
$\Delta_{vap}H\ (T_b)$: Vapor pressure (25°C):
$\Delta_{vap}H\ (25°C)$: Vapor pressure (100°C):

PROPERTIES AT 25°C AND 100 kPa

	Solid	Liquid	Gas		Solid
$\Delta_f H°$/kJ mol^{-1}:					d:
$S°$/J mol^{-1}K^{-1}:					η: N/A
C_p/J mol^{-1}K^{-1}:					k:
COMMENTS:					

Name: Heptanal
Synonyms: Heptaldehyde
        Enanthaldehyde
Mol. Form.: $C_7H_{14}O$

CAS RN: 111-71-7
Merck No.: 4578
Mol. Wt.: 114.188

PHYSICAL CONSTANTS

T_m: -43.3°C (229.8 K) T_c: μ:
T_b: 152.8°C (425.9 K) P_c: IP: 9.65 eV

TRANSITION PROPERTIES

$\Delta_{fus}H\ (T_m)$: Vapor pressure (0°C):
$\Delta_{vap}H\ (T_b)$: Vapor pressure (25°C):
$\Delta_{vap}H\ (25°C)$: 47.7 kJ/mol Vapor pressure (100°C): 19.0 kPa

PROPERTIES AT 25°C AND 100 kPa

	Solid	Liquid	Gas		Liquid
$\Delta_f H°$/kJ mol^{-1}:		-311.5	-263.8		d: 0.8132 g/mL
$S°$/J mol^{-1}K^{-1}:		335.4			η:
C_p/J mol^{-1}K^{-1}:		230.1			k:
COMMENTS:					

Name: Heptane

CAS RN: 142-82-5
Merck No.: 4580
Mol. Wt.: 100.204

Mol. Form.: C_7H_{16}

PHYSICAL CONSTANTS

T_m: -90.6°C (182.5 K) T_c: 267.2°C (540.3 K) μ:
T_b: 98.5°C (371.6 K) P_c: 2.756 MPa   IP: 9.92 eV

TRANSITION PROPERTIES

$\Delta_{fus}H$ (T_m): 14.16 kJ/mol   Vapor pressure (0°C): 1.52 kPa
$\Delta_{vap}H$ (T_b): 31.77 kJ/mol   Vapor pressure (25°C): 6.09 kPa
$\Delta_{vap}H$ (25°C): 36.57 kJ/mol   Vapor pressure (100°C): 106 kPa

PROPERTIES AT 25°C AND 100 kPa

	Solid	Liquid	Gas		Liquid
$\Delta_f H°$/kJ mol^{-1}:		-224.2	-187.7		d: 0.6795 g/mL
$S°$/J mol^{-1}K^{-1}:					η: 0.387 mPa s
C_p/J mol^{-1}K^{-1}:		224.7			k: 0.1228 W/m K

COMMENTS: TLV=400 ppm; flammable

Name: Heptanoic acid
Synonym: Enanthylic acid

CAS RN: 111-14-8
Merck No.: 4581
Mol. Wt.: 130.187

Mol. Form.: $C_7H_{14}O_2$

PHYSICAL CONSTANTS

T_m: -7.5°C (265.6 K) T_c: 406°C (679 K) μ:
T_b: 222.25°C (495.40 K) P_c: 2.90 MPa   IP:

TRANSITION PROPERTIES

$\Delta_{fus}H$ (T_m):   Vapor pressure (0°C):
$\Delta_{vap}H$ (T_b):   Vapor pressure (25°C):
$\Delta_{vap}H$ (25°C): 74.0 kJ/mol   Vapor pressure (100°C):

PROPERTIES AT 25°C AND 100 kPa

	Solid	Liquid	Gas		Liquid
$\Delta_f H°$/kJ mol^{-1}:		-610.2	-536.2		d: 0.9140 g/mL
$S°$/J mol^{-1}K^{-1}:					η: 3.84 mPa s
C_p/J mol^{-1}K^{-1}:		265.4			k:

COMMENTS:

Name: 1-Heptanol
Synonyms: Enanthic alcohol
          Heptyl alcohol
Mol. Form.: $C_7H_{16}O$

CAS RN: 111-70-6
Merck No.: 4582
Mol. Wt.: 116.203

PHYSICAL CONSTANTS

T_m: -34°C (239 K) T_c: 359.4°C (632.5 K) μ:
T_b: 176.45°C (449.60 K) P_c: 3.135 MPa   IP: 9.84 eV

TRANSITION PROPERTIES

$\Delta_{fus}H$ (T_m):   Vapor pressure (0°C):
$\Delta_{vap}H$ (T_b):   Vapor pressure (25°C):
$\Delta_{vap}H$ (25°C): 66.81 kJ/mol   Vapor pressure (100°C): 5.53 kPa

PROPERTIES AT 25°C AND 100 kPa

	Solid	Liquid	Gas		Liquid
$\Delta_f H°$/kJ mol^{-1}:		-403.3	-336.4		d: 0.8187 g/mL
$S°$/J mol^{-1}K^{-1}:					η: 5.81 mPa s
C_p/J mol^{-1}K^{-1}:		272.1			k: 0.159 W/m K

COMMENTS:

Name: 2-Heptanol

CAS RN: 543-49-7
Merck No.: 4583
Mol. Wt.: 116.203

Mol. Form.: $C_7H_{16}O$

PHYSICAL CONSTANTS

T_m: -30°C (243 K) T_c: 338.3°C (611.4 K)   μ:
T_b: 159°C (432 K) P_c: IP: 9.70 eV

TRANSITION PROPERTIES

$\Delta_{fus}H$ (T_m):   Vapor pressure (0°C):
$\Delta_{vap}H$ (T_b):   Vapor pressure (25°C): 0.100 kPa
$\Delta_{vap}H$ (25°C):   Vapor pressure (100°C): 12.5 kPa

PROPERTIES AT 25°C AND 100 kPa

	Solid	Liquid	Gas		Liquid
$\Delta_f H°$/kJ mol^{-1}:					d: 0.8139 g/mL
$S°$/J mol^{-1}K^{-1}:					η: 3.95 mPa s
C_p/J mol^{-1}K^{-1}:					k:
COMMENTS:					

Name: 2-Heptanone
Synonyms: Methyl pentyl ketone
        Amyl methyl ketone
Mol. Form.: $C_7H_{14}O$

CAS RN: 110-43-0
Merck No.: 4584
Mol. Wt.: 114.188

PHYSICAL CONSTANTS

T_m: -35°C (238 K) T_c: 338.4°C (611.5 K)   μ:
T_b: 151.05°C (424.20 K) P_c: 3.436 MPa   IP: 9.30 eV

TRANSITION PROPERTIES

$\Delta_{fus}H$ (T_m):   Vapor pressure (0°C):
$\Delta_{vap}H$ (T_b): 38.30 kJ/mol   Vapor pressure (25°C): 0.510 kPa
$\Delta_{vap}H$ (25°C): 47.24 kJ/mol   Vapor pressure (100°C): 20.2 kPa

PROPERTIES AT 25°C AND 100 kPa

	Solid	Liquid	Gas		Liquid
$\Delta_f H°$/kJ mol^{-1}:					d: 0.8111 g/mL
$S°$/J mol^{-1}K^{-1}:					η: 0.714 mPa s
C_p/J mol^{-1}K^{-1}:		232.6			k:
COMMENTS: TLV=50 ppm					

Name: 1-Heptene

CAS RN: 592-76-7
Merck No.:
Mol. Wt.: 98.188

Mol. Form.: C_7H_{14}

PHYSICAL CONSTANTS

T_m: -119.7°C (153.4 K) T_c: 264.2°C (537.3 K)   μ:
T_b: 93.64°C (366.79 K) P_c: 2.921 MPa   IP: 9.44 eV

TRANSITION PROPERTIES

$\Delta_{fus}H$ (T_m): 12.66 kJ/mol   Vapor pressure (0°C): 1.94 kPa
$\Delta_{vap}H$ (T_b):   Vapor pressure (25°C): 7.52 kPa
$\Delta_{vap}H$ (25°C): 35.49 kJ/mol   Vapor pressure (100°C): 122 kPa

PROPERTIES AT 25°C AND 100 kPa

	Solid	Liquid	Gas		Liquid
$\Delta_f H°$/kJ mol^{-1}:		-97.9	-62.3		d: 0.694 g/mL
$S°$/J mol^{-1}K^{-1}:		327.6			η: 0.340 mPa s
C_p/J mol^{-1}K^{-1}:		211.8			k:
COMMENTS: Flammable					

Name: Hexaborane
Synonym: Hexaboron decahydride

Mol. Form.: B_6H_{10}

CAS RN: 23777-80-2
Merck No.: 4598
Mol. Wt.: 74.945

PHYSICAL CONSTANTS

T_m: -62.3°C (210.8 K) T_c: μ: 2.50 D
T_b: 108°C (381 K) P_c: IP: 9.00 eV

TRANSITION PROPERTIES

$\Delta_{fus}H$ (T_m):   Vapor pressure (0°C):
$\Delta_{vap}H$ (T_b):   Vapor pressure (25°C):
$\Delta_{vap}H$ (25°C): 38.3 kJ/mol   Vapor pressure (100°C):

PROPERTIES AT 25°C AND 100 kPa

	Solid	Liquid	Gas		Liquid
$\Delta_f H°$/kJ mol^{-1}:		56.3	94.6		d: 0.67 g/mL
$S°$/J mol^{-1}K^{-1}:					η:
C_p/J mol^{-1}K^{-1}:					k:

COMMENTS: Highly toxic

Name: Hexachlorobenzene
Synonym: Perchlorobenzene

Mol. Form.: C_6Cl_6

CAS RN: 118-74-1
Merck No.: 4600
Mol. Wt.: 284.782

PHYSICAL CONSTANTS

T_m: 231.8°C (504.9 K) T_c: μ:
T_b: 325°C (598 K) P_c: IP: 8.98 eV

TRANSITION PROPERTIES

$\Delta_{fus}H$ (T_m): 23.85 kJ/mol   Vapor pressure (0°C): N/A
$\Delta_{vap}H$ (T_b):   Vapor pressure (25°C): N/A
$\Delta_{vap}H$ (25°C):   Vapor pressure (100°C): N/A

PROPERTIES AT 25°C AND 100 kPa

	Solid	Liquid	Gas		Solid
$\Delta_f H°$/kJ mol^{-1}:	-127.6		-35.5		d: 2.040 g/mL
$S°$/J mol^{-1}K^{-1}:	260.2				η: N/A
C_p/J mol^{-1}K^{-1}:	201.2				k:

COMMENTS: Highly toxic

Name: Hexachloro-1,3-butadiene
Synonym: Perchlorobutadiene

Mol. Form.: C_4Cl_6

CAS RN: 87-68-3
Merck No.:
Mol. Wt.: 260.760

PHYSICAL CONSTANTS

T_m: -21°C (252 K) T_c: μ:
T_b: 215°C (488 K) P_c: IP:

TRANSITION PROPERTIES

$\Delta_{fus}H$ (T_m):   Vapor pressure (0°C):
$\Delta_{vap}H$ (T_b):   Vapor pressure (25°C):
$\Delta_{vap}H$ (25°C):   Vapor pressure (100°C): 1.99 kPa

PROPERTIES AT 25°C AND 100 kPa

	Solid	Liquid	Gas		Liquid
$\Delta_f H°$/kJ mol^{-1}:					d: 1.556 g/mL
$S°$/J mol^{-1}K^{-1}:					η:
C_p/J mol^{-1}K^{-1}:					k:

COMMENTS: TLV=0.02 ppm; carcinogen; highly toxic

Name: Hexachloroethane
Synonym: Perchloroethane

Mol. Form.: C_2Cl_6

CAS RN: 67-72-1
Merck No.: 4601
Mol. Wt.: 236.738

PHYSICAL CONSTANTS

T_m: 187°C (460 K)
T_b: 187°C (460 K)

T_c:
P_c:

μ: 0 D
IP: 11.10 eV

TRANSITION PROPERTIES

$\Delta_{fus}H$ (T_m):
$\Delta_{vap}H$ (T_b):
$\Delta_{vap}H$ (25°C):

Vapor pressure (0°C):
Vapor pressure (25°C): 0.047 kPa
Vapor pressure (100°C): 4.89 kPa

PROPERTIES AT 25°C AND 100 kPa

	Solid	Liquid	Gas		Solid
$\Delta_f H°$/kJ mol^{-1}:	-202.8		-143.6	d:	2.080 g/mL
$S°$/J mol^{-1}K^{-1}:	237.3			η:	N/A
C_p/J mol^{-1}K^{-1}:	198.2			k:	
COMMENTS: TLV=1 ppm; carcinogen					

Name: Hexadecane
Synonym: Cetane

Mol. Form.: $C_{16}H_{34}$

CAS RN: 544-76-3
Merck No.:
Mol. Wt.: 226.446

PHYSICAL CONSTANTS

T_m: 18.19°C (291.34 K)
T_b: 286.86°C (560.01 K)

T_c: 449°C (722 K)
P_c: 1.435 MPa

μ:
IP:

TRANSITION PROPERTIES

$\Delta_{fus}H$ (T_m):
$\Delta_{vap}H$ (T_b):
$\Delta_{vap}H$ (25°C): 81.38 kJ/mol

Vapor pressure (0°C):
Vapor pressure (25°C):
Vapor pressure (100°C):

PROPERTIES AT 25°C AND 100 kPa

	Solid	Liquid	Gas		Liquid
$\Delta_f H°$/kJ mol^{-1}:		-456.1	-374.8	d:	0.769 g/mL
$S°$/J mol^{-1}K^{-1}:				η:	3.03 mPa s
C_p/J mol^{-1}K^{-1}:		501.6		k:	0.140 W/m K
COMMENTS:					

Name: Hexadecanoic acid
Synonym: Palmitic acid

Mol. Form.: $C_{16}H_{32}O_2$

CAS RN: 57-10-3
Merck No.: 6947
Mol. Wt.: 256.429

PHYSICAL CONSTANTS

T_m: 61.82°C (334.97 K)
T_b: 351.5°C (624.6 K)

T_c:
P_c:

μ:
IP:

TRANSITION PROPERTIES

$\Delta_{fus}H$ (T_m): 53.40 kJ/mol
$\Delta_{vap}H$ (T_b):
$\Delta_{vap}H$ (25°C): 101.0 kJ/mol

Vapor pressure (0°C):
Vapor pressure (25°C):
Vapor pressure (100°C):

PROPERTIES AT 25°C AND 100 kPa

	Solid	Liquid	Gas		Solid
$\Delta_f H°$/kJ mol^{-1}:	-891.5	-838.1	-737.1	d:	
$S°$/J mol^{-1}K^{-1}:	452.4			η:	N/A
C_p/J mol^{-1}K^{-1}:	460.7			k:	
COMMENTS:					

Name: 1-Hexadecanol

Synonyms: Cetyl alcohol

          Hexadecyl alcohol

Mol. Form.: $C_{16}H_{34}O$

CAS RN: 36653-82-4

Merck No.: 2020

Mol. Wt.: 242.445

PHYSICAL CONSTANTS

T_m: 49.3°C (322.4 K)	T_c: 497°C (770 K)	μ:
T_b: 334°C (607 K)	P_c: 1.61 MPa	IP:

TRANSITION PROPERTIES

$\Delta_{fus}H$ (T_m): 34.29 kJ/mol	Vapor pressure (0°C):
$\Delta_{vap}H$ (T_b):	Vapor pressure (25°C):
$\Delta_{vap}H$ (25°C):	Vapor pressure (100°C):

PROPERTIES AT 25°C AND 100 kPa

	Solid	Liquid	Gas		Solid
$\Delta_f H°$/kJ mol^{-1}:	-686.5		-517.0	d: 0.982 g/mL	
$S°$/J mol^{-1}K^{-1}:				η: N/A	
C_p/J mol^{-1}K^{-1}:	422.0			k:	
COMMENTS:					

Name: 1-Hexadecene

Synonym: 1-Cetene

Mol. Form.: $C_{16}H_{32}$

CAS RN: 629-73-2

Merck No.:

Mol. Wt.: 224.430

PHYSICAL CONSTANTS

T_m: 4.1°C (277.2 K)	T_c:	μ:
T_b: 284.9°C (558.0 K)	P_c:	IP:

TRANSITION PROPERTIES

$\Delta_{fus}H$ (T_m):	Vapor pressure (0°C):
$\Delta_{vap}H$ (T_b):	Vapor pressure (25°C):
$\Delta_{vap}H$ (25°C): 80.25 kJ/mol	Vapor pressure (100°C):

PROPERTIES AT 25°C AND 100 kPa

	Solid	Liquid	Gas		Liquid
$\Delta_f H°$/kJ mol^{-1}:		-328.7	-248.5	d: 0.777 g/mL	
$S°$/J mol^{-1}K^{-1}:		587.9		η:	
C_p/J mol^{-1}K^{-1}:		488.9		k:	
COMMENTS:					

Name: Hexafluorobenzene

Synonym: Perfluorobenzene

Mol. Form.: C_6F_6

CAS RN: 392-56-3

Merck No.:

Mol. Wt.: 186.056

PHYSICAL CONSTANTS

T_m: 5.35°C (278.50 K)	T_c: 243.58°C (516.73 K)	μ:
T_b: 80.26°C (353.41 K)	P_c: 3.273 MPa	IP: 9.91 eV

TRANSITION PROPERTIES

$\Delta_{fus}H$ (T_m): 11.59 kJ/mol	Vapor pressure (0°C):
$\Delta_{vap}H$ (T_b): 31.66 kJ/mol	Vapor pressure (25°C): 11.3 kPa
$\Delta_{vap}H$ (25°C): 35.71 kJ/mol	Vapor pressure (100°C): 184 kPa

PROPERTIES AT 25°C AND 100 kPa

	Solid	Liquid	Gas		Liquid
$\Delta_f H°$/kJ mol^{-1}:		-991.3	-955.4	d: 1.6144 g/mL	
$S°$/J mol^{-1}K^{-1}:		280.8		η: 2.79 mPa s	
C_p/J mol^{-1}K^{-1}:		221.6		k:	
COMMENTS:					

Name: Hexafluoroethane
Synonyms: Perfluoroethane
        Refrigerant 116
Mol. Form.: C_2F_6

CAS RN: 76-16-4
Merck No.:
Mol. Wt.: 138.012

PHYSICAL CONSTANTS

T_m: -100.7°C (172.4 K)
T_b: -78.1°C (195.0 K)

T_c: 20°C (293 K)
P_c:

μ: 0 D
IP: 13.40 eV

TRANSITION PROPERTIES

$\Delta_{fus}H$ (T_m):
$\Delta_{vap}H$ (T_b): 16.15 kJ/mol
$\Delta_{vap}H$ (25°C):

Vapor pressure (0°C):
Vapor pressure (25°C): N/A
Vapor pressure (100°C): N/A

PROPERTIES AT 25°C AND 100 kPa

	Solid	Liquid	Gas		Gas
$\Delta_f H°$/kJ mol^{-1}:			-1344.2	d:	5.641 g/L
$S°$/J mol^{-1}K^{-1}:			332.3	η:	
C_p/J mol^{-1}K^{-1}:			106.7	k:	
COMMENTS:					

Name: Hexamethylenediamine
Synonym: 1,6-Hexanediamine

Mol. Form.: $C_6H_{16}N_2$

CAS RN: 124-09-4
Merck No.: 4614
Mol. Wt.: 116.207

PHYSICAL CONSTANTS

T_m: 41.5°C (314.6 K)
T_b: 205°C (478 K)

T_c:
P_c:

μ:
IP:

TRANSITION PROPERTIES

$\Delta_{fus}H$ (T_m):
$\Delta_{vap}H$ (T_b):
$\Delta_{vap}H$ (25°C):

Vapor pressure (0°C):
Vapor pressure (25°C):
Vapor pressure (100°C):

PROPERTIES AT 25°C AND 100 kPa

	Solid	Liquid	Gas		Solid
$\Delta_f H°$/kJ mol^{-1}:				d:	
$S°$/J mol^{-1}K^{-1}:				η:	N/A
C_p/J mol^{-1}K^{-1}:				k:	
COMMENTS:					

Name: Hexanal
Synonyms: Caproaldehyde
        Hexaldehyde
Mol. Form.: $C_6H_{12}O$

CAS RN: 66-25-1
Merck No.: 1761
Mol. Wt.: 100.161

PHYSICAL CONSTANTS

T_m: -56°C (217 K)
T_b: 131°C (404 K)

T_c: 318°C (591 K)
P_c: 3.46 MPa

μ:
IP: 9.67 eV

TRANSITION PROPERTIES

$\Delta_{fus}H$ (T_m):
$\Delta_{vap}H$ (T_b):
$\Delta_{vap}H$ (25°C):

Vapor pressure (0°C):
Vapor pressure (25°C): 1.48 kPa
Vapor pressure (100°C): 42.2 kPa

PROPERTIES AT 25°C AND 100 kPa

	Solid	Liquid	Gas		Liquid
$\Delta_f H°$/kJ mol^{-1}:				d:	0.829 g/mL
$S°$/J mol^{-1}K^{-1}:				η:	
C_p/J mol^{-1}K^{-1}:				k:	
COMMENTS: Flammable					

Name: Hexane

Mol. Form.: C_6H_{14}

CAS RN: 110-54-3
Merck No.: 4613
Mol. Wt.: 86.177

PHYSICAL CONSTANTS

T_m: -95.3°C (177.8 K)	T_c: 234.6°C (507.7 K)	μ:
T_b: 68.73°C (341.88 K)	P_c: 3.010 MPa	IP: 10.13 eV

TRANSITION PROPERTIES

$\Delta_{fus}H$ (T_m): 13.08 kJ/mol
$\Delta_{vap}H$ (T_b): 28.85 kJ/mol
$\Delta_{vap}H$ (25°C): 31.56 kJ/mol

Vapor pressure (0°C): 6.05 kPa
Vapor pressure (25°C): 20.2 kPa
Vapor pressure (100°C): 246 kPa

PROPERTIES AT 25°C AND 100 kPa

	Solid	Liquid	Gas		Liquid
$\Delta_f H°$/kJ mol^{-1}:		-198.7	-167.1	d:	0.6548 g/mL
$S°$/J mol^{-1}K^{-1}:				η:	0.300 mPa s
C_p/J mol^{-1}K^{-1}:		195.6		k:	0.120 W/m K

COMMENTS: TLV=50 ppm; flammable

Name: 1,6-Hexanediol
Synonym: Hexamethylene glycol

Mol. Form.: $C_6H_{14}O_2$

CAS RN: 629-11-8
Merck No.: 4610
Mol. Wt.: 118.176

PHYSICAL CONSTANTS

T_m: 42.8°C (315.9 K)	T_c:	μ:
T_b: 208°C (481 K)	P_c:	IP:

TRANSITION PROPERTIES

$\Delta_{fus}H$ (T_m): 25.5 kJ/mol
$\Delta_{vap}H$ (T_b):
$\Delta_{vap}H$ (25°C): 83.2 kJ/mol

Vapor pressure (0°C):
Vapor pressure (25°C):
Vapor pressure (100°C):

PROPERTIES AT 25°C AND 100 kPa

	Solid	Liquid	Gas		Solid
$\Delta_f H°$/kJ mol^{-1}:	-569.9	-544.4	-461.2	d:	
$S°$/J mol^{-1}K^{-1}:				η:	N/A
C_p/J mol^{-1}K^{-1}:				k:	

COMMENTS:

Name: Hexanenitrile
Synonyms: Capronitrile
       Pentyl cyanide
Mol. Form.: $C_6H_{11}N$

CAS RN: 628-73-9
Merck No.:
Mol. Wt.: 97.160

PHYSICAL CONSTANTS

T_m: -80.3°C (192.8 K)	T_c: 360.7°C (633.8 K)	μ:
T_b: 163.65°C (436.80 K)	P_c: 3.30 MPa	IP:

TRANSITION PROPERTIES

$\Delta_{fus}H$ (T_m):
$\Delta_{vap}H$ (T_b):
$\Delta_{vap}H$ (25°C): 47.91 kJ/mol

Vapor pressure (0°C):
Vapor pressure (25°C): 0.355 kPa
Vapor pressure (100°C): 13.8 kPa

PROPERTIES AT 25°C AND 100 kPa

	Solid	Liquid	Gas		Liquid
$\Delta_f H°$/kJ mol^{-1}:				d:	0.801 g/mL
$S°$/J mol^{-1}K^{-1}:				η:	0.912 mPa s
C_p/J mol^{-1}K^{-1}:				k:	

COMMENTS: Highly toxic

Name: Hexanoic acid
Synonym: Caproic acid

Mol. Form.: $C_6H_{12}O_2$

CAS RN: 142-62-1
Merck No.: 1760
Mol. Wt.: 116.160

PHYSICAL CONSTANTS

T_m: -3°C (270 K)
T_b: 205.25°C (478.40 K)

T_c: 389°C (662 K)
P_c: 3.20 MPa

μ:
IP: 10.12 eV

TRANSITION PROPERTIES

$\Delta_{fus}H$ (T_m): 15.40 kJ/mol
$\Delta_{vap}H$ (T_b):
$\Delta_{vap}H$ (25°C): 71.9 kJ/mol

Vapor pressure (0°C):
Vapor pressure (25°C): 0.005 kPa
Vapor pressure (100°C):

PROPERTIES AT 25°C AND 100 kPa

	Solid	Liquid	Gas		Liquid
$\Delta_f H°$/kJ mol^{-1}:		-583.8	-511.9		d: 0.923 g/mL
$S°$/J mol^{-1}K^{-1}:					η: 2.83 mPa s
C_p/J mol^{-1}K^{-1}:		225.0			k:
COMMENTS:					

Name: 1-Hexanol
Synonyms: Caproyl alcohol
       Hexyl alcohol
Mol. Form.: $C_6H_{14}O$

CAS RN: 111-27-3
Merck No.: 4615
Mol. Wt.: 102.177

PHYSICAL CONSTANTS

T_m: -44.6°C (228.5 K)
T_b: 157.6°C (430.7 K)

T_c: 337.6°C (610.7 K)
P_c: 3.47 MPa

μ:
IP: 9.89 eV

TRANSITION PROPERTIES

$\Delta_{fus}H$ (T_m): 15.40 kJ/mol
$\Delta_{vap}H$ (T_b): 44.50 kJ/mol
$\Delta_{vap}H$ (25°C): 61.61 kJ/mol

Vapor pressure (0°C):
Vapor pressure (25°C): 0.110 kPa
Vapor pressure (100°C): 11.4 kPa

PROPERTIES AT 25°C AND 100 kPa

	Solid	Liquid	Gas		Liquid
$\Delta_f H°$/kJ mol^{-1}:		-377.5	-315.8		d: 0.8153 g/mL
$S°$/J mol^{-1}K^{-1}:		287.4			η: 4.58 mPa s
C_p/J mol^{-1}K^{-1}:		240.4			k: 0.150 W/m K
COMMENTS:					

Name: 2-Hexanol

Mol. Form.: $C_6H_{14}O$

CAS RN: 626-93-7
Merck No.:
Mol. Wt.: 102.177

PHYSICAL CONSTANTS

T_m:
T_b: 138°C (411 K)

T_c: 313.1°C (586.2 K)
P_c:

μ:
IP: 9.80 eV

TRANSITION PROPERTIES

$\Delta_{fus}H$ (T_m):
$\Delta_{vap}H$ (T_b): 41.01 kJ/mol
$\Delta_{vap}H$ (25°C): 58.46 kJ/mol

Vapor pressure (0°C):
Vapor pressure (25°C):
Vapor pressure (100°C): 23.2 kPa

PROPERTIES AT 25°C AND 100 kPa

	Solid	Liquid	Gas		Liquid
$\Delta_f H°$/kJ mol^{-1}:		-392.0			d: 0.8105 g/mL
$S°$/J mol^{-1}K^{-1}:					η:
C_p/J mol^{-1}K^{-1}:					k:
COMMENTS:					

Name: 2-Hexanone

Synonyms: Butyl methyl ketone

        Methyl butyl ketone

Mol. Form.: $C_6H_{12}O$

CAS RN: 591-78-6

Merck No.: 5955

Mol. Wt.: 100.161

PHYSICAL CONSTANTS

T_m: -55.5°C (217.6 K) T_c: 313.9°C (587.0 K) μ:

T_b: 127.6°C (400.7 K) P_c: 3.32 MPa   IP: 9.35 eV

TRANSITION PROPERTIES

$\Delta_{fus}H$ (T_m): 14.90 kJ/mol   Vapor pressure (0°C):

$\Delta_{vap}H$ (T_b): 36.35 kJ/mol   Vapor pressure (25°C): 1.54 kPa

$\Delta_{vap}H$ (25°C): 43.14 kJ/mol   Vapor pressure (100°C): 43.2 kPa

PROPERTIES AT 25°C AND 100 kPa

	Solid	Liquid	Gas		Liquid
$\Delta_f H°$/kJ mol^{-1}:		-322.0	-279.8		d: 0.8070 g/mL
$S°$/J mol^{-1}K^{-1}:					η: 0.583 mPa s
C_p/J mol^{-1}K^{-1}:		213.3			k: 0.139 W/m K

COMMENTS: TLV=5 ppm; flammable

Name: 1-Hexene

CAS RN: 592-41-6

Merck No.:

Mol. Wt.: 84.161

Mol. Form.: C_6H_{12}

PHYSICAL CONSTANTS

T_m: -139.76°C (133.39 K) T_c: 231.0°C (504.1 K) μ:

T_b: 63.48°C (336.63 K) P_c: 3.206 MPa   IP: 9.44 eV

TRANSITION PROPERTIES

$\Delta_{fus}H$ (T_m): 9.35 kJ/mol   Vapor pressure (0°C): 7.68 kPa

$\Delta_{vap}H$ (T_b): 28.28 kJ/mol   Vapor pressure (25°C): 24.8 kPa

$\Delta_{vap}H$ (25°C): 30.61 kJ/mol   Vapor pressure (100°C): 284 kPa

PROPERTIES AT 25°C AND 100 kPa

	Solid	Liquid	Gas		Liquid
$\Delta_f H°$/kJ mol^{-1}:		-74.2	-43.5		d: 0.6686 g/mL
$S°$/J mol^{-1}K^{-1}:		295.2			η: 0.252 mPa s
C_p/J mol^{-1}K^{-1}:		183.3			k: 0.121 W/m K

COMMENTS: Flammable

Name: *cis*-2-Hexene

CAS RN: 7688-21-3

Merck No.:

Mol. Wt.: 84.161

Mol. Form.: C_6H_{12}

PHYSICAL CONSTANTS

T_m: -141.1°C (132.0 K) T_c: μ:

T_b: 68.8°C (341.9 K) P_c: IP: 8.97 eV

TRANSITION PROPERTIES

$\Delta_{fus}H$ (T_m):   Vapor pressure (0°C): 6.04 kPa

$\Delta_{vap}H$ (T_b):   Vapor pressure (25°C): 20.0 kPa

$\Delta_{vap}H$ (25°C): 32.19 kJ/mol   Vapor pressure (100°C):

PROPERTIES AT 25°C AND 100 kPa

	Solid	Liquid	Gas		Liquid
$\Delta_f H°$/kJ mol^{-1}:		-83.9	-52.3		d: 0.6824 g/mL
$S°$/J mol^{-1}K^{-1}:					η:
C_p/J mol^{-1}K^{-1}:					k:

COMMENTS: Flammable

Name: *trans*-2-Hexene

CAS RN: 4050-45-7
Merck No.:
Mol. Wt.: 84.161

Mol. Form.: C_6H_{12}

PHYSICAL CONSTANTS

T_m: -133°C (140 K)	T_c:	μ:
T_b: 67.9°C (341.0 K)	P_c:	IP: 8.97 eV

TRANSITION PROPERTIES

$\Delta_{fus}H$ (T_m):   Vapor pressure (0°C): 6.20 kPa
$\Delta_{vap}H$ (T_b):   Vapor pressure (25°C): 20.7 kPa
$\Delta_{vap}H$ (25°C): 31.60 kJ/mol   Vapor pressure (100°C):

PROPERTIES AT 25°C AND 100 kPa

	Solid	Liquid	Gas		Liquid
$\Delta_f H°$/kJ mol^{-1}:		-85.5	-53.9	d:	0.6733 g/mL
$S°$/J mol^{-1}K^{-1}:				η:	
C_p/J mol^{-1}K^{-1}:				k:	
COMMENTS: Flammable					

Name: Hexylamine
Synonyms: 1-Hexanamine
          1-Aminohexane
Mol. Form.: $C_6H_{15}N$

CAS RN: 111-26-2
Merck No.:
Mol. Wt.: 101.192

PHYSICAL CONSTANTS

T_m: -22.9°C (250.2 K)	T_c:	μ:
T_b: 132.8°C (405.9 K)	P_c:	IP: 8.63 eV

TRANSITION PROPERTIES

$\Delta_{fus}H$ (T_m):   Vapor pressure (0°C):
$\Delta_{vap}H$ (T_b): 36.54 kJ/mol   Vapor pressure (25°C): 1.17 kPa
$\Delta_{vap}H$ (25°C): 45.10 kJ/mol   Vapor pressure (100°C): 35.8 kPa

PROPERTIES AT 25°C AND 100 kPa

	Solid	Liquid	Gas		Liquid
$\Delta_f H°$/kJ mol^{-1}:				d:	0.762 g/mL
$S°$/J mol^{-1}K^{-1}:				η:	
C_p/J mol^{-1}K^{-1}:		252.0		k:	
COMMENTS:					

Name: Hexylene glycol
Synonym: 2-Methyl-2,4-pentanediol

CAS RN: 107-41-5
Merck No.: 4631
Mol. Wt.: 118.176

Mol. Form.: $C_6H_{14}O_2$

PHYSICAL CONSTANTS

T_m: -50°C (223 K)	T_c:	μ:
T_b: 197.1°C (470.2 K)	P_c:	IP:

TRANSITION PROPERTIES

$\Delta_{fus}H$ (T_m):   Vapor pressure (0°C):
$\Delta_{vap}H$ (T_b): 57.30 kJ/mol   Vapor pressure (25°C): 2.06 kPa
$\Delta_{vap}H$ (25°C):   Vapor pressure (100°C):

PROPERTIES AT 25°C AND 100 kPa

	Solid	Liquid	Gas		Liquid
$\Delta_f H°$/kJ mol^{-1}:				d:	0.9182 g/mL
$S°$/J mol^{-1}K^{-1}:				η:	
C_p/J mol^{-1}K^{-1}:		336.0		k:	
COMMENTS: TLV=25 ppm					

Name: Hydracrylonitrile

Synonyms: 3-Hydroxypropanenitrile

2-Cyano-1-ethanol

Mol. Form.: C_3H_5NO

CAS RN: 109-78-4

Merck No.: 3751

Mol. Wt.: 71.079

PHYSICAL CONSTANTS

T_m: -46°C (227 K)

T_b: 221°C (494 K)

T_c:

P_c:

μ:

IP:

TRANSITION PROPERTIES

$\Delta_{fus}H$ (T_m):

$\Delta_{vap}H$ (T_b):

$\Delta_{vap}H$ (25°C):

Vapor pressure (0°C):

Vapor pressure (25°C): 0.010 kPa

Vapor pressure (100°C): 1.21 kPa

PROPERTIES AT 25°C AND 100 kPa

	Solid	Liquid	Gas		Liquid
$\Delta_f H°$/kJ mol^{-1}:					d: 1.0404 g/mL
$S°$/J mol^{-1}K^{-1}:					η:
C_p/J mol^{-1}K^{-1}:					k:

COMMENTS: Highly toxic

Name: Hydrazine

Mol. Form.: H_4N_2

CAS RN: 302-01-2

Merck No.: 4691

Mol. Wt.: 32.045

PHYSICAL CONSTANTS

T_m: 1.4°C (274.5 K)

T_b: 113.55°C (386.70 K)

T_c: 380°C (653 K)

P_c: 14.7 MPa

μ: 1.75 D

IP: 8.10 eV

TRANSITION PROPERTIES

$\Delta_{fus}H$ (T_m): 12.60 kJ/mol

$\Delta_{vap}H$ (T_b): 41.80 kJ/mol

$\Delta_{vap}H$ (25°C): 44.70 kJ/mol

Vapor pressure (0°C):

Vapor pressure (25°C): 1.91 kPa

Vapor pressure (100°C):

PROPERTIES AT 25°C AND 100 kPa

	Solid	Liquid	Gas		Liquid
$\Delta_f H°$/kJ mol^{-1}:		50.6	95.4		d: 1.0036 g/mL
$S°$/J mol^{-1}K^{-1}:		121.2	238.5		η: 0.876 mPa s
C_p/J mol^{-1}K^{-1}:		98.9	49.6		k:

COMMENTS: TLV=0.01 ppm; carcinogen; highly toxic; flammable

Name: Hydrazoic acid

Synonym: Hydrogen azide

Mol. Form.: HN_3

CAS RN: 7782-79-8

Merck No.: 4697

Mol. Wt.: 43.028

PHYSICAL CONSTANTS

T_m: -80°C (193 K)

T_b: 35.7°C (308.8 K)

T_c:

P_c:

μ: 1.70 D

IP: 10.72 eV

TRANSITION PROPERTIES

$\Delta_{fus}H$ (T_m):

$\Delta_{vap}H$ (T_b): 30.50 kJ/mol

$\Delta_{vap}H$ (25°C): 30.1 kJ/mol

Vapor pressure (0°C):

Vapor pressure (25°C):

Vapor pressure (100°C):

PROPERTIES AT 25°C AND 100 kPa

	Solid	Liquid	Gas		Liquid
$\Delta_f H°$/kJ mol^{-1}:		264.0	294.1		d:
$S°$/J mol^{-1}K^{-1}:		140.6	239.0		η:
C_p/J mol^{-1}K^{-1}:			43.7		k:

COMMENTS: Explosive

Name: Hydrogen (H$_2$)
Synonym: Dihydrogen

Mol. Form.: H$_2$

CAS RN: 1333-74-0
Merck No.: 4719
Mol. Wt.: 2.016

PHYSICAL CONSTANTS

T_m: -259.34°C (13.81 K)
T_b: -252.87°C (20.28 K)

T_c: -240.18°C (32.97 K)
P_c: 1.293 MPa

μ: 0 D
IP: 15.43 eV

TRANSITION PROPERTIES

$\Delta_{fus}H$ (T_m): 0.12 kJ/mol
$\Delta_{vap}H$ (T_b): 0.90 kJ/mol
$\Delta_{vap}H$ (25°C):

Vapor pressure (0°C): N/A
Vapor pressure (25°C): N/A
Vapor pressure (100°C): N/A

PROPERTIES AT 25°C AND 100 kPa

	Solid	Liquid	Gas	Gas
$\Delta_f H°$/kJ mol^{-1}:			0.0	d: 0.082 g/L
$S°$/J mol^{-1}K^{-1}:			130.7	η: 8.9 µPa s
C_p/J mol^{-1}K^{-1}:			28.8	k: 0.1869 W/m K

COMMENTS: Very flammable

Name: Hydrogen bromide
Synonym: Hydrobromic acid

Mol. Form.: BrH

CAS RN: 10035-10-6
Merck No.: 4720
Mol. Wt.: 80.912

PHYSICAL CONSTANTS

T_m: -86.81°C (186.34 K)
T_b: -66.38°C (206.77 K)

T_c: 90.1°C (363.2 K)
P_c: 8.55 MPa

μ: 0.827 D
IP: 11.66 eV

TRANSITION PROPERTIES

$\Delta_{fus}H$ (T_m): 2.41 kJ/mol
$\Delta_{vap}H$ (T_b):
$\Delta_{vap}H$ (25°C): 12.69 kJ/mol

Vapor pressure (0°C): 1296 kPa
Vapor pressure (25°C): 2454 kPa
Vapor pressure (100°C): N/A

PROPERTIES AT 25°C AND 100 kPa

	Solid	Liquid	Gas	Gas
$\Delta_f H°$/kJ mol^{-1}:			-36.3	d: 3.307 g/L
$S°$/J mol^{-1}K^{-1}:			198.7	η:
C_p/J mol^{-1}K^{-1}:			29.1	k:

COMMENTS: TLV=3 ppm; highly toxic

Name: Hydrogen chloride
Synonym: Hydrochloric acid

Mol. Form.: ClH

CAS RN: 7647-01-0
Merck No.: 4721
Mol. Wt.: 36.461

PHYSICAL CONSTANTS

T_m: -114.18°C (158.97 K)
T_b: -85°C (188 K)

T_c: 51.6°C (324.7 K)
P_c: 8.31 MPa

μ: 1.109 D
IP: 12.75 eV

TRANSITION PROPERTIES

$\Delta_{fus}H$ (T_m): 2.00 kJ/mol
$\Delta_{vap}H$ (T_b): 16.15 kJ/mol
$\Delta_{vap}H$ (25°C): 9.08 kJ/mol

Vapor pressure (0°C): 2559 kPa
Vapor pressure (25°C): 4718 kPa
Vapor pressure (100°C): N/A

PROPERTIES AT 25°C AND 100 kPa

	Solid	Liquid	Gas	Gas
$\Delta_f H°$/kJ mol^{-1}:			-92.3	d: 1.490 g/L
$S°$/J mol^{-1}K^{-1}:			186.9	η: 14.5 µPa s
C_p/J mol^{-1}K^{-1}:			29.1	k: 0.0145 W/m K

COMMENTS: TLV=5 ppm; highly toxic

Name: Hydrogen cyanide
Synonyms: Hydrocyanic acid
              Prussic acid
Mol. Form.: CHN

CAS RN: 74-90-8
Merck No.: 4722
Mol. Wt.: 27.026

PHYSICAL CONSTANTS

T_m: -13.4°C (259.7 K) T_c: 183.6°C (456.7 K) μ: 2.984 D
T_b: 26°C (299 K) P_c: 5.39 MPa IP: 13.60 eV

TRANSITION PROPERTIES

$\Delta_{fus}H$ (T_m): 8.41 kJ/mol Vapor pressure (0°C): 35.3 kPa
$\Delta_{vap}H$ (T_b): Vapor pressure (25°C): 98.8 kPa
$\Delta_{vap}H$ (25°C): 26.2 kJ/mol Vapor pressure (100°C): 936 kPa

PROPERTIES AT 25°C AND 100 kPa

	Solid	Liquid	Gas		Liquid
$\Delta_f H°$/kJ mol^{-1}:		108.9	135.1	d: 0.684 g/mL	
$S°$/J mol^{-1}K^{-1}:		112.8	201.8	η: 0.183 mPa s	
C_p/J mol^{-1}K^{-1}:		70.6	35.9	k:	

COMMENTS: TLV=10 ppm; highly toxic; very flammable

Name: Hydrogen fluoride
Synonym: Hydrofluoric acid

Mol. Form.: FH

CAS RN: 7664-39-3
Merck No.: 4723
Mol. Wt.: 20.006

PHYSICAL CONSTANTS

T_m: -83.36°C (189.79 K) T_c: 188°C (461 K) μ: 1.826 D
T_b: 20°C (293 K) P_c: 6.48 MPa IP: 16.04 eV

TRANSITION PROPERTIES

$\Delta_{fus}H$ (T_m): 4.58 kJ/mol Vapor pressure (0°C): 48.3 kPa
$\Delta_{vap}H$ (T_b): Vapor pressure (25°C): 123 kPa
$\Delta_{vap}H$ (25°C): 26.5 kJ/mol Vapor pressure (100°C):

PROPERTIES AT 25°C AND 100 kPa

	Solid	Liquid	Gas		Gas
$\Delta_f H°$/kJ mol^{-1}:		-299.8	-273.3	d: 0.818 g/L	
$S°$/J mol^{-1}K^{-1}:			173.8	η:	
C_p/J mol^{-1}K^{-1}:				k:	

COMMENTS: TLV=3 ppm; highly toxic

Name: Hydrogen iodide
Synonym: Hydroiodic acid

Mol. Form.: HI

CAS RN: 10034-85-2
Merck No.: 4724
Mol. Wt.: 127.912

PHYSICAL CONSTANTS

T_m: -50.77°C (222.38 K) T_c: 150.9°C (424.0 K) μ: 0.448 D
T_b: -35.55°C (237.60 K) P_c: 8.31 MPa IP: 10.39 eV

TRANSITION PROPERTIES

$\Delta_{fus}H$ (T_m): 2.87 kJ/mol Vapor pressure (0°C): 380 kPa
$\Delta_{vap}H$ (T_b): 19.76 kJ/mol Vapor pressure (25°C): 791 kPa
$\Delta_{vap}H$ (25°C): 17.36 kJ/mol Vapor pressure (100°C): 3870 kPa

PROPERTIES AT 25°C AND 100 kPa

	Solid	Liquid	Gas		Gas
$\Delta_f H°$/kJ mol^{-1}:			26.5	d: 5.228 g/L	
$S°$/J mol^{-1}K^{-1}:			206.6	η:	
C_p/J mol^{-1}K^{-1}:			29.2	k:	

COMMENTS:

Name: Hydrogen peroxide
Synonym: Dihydrogen peroxide

Mol. Form.: H_2O_2

CAS RN: 7722-84-1
Merck No.: 4725
Mol. Wt.: 34.015

PHYSICAL CONSTANTS

T_m: -0.43°C (272.72 K)	T_c:	μ: 1.573 D
T_b: 150.2°C (423.3 K)	P_c:	IP: 10.54 eV

TRANSITION PROPERTIES

$\Delta_{fus}H$ (T_m): 12.50 kJ/mol
$\Delta_{vap}H$ (T_b):
$\Delta_{vap}H$ (25°C): 51.60 kJ/mol

Vapor pressure (0°C):
Vapor pressure (25°C):
Vapor pressure (100°C): 16.2 kPa

PROPERTIES AT 25°C AND 100 kPa

	Solid	Liquid	Gas		Liquid
$\Delta_f H°$/kJ mol^{-1}:		-187.8	-136.3		d: 1.4 g/mL
$S°$/J mol^{-1}K^{-1}:		109.6	232.7		η:
C_p/J mol^{-1}K^{-1}:		89.1	43.1		k:

COMMENTS: TLV=1 ppm; explosive

Name: Hydrogen selenide
Synonym: Dihydrogen selenide

Mol. Form.: H_2Se

CAS RN: 7783-07-5
Merck No.: 4728
Mol. Wt.: 80.976

PHYSICAL CONSTANTS

T_m: -65.73°C (207.42 K)	T_c: 138°C (411 K)	μ:
T_b: -41.25°C (231.90 K)	P_c: 8.92 MPa	IP: 9.88 eV

TRANSITION PROPERTIES

$\Delta_{fus}H$ (T_m):
$\Delta_{vap}H$ (T_b): 19.70 kJ/mol
$\Delta_{vap}H$ (25°C):

Vapor pressure (0°C):
Vapor pressure (25°C):
Vapor pressure (100°C):

PROPERTIES AT 25°C AND 100 kPa

	Solid	Liquid	Gas		Gas
$\Delta_f H°$/kJ mol^{-1}:			29.7		d: 3.310 g/L
$S°$/J mol^{-1}K^{-1}:			219.0		η:
C_p/J mol^{-1}K^{-1}:			34.7		k:

COMMENTS: TLV=0.05 ppm; highly toxic

Name: Hydrogen sulfide
Synonym: Dihydrogen sulfide

Mol. Form.: H_2S

CAS RN: 7783-06-4
Merck No.: 4729
Mol. Wt.: 34.082

PHYSICAL CONSTANTS

T_m: -85.5°C (187.6 K)	T_c: 100.1°C (373.2 K)	μ: 0.97 D
T_b: -59.55°C (213.60 K)	P_c: 8.94 MPa	IP: 10.45 eV

TRANSITION PROPERTIES

$\Delta_{fus}H$ (T_m): 23.80 kJ/mol
$\Delta_{vap}H$ (T_b): 18.67 kJ/mol
$\Delta_{vap}H$ (25°C): 14.08 kJ/mol

Vapor pressure (0°C): 1065 kPa
Vapor pressure (25°C): 2085 kPa
Vapor pressure (100°C): 8909 kPa

PROPERTIES AT 25°C AND 100 kPa

	Solid	Liquid	Gas		Gas
$\Delta_f H°$/kJ mol^{-1}:			-20.6		d: 1.393 g/L
$S°$/J mol^{-1}K^{-1}:			205.8		η: 12.6 μPa s
C_p/J mol^{-1}K^{-1}:			34.2		k: 0.0146 W/m K

COMMENTS: TLV=10 ppm; highly toxic; very flammable

Name: Hydrogen telluride
Synonym: Dihydrogen telluride

Mol. Form.: H_2Te

CAS RN: 7783-09-7
Merck No.: 4730
Mol. Wt.: 129.616

PHYSICAL CONSTANTS

T_m: -49°C (224 K)	T_c:	μ:
T_b: -2°C (271 K)	P_c:	IP:

TRANSITION PROPERTIES

$\Delta_{fus}H$ (T_m):   Vapor pressure (0°C):
$\Delta_{vap}H$ (T_b): 19.20 kJ/mol   Vapor pressure (25°C):
$\Delta_{vap}H$ (25°C):   Vapor pressure (100°C):

PROPERTIES AT 25°C AND 100 kPa

	Solid	Liquid	Gas		Gas
$\Delta_f H°$/kJ mol^{-1}:			99.6	d:	5.298 g/L
$S°$/J mol^{-1}K^{-1}:				η:	
C_p/J mol^{-1}K^{-1}:				k:	
COMMENTS:					

Name: *p*-Hydroquinone
Synonyms: 1,4-Benzenediol
        Quinol
Mol. Form.: $C_6H_6O_2$

CAS RN: 123-31-9
Merck No.: 4738
Mol. Wt.: 110.112

PHYSICAL CONSTANTS

T_m: 172.3°C (445.4 K)	T_c:	μ:
T_b: 286°C (559 K)	P_c:	IP: 7.95 eV

TRANSITION PROPERTIES

$\Delta_{fus}H$ (T_m): 27.11 kJ/mol   Vapor pressure (0°C):
$\Delta_{vap}H$ (T_b):   Vapor pressure (25°C):
$\Delta_{vap}H$ (25°C):   Vapor pressure (100°C):

PROPERTIES AT 25°C AND 100 kPa

	Solid	Liquid	Gas		Solid
$\Delta_f H°$/kJ mol^{-1}:	-364.5		-265.3	d:	1.33 g/mL
$S°$/J mol^{-1}K^{-1}:				η:	N/A
C_p/J mol^{-1}K^{-1}:	136.0			k:	
COMMENTS: Highly toxic					

Name: Hydroxylamine (NH$_2$OH)

Mol. Form.: H_3NO

CAS RN: 7803-49-8
Merck No.: 4759
Mol. Wt.: 33.030

PHYSICAL CONSTANTS

T_m: 33.1°C (306.2 K)	T_c:	μ: 0.59 D
T_b: 58°C (331 K)	P_c:	IP: 10.00 eV

TRANSITION PROPERTIES

$\Delta_{fus}H$ (T_m):   Vapor pressure (0°C):
$\Delta_{vap}H$ (T_b):   Vapor pressure (25°C):
$\Delta_{vap}H$ (25°C):   Vapor pressure (100°C):

PROPERTIES AT 25°C AND 100 kPa

	Solid	Liquid	Gas		Solid
$\Delta_f H°$/kJ mol^{-1}:	-114.2			d:	1.21 g/mL
$S°$/J mol^{-1}K^{-1}:				η:	N/A
C_p/J mol^{-1}K^{-1}:				k:	
COMMENTS:					

Name: Hypophosphorous acid (H$_3$PO$_2$)

Synonym: Phosphonous acid

Mol. Form.: H$_3$O$_2$P

CAS RN: 6303-21-5

Merck No.: 4804

Mol. Wt.: 65.996

PHYSICAL CONSTANTS

T_m: 26.5°C (299.6 K)

T_b: 130°C (403 K)

T_c:

P_c:

μ:

IP:

TRANSITION PROPERTIES

$\Delta_{fus}H$ (T_m): 9.70 kJ/mol

$\Delta_{vap}H$ (T_b):

$\Delta_{vap}H$ (25°C):

Vapor pressure (0°C):

Vapor pressure (25°C):

Vapor pressure (100°C):

PROPERTIES AT 25°C AND 100 kPa

	Solid	Liquid	Gas	Solid
$\Delta_f H°$/kJ mol^{-1}:	-604.6	-595.4		d: 1.49 g/mL
$S°$/J mol^{-1}K^{-1}:				η: N/A
C_p/J mol^{-1}K^{-1}:				k:
COMMENTS:				

Name: 1*H*-Imidazole

Synonyms: 1,3-Diaza-2,4-cyclopentadiene

          1,3-Diazole

Mol. Form.: C$_3$H$_4$N$_2$

CAS RN: 288-32-4

Merck No.: 4828

Mol. Wt.: 68.078

PHYSICAL CONSTANTS

T_m: 90.5°C (363.6 K)

T_b: 257°C (530 K)

T_c:

P_c:

μ: 3.8 D

IP: 8.81 eV

TRANSITION PROPERTIES

$\Delta_{fus}H$ (T_m):

$\Delta_{vap}H$ (T_b):

$\Delta_{vap}H$ (25°C):

Vapor pressure (0°C):

Vapor pressure (25°C):

Vapor pressure (100°C):

PROPERTIES AT 25°C AND 100 kPa

	Solid	Liquid	Gas	Solid
$\Delta_f H°$/kJ mol^{-1}:	58.5			d:
$S°$/J mol^{-1}K^{-1}:				η: N/A
C_p/J mol^{-1}K^{-1}:				k:
COMMENTS:				

Name: Indan

Synonyms: 2,3-Dihydro-1*H*-indene

          1,2-Hydrindene

Mol. Form.: C$_9$H$_{10}$

CAS RN: 496-11-7

Merck No.: 4844

Mol. Wt.: 118.178

PHYSICAL CONSTANTS

T_m: -51.4°C (221.7 K)

T_b: 177.97°C (451.12 K)

T_c: 411.8°C (684.9 K)

P_c: 3.95 MPa

μ:

IP: 8.30 eV

TRANSITION PROPERTIES

$\Delta_{fus}H$ (T_m):

$\Delta_{vap}H$ (T_b): 39.63 kJ/mol

$\Delta_{vap}H$ (25°C): 48.79 kJ/mol

Vapor pressure (0°C):

Vapor pressure (25°C):

Vapor pressure (100°C): 9.16 kPa

PROPERTIES AT 25°C AND 100 kPa

	Solid	Liquid	Gas	Liquid
$\Delta_f H°$/kJ mol^{-1}:		11.5	60.7	d: 0.959 g/mL
$S°$/J mol^{-1}K^{-1}:		56.0		η: 1.36 mPa s
C_p/J mol^{-1}K^{-1}:		190.2		k:
COMMENTS:				

Name: Indene

Synonym: Indonaphthene

Mol. Form.: C_9H_8

CAS RN: 95-13-6

Merck No.: 4851

Mol. Wt.: 116.163

PHYSICAL CONSTANTS

T_m: -1.8°C (271.3 K)

T_b: 182°C (455 K)

T_c:

P_c:

μ:

IP: 8.14 eV

TRANSITION PROPERTIES

$\Delta_{fus}H$ (T_m):

$\Delta_{vap}H$ (T_b):

$\Delta_{vap}H$ (25°C): 52.8 kJ/mol

Vapor pressure (0°C):

Vapor pressure (25°C): 0.220 kPa

Vapor pressure (100°C): 7.74 kPa

PROPERTIES AT 25°C AND 100 kPa

	Solid	Liquid	Gas		Liquid
$\Delta_f H°$/kJ mol^{-1}:		110.6	163.4		d: 0.996 g/mL
$S°$/J mol^{-1}K^{-1}:		215.3			η:
C_p/J mol^{-1}K^{-1}:		186.9			k:

COMMENTS: TLV=10 ppm

Name: Indium

Mol. Form.: In

CAS RN: 7440-74-6

Merck No.: 4857

Mol. Wt.: 114.818

PHYSICAL CONSTANTS

T_m: 156.60°C (429.75 K)

T_b: 2072°C (2345 K)

T_c:

P_c:

μ:

IP: 5.79 eV

TRANSITION PROPERTIES

$\Delta_{fus}H$ (T_m): 3.28 kJ/mol

$\Delta_{vap}H$ (T_b):

$\Delta_{vap}H$ (25°C):

Vapor pressure (0°C):

Vapor pressure (25°C):

Vapor pressure (100°C):

PROPERTIES AT 25°C AND 100 kPa

	Solid	Liquid	Gas		Solid
$\Delta_f H°$/kJ mol^{-1}:	0.0		243.3		d: 7.31 g/mL
$S°$/J mol^{-1}K^{-1}:	57.8		173.8		η: N/A
C_p/J mol^{-1}K^{-1}:	26.7		20.8		k: 81.8 W/m K

COMMENTS:

Name: Iodine (I_2)

Synonym: Diiodine

Mol. Form.: I_2

CAS RN: 7553-56-2

Merck No.: 4908

Mol. Wt.: 253.809

PHYSICAL CONSTANTS

T_m: 113.7°C (386.8 K)

T_b: 184.4°C (457.5 K)

T_c: 546°C (819 K)

P_c:

μ: 0 D

IP: 9.40 eV

TRANSITION PROPERTIES

$\Delta_{fus}H$ (T_m): 15.52 kJ/mol

$\Delta_{vap}H$ (T_b): 41.57 kJ/mol

$\Delta_{vap}H$ (25°C):

Vapor pressure (0°C): 0.004 kPa

Vapor pressure (25°C): 0.041 kPa

Vapor pressure (100°C): 6.13 kPa

PROPERTIES AT 25°C AND 100 kPa

	Solid	Liquid	Gas		Solid
$\Delta_f H°$/kJ mol^{-1}:	0.0		62.4		d: 4.933 g/mL
$S°$/J mol^{-1}K^{-1}:	116.1		260.7		η: N/A
C_p/J mol^{-1}K^{-1}:	54.4		36.9		k:

COMMENTS: TLV=0.1 ppm

Name: Iodine bromide
Synonym: Iodine monobromide

Mol. Form.: BrI

CAS RN: 7789-33-5
Merck No.: 4910
Mol. Wt.: 206.809

PHYSICAL CONSTANTS

T_m: 40°C (313 K)
T_b: 116°C (389 K)

T_c: 446°C (719 K)
P_c:

μ: 0.726 D
IP: 9.79 eV

TRANSITION PROPERTIES

$\Delta_{fus}H$ (T_m):
$\Delta_{vap}H$ (T_b):
$\Delta_{vap}H$ (25°C):

Vapor pressure (0°C):
Vapor pressure (25°C):
Vapor pressure (100°C):

PROPERTIES AT 25°C AND 100 kPa

	Solid	Liquid	Gas	Solid
$\Delta_f H°$/kJ mol^{-1}:			40.8	d: 4.3 g/mL
$S°$/J mol^{-1}K^{-1}:			258.8	η: N/A
C_p/J mol^{-1}K^{-1}:			36.4	k:
COMMENTS:				

Name: Iodine pentafluoride
Synonym: Iodine(V) fluoride

Mol. Form.: F$_5$I

CAS RN: 7783-66-6
Merck No.: 4912
Mol. Wt.: 221.897

PHYSICAL CONSTANTS

T_m: 9.43°C (282.58 K)
T_b: 100.5°C (373.6 K)

T_c:
P_c:

μ: 2.18 D
IP: 12.94 eV

TRANSITION PROPERTIES

$\Delta_{fus}H$ (T_m):
$\Delta_{vap}H$ (T_b): 41.30 kJ/mol
$\Delta_{vap}H$ (25°C): 42.3 kJ/mol

Vapor pressure (0°C):
Vapor pressure (25°C):
Vapor pressure (100°C):

PROPERTIES AT 25°C AND 100 kPa

	Solid	Liquid	Gas	Liquid
$\Delta_f H°$/kJ mol^{-1}:		-864.8	-822.5	d: 3.19 g/mL
$S°$/J mol^{-1}K^{-1}:			327.7	η:
C_p/J mol^{-1}K^{-1}:			99.2	k:
COMMENTS:				

Name: Iodobenzene
Synonym: Phenyl iodide

Mol. Form.: C$_6$H$_5$I

CAS RN: 591-50-4
Merck No.: 4922
Mol. Wt.: 204.010

PHYSICAL CONSTANTS

T_m: -31.3°C (241.8 K)
T_b: 188.45°C (461.60 K)

T_c: 448°C (721 K)
P_c: 4.52 MPa

μ: 1.70 D
IP: 8.69 eV

TRANSITION PROPERTIES

$\Delta_{fus}H$ (T_m): 9.76 kJ/mol
$\Delta_{vap}H$ (T_b): 39.50 kJ/mol
$\Delta_{vap}H$ (25°C): 49.58 kJ/mol

Vapor pressure (0°C):
Vapor pressure (25°C): 0.133 kPa
Vapor pressure (100°C): 6.67 kPa

PROPERTIES AT 25°C AND 100 kPa

	Solid	Liquid	Gas	Liquid
$\Delta_f H°$/kJ mol^{-1}:		117.2	164.9	d: 1.8229 g/mL
$S°$/J mol^{-1}K^{-1}:		205.4		η: 1.55 mPa s
C_p/J mol^{-1}K^{-1}:		158.7		k:
COMMENTS:				

Name: 1-Iodobutane
Synonym: Butyl iodide

Mol. Form.: C_4H_9I

CAS RN: 542-69-8
Merck No.: 1572
Mol. Wt.: 184.020

PHYSICAL CONSTANTS

T_m: -103°C (170 K)
T_b: 130.6°C (403.7 K)

T_c:
P_c:

μ:
IP: 9.23 eV

TRANSITION PROPERTIES

$\Delta_{fus}H$ (T_m):
$\Delta_{vap}H$ (T_b): 34.66 kJ/mol
$\Delta_{vap}H$ (25°C): 40.63 kJ/mol

Vapor pressure (0°C):
Vapor pressure (25°C): 1.85 kPa
Vapor pressure (100°C): 42.0 kPa

PROPERTIES AT 25°C AND 100 kPa

	Solid	Liquid	Gas		Liquid
$\Delta_f H°$/kJ mol^{-1}:					d: 1.6072 g/mL
$S°$/J mol^{-1}K^{-1}:					η: 0.826 mPa s
C_p/J mol^{-1}K^{-1}:					k:
COMMENTS:					

Name: Iodoethane
Synonym: Ethyl iodide

Mol. Form.: C_2H_5I

CAS RN: 75-03-6
Merck No.: 3769
Mol. Wt.: 155.966

PHYSICAL CONSTANTS

T_m: -111.1°C (162.0 K)
T_b: 72.5°C (345.6 K)

T_c:
P_c:

μ: 1.91 D
IP: 9.35 eV

TRANSITION PROPERTIES

$\Delta_{fus}H$ (T_m):
$\Delta_{vap}H$ (T_b): 29.44 kJ/mol
$\Delta_{vap}H$ (25°C): 31.93 kJ/mol

Vapor pressure (0°C): 5.46 kPa
Vapor pressure (25°C): 18.2 kPa
Vapor pressure (100°C):

PROPERTIES AT 25°C AND 100 kPa

	Solid	Liquid	Gas		Liquid
$\Delta_f H°$/kJ mol^{-1}:		-40.2	-7.7		d: 1.9244 g/mL
$S°$/J mol^{-1}K^{-1}:		211.7	306.0		η: 0.556 mPa s
C_p/J mol^{-1}K^{-1}:		115.1	66.9		k:
COMMENTS:					

Name: Iodomethane
Synonym: Methyl iodide

Mol. Form.: CH_3I

CAS RN: 74-88-4
Merck No.: 6002
Mol. Wt.: 141.939

PHYSICAL CONSTANTS

T_m: -66.45°C (206.70 K)
T_b: 42.55°C (315.70 K)

T_c: 255°C (528 K)
P_c:

μ: 1.62 D
IP: 9.54 eV

TRANSITION PROPERTIES

$\Delta_{fus}H$ (T_m):
$\Delta_{vap}H$ (T_b): 27.34 kJ/mol
$\Delta_{vap}H$ (25°C): 27.97 kJ/mol

Vapor pressure (0°C): 18.6 kPa
Vapor pressure (25°C): 53.9 kPa
Vapor pressure (100°C):

PROPERTIES AT 25°C AND 100 kPa

	Solid	Liquid	Gas		Liquid
$\Delta_f H°$/kJ mol^{-1}:		-12.3	14.7		d: 2.2650 g/mL
$S°$/J mol^{-1}K^{-1}:		163.2	254.1		η: 0.469 mPa s
C_p/J mol^{-1}K^{-1}:		126.0	44.1		k:
COMMENTS: TLV=2 ppm; carcinogen; highly toxic					

Name: 1-Iodo-2-methylpropane
Synonym: Isobutyl iodide

Mol. Form.: C_4H_9I

CAS RN: 513-38-2
Merck No.: 5027
Mol. Wt.: 184.020

PHYSICAL CONSTANTS

T_m: -90.7°C (182.4 K)	T_c:	μ:
T_b: 121.1°C (394.2 K)	P_c:	IP: 9.20 eV

TRANSITION PROPERTIES

$\Delta_{fus}H$ (T_m):
$\Delta_{vap}H$ (T_b): 33.54 kJ/mol
$\Delta_{vap}H$ (25°C): 38.83 kJ/mol

Vapor pressure (0°C): 0.468 kPa
Vapor pressure (25°C): 2.10 kPa
Vapor pressure (100°C): 52.8 kPa

PROPERTIES AT 25°C AND 100 kPa

	Solid	Liquid	Gas		Gas
$\Delta_fH°$/kJ mol^{-1}:				d:	1.5952 g/mL
$S°$/J mol^{-1}K^{-1}:				η:	
C_p/J mol^{-1}K^{-1}:				k:	
COMMENTS:					

Name: 1-Iodopropane
Synonym: Propyl iodide

Mol. Form.: C_3H_7I

CAS RN: 107-08-4
Merck No.: 7875
Mol. Wt.: 169.993

PHYSICAL CONSTANTS

T_m: -101.3°C (171.8 K)	T_c:	μ: 2.04 D
T_b: 102.6°C (375.7 K)	P_c:	IP: 9.27 eV

TRANSITION PROPERTIES

$\Delta_{fus}H$ (T_m):
$\Delta_{vap}H$ (T_b): 32.08 kJ/mol
$\Delta_{vap}H$ (25°C): 36.25 kJ/mol

Vapor pressure (0°C): 1.48 kPa
Vapor pressure (25°C): 5.75 kPa
Vapor pressure (100°C): 94.5 kPa

PROPERTIES AT 25°C AND 100 kPa

	Solid	Liquid	Gas		Liquid
$\Delta_fH°$/kJ mol^{-1}:		-66.0	-30.3	d:	1.740 g/mL
$S°$/J mol^{-1}K^{-1}:				η:	0.703 mPa s
C_p/J mol^{-1}K^{-1}:				k:	
COMMENTS:					

Name: 2-Iodopropane
Synonym: Isopropyl iodide

Mol. Form.: C_3H_7I

CAS RN: 75-30-9
Merck No.: 5102
Mol. Wt.: 169.993

PHYSICAL CONSTANTS

T_m: -90°C (183 K)	T_c:	μ:
T_b: 89.5°C (362.6 K)	P_c:	IP: 9.18 eV

TRANSITION PROPERTIES

$\Delta_{fus}H$ (T_m):
$\Delta_{vap}H$ (T_b): 30.68 kJ/mol
$\Delta_{vap}H$ (25°C): 34.06 kJ/mol

Vapor pressure (0°C): 2.65 kPa
Vapor pressure (25°C): 9.36 kPa
Vapor pressure (100°C): 137 kPa

PROPERTIES AT 25°C AND 100 kPa

	Solid	Liquid	Gas		Gas
$\Delta_fH°$/kJ mol^{-1}:		-74.8	-40.3	d:	1.6946 g/mL
$S°$/J mol^{-1}K^{-1}:				η:	0.653 mPa s
C_p/J mol^{-1}K^{-1}:		91.0		k:	
COMMENTS:					

Name: Iridium

CAS RN: 7439-88-5
Merck No.: 4968
Mol. Wt.: 192.220

Mol. Form.: Ir

PHYSICAL CONSTANTS

T_m: 2446°C (2719 K) T_c: μ:
T_b: 4428°C (4701 K) P_c: IP: 9.10 eV

TRANSITION PROPERTIES

$\Delta_{fus}H$ (T_m): 41.12 kJ/mol   Vapor pressure (0°C): N/A
$\Delta_{vap}H$ (T_b):   Vapor pressure (25°C): N/A
$\Delta_{vap}H$ (25°C):   Vapor pressure (100°C): N/A

PROPERTIES AT 25°C AND 100 kPa

	Solid	Liquid	Gas	Solid
$\Delta_f H°$/kJ mol^{-1}:	0.0		665.3	d: 22.5 g/mL
$S°$/J mol^{-1}K^{-1}:	35.5		193.6	η: N/A
C_p/J mol^{-1}K^{-1}:	25.1		20.8	k: 147 W/m K
COMMENTS:				

Name: Iron

CAS RN: 7439-89-6
Merck No.: 4975
Mol. Wt.: 55.847

Mol. Form.: Fe

PHYSICAL CONSTANTS

T_m: 1538°C (1811 K) T_c: μ:
T_b: 2861°C (3134 K) P_c: IP: 7.90 eV

TRANSITION PROPERTIES

$\Delta_{fus}H$ (T_m): 13.81 kJ/mol   Vapor pressure (0°C): N/A
$\Delta_{vap}H$ (T_b):   Vapor pressure (25°C): N/A
$\Delta_{vap}H$ (25°C):   Vapor pressure (100°C): N/A

PROPERTIES AT 25°C AND 100 kPa

	Solid	Liquid	Gas	Solid
$\Delta_f H°$/kJ mol^{-1}:	0.0		416.3	d: 7.87 g/mL
$S°$/J mol^{-1}K^{-1}:	27.3		180.5	η: N/A
C_p/J mol^{-1}K^{-1}:	25.1		25.7	k: 80.2 W/m K
COMMENTS:				

Name: Iron oxide (FeO)
Synonyms: Ferrous oxide
          Iron(II) oxide
Mol. Form.: FeO

CAS RN: 1345-25-1
Merck No.: 4001
Mol. Wt.: 71.846

PHYSICAL CONSTANTS

T_m: 1377°C (1650 K) T_c: μ:
T_b: P_c: IP:

TRANSITION PROPERTIES

$\Delta_{fus}H$ (T_m): 24.00 kJ/mol   Vapor pressure (0°C): N/A
$\Delta_{vap}H$ (T_b):   Vapor pressure (25°C): N/A
$\Delta_{vap}H$ (25°C):   Vapor pressure (100°C): N/A

PROPERTIES AT 25°C AND 100 kPa

	Solid	Liquid	Gas	Solid
$\Delta_f H°$/kJ mol^{-1}:	-272.0			d: 6.0 g/mL
$S°$/J mol^{-1}K^{-1}:				η: N/A
C_p/J mol^{-1}K^{-1}:				k:
COMMENTS:				

Name: Iron oxide (Fe$_3$O$_4$)   CAS RN: 1317-61-9
Synonyms: Triiron tetraoxide   Merck No.: 3988
          Iron(II,III) oxide   Mol. Wt.: 231.539
Mol. Form.: Fe$_3$O$_4$

PHYSICAL CONSTANTS

T_m: 1597°C (1870 K) T_c: μ:
T_b: P_c: IP:

TRANSITION PROPERTIES

$\Delta_{fus}H$ (T_m): 138.00 kJ/mol   Vapor pressure (0°C): N/A
$\Delta_{vap}H$ (T_b):   Vapor pressure (25°C): N/A
$\Delta_{vap}H$ (25°C):   Vapor pressure (100°C): N/A

PROPERTIES AT 25°C AND 100 kPa

	Solid	Liquid	Gas		Solid
$\Delta_f H°$/kJ mol^{-1}:	-1118.4			d:	5.2 g/mL
$S°$/J mol^{-1}K^{-1}:	146.4			η:	N/A
C_p/J mol^{-1}K^{-1}:	143.4			k:	

COMMENTS:

Name: Iron sulfide (FeS)   CAS RN: 1317-37-9
Synonyms: Ferrous sulfide   Merck No.: 4007
          Iron(II) sulfide   Mol. Wt.: 87.913
Mol. Form.: FeS

PHYSICAL CONSTANTS

T_m: 1190°C (1463 K) T_c: μ:
T_b: P_c: IP:

TRANSITION PROPERTIES

$\Delta_{fus}H$ (T_m): 31.50 kJ/mol   Vapor pressure (0°C): N/A
$\Delta_{vap}H$ (T_b):   Vapor pressure (25°C): N/A
$\Delta_{vap}H$ (25°C):   Vapor pressure (100°C): N/A

PROPERTIES AT 25°C AND 100 kPa

	Solid	Liquid	Gas		Solid
$\Delta_f H°$/kJ mol^{-1}:	-100.0			d:	4.3 g/mL
$S°$/J mol^{-1}K^{-1}:	60.3			η:	N/A
C_p/J mol^{-1}K^{-1}:	50.5			k:	

COMMENTS:

Name: Isobutanal   CAS RN: 78-84-2
Synonyms: 2-Methylpropanal   Merck No.: 5038
          Isobutyraldehyde   Mol. Wt.: 72.107
Mol. Form.: C$_4$H$_8$O

PHYSICAL CONSTANTS

T_m: -65.9°C (207.2 K) T_c: μ:
T_b: 64.5°C (337.6 K) P_c: IP: 9.71 eV

TRANSITION PROPERTIES

$\Delta_{fus}H$ (T_m):   Vapor pressure (0°C): 6.66 kPa
$\Delta_{vap}H$ (T_b):   Vapor pressure (25°C): 23.0 kPa
$\Delta_{vap}H$ (25°C): 31.6 kJ/mol   Vapor pressure (100°C):

PROPERTIES AT 25°C AND 100 kPa

	Solid	Liquid	Gas		Liquid
$\Delta_f H°$/kJ mol^{-1}:		-247.4	-215.8	d:	0.7836 g/mL
$S°$/J mol^{-1}K^{-1}:				η:	
C_p/J mol^{-1}K^{-1}:				k:	

COMMENTS: Highly toxic; flammable

Name: Isobutane
Synonym: 2-Methylpropane

Mol. Form.: C_4H_{10}

CAS RN: 75-28-5
Merck No.:
Mol. Wt.: 58.123

PHYSICAL CONSTANTS

T_m: -138.3°C (134.8 K)
T_b: -11.73°C (261.42 K)

T_c: 134.70°C (407.85 K)
P_c: 3.630 MPa

μ: 0.132 D
IP: 10.57 eV

TRANSITION PROPERTIES

$\Delta_{fus}H$ (T_m): 4.66 kJ/mol
$\Delta_{vap}H$ (T_b): 21.30 kJ/mol
$\Delta_{vap}H$ (25°C): 19.23 kJ/mol

Vapor pressure (0°C): 156 kPa
Vapor pressure (25°C): 348 kPa
Vapor pressure (100°C): 1928 kPa

PROPERTIES AT 25°C AND 100 kPa

	Solid	Liquid	Gas		Gas
$\Delta_f H°$/kJ mol^{-1}:		-153.5	-134.2	*d*:	2.376 g/L
$S°$/J mol^{-1}K^{-1}:			295.5	η:	7.5 µPa s
C_p/J mol^{-1}K^{-1}:		143.0	98.7	*k*:	0.0161 W/m K
COMMENTS: Very flammable					

Name: Isobutene
Synonyms: 2-Methyl-1-propene
         Isobutylene
Mol. Form.: C_4H_8

CAS RN: 115-11-7
Merck No.: 5024
Mol. Wt.: 56.107

PHYSICAL CONSTANTS

T_m: -140.4°C (132.7 K)
T_b: -6.95°C (266.20 K)

T_c: 144.8°C (417.9 K)
P_c: 4.000 MPa

μ: 0.503 D
IP: 9.24 eV

TRANSITION PROPERTIES

$\Delta_{fus}H$ (T_m): 5.93 kJ/mol
$\Delta_{vap}H$ (T_b):
$\Delta_{vap}H$ (25°C): 20.6 kJ/mol

Vapor pressure (0°C): 132 kPa
Vapor pressure (25°C):
Vapor pressure (100°C):

PROPERTIES AT 25°C AND 100 kPa

	Solid	Liquid	Gas		Gas
$\Delta_f H°$/kJ mol^{-1}:		-37.5	-16.9	*d*:	2.293 g/L
$S°$/J mol^{-1}K^{-1}:				η:	
C_p/J mol^{-1}K^{-1}:				*k*:	
COMMENTS: Very flammable					

Name: Isobutyl acetate
Synonyms: 2-Methylpropyl acetate
         Isobutyl ethanoate
Mol. Form.: $C_6H_{12}O_2$

CAS RN: 110-19-0
Merck No.: 5014
Mol. Wt.: 116.160

PHYSICAL CONSTANTS

T_m: -98.85°C (174.30 K)
T_b: 116.55°C (389.70 K)

T_c: 288°C (561 K)
P_c: 3.16 MPa

μ:
IP:

TRANSITION PROPERTIES

$\Delta_{fus}H$ (T_m):
$\Delta_{vap}H$ (T_b): 35.85 kJ/mol
$\Delta_{vap}H$ (25°C): 39.20 kJ/mol

Vapor pressure (0°C):
Vapor pressure (25°C): 2.37 kPa
Vapor pressure (100°C): 60.3 kPa

PROPERTIES AT 25°C AND 100 kPa

	Solid	Liquid	Gas		Liquid
$\Delta_f H°$/kJ mol^{-1}:				*d*:	0.8695 g/mL
$S°$/J mol^{-1}K^{-1}:				η:	0.676 mPa s
C_p/J mol^{-1}K^{-1}:		233.8		*k*:	
COMMENTS: TLV=150 ppm; flammable					

Name: Isobutylamine   CAS RN: 78-81-9
Synonyms: 2-Methyl-1-propanamine   Merck No.: 5016
　　　　　1-Amino-2-methylpropane   Mol. Wt.: 73.138
Mol. Form.: $C_4H_{11}N$

PHYSICAL CONSTANTS

T_m: -86.7°C (186.4 K) T_c: 246°C (519 K)   μ:
T_b: 67.75°C (340.90 K) P_c: 4.07 MPa   IP: 8.70 eV

TRANSITION PROPERTIES

$\Delta_{fus}H$ (T_m):   Vapor pressure (0°C): 5.14 kPa
$\Delta_{vap}H$ (T_b): 30.61 kJ/mol   Vapor pressure (25°C): 19.0 kPa
$\Delta_{vap}H$ (25°C): 33.85 kJ/mol   Vapor pressure (100°C):

PROPERTIES AT 25°C AND 100 kPa

	Solid	Liquid	Gas		Liquid
$\Delta_fH°$/kJ mol^{-1}:		-132.6	-98.7		d: 0.7297 g/mL
$S°$/J mol^{-1}K^{-1}:					η: 0.571 mPa s
C_p/J mol^{-1}K^{-1}:		183.2			k:
COMMENTS: Flammable					

Name: Isobutylbenzene   CAS RN: 538-93-2
Synonym: (2-Methylpropyl)benzene   Merck No.: 5018
                                                                         Mol. Wt.: 134.221
Mol. Form.: $C_{10}H_{14}$

PHYSICAL CONSTANTS

T_m: -51.45°C (221.70 K) T_c: 377°C (650 K)   μ:
T_b: 172.79°C (445.94 K) P_c: 3.05 MPa   IP: 8.68 eV

TRANSITION PROPERTIES

$\Delta_{fus}H$ (T_m): 12.51 kJ/mol   Vapor pressure (0°C):
$\Delta_{vap}H$ (T_b): 37.82 kJ/mol   Vapor pressure (25°C): 0.257 kPa
$\Delta_{vap}H$ (25°C): 47.86 kJ/mol   Vapor pressure (100°C): 10.9 kPa

PROPERTIES AT 25°C AND 100 kPa

	Solid	Liquid	Gas		Liquid
$\Delta_fH°$/kJ mol^{-1}:		-69.8	-21.5		d: 0.8491 g/mL
$S°$/J mol^{-1}K^{-1}:					η:
C_p/J mol^{-1}K^{-1}:		240.6			k:
COMMENTS:					

Name: Isobutyl butanoate   CAS RN: 539-90-2
Synonyms: 2-Methylpropyl butanoate   Merck No.: 5020
　　　　　Isobutyl butyrate   Mol. Wt.: 144.214
Mol. Form.: $C_8H_{16}O_2$

PHYSICAL CONSTANTS

T_m: T_c: 338°C (611 K)   μ:
T_b: 156.95°C (430.10 K) P_c: IP:

TRANSITION PROPERTIES

$\Delta_{fus}H$ (T_m):   Vapor pressure (0°C): 0.096 kPa
$\Delta_{vap}H$ (T_b):   Vapor pressure (25°C): 0.500 kPa
$\Delta_{vap}H$ (25°C):   Vapor pressure (100°C): 16.7 kPa

PROPERTIES AT 25°C AND 100 kPa

	Solid	Liquid	Gas		Liquid
$\Delta_fH°$/kJ mol^{-1}:					d: 0.866 g/mL
$S°$/J mol^{-1}K^{-1}:					η:
C_p/J mol^{-1}K^{-1}:					k:
COMMENTS:					

Name: Isobutyl formate

Synonyms: 2-Methylpropyl methanoate

         Isobutyl methanoate

Mol. Form.: $C_5H_{10}O_2$

CAS RN: 542-55-2

Merck No.: 5026

Mol. Wt.: 102.133

PHYSICAL CONSTANTS

T_m: -95.8°C (177.3 K)	T_c: 278°C (551 K)	μ:
T_b: 98.25°C (371.40 K)	P_c: 3.88 MPa	IP:

TRANSITION PROPERTIES

$\Delta_{fus}H$ (T_m):

$\Delta_{vap}H$ (T_b): 33.55 kJ/mol

$\Delta_{vap}H$ (25°C):

Vapor pressure (0°C):

Vapor pressure (25°C): 5.34 kPa

Vapor pressure (100°C): 107 kPa

PROPERTIES AT 25°C AND 100 kPa

	Solid	Liquid	Gas		Liquid
$\Delta_f H°$/kJ mol^{-1}:					d: 0.8732 g/mL
$S°$/J mol^{-1}K^{-1}:					η: 0.6 mPa s
C_p/J mol^{-1}K^{-1}:					k:
COMMENTS: Flammable					

Name: Isobutyl 3-methylbutanoate

Synonyms: 2-Methylpropyl isovalerate

         Isobutyl 3-methylbutyrate

Mol. Form.: $C_9H_{18}O_2$

CAS RN: 589-59-3

Merck No.: 5029

Mol. Wt.: 158.241

PHYSICAL CONSTANTS

T_m:	T_c: 348°C (621 K)	μ:
T_b: 168.55°C (441.70 K)	P_c:	IP:

TRANSITION PROPERTIES

$\Delta_{fus}H$ (T_m):

$\Delta_{vap}H$ (T_b):

$\Delta_{vap}H$ (25°C):

Vapor pressure (0°C):

Vapor pressure (25°C): 0.253 kPa

Vapor pressure (100°C): 10.8 kPa

PROPERTIES AT 25°C AND 100 kPa

	Solid	Liquid	Gas		Liquid
$\Delta_f H°$/kJ mol^{-1}:					d: 0.849 g/mL
$S°$/J mol^{-1}K^{-1}:					η:
C_p/J mol^{-1}K^{-1}:					k:
COMMENTS:					

Name: Isobutyl 2-methylpropanoate

Synonyms: Isobutyl isobutanoate

         Isobutyl isobutyrate

Mol. Form.: $C_8H_{16}O_2$

CAS RN: 97-85-8

Merck No.: 5028

Mol. Wt.: 144.214

PHYSICAL CONSTANTS

T_m: -80.7°C (192.4 K)	T_c: 329°C (602 K)	μ:
T_b: 148.65°C (421.80 K)	P_c:	IP:

TRANSITION PROPERTIES

$\Delta_{fus}H$ (T_m):

$\Delta_{vap}H$ (T_b): 38.20 kJ/mol

$\Delta_{vap}H$ (25°C): 46.40 kJ/mol

Vapor pressure (0°C): 0.097 kPa

Vapor pressure (25°C): 0.552 kPa

Vapor pressure (100°C): 21.2 kPa

PROPERTIES AT 25°C AND 100 kPa

	Solid	Liquid	Gas		Liquid
$\Delta_f H°$/kJ mol^{-1}:					d: 0.850 g/mL
$S°$/J mol^{-1}K^{-1}:					η: 0.83 mPa s
C_p/J mol^{-1}K^{-1}:					k:
COMMENTS:					

Name: Isobutyl propanoate

Synonyms: 2-Methylpropyl propanoate

Isobutyl propionate

Mol. Form.: $C_7H_{14}O_2$

CAS RN: 540-42-1

Merck No.: 5033

Mol. Wt.: 130.187

PHYSICAL CONSTANTS

T_m: -71.4°C (201.7 K)	T_c: 319°C (592 K)	μ:
T_b: 137°C (410 K)	P_c:	IP:

TRANSITION PROPERTIES

$\Delta_{fus}H$ (T_m):   Vapor pressure (0°C):

$\Delta_{vap}H$ (T_b):   Vapor pressure (25°C):

$\Delta_{vap}H$ (25°C):   Vapor pressure (100°C): 31.5 kPa

PROPERTIES AT 25°C AND 100 kPa

	Solid	Liquid	Gas	Liquid
$\Delta_f H°$/kJ mol^{-1}:				d: 0.87 g/mL
$S°$/J mol^{-1}K^{-1}:				η:
C_p/J mol^{-1}K^{-1}:				k:
COMMENTS:				

Name: Isopentane

Synonym: 2-Methylbutane

Mol. Form.: C_5H_{12}

CAS RN: 78-78-4

Merck No.:

Mol. Wt.: 72.150

PHYSICAL CONSTANTS

T_m: -159.9°C (113.2 K)	T_c: 187.28°C (460.43 K)	μ: 0.13 D
T_b: 27.88°C (301.03 K)	P_c: 3.381 MPa	IP: 10.22 eV

TRANSITION PROPERTIES

$\Delta_{fus}H$ (T_m): 5.15 kJ/mol   Vapor pressure (0°C): 34.6 kPa

$\Delta_{vap}H$ (T_b): 24.69 kJ/mol   Vapor pressure (25°C): 91.7 kPa

$\Delta_{vap}H$ (25°C): 24.85 kJ/mol   Vapor pressure (100°C): 723 kPa

PROPERTIES AT 25°C AND 100 kPa

	Solid	Liquid	Gas	Liquid
$\Delta_f H°$/kJ mol^{-1}:		-178.5	-153.7	d: 0.6142 g/mL
$S°$/J mol^{-1}K^{-1}:		260.4		η: 0.214 mPa s
C_p/J mol^{-1}K^{-1}:		164.8		k:
COMMENTS: Very flammable				

Name: Isopentyl acetate

Synonyms: Isoamyl acetate

3-Methylbutyl ethanoate

Mol. Form.: $C_7H_{14}O_2$

CAS RN: 123-92-2

Merck No.: 4993

Mol. Wt.: 130.187

PHYSICAL CONSTANTS

T_m: -78.5°C (194.6 K)	T_c: 326°C (599 K)	μ:
T_b: 142.55°C (415.70 K)	P_c:	IP:

TRANSITION PROPERTIES

$\Delta_{fus}H$ (T_m):   Vapor pressure (0°C):

$\Delta_{vap}H$ (T_b): 37.53 kJ/mol   Vapor pressure (25°C): 0.728 kPa

$\Delta_{vap}H$ (25°C):   Vapor pressure (100°C): 25.7 kPa

PROPERTIES AT 25°C AND 100 kPa

	Solid	Liquid	Gas	Liquid
$\Delta_f H°$/kJ mol^{-1}:				d: 0.8666 g/mL
$S°$/J mol^{-1}K^{-1}:				η: 0.790 mPa s
C_p/J mol^{-1}K^{-1}:		248.5		k:
COMMENTS: TLV=100 ppm; flammable				

Name: Isopentyl butanoate

Synonyms: 3-Methylbutyl butanoate

    Isoamyl butyrate

Mol. Form.: $C_9H_{18}O_2$

CAS RN: 106-27-4

Merck No.: 4997

Mol. Wt.: 158.241

PHYSICAL CONSTANTS

T_m:

T_b: 179°C (452 K)

T_c: 346°C (619 K)

P_c:

μ:

IP:

TRANSITION PROPERTIES

$\Delta_{fus}H$ (T_m):

$\Delta_{vap}H$ (T_b):

$\Delta_{vap}H$ (25°C):

Vapor pressure (0°C):

Vapor pressure (25°C): 0.160 kPa

Vapor pressure (100°C): 8.01 kPa

PROPERTIES AT 25°C AND 100 kPa

	Solid	Liquid	Gas	Liquid
$\Delta_f H°$/kJ mol^{-1}:				d: 0.860 g/mL
$S°$/J mol^{-1}K^{-1}:				η:
C_p/J mol^{-1}K^{-1}:				k:
COMMENTS:				

Name: Isopentyl formate

Synonyms: Isoamyl formate

    3-Methylbutyl methanoate

Mol. Form.: $C_6H_{12}O_2$

CAS RN: 110-45-2

Merck No.: 5001

Mol. Wt.: 116.160

PHYSICAL CONSTANTS

T_m: -93.5°C (179.6 K)

T_b: 123.55°C (396.70 K)

T_c: 305°C (578 K)

P_c:

μ:

IP:

TRANSITION PROPERTIES

$\Delta_{fus}H$ (T_m):

$\Delta_{vap}H$ (T_b):

$\Delta_{vap}H$ (25°C):

Vapor pressure (0°C):

Vapor pressure (25°C): 1.79 kPa

Vapor pressure (100°C): 47.9 kPa

PROPERTIES AT 25°C AND 100 kPa

	Solid	Liquid	Gas	Liquid
$\Delta_f H°$/kJ mol^{-1}:				d: 0.873 g/mL
$S°$/J mol^{-1}K^{-1}:				η:
C_p/J mol^{-1}K^{-1}:				k:
COMMENTS:				

Name: Isopentyl isopentanoate

Synonyms: 3-Methylbutyl 3-methylbutanoate

    Isoamyl isovalerate

Mol. Form.: $C_{10}H_{20}O_2$

CAS RN: 659-70-1

Merck No.: 5003

Mol. Wt.: 172.268

PHYSICAL CONSTANTS

T_m:

T_b: 190.4°C (463.5 K)

T_c:

P_c:

μ:

IP:

TRANSITION PROPERTIES

$\Delta_{fus}H$ (T_m):

$\Delta_{vap}H$ (T_b): 45.86 kJ/mol

$\Delta_{vap}H$ (25°C): 46.74 kJ/mol

Vapor pressure (0°C):

Vapor pressure (25°C):

Vapor pressure (100°C): 5.25 kPa

PROPERTIES AT 25°C AND 100 kPa

	Solid	Liquid	Gas	Liquid
$\Delta_f H°$/kJ mol^{-1}:				d: 0.853 g/mL
$S°$/J mol^{-1}K^{-1}:				η:
C_p/J mol^{-1}K^{-1}:				k:
COMMENTS:				

Name: Isophorone   CAS RN: 78-59-1
Synonym: 3,5,5-Trimethyl-2-cyclohexen-1-one   Merck No.:
                                                                          Mol. Wt.: 138.210
Mol. Form.: $C_9H_{14}O$

PHYSICAL CONSTANTS

T_m: -8.1°C (265.0 K)	T_c:	μ:
T_b: 215.2°C (488.3 K)	P_c:	IP: 9.07 eV

TRANSITION PROPERTIES

$\Delta_{fus}H$ (T_m):   Vapor pressure (0°C):
$\Delta_{vap}H$ (T_b):   Vapor pressure (25°C): 0.060 kPa
$\Delta_{vap}H$ (25°C):   Vapor pressure (100°C): 2.99 kPa

PROPERTIES AT 25°C AND 100 kPa

	Solid	Liquid	Gas		Liquid
$\Delta_fH°$/kJ mol^{-1}:				d: 0.9196 g/mL	
$S°$/J mol^{-1}K^{-1}:				η: 2.33 mPa s	
C_p/J mol^{-1}K^{-1}:		253.5		k:	

COMMENTS: TLV=5 ppm

Name: Isophthalic acid   CAS RN: 121-91-5
Synonym: 1,3-Benzenedicarboxylic acid   Merck No.: 5083
                                                                          Mol. Wt.: 166.133
Mol. Form.: $C_8H_6O_4$

PHYSICAL CONSTANTS

T_m: 347°C (620 K)	T_c:	μ:
T_b:	P_c:	IP: 9.98 eV

TRANSITION PROPERTIES

$\Delta_{fus}H$ (T_m):   Vapor pressure (0°C): N/A
$\Delta_{vap}H$ (T_b):   Vapor pressure (25°C): N/A
$\Delta_{vap}H$ (25°C):   Vapor pressure (100°C): N/A

PROPERTIES AT 25°C AND 100 kPa

	Solid	Liquid	Gas		Solid
$\Delta_fH°$/kJ mol^{-1}:	-803.0		-696.3	d:	
$S°$/J mol^{-1}K^{-1}:				η: N/A	
C_p/J mol^{-1}K^{-1}:				k:	

COMMENTS:

Name: Isopropyl acetate   CAS RN: 108-21-4
Synonyms: 1-Methylethyl acetate   Merck No.: 5093
          Isopropyl ethanoate   Mol. Wt.: 102.133
Mol. Form.: $C_5H_{10}O_2$

PHYSICAL CONSTANTS

T_m: -73.4°C (199.7 K)	T_c: 258°C (531 K)	μ:
T_b: 88.6°C (361.7 K)	P_c:	IP: 9.99 eV

TRANSITION PROPERTIES

$\Delta_{fus}H$ (T_m):   Vapor pressure (0°C):
$\Delta_{vap}H$ (T_b): 32.93 kJ/mol   Vapor pressure (25°C): 8.05 kPa
$\Delta_{vap}H$ (25°C): 37.20 kJ/mol   Vapor pressure (100°C):

PROPERTIES AT 25°C AND 100 kPa

	Solid	Liquid	Gas		Liquid
$\Delta_fH°$/kJ mol^{-1}:		-518.9	-481.7	d: 0.8711 g/mL	
$S°$/J mol^{-1}K^{-1}:				η: 0.52 mPa s	
C_p/J mol^{-1}K^{-1}:		199.4		k:	

COMMENTS: TLV=250 ppm; flammable

Name: Isopropylamine
Synonyms: 2-Propanamine
           2-Aminopropane
Mol. Form.: C_3H_9N

CAS RN: 75-31-0
Merck No.: 5097
Mol. Wt.: 59.111

PHYSICAL CONSTANTS

T_m: -95.14°C (178.01 K)
T_b: 31.76°C (304.91 K)

T_c: 198.7°C (471.8 K)
P_c: 4.54 MPa

μ: 1.19 D
IP: 8.72 eV

TRANSITION PROPERTIES

$\Delta_{fus}H$ (T_m): 7.33 kJ/mol
$\Delta_{vap}H$ (T_b): 27.83 kJ/mol
$\Delta_{vap}H$ (25°C): 28.36 kJ/mol

Vapor pressure (0°C):
Vapor pressure (25°C): 78.0 kPa
Vapor pressure (100°C):

PROPERTIES AT 25°C AND 100 kPa

	Solid	Liquid	Gas		Liquid
$\Delta_f H°$/kJ mol^{-1}:		-112.3	-83.7	d:	0.6821 g/mL
$S°$/J mol^{-1}K^{-1}:		218.3	312.2	η:	0.325 mPa s
C_p/J mol^{-1}K^{-1}:		163.8	97.5	k:	

COMMENTS: TLV=5 ppm; very flammable

Name: Isoquinoline
Synonym: 2-Azanaphthalene

Mol. Form.: C_9H_7N

CAS RN: 119-65-3
Merck No.: 5110
Mol. Wt.: 129.161

PHYSICAL CONSTANTS

T_m: 26.47°C (299.62 K)
T_b: 243.22°C (516.37 K)

T_c: 530°C (803 K)
P_c: 5.10 MPa

μ: 2.73 D
IP: 8.53 eV

TRANSITION PROPERTIES

$\Delta_{fus}H$ (T_m): 7.45 kJ/mol
$\Delta_{vap}H$ (T_b): 48.96 kJ/mol
$\Delta_{vap}H$ (25°C): 60.26 kJ/mol

Vapor pressure (0°C):
Vapor pressure (25°C):
Vapor pressure (100°C):

PROPERTIES AT 25°C AND 100 kPa

	Solid	Liquid	Gas		Solid
$\Delta_f H°$/kJ mol^{-1}:		144.5	204.6	d:	
$S°$/J mol^{-1}K^{-1}:		216.0		η:	N/A
C_p/J mol^{-1}K^{-1}:		196.2		k:	

COMMENTS:

Name: Ketene
Synonym: Ethenone

Mol. Form.: C_2H_2O

CAS RN: 463-51-4
Merck No.: 5177
Mol. Wt.: 42.037

PHYSICAL CONSTANTS

T_m: -151°C (122 K)
T_b: -49.81°C (223.34 K)

T_c:
P_c:

μ: 1.422 D
IP: 9.61 eV

TRANSITION PROPERTIES

$\Delta_{fus}H$ (T_m):
$\Delta_{vap}H$ (T_b): 20.4 kJ/mol
$\Delta_{vap}H$ (25°C):

Vapor pressure (0°C):
Vapor pressure (25°C):
Vapor pressure (100°C):

PROPERTIES AT 25°C AND 100 kPa

	Solid	Liquid	Gas		Gas
$\Delta_f H°$/kJ mol^{-1}:		-67.9	-47.5	d:	1.718 g/L
$S°$/J mol^{-1}K^{-1}:			247.6	η:	
C_p/J mol^{-1}K^{-1}:			51.8	k:	

COMMENTS: TLV=0.5 ppm

Name: Krypton

CAS RN: 7439-90-9
Merck No.: 5202
Mol. Wt.: 83.800

Mol. Form.: Kr

PHYSICAL CONSTANTS

T_m: -157.36°C (115.79 K) T_c: -63.74°C (209.41 K) μ: 0 D
T_b: -153.22°C (119.93 K) P_c: 5.50 MPa   IP: 14.00 eV

TRANSITION PROPERTIES

$\Delta_{fus}H$ (T_m): 1.37 kJ/mol   Vapor pressure (0°C): N/A
$\Delta_{vap}H$ (T_b): 9.08 kJ/mol   Vapor pressure (25°C): N/A
$\Delta_{vap}H$ (25°C):   Vapor pressure (100°C): N/A

PROPERTIES AT 25°C AND 100 kPa

	Solid	Liquid	Gas		Gas
$\Delta_fH°$/kJ mol^{-1}:			0.0	d: 3.425 g/L	
$S°$/J mol^{-1}K^{-1}:			164.1	η: 25.4 µPa s	
C_p/J mol^{-1}K^{-1}:			20.8	k: 0.0095 W/m K	
COMMENTS:					

Name: Lead

CAS RN: 7439-92-1
Merck No.: 5267
Mol. Wt.: 207.200

Mol. Form.: Pb

PHYSICAL CONSTANTS

T_m: 327.46°C (600.61 K) T_c: μ:
T_b: 1749°C (2022 K) P_c: IP: 7.42 eV

TRANSITION PROPERTIES

$\Delta_{fus}H$ (T_m): 4.77 kJ/mol   Vapor pressure (0°C): N/A
$\Delta_{vap}H$ (T_b): 179.50 kJ/mol   Vapor pressure (25°C): N/A
$\Delta_{vap}H$ (25°C):   Vapor pressure (100°C): N/A

PROPERTIES AT 25°C AND 100 kPa

	Solid	Liquid	Gas		Solid
$\Delta_fH°$/kJ mol^{-1}:	0.0		195.2	d: 11.3 g/mL	
$S°$/J mol^{-1}K^{-1}:	64.8		175.4	η: N/A	
C_p/J mol^{-1}K^{-1}:	26.4		20.8	k: 35.3 W/m K	
COMMENTS: Highly toxic					

Name: Lead bromide (PbBr$_2$)
Synonym: Lead(II) bromide

CAS RN: 10031-22-8
Merck No.: 5275
Mol. Wt.: 367.008

Mol. Form.: Br$_2$Pb

PHYSICAL CONSTANTS

T_m: 371°C (644 K) T_c: μ:
T_b: 892°C (1165 K) P_c: IP:

TRANSITION PROPERTIES

$\Delta_{fus}H$ (T_m): 16.44 kJ/mol   Vapor pressure (0°C): N/A
$\Delta_{vap}H$ (T_b): 133.00 kJ/mol   Vapor pressure (25°C): N/A
$\Delta_{vap}H$ (25°C):   Vapor pressure (100°C): N/A

PROPERTIES AT 25°C AND 100 kPa

	Solid	Liquid	Gas		Solid
$\Delta_fH°$/kJ mol^{-1}:	-278.7			d: 6.69 g/mL	
$S°$/J mol^{-1}K^{-1}:	161.5			η: N/A	
C_p/J mol^{-1}K^{-1}:	80.1			k:	
COMMENTS: Highly toxic					

Name: Lead chloride (PbCl$_2$)
Synonyms: Lead dichloride
       Lead(II) chloride
Mol. Form.: Cl$_2$Pb

CAS RN: 7758-95-4
Merck No.: 5278
Mol. Wt.: 278.105

PHYSICAL CONSTANTS

T_m: 501°C (774 K) T_c: μ:
T_b: 951°C (1224 K) P_c: IP: 10.00 eV

TRANSITION PROPERTIES

$\Delta_{fus}H$ (T_m): 21.90 kJ/mol   Vapor pressure (0°C): N/A
$\Delta_{vap}H$ (T_b): 127.00 kJ/mol   Vapor pressure (25°C): N/A
$\Delta_{vap}H$ (25°C):   Vapor pressure (100°C): N/A

PROPERTIES AT 25°C AND 100 kPa

	Solid	Liquid	Gas		Solid
$\Delta_f H°$/kJ mol^{-1}:	-359.4			d:	5.98 g/mL
$S°$/J mol^{-1}K^{-1}:	136.0			η:	N/A
C_p/J mol^{-1}K^{-1}:				k:	

COMMENTS: Highly toxic

Name: Lead fluoride (PbF$_2$)
Synonyms: Lead difluoride
       Lead(II) fluoride
Mol. Form.: F$_2$Pb

CAS RN: 7783-46-2
Merck No.: 5282
Mol. Wt.: 245.197

PHYSICAL CONSTANTS

T_m: 830°C (1103 K) T_c: μ:
T_b: 1293°C (1566 K) P_c: IP: 11.50 eV

TRANSITION PROPERTIES

$\Delta_{fus}H$ (T_m): 14.70 kJ/mol   Vapor pressure (0°C): N/A
$\Delta_{vap}H$ (T_b): 160.40 kJ/mol   Vapor pressure (25°C): N/A
$\Delta_{vap}H$ (25°C):   Vapor pressure (100°C): N/A

PROPERTIES AT 25°C AND 100 kPa

	Solid	Liquid	Gas		Solid
$\Delta_f H°$/kJ mol^{-1}:	-664.0			d:	8.44 g/mL
$S°$/J mol^{-1}K^{-1}:	110.5			η:	N/A
C_p/J mol^{-1}K^{-1}:				k:	

COMMENTS: Highly toxic

Name: Lead iodide (PbI$_2$)
Synonyms: Lead diiodide
       Lead(II) iodide
Mol. Form.: I$_2$Pb

CAS RN: 10101-63-0
Merck No.: 5288
Mol. Wt.: 461.009

PHYSICAL CONSTANTS

T_m: 410°C (683 K) T_c: μ:
T_b: 872°C (1145 K) P_c: IP:

TRANSITION PROPERTIES

$\Delta_{fus}H$ (T_m): 23.40 kJ/mol   Vapor pressure (0°C): N/A
$\Delta_{vap}H$ (T_b): 104.00 kJ/mol   Vapor pressure (25°C): N/A
$\Delta_{vap}H$ (25°C):   Vapor pressure (100°C): N/A

PROPERTIES AT 25°C AND 100 kPa

	Solid	Liquid	Gas		Solid
$\Delta_f H°$/kJ mol^{-1}:	-175.5			d:	6.16 g/mL
$S°$/J mol^{-1}K^{-1}:	174.9			η:	N/A
C_p/J mol^{-1}K^{-1}:	77.4			k:	

COMMENTS: Highly toxic

Name: Lead sulfide (PbS)

Synonym: Lead(II) sulfide

Mol. Form.: PbS

CAS RN: 1314-87-0
Merck No.: 5303
Mol. Wt.: 239.266

PHYSICAL CONSTANTS

T_m: 1113.4°C (1386.5 K)

T_b:

T_c:

P_c:

μ: 3.59 D

IP: 8.50 eV

TRANSITION PROPERTIES

$\Delta_{fus}H$ (T_m): 19.00 kJ/mol

$\Delta_{vap}H$ (T_b):

$\Delta_{vap}H$ (25°C):

Vapor pressure (0°C): N/A

Vapor pressure (25°C): N/A

Vapor pressure (100°C): N/A

PROPERTIES AT 25°C AND 100 kPa

	Solid	Liquid	Gas		Solid
$\Delta_f H°$/kJ mol^{-1}:	-100.4			d:	7.60 g/mL
$S°$/J mol^{-1}K^{-1}:	91.2			η:	N/A
C_p/J mol^{-1}K^{-1}:	49.5			k:	

COMMENTS: Highly toxic

Name: Lithium

Mol. Form.: Li

CAS RN: 7439-93-2
Merck No.: 5395
Mol. Wt.: 6.941

PHYSICAL CONSTANTS

T_m: 180.5°C (453.6 K)

T_b: 1342°C (1615 K)

T_c:

P_c:

μ:

IP: 5.39 eV

TRANSITION PROPERTIES

$\Delta_{fus}H$ (T_m): 3.00 kJ/mol

$\Delta_{vap}H$ (T_b):

$\Delta_{vap}H$ (25°C):

Vapor pressure (0°C):

Vapor pressure (25°C):

Vapor pressure (100°C):

PROPERTIES AT 25°C AND 100 kPa

	Solid	Liquid	Gas		Solid
$\Delta_f H°$/kJ mol^{-1}:	0.0		159.3	d:	0.534 g/mL
$S°$/J mol^{-1}K^{-1}:	29.1		138.8	η:	N/A
C_p/J mol^{-1}K^{-1}:	24.8		20.8	k:	84.8 W/m K

COMMENTS: Flammable

Name: Lithium bromide

Mol. Form.: BrLi

CAS RN: 7550-35-8
Merck No.: 5403
Mol. Wt.: 86.845

PHYSICAL CONSTANTS

T_m: 552°C (825 K)

T_b:

T_c:

P_c:

μ: 7.268 D

IP: 8.70 eV

TRANSITION PROPERTIES

$\Delta_{fus}H$ (T_m): 17.60 kJ/mol

$\Delta_{vap}H$ (T_b):

$\Delta_{vap}H$ (25°C):

Vapor pressure (0°C): N/A

Vapor pressure (25°C): N/A

Vapor pressure (100°C): N/A

PROPERTIES AT 25°C AND 100 kPa

	Solid	Liquid	Gas		Solid
$\Delta_f H°$/kJ mol^{-1}:	-351.2			d:	3.464 g/mL
$S°$/J mol^{-1}K^{-1}:	74.3			η:	N/A
C_p/J mol^{-1}K^{-1}:				k:	

COMMENTS:

Name: Lithium chloride

CAS RN: 7447-41-8
Merck No.: 5405
Mol. Wt.: 42.394

Mol. Form.: ClLi

PHYSICAL CONSTANTS

T_m: 610°C (883 K) T_c: μ: 7.129 D
T_b: 1383°C (1656 K) P_c: IP: 9.57 eV

TRANSITION PROPERTIES

$\Delta_{fus}H$ (T_m): 19.90 kJ/mol Vapor pressure (0°C): N/A
$\Delta_{vap}H$ (T_b): Vapor pressure (25°C): N/A
$\Delta_{vap}H$ (25°C): Vapor pressure (100°C): N/A

PROPERTIES AT 25°C AND 100 kPa

	Solid	**Liquid**	**Gas**	**Solid**
$\Delta_f H°$/kJ mol^{-1}:	-408.6			d: 2.07 g/mL
$S°$/J mol^{-1}K^{-1}:	59.3			η: N/A
C_p/J mol^{-1}K^{-1}:	48.0			k:
COMMENTS:				

Name: Lithium fluoride

CAS RN: 7789-24-4
Merck No.: 5409
Mol. Wt.: 25.939

Mol. Form.: FLi

PHYSICAL CONSTANTS

T_m: 848.2°C (1121.3 K) T_c: μ: 6.326 D
T_b: 1673°C (1946 K) P_c: IP:

TRANSITION PROPERTIES

$\Delta_{fus}H$ (T_m): 27.09 kJ/mol Vapor pressure (0°C): N/A
$\Delta_{vap}H$ (T_b): 147.00 kJ/mol Vapor pressure (25°C): N/A
$\Delta_{vap}H$ (25°C): Vapor pressure (100°C): N/A

PROPERTIES AT 25°C AND 100 kPa

	Solid	**Liquid**	**Gas**	**Solid**
$\Delta_f H°$/kJ mol^{-1}:	-616.0			d: 2.640 g/mL
$S°$/J mol^{-1}K^{-1}:	35.7			η: N/A
C_p/J mol^{-1}K^{-1}:	41.6			k:
COMMENTS: Highly toxic				

Name: Lithium hydroxide

CAS RN: 1310-65-2
Merck No.: 5412
Mol. Wt.: 23.948

Mol. Form.: HLiO

PHYSICAL CONSTANTS

T_m: 471.2°C (744.3 K) T_c: μ: 4.754 D
T_b: 1626°C (1899 K) P_c: IP:

TRANSITION PROPERTIES

$\Delta_{fus}H$ (T_m): 20.88 kJ/mol Vapor pressure (0°C): N/A
$\Delta_{vap}H$ (T_b): 188.00 kJ/mol Vapor pressure (25°C): N/A
$\Delta_{vap}H$ (25°C): Vapor pressure (100°C): N/A

PROPERTIES AT 25°C AND 100 kPa

	Solid	**Liquid**	**Gas**	**Solid**
$\Delta_f H°$/kJ mol^{-1}:	-484.9			d: 1.5 g/mL
$S°$/J mol^{-1}K^{-1}:	42.8			η: N/A
C_p/J mol^{-1}K^{-1}:	49.7			k:
COMMENTS:				

Name: Lithium iodide

CAS RN: 10377-51-2
Merck No.: 5413
Mol. Wt.: 133.846

Mol. Form.: ILi

PHYSICAL CONSTANTS

T_m: 469°C (742 K) T_c: μ: 7.428 D
T_b: 1171°C (1444 K) P_c: IP: 7.50 eV

TRANSITION PROPERTIES

$\Delta_{fus}H$ (T_m): 14.60 kJ/mol   Vapor pressure (0°C): N/A
$\Delta_{vap}H$ (T_b):   Vapor pressure (25°C): N/A
$\Delta_{vap}H$ (25°C):   Vapor pressure (100°C): N/A

PROPERTIES AT 25°C AND 100 kPa

	Solid	Liquid	Gas	Solid
$\Delta_f H°$/kJ mol^{-1}:	-270.4			d: 4.06 g/mL
$S°$/J mol^{-1}K^{-1}:	86.8			η: N/A
C_p/J mol^{-1}K^{-1}:	51.0			k:
COMMENTS:				

Name: Lithium nitrate

CAS RN: 7790-69-4
Merck No.: 5414
Mol. Wt.: 68.946

Mol. Form.: LiNO$_3$

PHYSICAL CONSTANTS

T_m: 253°C (526 K) T_c: μ:
T_b: P_c: IP:

TRANSITION PROPERTIES

$\Delta_{fus}H$ (T_m): 24.90 kJ/mol   Vapor pressure (0°C): N/A
$\Delta_{vap}H$ (T_b):   Vapor pressure (25°C): N/A
$\Delta_{vap}H$ (25°C):   Vapor pressure (100°C): N/A

PROPERTIES AT 25°C AND 100 kPa

	Solid	Liquid	Gas	Solid
$\Delta_f H°$/kJ mol^{-1}:	-483.1			d: 2.38 g/mL
$S°$/J mol^{-1}K^{-1}:	90.0			η: N/A
C_p/J mol^{-1}K^{-1}:				k:
COMMENTS:				

Name: Magnesium

CAS RN: 7439-95-4
Merck No.: 5529
Mol. Wt.: 24.305

Mol. Form.: Mg

PHYSICAL CONSTANTS

T_m: 650°C (923 K) T_c: μ:
T_b: 1090°C (1363 K) P_c: IP: 7.65 eV

TRANSITION PROPERTIES

$\Delta_{fus}H$ (T_m): 8.48 kJ/mol   Vapor pressure (0°C): N/A
$\Delta_{vap}H$ (T_b):   Vapor pressure (25°C): N/A
$\Delta_{vap}H$ (25°C):   Vapor pressure (100°C): N/A

PROPERTIES AT 25°C AND 100 kPa

	Solid	Liquid	Gas	Solid
$\Delta_f H°$/kJ mol^{-1}:	0.0		147.1	d: 1.74 g/mL
$S°$/J mol^{-1}K^{-1}:	32.7		148.6	η: N/A
C_p/J mol^{-1}K^{-1}:	24.9		20.8	k: 156 W/m K
COMMENTS:				

Name: Magnesium oxide
Synonyms: Magnesia
           Periclase
Mol. Form.: MgO

CAS RN: 1309-48-4
Merck No.: 5555
Mol. Wt.: 40.304

PHYSICAL CONSTANTS

T_m: 2826°C (3099 K)	T_c:	μ:
T_b:	P_c:	IP: 9.70 eV

TRANSITION PROPERTIES

$\Delta_{fus}H$ (T_m): 78.00 kJ/mol	Vapor pressure (0°C): N/A
$\Delta_{vap}H$ (T_b):	Vapor pressure (25°C): N/A
$\Delta_{vap}H$ (25°C):	Vapor pressure (100°C): N/A

PROPERTIES AT 25°C AND 100 kPa

	Solid	Liquid	Gas		Solid
$\Delta_f H°$/kJ mol^{-1}:	-601.6			d:	3.6 g/mL
$S°$/J mol^{-1}K^{-1}:	27.0			η:	N/A
C_p/J mol^{-1}K^{-1}:	37.2			k:	50 W/m K
COMMENTS:					

Name: Maleic acid
Synonym: *cis*-2-Butenedioic acid

Mol. Form.: C$_4$H$_4$O$_4$

CAS RN: 110-16-7
Merck No.: 5585
Mol. Wt.: 116.073

PHYSICAL CONSTANTS

T_m: 130.5°C (403.6 K)	T_c:	μ:
T_b:	P_c:	IP:

TRANSITION PROPERTIES

$\Delta_{fus}H$ (T_m):	Vapor pressure (0°C):
$\Delta_{vap}H$ (T_b):	Vapor pressure (25°C):
$\Delta_{vap}H$ (25°C):	Vapor pressure (100°C):

PROPERTIES AT 25°C AND 100 kPa

	Solid	Liquid	Gas		Solid
$\Delta_f H°$/kJ mol^{-1}:	-789.4		-679.4	d:	
$S°$/J mol^{-1}K^{-1}:	160.8			η:	N/A
C_p/J mol^{-1}K^{-1}:	137.0			k:	
COMMENTS:					

Name: Maleic anhydride
Synonyms: *cis*-Butenedioic anhydride
           2,5-Furandione
Mol. Form.: C$_4$H$_2$O$_3$

CAS RN: 108-31-6
Merck No.: 5586
Mol. Wt.: 98.058

PHYSICAL CONSTANTS

T_m: 52.8°C (325.9 K)	T_c:	μ:
T_b: 202°C (475 K)	P_c:	IP: 10.80 eV

TRANSITION PROPERTIES

$\Delta_{fus}H$ (T_m):	Vapor pressure (0°C):
$\Delta_{vap}H$ (T_b):	Vapor pressure (25°C):
$\Delta_{vap}H$ (25°C):	Vapor pressure (100°C): 3.32 kPa

PROPERTIES AT 25°C AND 100 kPa

	Solid	Liquid	Gas		Solid
$\Delta_f H°$/kJ mol^{-1}:	-469.8		-398.3	d:	1.5 g/mL
$S°$/J mol^{-1}K^{-1}:				η:	N/A
C_p/J mol^{-1}K^{-1}:				k:	
COMMENTS: TLV=0.25 ppm; highly toxic					

Name: Manganese

Mol. Form.: Mn

CAS RN: 7439-96-5
Merck No.: 5604
Mol. Wt.: 54.938

PHYSICAL CONSTANTS

T_m: 1246°C (1519 K) T_c: μ:
T_b: 2061°C (2334 K) P_c: IP: 7.43 eV

TRANSITION PROPERTIES

$\Delta_{fus}H$ (T_m): 12.91 kJ/mol   Vapor pressure (0°C): N/A
$\Delta_{vap}H$ (T_b):   Vapor pressure (25°C): N/A
$\Delta_{vap}H$ (25°C):   Vapor pressure (100°C): N/A

PROPERTIES AT 25°C AND 100 kPa

	Solid	Liquid	Gas		Solid
$\Delta_f H°$/kJ mol^{-1}:	0.0		280.7		d: 7.3 g/mL
$S°$/J mol^{-1}K^{-1}:	32.0		173.7		η: N/A
C_p/J mol^{-1}K^{-1}:	26.3		20.8		k: 7.81 W/m K
COMMENTS:					

Name: Menthol
Synonym: 2-Isopropyl-5-methylcyclohexanol

Mol. Form.: $C_{10}H_{20}O$

CAS RN: 89-78-1
Merck No.: 5723
Mol. Wt.: 156.268

PHYSICAL CONSTANTS

T_m: 42°C (315 K) T_c: 421°C (694 K) μ:
T_b: 216.35°C (489.50 K) P_c: IP:

TRANSITION PROPERTIES

$\Delta_{fus}H$ (T_m):   Vapor pressure (0°C):
$\Delta_{vap}H$ (T_b):   Vapor pressure (25°C):
$\Delta_{vap}H$ (25°C):   Vapor pressure (100°C):

PROPERTIES AT 25°C AND 100 kPa

	Solid	Liquid	Gas		Solid
$\Delta_f H°$/kJ mol^{-1}:					d: 0.89 g/mL
$S°$/J mol^{-1}K^{-1}:					η: N/A
C_p/J mol^{-1}K^{-1}:					k:
COMMENTS:					

Name: Mercury

Mol. Form.: Hg

CAS RN: 7439-97-6
Merck No.: 5801
Mol. Wt.: 200.590

PHYSICAL CONSTANTS

T_m: -38.83°C (234.32 K) T_c: 1477°C (1750 K) μ:
T_b: 356.73°C (629.88 K) P_c: 172.00 MPa IP: 10.44 eV

TRANSITION PROPERTIES

$\Delta_{fus}H$ (T_m): 2.29 kJ/mol   Vapor pressure (0°C):
$\Delta_{vap}H$ (T_b): 59.11 kJ/mol   Vapor pressure (25°C):
$\Delta_{vap}H$ (25°C):   Vapor pressure (100°C): 0.037 kPa

PROPERTIES AT 25°C AND 100 kPa

	Solid	Liquid	Gas		Liquid
$\Delta_f H°$/kJ mol^{-1}:		0.0	61.4		d: 13.5336 g/mL
$S°$/J mol^{-1}K^{-1}:		75.9	175.0		η: 1.53 mPa s
C_p/J mol^{-1}K^{-1}:		28.0	20.8		k: 8.30 W/m K
COMMENTS: Highly toxic					

Name: Mercury bromide (HgBr$_2$)
Synonyms: Mercuric bromide
        Mercury(II) bromide
Mol. Form.: Br$_2$Hg

CAS RN: 7789-47-1
Merck No.: 5769
Mol. Wt.: 360.398

PHYSICAL CONSTANTS

T_m: 236°C (509 K) T_c: 739°C (1012 K) μ:
T_b: 322°C (595 K) P_c: IP: 10.56 eV

TRANSITION PROPERTIES

$\Delta_{fus}H$ (T_m): 17.90 kJ/mol Vapor pressure (0°C): N/A
$\Delta_{vap}H$ (T_b): 58.89 kJ/mol Vapor pressure (25°C): N/A
$\Delta_{vap}H$ (25°C): Vapor pressure (100°C): N/A

PROPERTIES AT 25°C AND 100 kPa

	Solid	Liquid	Gas		Solid
$\Delta_f H°$/kJ mol^{-1}:	-170.7			d:	6.05 g/mL
$S°$/J mol^{-1}K^{-1}:	172.0			η:	N/A
C_p/J mol^{-1}K^{-1}:				k:	

COMMENTS: Highly toxic

Name: Mercury chloride (HgCl$_2$)
Synonyms: Mercuric chloride
        Mercury(II) chloride
Mol. Form.: Cl$_2$Hg

CAS RN: 7487-94-7
Merck No.: 5770
Mol. Wt.: 271.495

PHYSICAL CONSTANTS

T_m: 276°C (549 K) T_c: 700°C (973 K) μ:
T_b: 304°C (577 K) P_c: IP: 11.38 eV

TRANSITION PROPERTIES

$\Delta_{fus}H$ (T_m): 19.41 kJ/mol Vapor pressure (0°C): N/A
$\Delta_{vap}H$ (T_b): 58.90 kJ/mol Vapor pressure (25°C): N/A
$\Delta_{vap}H$ (25°C): Vapor pressure (100°C): N/A

PROPERTIES AT 25°C AND 100 kPa

	Solid	Liquid	Gas		Solid
$\Delta_f H°$/kJ mol^{-1}:	-224.3			d:	5.6 g/mL
$S°$/J mol^{-1}K^{-1}:	146.0			η:	N/A
C_p/J mol^{-1}K^{-1}:				k:	

COMMENTS: Highly toxic

Name: Mercury chloride (Hg$_2$Cl$_2$)
Synonyms: Mercurous chloride
        Dimercury dichloride
Mol. Form.: Cl$_2$Hg$_2$

CAS RN: 10112-91-1
Merck No.: 5795
Mol. Wt.: 472.085

PHYSICAL CONSTANTS

T_m: 543°C (816 K) T_c: μ:
T_b: P_c: IP:

TRANSITION PROPERTIES

$\Delta_{fus}H$ (T_m): Vapor pressure (0°C): N/A
$\Delta_{vap}H$ (T_b): Vapor pressure (25°C): N/A
$\Delta_{vap}H$ (25°C): Vapor pressure (100°C): N/A

PROPERTIES AT 25°C AND 100 kPa

	Solid	Liquid	Gas		Solid
$\Delta_f H°$/kJ mol^{-1}:	-265.4			d:	6.97 g/mL
$S°$/J mol^{-1}K^{-1}:	191.6			η:	N/A
C_p/J mol^{-1}K^{-1}:				k:	

COMMENTS: Highly toxic

Name: Mercury iodide (HgI$_2$)   CAS RN: 7774-29-0
Synonyms: Mercuric iodide   Merck No.: 5776
          Mercury(II) iodide   Mol. Wt.: 454.399
Mol. Form.: HgI$_2$

PHYSICAL CONSTANTS

T_m: 259°C (532 K) T_c: 799°C (1072 K)   μ:
T_b: 354°C (627 K) P_c: IP: 9.51 eV

TRANSITION PROPERTIES

$\Delta_{fus}H$ (T_m): 18.90 kJ/mol   Vapor pressure (0°C): N/A
$\Delta_{vap}H$ (T_b): 59.20 kJ/mol   Vapor pressure (25°C): N/A
$\Delta_{vap}H$ (25°C):   Vapor pressure (100°C): N/A

PROPERTIES AT 25°C AND 100 kPa

	Solid	Liquid	Gas		Solid
$\Delta_f H°$/kJ mol^{-1}:	-105.4			d:	6.3 g/mL
$S°$/J mol^{-1}K^{-1}:	180.0			η:	N/A
C_p/J mol^{-1}K^{-1}:				k:	
COMMENTS:					

Name: Mesitylene   CAS RN: 108-67-8
Synonym: 1,3,5-Trimethylbenzene   Merck No.: 5810
                                                                        Mol. Wt.: 120.194

Mol. Form.: C$_9$H$_{12}$

PHYSICAL CONSTANTS

T_m: -44.7°C (228.4 K) T_c: 364.10°C (637.25 K)   μ: 0 D
T_b: 164.74°C (437.89 K) P_c: 3.127 MPa   IP: 8.41 eV

TRANSITION PROPERTIES

$\Delta_{fus}H$ (T_m): 9.51 kJ/mol   Vapor pressure (0°C):
$\Delta_{vap}H$ (T_b): 39.04 kJ/mol   Vapor pressure (25°C): 0.330 kPa
$\Delta_{vap}H$ (25°C): 47.50 kJ/mol   Vapor pressure (100°C): 13.5 kPa

PROPERTIES AT 25°C AND 100 kPa

	Solid	Liquid	Gas		Liquid
$\Delta_f H°$/kJ mol^{-1}:		-63.4	-15.9	d:	0.8614 g/mL
$S°$/J mol^{-1}K^{-1}:				η:	1.05 mPa s
C_p/J mol^{-1}K^{-1}:		209.3		k:	0.136 W/m K
COMMENTS: TLV=25 ppm					

Name: Mesityl oxide   CAS RN: 141-79-7
Synonyms: 4-Methyl-3-penten-2-one   Merck No.: 5811
          Isobutenyl methyl ketone   Mol. Wt.: 98.145
Mol. Form.: C$_6$H$_{10}$O

PHYSICAL CONSTANTS

T_m: -59°C (214 K) T_c: μ:
T_b: 130°C (403 K) P_c: IP: 9.08 eV

TRANSITION PROPERTIES

$\Delta_{fus}H$ (T_m):   Vapor pressure (0°C):
$\Delta_{vap}H$ (T_b): 36.11 kJ/mol   Vapor pressure (25°C): 1.47 kPa
$\Delta_{vap}H$ (25°C): 43.40 kJ/mol   Vapor pressure (100°C): 40.5 kPa

PROPERTIES AT 25°C AND 100 kPa

	Solid	Liquid	Gas		Liquid
$\Delta_f H°$/kJ mol^{-1}:				d:	0.7478 g/mL
$S°$/J mol^{-1}K^{-1}:				η:	0.602 mPa s
C_p/J mol^{-1}K^{-1}:		212.5		k:	0.156 W/m K
COMMENTS: TLV=15 ppm; highly toxic					

Name: Methacrylic acid
Synonym: 2-Methylpropenoic acid

Mol. Form.: $C_4H_6O_2$

CAS RN: 79-41-4
Merck No.: 5849
Mol. Wt.: 86.090

PHYSICAL CONSTANTS

T_m: 16°C (289 K) T_c: μ:
T_b: 162.5°C (435.6 K) P_c: IP: 10.15 eV

TRANSITION PROPERTIES

$\Delta_{fus}H$ (T_m):   Vapor pressure (0°C):
$\Delta_{vap}H$ (T_b):   Vapor pressure (25°C):
$\Delta_{vap}H$ (25°C):   Vapor pressure (100°C): 10.1 kPa

PROPERTIES AT 25°C AND 100 kPa

	Solid	Liquid	Gas		Liquid
$\Delta_f H°$/kJ mol^{-1}:					d: 1.015 g/mL
$S°$/J mol^{-1}K^{-1}:					η:
C_p/J mol^{-1}K^{-1}:					k:
COMMENTS: TLV=20 ppm; highly toxic					

Name: Methane
Synonym: Refrigerant 50

Mol. Form.: CH_4

CAS RN: 74-82-8
Merck No.: 5863
Mol. Wt.: 16.043

PHYSICAL CONSTANTS

T_m: -182.46°C (90.69 K) T_c: -82.62°C (190.53 K)   μ: 0 D
T_b: -161.48°C (111.67 K) P_c: 4.604 MPa   IP: 12.51 eV

TRANSITION PROPERTIES

$\Delta_{fus}H$ (T_m): 0.94 kJ/mol   Vapor pressure (0°C): N/A
$\Delta_{vap}H$ (T_b): 8.19 kJ/mol   Vapor pressure (25°C): N/A
$\Delta_{vap}H$ (25°C):   Vapor pressure (100°C): N/A

PROPERTIES AT 25°C AND 100 kPa

	Solid	Liquid	Gas		Gas
$\Delta_f H°$/kJ mol^{-1}:			-74.4		d: 0.656 g/L
$S°$/J mol^{-1}K^{-1}:			186.3		η: 11.1 μPa s
C_p/J mol^{-1}K^{-1}:			35.3		k: 0.0341 W/m K
COMMENTS: Very flammable					

Name: Methanethiol
Synonym: Methyl mercaptan

Mol. Form.: CH_4S

CAS RN: 74-93-1
Merck No.: 5867
Mol. Wt.: 48.109

PHYSICAL CONSTANTS

T_m: -123°C (150 K) T_c: 196.9°C (470.0 K)   μ: 1.52 D
T_b: 5.96°C (279.11 K) P_c: 7.23 MPa   IP: 9.44 eV

TRANSITION PROPERTIES

$\Delta_{fus}H$ (T_m): 5.91 kJ/mol   Vapor pressure (0°C): 79.8 kPa
$\Delta_{vap}H$ (T_b):   Vapor pressure (25°C):
$\Delta_{vap}H$ (25°C): 24.1 kJ/mol   Vapor pressure (100°C):

PROPERTIES AT 25°C AND 100 kPa

	Solid	Liquid	Gas		Gas
$\Delta_f H°$/kJ mol^{-1}:		-46.4	-22.3		d: 1.966 g/L
$S°$/J mol^{-1}K^{-1}:		169.2	255.2		η:
C_p/J mol^{-1}K^{-1}:		90.5	50.3		k:
COMMENTS: TLV=0.5 ppm; highly toxic; very flammable					

Name: Methanol
Synonyms: Methyl alcohol
        Carbinol
Mol. Form.: CH_4O

CAS RN: 67-56-1
Merck No.: 5868
Mol. Wt.: 32.042

PHYSICAL CONSTANTS

T_m: -97.68°C (175.47 K) T_c: 239.49°C (512.64 K) μ: 1.70 D
T_b: 64.6°C (337.7 K) P_c: 8.092 MPa IP: 10.85 eV

TRANSITION PROPERTIES

$\Delta_{fus}H$ (T_m): 3.18 kJ/mol Vapor pressure (0°C): 4.03 kPa
$\Delta_{vap}H$ (T_b): 35.21 kJ/mol Vapor pressure (25°C): 16.9 kPa
$\Delta_{vap}H$ (25°C): 37.43 kJ/mol Vapor pressure (100°C): 353 kPa

PROPERTIES AT 25°C AND 100 kPa

	Solid	Liquid	Gas		Liquid
$\Delta_f H°$/kJ mol^{-1}:		-239.1	-201.5		d: 0.7866 g/mL
$S°$/J mol^{-1}K^{-1}:		126.8	239.8		η: 0.549 mPa s
C_p/J mol^{-1}K^{-1}:		81.1	43.9		k: 0.200 W/m K

COMMENTS: TLV=200 ppm; highly toxic; flammable

Name: 2-Methoxyethanol
Synonyms: Ethylene glycol monomethyl ether
        Methyl cellosolve
Mol. Form.: $C_3H_8O_2$

CAS RN: 109-86-4
Merck No.: 5961
Mol. Wt.: 76.095

PHYSICAL CONSTANTS

T_m: -85.1°C (188.0 K) T_c: μ: 2.36 D
T_b: 124.1°C (397.2 K) P_c: IP: 9.60 eV

TRANSITION PROPERTIES

$\Delta_{fus}H$ (T_m): Vapor pressure (0°C):
$\Delta_{vap}H$ (T_b): 37.54 kJ/mol Vapor pressure (25°C): 1.31 kPa
$\Delta_{vap}H$ (25°C): 45.17 kJ/mol Vapor pressure (100°C): 43.9 kPa

PROPERTIES AT 25°C AND 100 kPa

	Solid	Liquid	Gas		Liquid
$\Delta_f H°$/kJ mol^{-1}:					d: 0.9598 g/mL
$S°$/J mol^{-1}K^{-1}:					η: 1.60 mPa s
C_p/J mol^{-1}K^{-1}:		171.1			k:

COMMENTS: TLV=5 ppm; highly toxic

Name: 2-(2-Methoxyethoxy)ethanol
Synonyms: Diethylene glycol monomethyl ether
        3,6-Dioxa-1-heptanol
Mol. Form.: $C_5H_{12}O_3$

CAS RN: 111-77-3
Merck No.: 5959
Mol. Wt.: 120.149

PHYSICAL CONSTANTS

T_m: <-84°C (<189 K) T_c: μ:
T_b: 193°C (466 K) P_c: IP:

TRANSITION PROPERTIES

$\Delta_{fus}H$ (T_m): Vapor pressure (0°C):
$\Delta_{vap}H$ (T_b): 46.57 kJ/mol Vapor pressure (25°C): 0.024 kPa
$\Delta_{vap}H$ (25°C): Vapor pressure (100°C): 3.65 kPa

PROPERTIES AT 25°C AND 100 kPa

	Solid	Liquid	Gas		Liquid
$\Delta_f H°$/kJ mol^{-1}:					d: 1.030 g/mL
$S°$/J mol^{-1}K^{-1}:					η: 3.48 mPa s
C_p/J mol^{-1}K^{-1}:		271.1			k:

COMMENTS:

Name: 2-Methoxyethyl acetate

Synonyms: Methyl cellosolve acetate

          Ethylene glycol monomethyl ether acetate

Mol. Form.: $C_5H_{10}O_3$

CAS RN: 110-49-6

Merck No.: 5962

Mol. Wt.: 118.133

PHYSICAL CONSTANTS

T_m: -70°C (203 K)

T_b: 143°C (416 K)

T_c:

P_c:

μ:

IP:

TRANSITION PROPERTIES

$\Delta_{fus}H$ (T_m):

$\Delta_{vap}H$ (T_b): 43.90 kJ/mol

$\Delta_{vap}H$ (25°C): 50.27 kJ/mol

Vapor pressure (0°C):

Vapor pressure (25°C): 0.670 kPa

Vapor pressure (100°C): 23.0 kPa

PROPERTIES AT 25°C AND 100 kPa

	Solid	Liquid	Gas	Liquid
$\Delta_f H°$/kJ mol^{-1}:				d: 1.0049 g/mL
$S°$/J mol^{-1}K^{-1}:				η:
C_p/J mol^{-1}K^{-1}:		310.0		k:

COMMENTS: TLV=5 ppm

Name: Methyl acetate

Synonym: Methyl ethanoate

Mol. Form.: $C_3H_6O_2$

CAS RN: 79-20-9

Merck No.: 5932

Mol. Wt.: 74.079

PHYSICAL CONSTANTS

T_m: -98°C (175 K)

T_b: 56.87°C (330.02 K)

T_c: 233.40°C (506.55 K)

P_c: 4.75 MPa

μ: 1.72 D

IP: 10.27 eV

TRANSITION PROPERTIES

$\Delta_{fus}H$ (T_m):

$\Delta_{vap}H$ (T_b): 30.32 kJ/mol

$\Delta_{vap}H$ (25°C): 32.29 kJ/mol

Vapor pressure (0°C): 8.37 kPa

Vapor pressure (25°C): 28.8 kPa

Vapor pressure (100°C): 371 kPa

PROPERTIES AT 25°C AND 100 kPa

	Solid	Liquid	Gas	Liquid
$\Delta_f H°$/kJ mol^{-1}:		-445.8	-411.9	d: 0.9279 g/mL
$S°$/J mol^{-1}K^{-1}:			324.4	η: 0.364 mPa s
C_p/J mol^{-1}K^{-1}:		141.9	86.0	k: 0.153 W/m K

COMMENTS: TLV=200 ppm; flammable

Name: Methyl acrylate

Synonym: Methyl propenoate

Mol. Form.: $C_4H_6O_2$

CAS RN: 96-33-3

Merck No.: 5935

Mol. Wt.: 86.090

PHYSICAL CONSTANTS

T_m: -76.5°C (196.6 K)

T_b: 80.7°C (353.8 K)

T_c:

P_c:

μ:

IP: 9.90 eV

TRANSITION PROPERTIES

$\Delta_{fus}H$ (T_m):

$\Delta_{vap}H$ (T_b): 33.10 kJ/mol

$\Delta_{vap}H$ (25°C): 29.20 kJ/mol

Vapor pressure (0°C):

Vapor pressure (25°C): 11.0 kPa

Vapor pressure (100°C):

PROPERTIES AT 25°C AND 100 kPa

	Solid	Liquid	Gas	Liquid
$\Delta_f H°$/kJ mol^{-1}:		-362.2	-333.0	d: 0.949 g/mL
$S°$/J mol^{-1}K^{-1}:		239.5		η:
C_p/J mol^{-1}K^{-1}:		158.8		k:

COMMENTS: TLV=10 ppm; highly toxic; flammable

Name: Methylacrylonitrile   CAS RN: 126-98-7
Synonyms: 2-Methylpropenenitrile   Merck No.: 5850
            2-Cyano-1-propene   Mol. Wt.: 67.090
Mol. Form.: C_4H_5N

PHYSICAL CONSTANTS

T_m: -35.8°C (237.3 K)	T_c:	μ: 3.69 D
T_b: 90.3°C (363.4 K)	P_c:	IP: 10.34 eV

TRANSITION PROPERTIES

$\Delta_{fus}H$ (T_m):	Vapor pressure (0°C): 2.13 kPa
$\Delta_{vap}H$ (T_b): 31.80 kJ/mol	Vapor pressure (25°C): 8.26 kPa
$\Delta_{vap}H$ (25°C):	Vapor pressure (100°C): 135 kPa

PROPERTIES AT 25°C AND 100 kPa

	Solid	Liquid	Gas	Liquid
$\Delta_fH°$/kJ mol^{-1}:				d: 0.7948 g/mL
$S°$/J mol^{-1}K^{-1}:				η:
C_p/J mol^{-1}K^{-1}:				k:

COMMENTS: TLV=1 ppm; highly toxic

Name: Methylamine   CAS RN: 74-89-5
Synonyms: Methanamine   Merck No.: 5938
            Aminomethane   Mol. Wt.: 31.057
Mol. Form.: CH_5N

PHYSICAL CONSTANTS

T_m: -93.44°C (179.71 K)	T_c: 157.6°C (430.7 K)	μ: 1.31 D
T_b: -6.32°C (266.83 K)	P_c: 7.614 MPa	IP: 8.97 eV

TRANSITION PROPERTIES

$\Delta_{fus}H$ (T_m): 6.13 kJ/mol	Vapor pressure (0°C):
$\Delta_{vap}H$ (T_b): 25.60 kJ/mol	Vapor pressure (25°C): 353 kPa
$\Delta_{vap}H$ (25°C): 23.37 kJ/mol	Vapor pressure (100°C):

PROPERTIES AT 25°C AND 100 kPa

	Solid	Liquid	Gas	Gas
$\Delta_fH°$/kJ mol^{-1}:		-47.3	-22.5	d: 1.269 g/L
$S°$/J mol^{-1}K^{-1}:		150.2	242.9	η:
C_p/J mol^{-1}K^{-1}:		102.1	50.1	k:

COMMENTS: TLV=5 ppm; highly toxic; very flammable

Name: *N*-Methylaniline   CAS RN: 100-61-8
Synonym: *N*-Methylbenzenamine   Merck No.: 5941
                                                                                       Mol. Wt.: 107.155
Mol. Form.: C_7H_9N

PHYSICAL CONSTANTS

T_m: -57°C (216 K)	T_c: 428°C (701 K)	μ:
T_b: 196.25°C (469.40 K)	P_c: 5.20 MPa	IP: 7.33 eV

TRANSITION PROPERTIES

$\Delta_{fus}H$ (T_m):	Vapor pressure (0°C):
$\Delta_{vap}H$ (T_b):	Vapor pressure (25°C):
$\Delta_{vap}H$ (25°C): 53.10 kJ/mol	Vapor pressure (100°C): 4.23 kPa

PROPERTIES AT 25°C AND 100 kPa

	Solid	Liquid	Gas	Liquid
$\Delta_fH°$/kJ mol^{-1}:				d: 0.9822 g/mL
$S°$/J mol^{-1}K^{-1}:				η: 2.04 mPa s
C_p/J mol^{-1}K^{-1}:		207.1		k:

COMMENTS: TLV=0.5 ppm

Name: *o*-Methylaniline
Synonyms: *o*-Toluidine
2-Methylbenzenamine
Mol. Form.: C_7H_9N

CAS RN: 95-53-4
Merck No.: 9462
Mol. Wt.: 107.155

PHYSICAL CONSTANTS

T_m: -16.35°C (256.80 K)
T_b: 200.34°C (473.49 K)

T_c: 434°C (707 K)
P_c: 4.37 MPa

μ:
IP: 7.44 eV

TRANSITION PROPERTIES

$\Delta_{fus}H$ (T_m):
$\Delta_{vap}H$ (T_b): 44.60 kJ/mol
$\Delta_{vap}H$ (25°C): 56.74 kJ/mol

Vapor pressure (0°C):
Vapor pressure (25°C): 0.043 kPa
Vapor pressure (100°C):

PROPERTIES AT 25°C AND 100 kPa

	Solid	Liquid	Gas		Liquid
$\Delta_f H°$/kJ mol^{-1}:		-6.3	56.4		*d*: 0.9947 g/mL
$S°$/J mol^{-1}K^{-1}:			351.0		η: 3.82 mPa s
C_p/J mol^{-1}K^{-1}:			130.2		*k*:

COMMENTS: TLV=2 ppm; carcinogen; highly toxic

Name: *m*-Methylaniline
Synonyms: *m*-Toluidine
3-Methylbenzenamine
Mol. Form.: C_7H_9N

CAS RN: 108-44-1
Merck No.: 9462
Mol. Wt.: 107.155

PHYSICAL CONSTANTS

T_m: -31.25°C (241.90 K)
T_b: 203.37°C (476.52 K)

T_c: 434°C (707 K)
P_c: 4.28 MPa

μ:
IP: 7.50 eV

TRANSITION PROPERTIES

$\Delta_{fus}H$ (T_m): 3.89 kJ/mol
$\Delta_{vap}H$ (T_b): 44.85 kJ/mol
$\Delta_{vap}H$ (25°C): 57.28 kJ/mol

Vapor pressure (0°C):
Vapor pressure (25°C): 0.036 kPa
Vapor pressure (100°C):

PROPERTIES AT 25°C AND 100 kPa

	Solid	Liquid	Gas		Liquid
$\Delta_f H°$/kJ mol^{-1}:		-8.1	54.6		*d*: 0.9850 g/mL
$S°$/J mol^{-1}K^{-1}:			352.5		η: 3.31 mPa s
C_p/J mol^{-1}K^{-1}:		227.0	125.5		*k*:

COMMENTS: TLV=2 ppm; carcinogen; highly toxic

Name: *p*-Methylaniline
Synonyms: *p*-Toluidine
4-Methylbenzenamine
Mol. Form.: C_7H_9N

CAS RN: 106-49-0
Merck No.: 9462
Mol. Wt.: 107.155

PHYSICAL CONSTANTS

T_m: 43.75°C (316.90 K)
T_b: 200.42°C (473.57 K)

T_c: 433°C (706 K)
P_c: 4.58 MPa

μ:
IP: 7.24 eV

TRANSITION PROPERTIES

$\Delta_{fus}H$ (T_m): 18.22 kJ/mol
$\Delta_{vap}H$ (T_b): 44.27 kJ/mol
$\Delta_{vap}H$ (25°C): 56.20 kJ/mol

Vapor pressure (0°C): 0.638 kPa
Vapor pressure (25°C): 1.74 kPa
Vapor pressure (100°C): 15.8 kPa

PROPERTIES AT 25°C AND 100 kPa

	Solid	Liquid	Gas		Solid
$\Delta_f H°$/kJ mol^{-1}:	-23.5		55.3		*d*: 0.957 g/mL
$S°$/J mol^{-1}K^{-1}:			347.0		η: N/A
C_p/J mol^{-1}K^{-1}:			126.2		*k*:

COMMENTS: TLV=2 ppm; highly toxic

Name: Methyl benzoate

CAS RN: 93-58-3
Merck No.: 5947
Mol. Wt.: 136.150

Mol. Form.: $C_8H_8O_2$

PHYSICAL CONSTANTS

T_m: -15°C (258 K) T_c: μ:
T_b: 199°C (472 K) P_c: IP: 9.32 eV

TRANSITION PROPERTIES

$\Delta_{fus}H$ (T_m):   Vapor pressure (0°C):
$\Delta_{vap}H$ (T_b): 43.18 kJ/mol   Vapor pressure (25°C): 0.052 kPa
$\Delta_{vap}H$ (25°C): 55.57 kJ/mol   Vapor pressure (100°C):

PROPERTIES AT 25°C AND 100 kPa

	Solid	Liquid	Gas		Liquid
$\Delta_f H°$/kJ mol^{-1}:		-343.5	-287.9		d: 1.0838 g/mL
$S°$/J mol^{-1}K^{-1}:					η: 1.86 mPa s
C_p/J mol^{-1}K^{-1}:		221.3			k:
COMMENTS:					

Name: 2-Methyl-1,3-butadiene
Synonym: Isoprene

CAS RN: 78-79-5
Merck No.: 5087
Mol. Wt.: 68.119

Mol. Form.: C_5H_8

PHYSICAL CONSTANTS

T_m: -145.9°C (127.2 K) T_c: μ: 0.25 D
T_b: 34.06°C (307.21 K) P_c: IP: 8.84 eV

TRANSITION PROPERTIES

$\Delta_{fus}H$ (T_m): 4.79 kJ/mol   Vapor pressure (0°C): 26.4 kPa
$\Delta_{vap}H$ (T_b):   Vapor pressure (25°C): 73.4 kPa
$\Delta_{vap}H$ (25°C): 27.3 kJ/mol   Vapor pressure (100°C):

PROPERTIES AT 25°C AND 100 kPa

	Solid	Liquid	Gas		Liquid
$\Delta_f H°$/kJ mol^{-1}:		48.2	75.5		d: 0.676 g/mL
$S°$/J mol^{-1}K^{-1}:		229.3			η:
C_p/J mol^{-1}K^{-1}:		152.6			k:
COMMENTS: Very flammable					

Name: Methyl butanoate
Synonym: Methyl butyrate

CAS RN: 623-42-7
Merck No.: 5957
Mol. Wt.: 102.133

Mol. Form.: $C_5H_{10}O_2$

PHYSICAL CONSTANTS

T_m: -85.8°C (187.3 K) T_c: 281.3°C (554.4 K)   μ:
T_b: 102.8°C (375.9 K) P_c: 3.47 MPa   IP: 10.07 eV

TRANSITION PROPERTIES

$\Delta_{fus}H$ (T_m):   Vapor pressure (0°C): 0.939 kPa
$\Delta_{vap}H$ (T_b): 33.79 kJ/mol   Vapor pressure (25°C): 4.30 kPa
$\Delta_{vap}H$ (25°C): 39.28 kJ/mol   Vapor pressure (100°C): 93.2 kPa

PROPERTIES AT 25°C AND 100 kPa

	Solid	Liquid	Gas		Liquid
$\Delta_f H°$/kJ mol^{-1}:					d: 0.8926 g/mL
$S°$/J mol^{-1}K^{-1}:					η: 0.541 mPa s
C_p/J mol^{-1}K^{-1}:					k:
COMMENTS: Flammable					

Name: 3-Methylbutanoic acid
Synonyms: Isovaleric acid
          Delphinic acid
Mol. Form.: $C_5H_{10}O_2$

CAS RN: 503-74-2
Merck No.: 5120
Mol. Wt.: 102.133

PHYSICAL CONSTANTS

T_m: -29.3°C (243.8 K) T_c: 356°C (629 K)   μ:
T_b: 176.55°C (449.70 K) P_c: 3.40 MPa   IP: 10.51 eV

TRANSITION PROPERTIES

$\Delta_{fus}H$ (T_m):   Vapor pressure (0°C):
$\Delta_{vap}H$ (T_b):   Vapor pressure (25°C):
$\Delta_{vap}H$ (25°C): 51.6 kJ/mol   Vapor pressure (100°C): 5.99 kPa

PROPERTIES AT 25°C AND 100 kPa

	Solid	Liquid	Gas	Liquid
$\Delta_fH°$/kJ mol^{-1}:		-561.6	-510.0	d: 0.926 g/mL
$S°$/J mol^{-1}K^{-1}:				η:
C_p/J mol^{-1}K^{-1}:				k:
COMMENTS:				

Name: 2-Methyl-1-butanol
Synonym: *sec*-Butylcarbinol

Mol. Form.: $C_5H_{12}O$

CAS RN: 137-32-6
Merck No.: 5952
Mol. Wt.: 88.150

PHYSICAL CONSTANTS

T_m: <-70°C (<-203 K) T_c: μ:
T_b: 128°C (401 K) P_c: IP: 9.86 eV

TRANSITION PROPERTIES

$\Delta_{fus}H$ (T_m):   Vapor pressure (0°C):
$\Delta_{vap}H$ (T_b): 45.20 kJ/mol   Vapor pressure (25°C): 0.416 kPa
$\Delta_{vap}H$ (25°C): 55.16 kJ/mol   Vapor pressure (100°C): 34.5 kPa

PROPERTIES AT 25°C AND 100 kPa

	Solid	Liquid	Gas	Liquid
$\Delta_fH°$/kJ mol^{-1}:		-356.6	-302.0	d: 0.8150 g/mL
$S°$/J mol^{-1}K^{-1}:				η: 4.45 mPa s
C_p/J mol^{-1}K^{-1}:		220.1		k:
COMMENTS:				

Name: 2-Methyl-2-butanol
Synonyms: *tert*-Amyl alcohol
          tert-Pentyl alcohol
Mol. Form.: $C_5H_{12}O$

CAS RN: 75-85-4
Merck No.: 7096
Mol. Wt.: 88.150

PHYSICAL CONSTANTS

T_m: -8.8°C (264.3 K) T_c: 272°C (545 K)   μ:
T_b: 102.4°C (375.5 K) P_c: IP: 9.80 eV

TRANSITION PROPERTIES

$\Delta_{fus}H$ (T_m): 4.45 kJ/mol   Vapor pressure (0°C):
$\Delta_{vap}H$ (T_b): 39.04 kJ/mol   Vapor pressure (25°C): 2.19 kPa
$\Delta_{vap}H$ (25°C): 50.10 kJ/mol   Vapor pressure (100°C): 94.0 kPa

PROPERTIES AT 25°C AND 100 kPa

	Solid	Liquid	Gas	Liquid
$\Delta_fH°$/kJ mol^{-1}:		-379.5	-330.8	d: 0.8050 g/mL
$S°$/J mol^{-1}K^{-1}:				η: 3.55 mPa s
C_p/J mol^{-1}K^{-1}:		247.1		k:
COMMENTS:				

Name: 3-Methyl-1-butanol   CAS RN: 123-51-3
Synonyms: Isoamyl alcohol   Merck No.: 5081
          Isopentyl alcohol   Mol. Wt.: 88.150
Mol. Form.: $C_5H_{12}O$

PHYSICAL CONSTANTS

T_m: -117.2°C (155.9 K) T_c: 306.3°C (579.4 K)   μ:
T_b: 131.1°C (404.2 K) P_c: IP:

TRANSITION PROPERTIES

$\Delta_{fus}H$ (T_m):   Vapor pressure (0°C):
$\Delta_{vap}H$ (T_b): 44.07 kJ/mol   Vapor pressure (25°C): 0.315 kPa
$\Delta_{vap}H$ (25°C): 55.61 kJ/mol   Vapor pressure (100°C): 31.7 kPa

PROPERTIES AT 25°C AND 100 kPa

	Solid	Liquid	Gas	Liquid
$\Delta_fH°$/kJ mol^{-1}:		-356.4	-301.3	d: 0.8071 g/mL
$S°$/J mol^{-1}K^{-1}:				η: 3.69 mPa s
C_p/J mol^{-1}K^{-1}:		210.0		k:
COMMENTS:				

Name: 3-Methyl-2-butanol   CAS RN: 598-75-4
Synonym: Isopropylethanol   Merck No.: 5953
                                                                 Mol. Wt.: 88.150
Mol. Form.: $C_5H_{12}O$

PHYSICAL CONSTANTS

T_m: T_c: μ:
T_b: 113.5°C (386.6 K) P_c: IP: 10.01 eV

TRANSITION PROPERTIES

$\Delta_{fus}H$ (T_m):   Vapor pressure (0°C):
$\Delta_{vap}H$ (T_b): 41.80 kJ/mol   Vapor pressure (25°C): 1.20 kPa
$\Delta_{vap}H$ (25°C): 53.00 kJ/mol   Vapor pressure (100°C): 66.8 kPa

PROPERTIES AT 25°C AND 100 kPa

	Solid	Liquid	Gas	Liquid
$\Delta_fH°$/kJ mol^{-1}:		-366.6	-315.2	d: 0.814 g/mL
$S°$/J mol^{-1}K^{-1}:				η: 3.51 mPa s
C_p/J mol^{-1}K^{-1}:				k:
COMMENTS:				

Name: 3-Methyl-2-butanone   CAS RN: 563-80-4
Synonyms: Methyl isopropyl ketone   Merck No.:
          Isopropyl methyl ketone   Mol. Wt.: 86.134
Mol. Form.: $C_5H_{10}O$

PHYSICAL CONSTANTS

T_m: -92°C (181 K) T_c: 280.3°C (553.4 K)   μ:
T_b: 94.33°C (367.48 K) P_c: 3.85 MPa   IP: 9.30 eV

TRANSITION PROPERTIES

$\Delta_{fus}H$ (T_m):   Vapor pressure (0°C):
$\Delta_{vap}H$ (T_b): 32.35 kJ/mol   Vapor pressure (25°C):
$\Delta_{vap}H$ (25°C): 36.78 kJ/mol   Vapor pressure (100°C): 120 kPa

PROPERTIES AT 25°C AND 100 kPa

	Solid	Liquid	Gas	Liquid
$\Delta_fH°$/kJ mol^{-1}:		-299.4	-262.5	d: 0.801 g/mL
$S°$/J mol^{-1}K^{-1}:		268.5		η:
C_p/J mol^{-1}K^{-1}:		179.9		k:
COMMENTS:				

Name: 2-Methyl-1-butene
Synonym: γ-Isoamylene

Mol. Form.: C_5H_{10}

CAS RN: 563-46-2
Merck No.:
Mol. Wt.: 70.134

PHYSICAL CONSTANTS

T_m: -137.57°C (135.58 K)
T_b: 31.2°C (304.3 K)

T_c: 197°C (470 K)
P_c: 3.8 MPa

μ:
IP: 9.13 eV

TRANSITION PROPERTIES

$\Delta_{fus}H$ (T_m):
$\Delta_{vap}H$ (T_b): 25.50 kJ/mol
$\Delta_{vap}H$ (25°C): 25.86 kJ/mol

Vapor pressure (0°C): 29.6 kPa
Vapor pressure (25°C): 81.4 kPa
Vapor pressure (100°C): 679 kPa

PROPERTIES AT 25°C AND 100 kPa

	Solid	Liquid	Gas		Liquid
$\Delta_f H°$/kJ mol^{-1}:		-61.0	-35.3		d: 0.6451 g/mL
$S°$/J mol^{-1}K^{-1}:		254.0			η:
C_p/J mol^{-1}K^{-1}:		157.2			k:
COMMENTS: Flammable					

Name: 2-Methyl-2-butene
Synonym: β-Isoamylene

Mol. Form.: C_5H_{10}

CAS RN: 513-35-9
Merck No.: 644
Mol. Wt.: 70.134

PHYSICAL CONSTANTS

T_m: -133.76°C (139.39 K)
T_b: 38.56°C (311.71 K)

T_c: 208°C (481 K)
P_c: 3.91 MPa

μ:
IP: 8.68 eV

TRANSITION PROPERTIES

$\Delta_{fus}H$ (T_m):
$\Delta_{vap}H$ (T_b): 26.31 kJ/mol
$\Delta_{vap}H$ (25°C): 27.06 kJ/mol

Vapor pressure (0°C): 21.9 kPa
Vapor pressure (25°C): 62.2 kPa
Vapor pressure (100°C): 568 kPa

PROPERTIES AT 25°C AND 100 kPa

	Solid	Liquid	Gas		Liquid
$\Delta_f H°$/kJ mol^{-1}:		-68.6	-41.8		d: 0.6570 g/mL
$S°$/J mol^{-1}K^{-1}:		251.0			η: 0.203 mPa s
C_p/J mol^{-1}K^{-1}:		152.8			k:
COMMENTS: Flammable					

Name: 3-Methyl-1-butene
Synonym: α-Isoamylene

Mol. Form.: C_5H_{10}

CAS RN: 563-45-1
Merck No.:
Mol. Wt.: 70.134

PHYSICAL CONSTANTS

T_m:
T_b: 20.1°C (293.2 K)

T_c:
P_c:

μ: 0.320 D
IP: 9.52 eV

TRANSITION PROPERTIES

$\Delta_{fus}H$ (T_m):
$\Delta_{vap}H$ (T_b):
$\Delta_{vap}H$ (25°C): 23.80 kJ/mol

Vapor pressure (0°C): 47.0 kPa
Vapor pressure (25°C): 120 kPa
Vapor pressure (100°C):

PROPERTIES AT 25°C AND 100 kPa

	Solid	Liquid	Gas		Gas
$\Delta_f H°$/kJ mol^{-1}:		-51.5	-27.6		d: 2.867 g/L
$S°$/J mol^{-1}K^{-1}:		253.3			η:
C_p/J mol^{-1}K^{-1}:		156.1			k:
COMMENTS: Flammable					

Name: Methyl chloroacetate
Synonym: Methyl chloroethanoate

Mol. Form.: $C_3H_5ClO_2$

CAS RN: 96-34-4
Merck No.: 5965
Mol. Wt.: 108.524

PHYSICAL CONSTANTS

T_m: -32.12°C (241.03 K)
T_b: 129.5°C (402.6 K)

T_c:
P_c:

μ:
IP: 10.30 eV

TRANSITION PROPERTIES

$\Delta_{fus}H\ (T_m)$:
$\Delta_{vap}H\ (T_b)$: 39.23 kJ/mol
$\Delta_{vap}H$ (25°C): 46.73 kJ/mol

Vapor pressure (0°C):
Vapor pressure (25°C):
Vapor pressure (100°C): 38.5 kPa

PROPERTIES AT 25°C AND 100 kPa

	Solid	Liquid	Gas	Liquid
$\Delta_f H°$/kJ mol^{-1}:				d: 1.228 g/mL
$S°$/J mol^{-1}K^{-1}:				η:
C_p/J mol^{-1}K^{-1}:				k:
COMMENTS:				

Name: Methyl cyanoacetate
Synonym: Malonomononitrile, monomethyl ester

Mol. Form.: $C_4H_5NO_2$

CAS RN: 105-34-0
Merck No.:
Mol. Wt.: 99.089

PHYSICAL CONSTANTS

T_m: -22.5°C (250.6 K)
T_b: 200.5°C (473.6 K)

T_c:
P_c:

μ:
IP:

TRANSITION PROPERTIES

$\Delta_{fus}H\ (T_m)$:
$\Delta_{vap}H\ (T_b)$: 48.15 kJ/mol
$\Delta_{vap}H$ (25°C): 61.67 kJ/mol

Vapor pressure (0°C):
Vapor pressure (25°C): 0.019 kPa
Vapor pressure (100°C): 2.20 kPa

PROPERTIES AT 25°C AND 100 kPa

	Solid	Liquid	Gas	Liquid
$\Delta_f H°$/kJ mol^{-1}:				d: 1.1225 g/mL
$S°$/J mol^{-1}K^{-1}:				η: 2.5 mPa s
C_p/J mol^{-1}K^{-1}:				k:
COMMENTS: Highly toxic				

Name: Methylcyclohexane
Synonym: Hexahydrotoluene

Mol. Form.: C_7H_{14}

CAS RN: 108-87-2
Merck No.:
Mol. Wt.: 98.188

PHYSICAL CONSTANTS

T_m: -126.6°C (146.5 K)
T_b: 100.93°C (374.08 K)

T_c: 299.1°C (572.2 K)
P_c: 3.471 MPa

μ:
IP: 9.64 eV

TRANSITION PROPERTIES

$\Delta_{fus}H\ (T_m)$: 6.75 kJ/mol
$\Delta_{vap}H\ (T_b)$: 31.27 kJ/mol
$\Delta_{vap}H$ (25°C): 35.36 kJ/mol

Vapor pressure (0°C):
Vapor pressure (25°C): 6.18 kPa
Vapor pressure (100°C): 98.7 kPa

PROPERTIES AT 25°C AND 100 kPa

	Solid	Liquid	Gas	Liquid
$\Delta_f H°$/kJ mol^{-1}:		-190.1	-154.7	d: 0.7651 g/mL
$S°$/J mol^{-1}K^{-1}:				η: 0.679 mPa s
C_p/J mol^{-1}K^{-1}:		184.8		k:
COMMENTS: TLV=400 ppm; flammable				

Name: *cis*-2-Methylcyclohexanol

CAS RN: 7443-70-1
Merck No.:
Mol. Wt.: 114.188

Mol. Form.: $C_7H_{14}O$

PHYSICAL CONSTANTS

T_m: 7°C (280 K) T_c: μ:
T_b: 165°C (438 K) P_c: IP:

TRANSITION PROPERTIES

$\Delta_{fus}H$ (T_m):   Vapor pressure (0°C):
$\Delta_{vap}H$ (T_b): 48.50 kJ/mol   Vapor pressure (25°C):
$\Delta_{vap}H$ (25°C): 63.2 kJ/mol   Vapor pressure (100°C):

PROPERTIES AT 25°C AND 100 kPa

	Solid	Liquid	Gas		Liquid
$\Delta_f H°$/kJ mol^{-1}:		-390.2	-327.0		d: 0.9318 g/mL
$S°$/J mol^{-1}K^{-1}:					η: 18.1 mPa s
C_p/J mol^{-1}K^{-1}:					k:
COMMENTS:					

Name: *trans*-2-Methylcyclohexanol

CAS RN: 7443-52-9
Merck No.:
Mol. Wt.: 114.188

Mol. Form.: $C_7H_{14}O$

PHYSICAL CONSTANTS

T_m: -4°C (269 K) T_c: μ:
T_b: 166°C (439 K) P_c: IP:

TRANSITION PROPERTIES

$\Delta_{fus}H$ (T_m):   Vapor pressure (0°C):
$\Delta_{vap}H$ (T_b): 53.00 kJ/mol   Vapor pressure (25°C):
$\Delta_{vap}H$ (25°C): 63.2 kJ/mol   Vapor pressure (100°C):

PROPERTIES AT 25°C AND 100 kPa

	Solid	Liquid	Gas		Liquid
$\Delta_f H°$/kJ mol^{-1}:		-415.7	-352.5		d: 0.9208 g/mL
$S°$/J mol^{-1}K^{-1}:					η: 37 mPa s
C_p/J mol^{-1}K^{-1}:					k:
COMMENTS:					

Name: Methylcyclopentane

CAS RN: 96-37-7
Merck No.:
Mol. Wt.: 84.161

Mol. Form.: C_6H_{12}

PHYSICAL CONSTANTS

T_m: -142.5°C (130.6 K) T_c: 259.58°C (532.73 K)   μ:
T_b: 71.8°C (344.9 K) P_c: 3.784 MPa   IP: 9.85 eV

TRANSITION PROPERTIES

$\Delta_{fus}H$ (T_m): 6.93 kJ/mol   Vapor pressure (0°C): 5.50 kPa
$\Delta_{vap}H$ (T_b): 29.08 kJ/mol   Vapor pressure (25°C): 18.3 kPa
$\Delta_{vap}H$ (25°C): 31.64 kJ/mol   Vapor pressure (100°C):

PROPERTIES AT 25°C AND 100 kPa

	Solid	Liquid	Gas		Liquid
$\Delta_f H°$/kJ mol^{-1}:		-137.9	-106.2		d: 0.745 g/mL
$S°$/J mol^{-1}K^{-1}:					η: 0.479 mPa s
C_p/J mol^{-1}K^{-1}:					k:
COMMENTS: Flammable					

Name: *N*-Methylformamide
Synonym: *N*-Methylmethanamide

Mol. Form.: C_2H_5NO

CAS RN: 123-39-7
Merck No.:
Mol. Wt.: 59.068

PHYSICAL CONSTANTS

T_m: -3.8°C (269.3 K)	T_c:	μ: 3.83 D
T_b: 199.51°C (472.66 K)	P_c:	IP: 9.79 eV

TRANSITION PROPERTIES

$\Delta_{fus}H$ (T_m):
$\Delta_{vap}H$ (T_b):
$\Delta_{vap}H$ (25°C): 56.19 kJ/mol

Vapor pressure (0°C):
Vapor pressure (25°C):
Vapor pressure (100°C): 3.04 kPa

PROPERTIES AT 25°C AND 100 kPa

	Solid	Liquid	Gas	Liquid
$\Delta_f H°$/kJ mol^{-1}:				d: 1.00 g/mL
$S°$/J mol^{-1}K^{-1}:				η: 1.68 mPa s
C_p/J mol^{-1}K^{-1}:		123.8		k: 0.203 W/m K
COMMENTS:				

Name: Methyl formate
Synonym: Methyl methanoate

Mol. Form.: $C_2H_4O_2$

CAS RN: 107-31-3
Merck No.: 5994
Mol. Wt.: 60.053

PHYSICAL CONSTANTS

T_m: -99°C (174 K)	T_c: 214.1°C (487.2 K)	μ: 1.77 D
T_b: 31.7°C (304.8 K)	P_c: 5.998 MPa	IP: 10.82 eV

TRANSITION PROPERTIES

$\Delta_{fus}H$ (T_m): 7.45 kJ/mol
$\Delta_{vap}H$ (T_b): 27.92 kJ/mol
$\Delta_{vap}H$ (25°C): 28.35 kJ/mol

Vapor pressure (0°C): 26.0 kPa
Vapor pressure (25°C): 78.1 kPa
Vapor pressure (100°C): 779 kPa

PROPERTIES AT 25°C AND 100 kPa

	Solid	Liquid	Gas	Liquid
$\Delta_f H°$/kJ mol^{-1}:		-386.1	-355.5	d: 0.9664 g/mL
$S°$/J mol^{-1}K^{-1}:			285.3	η: 0.325 mPa s
C_p/J mol^{-1}K^{-1}:		119.1	64.4	k:
COMMENTS: TLV=100 ppm; very flammable				

Name: 2-Methylheptane

Mol. Form.: C_8H_{18}

CAS RN: 592-27-8
Merck No.:
Mol. Wt.: 114.231

PHYSICAL CONSTANTS

T_m: -108.99°C (164.16 K)	T_c: 286.6°C (559.7 K)	μ:
T_b: 117.66°C (390.81 K)	P_c: 2.484 MPa	IP: 9.84 eV

TRANSITION PROPERTIES

$\Delta_{fus}H$ (T_m):
$\Delta_{vap}H$ (T_b): 33.26 kJ/mol
$\Delta_{vap}H$ (25°C): 39.67 kJ/mol

Vapor pressure (0°C):
Vapor pressure (25°C):
Vapor pressure (100°C): 60.1 kPa

PROPERTIES AT 25°C AND 100 kPa

	Solid	Liquid	Gas	Liquid
$\Delta_f H°$/kJ mol^{-1}:		-255.0	-215.4	d: 0.6939 g/mL
$S°$/J mol^{-1}K^{-1}:		356.4		η:
C_p/J mol^{-1}K^{-1}:		252.0		k:
COMMENTS: Flammable				

Name: 3-Methylheptane

CAS RN: 589-81-1
Merck No.:
Mol. Wt.: 114.231

Mol. Form.: C_8H_{18}

PHYSICAL CONSTANTS

T_m: -120.5°C (152.6 K)
T_b: 118.94°C (392.09 K)

T_c: 290.6°C (563.7 K)
P_c: 2.546 MPa

μ:
IP:

TRANSITION PROPERTIES

$\Delta_{fus}H$ (T_m): 11.38 kJ/mol
$\Delta_{vap}H$ (T_b): 33.66 kJ/mol
$\Delta_{vap}H$ (25°C): 39.83 kJ/mol

Vapor pressure (0°C):
Vapor pressure (25°C):
Vapor pressure (100°C): 57.8 kPa

PROPERTIES AT 25°C AND 100 kPa

	Solid	Liquid	Gas	Liquid
$\Delta_fH°$/kJ mol^{-1}:		-252.3	-212.5	d: 0.7018 g/mL
$S°$/J mol^{-1}K^{-1}:		362.6		η:
C_p/J mol^{-1}K^{-1}:		250.2		k:

COMMENTS: Flammable

Name: 4-Methylheptane

CAS RN: 589-53-7
Merck No.:
Mol. Wt.: 114.231

Mol. Form.: C_8H_{18}

PHYSICAL CONSTANTS

T_m: -121°C (152 K)
T_b: 117.72°C (390.87 K)

T_c: 288.7°C (561.8 K)
P_c: 2.542 MPa

μ:
IP:

TRANSITION PROPERTIES

$\Delta_{fus}H$ (T_m): 10.84 kJ/mol
$\Delta_{vap}H$ (T_b): 33.35 kJ/mol
$\Delta_{vap}H$ (25°C): 39.69 kJ/mol

Vapor pressure (0°C):
Vapor pressure (25°C): 2.74 kPa
Vapor pressure (100°C): 60.0 kPa

PROPERTIES AT 25°C AND 100 kPa

	Solid	Liquid	Gas	Liquid
$\Delta_fH°$/kJ mol^{-1}:		-251.6	-212.0	d: 0.7006 g/mL
$S°$/J mol^{-1}K^{-1}:				η:
C_p/J mol^{-1}K^{-1}:		251.1		k:

COMMENTS: Flammable

Name: 4-Methyl-3-heptanol

CAS RN: 14979-39-6
Merck No.:
Mol. Wt.: 130.230

Mol. Form.: $C_8H_{18}O$

PHYSICAL CONSTANTS

T_m: -123°C (150 K)
T_b: 170°C (443 K)

T_c: 350.4°C (623.5 K)
P_c:

μ:
IP:

TRANSITION PROPERTIES

$\Delta_{fus}H$ (T_m):
$\Delta_{vap}H$ (T_b):
$\Delta_{vap}H$ (25°C):

Vapor pressure (0°C):
Vapor pressure (25°C):
Vapor pressure (100°C): 16.3 kPa

PROPERTIES AT 25°C AND 100 kPa

	Solid	Liquid	Gas	Liquid
$\Delta_fH°$/kJ mol^{-1}:				d: 0.7940 g/mL
$S°$/J mol^{-1}K^{-1}:				η: 1.09 mPa s
C_p/J mol^{-1}K^{-1}:				k:

COMMENTS:

Name: 5-Methyl-3-heptanol

CAS RN: 18720-65-5
Merck No.:
Mol. Wt.: 130.230

Mol. Form.: $C_8H_{18}O$

PHYSICAL CONSTANTS

T_m: -91.2°C (181.9 K) T_c: 348.1°C (621.2 K)   μ:
T_b: 172°C (445 K) P_c: IP:

TRANSITION PROPERTIES

$\Delta_{fus}H$ (T_m):   Vapor pressure (0°C):
$\Delta_{vap}H$ (T_b):   Vapor pressure (25°C):
$\Delta_{vap}H$ (25°C):   Vapor pressure (100°C): 18.4 kPa

PROPERTIES AT 25°C AND 100 kPa

	Solid	Liquid	Gas	Liquid
$\Delta_f H°$/kJ mol^{-1}:				d: 0.8143 g/mL
$S°$/J mol^{-1}K^{-1}:				η: 1.18 mPa s
C_p/J mol^{-1}K^{-1}:				k:
COMMENTS:				

Name: 2-Methylhexane

CAS RN: 591-76-4
Merck No.:
Mol. Wt.: 100.204

Mol. Form.: C_7H_{16}

PHYSICAL CONSTANTS

T_m: -118.2°C (154.9 K) T_c: 257.3°C (530.4 K)   μ:
T_b: 90.04°C (363.19 K) P_c: 2.734 MPa   IP:

TRANSITION PROPERTIES

$\Delta_{fus}H$ (T_m): 8.87 kJ/mol   Vapor pressure (0°C): 2.34 kPa
$\Delta_{vap}H$ (T_b): 30.62 kJ/mol   Vapor pressure (25°C): 8.78 kPa
$\Delta_{vap}H$ (25°C): 34.87 kJ/mol   Vapor pressure (100°C): 135 kPa

PROPERTIES AT 25°C AND 100 kPa

	Solid	Liquid	Gas	Liquid
$\Delta_f H°$/kJ mol^{-1}:		-229.5	-194.6	d: 0.6744 g/mL
$S°$/J mol^{-1}K^{-1}:		323.3		η:
C_p/J mol^{-1}K^{-1}:		222.9		k:
COMMENTS: Flammable				

Name: 3-Methylhexane

CAS RN: 589-34-4
Merck No.:
Mol. Wt.: 100.204

Mol. Form.: C_7H_{16}

PHYSICAL CONSTANTS

T_m: -119.4°C (153.7 K) T_c: 262.2°C (535.3 K)   μ:
T_b: 91.84°C (364.99 K) P_c: 2.814 MPa   IP:

TRANSITION PROPERTIES

$\Delta_{fus}H$ (T_m):   Vapor pressure (0°C): 2.17 kPa
$\Delta_{vap}H$ (T_b): 30.89 kJ/mol   Vapor pressure (25°C): 8.22 kPa
$\Delta_{vap}H$ (25°C): 35.06 kJ/mol   Vapor pressure (100°C): 128 kPa

PROPERTIES AT 25°C AND 100 kPa

	Solid	Liquid	Gas	Liquid
$\Delta_f H°$/kJ mol^{-1}:		-226.4	-191.3	d: 0.6829 g/mL
$S°$/J mol^{-1}K^{-1}:				η: 0.350 mPa s
C_p/J mol^{-1}K^{-1}:		214.2		k:
COMMENTS: Flammable				

Name: 5-Methyl-2-hexanone

Synonyms: Methyl isopentyl ketone

          Isoamyl methyl ketone

Mol. Form.: $C_7H_{14}O$

CAS RN: 110-12-3

Merck No.:

Mol. Wt.: 114.188

PHYSICAL CONSTANTS

T_m: -73.9°C (199.2 K)	T_c:	μ:
T_b: 144°C (417 K)	P_c:	IP: 9.28 eV

TRANSITION PROPERTIES

$\Delta_{fus}H$ (T_m):

$\Delta_{vap}H$ (T_b):

$\Delta_{vap}H$ (25°C):

Vapor pressure (0°C):

Vapor pressure (25°C): 0.691 kPa

Vapor pressure (100°C): 24.8 kPa

PROPERTIES AT 25°C AND 100 kPa

	Solid	Liquid	Gas		Liquid
$\Delta_fH°$/kJ mol^{-1}:				d:	0.808 g/mL
$S°$/J mol^{-1}K^{-1}:				η:	
C_p/J mol^{-1}K^{-1}:				k:	

COMMENTS: TLV=50 ppm

Name: Methylhydrazine

Mol. Form.: CH_6N_2

CAS RN: 60-34-4

Merck No.: 6001

Mol. Wt.: 46.072

PHYSICAL CONSTANTS

T_m: -52.4°C (220.7 K)	T_c: 294°C (567 K)	μ:
T_b: 87.5°C (360.6 K)	P_c: 8.24 MPa	IP: 7.67 eV

TRANSITION PROPERTIES

$\Delta_{fus}H$ (T_m):

$\Delta_{vap}H$ (T_b): 36.12 kJ/mol

$\Delta_{vap}H$ (25°C): 40.37 kJ/mol

Vapor pressure (0°C):

Vapor pressure (25°C): 6.61 kPa

Vapor pressure (100°C):

PROPERTIES AT 25°C AND 100 kPa

	Solid	Liquid	Gas		Liquid
$\Delta_fH°$/kJ mol^{-1}:		54.0	94.3	d:	
$S°$/J mol^{-1}K^{-1}:		165.9	278.8	η:	
C_p/J mol^{-1}K^{-1}:		134.9	71.1	k:	

COMMENTS: TLV=0.01 ppm; carcinogen; highly toxic; flammable

Name: Methyl isobutanoate

Synonym: Methyl 2-methylpropanoate

Mol. Form.: $C_5H_{10}O_2$

CAS RN: 547-63-7

Merck No.: 6003

Mol. Wt.: 102.133

PHYSICAL CONSTANTS

T_m: -84.7°C (188.4 K)	T_c: 267.7°C (540.8 K)	μ:
T_b: 92.5°C (365.6 K)	P_c: 3.43 MPa	IP:

TRANSITION PROPERTIES

$\Delta_{fus}H$ (T_m):

$\Delta_{vap}H$ (T_b): 32.61 kJ/mol

$\Delta_{vap}H$ (25°C): 37.32 kJ/mol

Vapor pressure (0°C):

Vapor pressure (25°C):

Vapor pressure (100°C): 127 kPa

PROPERTIES AT 25°C AND 100 kPa

	Solid	Liquid	Gas		Liquid
$\Delta_fH°$/kJ mol^{-1}:				d:	0.8854 g/mL
$S°$/J mol^{-1}K^{-1}:				η:	0.488 mPa s
C_p/J mol^{-1}K^{-1}:				k:	

COMMENTS:

Name: Methylisocyanate   CAS RN: 624-83-9
Synonym: Isocyanatomethane   Merck No.: 6004
                                                               Mol. Wt.: 57.052

Mol. Form.: C_2H_3NO

PHYSICAL CONSTANTS

T_m: -45°C (228 K) T_c: μ: 2.8 D
T_b: 39.5°C (312.6 K) P_c: IP: 10.67 eV

TRANSITION PROPERTIES

$\Delta_{fus}H$ (T_m):   Vapor pressure (0°C): 17.7 kPa
$\Delta_{vap}H$ (T_b):   Vapor pressure (25°C): 57.7 kPa
$\Delta_{vap}H$ (25°C):   Vapor pressure (100°C):

PROPERTIES AT 25°C AND 100 kPa

	Solid	Liquid	Gas		Liquid
$\Delta_f H°$/kJ mol^{-1}:		-92.0		d:	
$S°$/J mol^{-1}K^{-1}:				η:	
C_p/J mol^{-1}K^{-1}:				k:	

COMMENTS: TLV=0.02 ppm; highly toxic; flammable

Name: Methyl methacrylate   CAS RN: 80-62-6
Synonym: Methyl 2-methylpropenoate   Merck No.: 5849
                                                               Mol. Wt.: 100.117

Mol. Form.: $C_5H_8O_2$

PHYSICAL CONSTANTS

T_m: -48°C (225 K) T_c: μ:
T_b: 100.5°C (373.6 K) P_c: IP: 9.70 eV

TRANSITION PROPERTIES

$\Delta_{fus}H$ (T_m):   Vapor pressure (0°C):
$\Delta_{vap}H$ (T_b): 36.00 kJ/mol   Vapor pressure (25°C): 5.10 kPa
$\Delta_{vap}H$ (25°C): 40.70 kJ/mol   Vapor pressure (100°C): 100 kPa

PROPERTIES AT 25°C AND 100 kPa

	Solid	Liquid	Gas		Liquid
$\Delta_f H°$/kJ mol^{-1}:				d:	0.939 g/mL
$S°$/J mol^{-1}K^{-1}:				η:	
C_p/J mol^{-1}K^{-1}:		191.2		k:	

COMMENTS: TLV=100 ppm; highly toxic; flammable

Name: 1-Methylnaphthalene   CAS RN: 90-12-0
                                                               Merck No.:
                                                               Mol. Wt.: 142.200

Mol. Form.: $C_{11}H_{10}$

PHYSICAL CONSTANTS

T_m: -30.48°C (242.67 K) T_c: 499°C (772 K)   μ:
T_b: 244.74°C (517.89 K) P_c: IP: 7.85 eV

TRANSITION PROPERTIES

$\Delta_{fus}H$ (T_m): 6.94 kJ/mol   Vapor pressure (0°C):
$\Delta_{vap}H$ (T_b): 45.48 kJ/mol   Vapor pressure (25°C): 0.009 kPa
$\Delta_{vap}H$ (25°C): 60.07 kJ/mol   Vapor pressure (100°C):

PROPERTIES AT 25°C AND 100 kPa

	Solid	Liquid	Gas		Liquid
$\Delta_f H°$/kJ mol^{-1}:		56.3		d:	1.015 g/mL
$S°$/J mol^{-1}K^{-1}:		254.8		η:	2.8 mPa s
C_p/J mol^{-1}K^{-1}:		224.4		k:	

COMMENTS:

Name: 2-Methylnaphthalene

CAS RN: 91-57-6
Merck No.:
Mol. Wt.: 142.200

Mol. Form.: $C_{11}H_{10}$

PHYSICAL CONSTANTS

T_m: 34.4°C (307.5 K) T_c: 488°C (761 K) μ:
T_b: 241.11°C (514.26 K) P_c: IP: 7.80 eV

TRANSITION PROPERTIES

$\Delta_{fus}H$ (T_m): 11.97 kJ/mol Vapor pressure (0°C):
$\Delta_{vap}H$ (T_b): Vapor pressure (25°C):
$\Delta_{vap}H$ (25°C): Vapor pressure (100°C):

PROPERTIES AT 25°C AND 100 kPa

	Solid	Liquid	Gas		Solid
$\Delta_f H°$/kJ mol^{-1}:	44.9		106.7	d:	1.001 g/mL
$S°$/J mol^{-1}K^{-1}:	220.0			η:	N/A
C_p/J mol^{-1}K^{-1}:	196.0			k:	
COMMENTS: Highly toxic					

Name: 2-Methyloctane

CAS RN: 3221-61-2
Merck No.:
Mol. Wt.: 128.258

Mol. Form.: C_9H_{20}

PHYSICAL CONSTANTS

T_m: -80.37°C (192.78 K) T_c: 313.9°C (587.0 K) μ:
T_b: 143.28°C (416.43 K) P_c: 2.310 MPa IP:

TRANSITION PROPERTIES

$\Delta_{fus}H$ (T_m): Vapor pressure (0°C):
$\Delta_{vap}H$ (T_b): Vapor pressure (25°C):
$\Delta_{vap}H$ (25°C): Vapor pressure (100°C): 27.0 kPa

PROPERTIES AT 25°C AND 100 kPa

	Solid	Liquid	Gas		Liquid
$\Delta_f H°$/kJ mol^{-1}:				d:	0.705 g/mL
$S°$/J mol^{-1}K^{-1}:				η:	
C_p/J mol^{-1}K^{-1}:				k:	
COMMENTS: Flammable					

Name: Methyl oleate
Synonym: Methyl *cis*-9-octadecenoate

CAS RN: 112-62-9
Merck No.: 6788
Mol. Wt.: 296.494

Mol. Form.: $C_{19}H_{36}O_2$

PHYSICAL CONSTANTS

T_m: -19.9°C (253.2 K) T_c: μ:
T_b: P_c: IP:

TRANSITION PROPERTIES

$\Delta_{fus}H$ (T_m): Vapor pressure (0°C):
$\Delta_{vap}H$ (T_b): Vapor pressure (25°C):
$\Delta_{vap}H$ (25°C): 106.80 kJ/mol Vapor pressure (100°C):

PROPERTIES AT 25°C AND 100 kPa

	Solid	Liquid	Gas		Liquid
$\Delta_f H°$/kJ mol^{-1}:		-734.5	-649.9	d:	0.872 g/mL
$S°$/J mol^{-1}K^{-1}:				η:	5.0 mPa s
C_p/J mol^{-1}K^{-1}:				k:	
COMMENTS:					

Name: Methyloxirane   CAS RN: 75-56-9
Synonyms: 1,2-Propylene oxide   Merck No.: 7869
          Epoxypropane   Mol. Wt.: 58.080
Mol. Form.: C_3H_6O

PHYSICAL CONSTANTS

T_m: -111.93°C (161.22 K) T_c: 209.1°C (482.2 K) μ: 2.01 D
T_b: 35°C (308 K) P_c: 4.92 MPa   IP: 10.22 eV

TRANSITION PROPERTIES

$\Delta_{fus}H$ (T_m): 6.53 kJ/mol   Vapor pressure (0°C): 24.4 kPa
$\Delta_{vap}H$ (T_b): 27.35 kJ/mol   Vapor pressure (25°C): 71.7 kPa
$\Delta_{vap}H$ (25°C): 27.89 kJ/mol   Vapor pressure (100°C):

PROPERTIES AT 25°C AND 100 kPa

	Solid	Liquid	Gas		Liquid
$\Delta_f H°$/kJ mol^{-1}:		-122.6	-94.7		d: 0.8209 g/mL
$S°$/J mol^{-1}K^{-1}:		196.5	286.9		η:
C_p/J mol^{-1}K^{-1}:		120.4	72.6		k:

COMMENTS: TLV=20 ppm; carcinogen; highly toxic; very flammable

---

Name: 2-Methylpentane   CAS RN: 107-83-5
Synonym: Isohexane   Merck No.:
                                                                     Mol. Wt.: 86.177
Mol. Form.: C_6H_{14}

PHYSICAL CONSTANTS

T_m: -153.7°C (119.4 K) T_c: 224.6°C (497.7 K) μ:
T_b: 60.26°C (333.41 K) P_c: 3.031 MPa   IP: 10.12 eV

TRANSITION PROPERTIES

$\Delta_{fus}H$ (T_m): 6.27 kJ/mol   Vapor pressure (0°C): 8.99 kPa
$\Delta_{vap}H$ (T_b): 27.79 kJ/mol   Vapor pressure (25°C): 28.2 kPa
$\Delta_{vap}H$ (25°C): 29.89 kJ/mol   Vapor pressure (100°C): 308 kPa

PROPERTIES AT 25°C AND 100 kPa

	Solid	Liquid	Gas		Liquid
$\Delta_f H°$/kJ mol^{-1}:		-204.6	-174.8		d: 0.650 g/mL
$S°$/J mol^{-1}K^{-1}:		290.6			η: 0.286 mPa s
C_p/J mol^{-1}K^{-1}:		193.7			k:

COMMENTS: Flammable

Name: 3-Methylpentane   CAS RN: 96-14-0
                                                                     Merck No.:
                                                                     Mol. Wt.: 86.177
Mol. Form.: C_6H_{14}

PHYSICAL CONSTANTS

T_m: -162.9°C (110.2 K) T_c: 231.4°C (504.5 K) μ:
T_b: 63.27°C (336.42 K) P_c: 3.126 MPa   IP: 10.08 eV

TRANSITION PROPERTIES

$\Delta_{fus}H$ (T_m): 5.30 kJ/mol   Vapor pressure (0°C): 7.93 kPa
$\Delta_{vap}H$ (T_b): 28.06 kJ/mol   Vapor pressure (25°C): 25.3 kPa
$\Delta_{vap}H$ (25°C): 30.28 kJ/mol   Vapor pressure (100°C): 282 kPa

PROPERTIES AT 25°C AND 100 kPa

	Solid	Liquid	Gas		Liquid
$\Delta_f H°$/kJ mol^{-1}:		-202.4	-172.1		d: 0.661 g/mL
$S°$/J mol^{-1}K^{-1}:		292.5			η: 0.306 mPa s
C_p/J mol^{-1}K^{-1}:		190.7			k:

COMMENTS: Flammable

Name: 4-Methyl-2-pentanol
Synonym: 4-Methyl-2-pentyl alcohol

Mol. Form.: $C_6H_{14}O$

CAS RN: 108-11-2
Merck No.:
Mol. Wt.: 102.177

PHYSICAL CONSTANTS

T_m: -90°C (183 K)
T_b: 131.65°C (404.80 K)

T_c: 301.3°C (574.4 K)
P_c:

μ:
IP:

TRANSITION PROPERTIES

$\Delta_{fus}H$ (T_m):
$\Delta_{vap}H$ (T_b): 44.22 kJ/mol
$\Delta_{vap}H$ (25°C): 58.70 kJ/mol

Vapor pressure (0°C):
Vapor pressure (25°C): 0.700 kPa
Vapor pressure (100°C): 33.2 kPa

PROPERTIES AT 25°C AND 100 kPa

	Solid	Liquid	Gas	Liquid
$\Delta_f H°$/kJ mol^{-1}:		-394.7		d: 0.8033 g/mL
$S°$/J mol^{-1}K^{-1}:				η: 4.07 mPa s
C_p/J mol^{-1}K^{-1}:		273.0		k:
COMMENTS:				

Name: 4-Methyl-2-pentanone
Synonyms: Isobutyl methyl ketone
Methyl isobutyl ketone
Mol. Form.: $C_6H_{12}O$

CAS RN: 108-10-1
Merck No.: 5095
Mol. Wt.: 100.161

PHYSICAL CONSTANTS

T_m: -84°C (189 K)
T_b: 116.5°C (389.6 K)

T_c: 298°C (571 K)
P_c: 3.27 MPa

μ:
IP: 9.30 eV

TRANSITION PROPERTIES

$\Delta_{fus}H$ (T_m):
$\Delta_{vap}H$ (T_b): 34.49 kJ/mol
$\Delta_{vap}H$ (25°C): 40.61 kJ/mol

Vapor pressure (0°C): 0.539 kPa
Vapor pressure (25°C): 2.64 kPa
Vapor pressure (100°C): 62.0 kPa

PROPERTIES AT 25°C AND 100 kPa

	Solid	Liquid	Gas	Liquid
$\Delta_f H°$/kJ mol^{-1}:				d: 0.7962 g/mL
$S°$/J mol^{-1}K^{-1}:				η: 0.545 mPa s
C_p/J mol^{-1}K^{-1}:		213.3		k:
COMMENTS: TLV=50 ppm; flammable				

Name: 2-Methyl-1-pentene

Mol. Form.: C_6H_{12}

CAS RN: 763-29-1
Merck No.:
Mol. Wt.: 84.161

PHYSICAL CONSTANTS

T_m: -136°C (137 K)
T_b: 62.1°C (335.2 K)

T_c:
P_c:

μ:
IP: 9.08 eV

TRANSITION PROPERTIES

$\Delta_{fus}H$ (T_m):
$\Delta_{vap}H$ (T_b):
$\Delta_{vap}H$ (25°C): 30.48 kJ/mol

Vapor pressure (0°C): 8.10 kPa
Vapor pressure (25°C): 26.0 kPa
Vapor pressure (100°C):

PROPERTIES AT 25°C AND 100 kPa

	Solid	Liquid	Gas	Liquid
$\Delta_f H°$/kJ mol^{-1}:		-90.0	-59.4	d: 0.6749 g/mL
$S°$/J mol^{-1}K^{-1}:				η:
C_p/J mol^{-1}K^{-1}:				k:
COMMENTS: Flammable				

Name: 2-Methyl-2-pentene

CAS RN: 625-27-4
Merck No.:
Mol. Wt.: 84.161

Mol. Form.: C_6H_{12}

PHYSICAL CONSTANTS

T_m: -135°C (138 K) T_c: μ:
T_b: 67.3°C (340.4 K) P_c: IP: 8.58 eV

TRANSITION PROPERTIES

$\Delta_{fus}H$ (T_m):   Vapor pressure (0°C): 6.30 kPa
$\Delta_{vap}H$ (T_b):   Vapor pressure (25°C): 21.0 kPa
$\Delta_{vap}H$ (25°C): 31.60 kJ/mol   Vapor pressure (100°C):

PROPERTIES AT 25°C AND 100 kPa

	Solid	Liquid	Gas		Liquid
$\Delta_f H°$/kJ mol^{-1}:		-98.5	-66.9		d: 0.6819 g/mL
$S°$/J mol^{-1}K^{-1}:					η:
C_p/J mol^{-1}K^{-1}:					k:

COMMENTS: Flammable

Name: 4-Methyl-1-pentene
Synonym: Isobutylethene

CAS RN: 691-37-2
Merck No.:
Mol. Wt.: 84.161

Mol. Form.: C_6H_{12}

PHYSICAL CONSTANTS

T_m: -154°C (119 K) T_c: μ:
T_b: 53.9°C (327.0 K) P_c: IP: 9.45 eV

TRANSITION PROPERTIES

$\Delta_{fus}H$ (T_m):   Vapor pressure (0°C): 12.0 kPa
$\Delta_{vap}H$ (T_b):   Vapor pressure (25°C): 36.1 kPa
$\Delta_{vap}H$ (25°C): 28.71 kJ/mol   Vapor pressure (100°C):

PROPERTIES AT 25°C AND 100 kPa

	Solid	Liquid	Gas		Liquid
$\Delta_f H°$/kJ mol^{-1}:		-80.0	-51.3		d: 0.6589 g/mL
$S°$/J mol^{-1}K^{-1}:					η:
C_p/J mol^{-1}K^{-1}:					k:

COMMENTS: Flammable

Name: 4-Methyl-*cis*-2-pentene

CAS RN: 691-38-3
Merck No.:
Mol. Wt.: 84.161

Mol. Form.: C_6H_{12}

PHYSICAL CONSTANTS

T_m: -134.8°C (138.3 K) T_c: μ:
T_b: 56.3°C (329.4 K) P_c: IP: 8.98 eV

TRANSITION PROPERTIES

$\Delta_{fus}H$ (T_m):   Vapor pressure (0°C): 10.5 kPa
$\Delta_{vap}H$ (T_b):   Vapor pressure (25°C): 32.5 kPa
$\Delta_{vap}H$ (25°C): 29.48 kJ/mol   Vapor pressure (100°C):

PROPERTIES AT 25°C AND 100 kPa

	Solid	Liquid	Gas		Liquid
$\Delta_f H°$/kJ mol^{-1}:		-87.0	-57.5		d: 0.6645 g/mL
$S°$/J mol^{-1}K^{-1}:					η:
C_p/J mol^{-1}K^{-1}:					k:

COMMENTS: Flammable

Name: 4-Methyl-*trans*-2-pentene

CAS RN: 674-76-0
Merck No.:
Mol. Wt.: 84.161

Mol. Form.: C_6H_{12}

PHYSICAL CONSTANTS

T_m: -140.8°C (132.3 K)	T_c:	μ:
T_b: 58.6°C (331.7 K)	P_c:	IP: 8.97 eV

TRANSITION PROPERTIES

$\Delta_{fus}H$ (T_m):   Vapor pressure (0°C): 9.41 kPa
$\Delta_{vap}H$ (T_b):   Vapor pressure (25°C): 29.7 kPa
$\Delta_{vap}H$ (25°C): 29.97 kJ/mol   Vapor pressure (100°C):

PROPERTIES AT 25°C AND 100 kPa

	Solid	Liquid	Gas	Liquid
$\Delta_f H°$/kJ mol^{-1}:		-91.5	-61.5	d: 0.6637 g/mL
$S°$/J mol^{-1}K^{-1}:				η:
C_p/J mol^{-1}K^{-1}:				k:

COMMENTS: Flammable

Name: 2-Methylpropanenitrile
Synonyms: Isobutyronitrile
          Isopropyl cyanide
Mol. Form.: C_4H_7N

CAS RN: 78-82-0
Merck No.:
Mol. Wt.: 69.106

PHYSICAL CONSTANTS

T_m: -71.5°C (201.6 K)	T_c:	μ: 4.29 D
T_b: 103.9°C (377.0 K)	P_c:	IP: 11.30 eV

TRANSITION PROPERTIES

$\Delta_{fus}H$ (T_m):   Vapor pressure (0°C):
$\Delta_{vap}H$ (T_b): 32.39 kJ/mol   Vapor pressure (25°C):
$\Delta_{vap}H$ (25°C): 37.13 kJ/mol   Vapor pressure (100°C): 90.9 kPa

PROPERTIES AT 25°C AND 100 kPa

	Solid	Liquid	Gas	Liquid
$\Delta_f H°$/kJ mol^{-1}:		-13.8	23.3	d: 0.7656 g/mL
$S°$/J mol^{-1}K^{-1}:				η: 0.49 mPa s
C_p/J mol^{-1}K^{-1}:				k:

COMMENTS: Highly toxic; flammable

Name: 2-Methyl-2-propanethiol
Synonym: *tert*-Butyl mercaptan

CAS RN: 75-66-1
Merck No.: 1577
Mol. Wt.: 90.189

Mol. Form.: $C_4H_{10}S$

PHYSICAL CONSTANTS

T_m: -0.5°C (272.6 K)	T_c:	μ: 1.66 D
T_b: 64.3°C (337.4 K)	P_c:	IP: 9.03 eV

TRANSITION PROPERTIES

$\Delta_{fus}H$ (T_m):   Vapor pressure (0°C): 7.45 kPa
$\Delta_{vap}H$ (T_b): 28.45 kJ/mol   Vapor pressure (25°C): 24.2 kPa
$\Delta_{vap}H$ (25°C): 30.78 kJ/mol   Vapor pressure (100°C): 276 kPa

PROPERTIES AT 25°C AND 100 kPa

	Solid	Liquid	Gas	Liquid
$\Delta_f H°$/kJ mol^{-1}:		-140.5	-109.6	d: 0.7943 g/mL
$S°$/J mol^{-1}K^{-1}:				η:
C_p/J mol^{-1}K^{-1}:				k:

COMMENTS:

Name: Methyl propanoate
Synonym: Methyl propionate

Mol. Form.: $C_4H_8O_2$

CAS RN: 554-12-1
Merck No.: 6030
Mol. Wt.: 88.106

PHYSICAL CONSTANTS

T_m: -87.5°C (185.6 K)
T_b: 79.8°C (352.9 K)

T_c: 257.5°C (530.6 K)
P_c: 4.004 MPa

μ:
IP: 10.15 eV

TRANSITION PROPERTIES

$\Delta_{fus}H$ (T_m):
$\Delta_{vap}H$ (T_b): 32.24 kJ/mol
$\Delta_{vap}H$ (25°C): 35.85 kJ/mol

Vapor pressure (0°C): 2.86 kPa
Vapor pressure (25°C): 11.5 kPa
Vapor pressure (100°C): 188 kPa

PROPERTIES AT 25°C AND 100 kPa

	Solid	Liquid	Gas	Liquid
$\Delta_f H°$/kJ mol^{-1}:				d: 0.9090 g/mL
$S°$/J mol^{-1}K^{-1}:				η: 0.431 mPa s
C_p/J mol^{-1}K^{-1}:				k:

COMMENTS: Flammable

Name: 2-Methylpropanoic acid
Synonym: Isobutyric acid

Mol. Form.: $C_4H_8O_2$

CAS RN: 79-31-2
Merck No.: 5039
Mol. Wt.: 88.106

PHYSICAL CONSTANTS

T_m: -46°C (227 K)
T_b: 154.45°C (427.60 K)

T_c: 332°C (605 K)
P_c: 3.7 MPa

μ:
IP: 10.33 eV

TRANSITION PROPERTIES

$\Delta_{fus}H$ (T_m):
$\Delta_{vap}H$ (T_b):
$\Delta_{vap}H$ (25°C): 35.30 kJ/mol

Vapor pressure (0°C):
Vapor pressure (25°C):
Vapor pressure (100°C): 16.5 kPa

PROPERTIES AT 25°C AND 100 kPa

	Solid	Liquid	Gas	Liquid
$\Delta_f H°$/kJ mol^{-1}:				d: 0.9431 g/mL
$S°$/J mol^{-1}K^{-1}:				η: 1.23 mPa s
C_p/J mol^{-1}K^{-1}:				k:

COMMENTS:

Name: 2-Methyl-2-propanol
Synonyms: *tert*-Butyl alcohol
             Trimethylcarbinol
Mol. Form.: $C_4H_{10}O$

CAS RN: 75-65-0
Merck No.: 1542
Mol. Wt.: 74.123

PHYSICAL CONSTANTS

T_m: 25.4°C (298.5 K)
T_b: 82.4°C (355.5 K)

T_c: 233.06°C (506.21 K)
P_c: 3.973 MPa

μ:
IP: 9.97 eV

TRANSITION PROPERTIES

$\Delta_{fus}H$ (T_m): 6.79 kJ/mol
$\Delta_{vap}H$ (T_b): 39.07 kJ/mol
$\Delta_{vap}H$ (25°C): 46.69 kJ/mol

Vapor pressure (0°C):
Vapor pressure (25°C): 5.52 kPa
Vapor pressure (100°C): 191 kPa

PROPERTIES AT 25°C AND 100 kPa

	Solid	Liquid	Gas	Solid
$\Delta_f H°$/kJ mol^{-1}:		-359.2	-312.5	d: 0.7812 g/mL
$S°$/J mol^{-1}K^{-1}:		193.3	326.7	η: N/A
C_p/J mol^{-1}K^{-1}:	146.1	218.6	113.6	k:

COMMENTS: TLV=100 ppm; highly toxic; flammable

Name: 2-Methyl-1-propanol
Synonyms: Isobutyl alcohol
        Isopropylcarbinol
Mol. Form.: $C_4H_{10}O$

CAS RN: 78-83-1
Merck No.: 5015
Mol. Wt.: 74.123

PHYSICAL CONSTANTS

T_m: -108°C (165 K)
T_b: 107.89°C (381.04 K)

T_c: 274.63°C (547.78 K)
P_c: 4.300 MPa

μ: 1.64 D
IP: 10.12 eV

TRANSITION PROPERTIES

$\Delta_{fus}H$ (T_m): 6.32 kJ/mol
$\Delta_{vap}H$ (T_b): 41.82 kJ/mol
$\Delta_{vap}H$ (25°C): 50.82 kJ/mol

Vapor pressure (0°C):
Vapor pressure (25°C): 1.39 kPa
Vapor pressure (100°C): 75.4 kPa

PROPERTIES AT 25°C AND 100 kPa

	Solid	Liquid	Gas		Liquid
$\Delta_f H°$/kJ mol^{-1}:		-334.7	-283.9	d:	0.7978 g/mL
$S°$/J mol^{-1}K^{-1}:		214.7		η:	3.33 mPa s
C_p/J mol^{-1}K^{-1}:		181.5	111.3	k:	

COMMENTS: TLV=50 ppm; flammable

Name: 2-Methylpropenal
Synonyms: Methacrolein
        Isobutenal
Mol. Form.: C_4H_6O

CAS RN: 78-85-3
Merck No.:
Mol. Wt.: 70.091

PHYSICAL CONSTANTS

T_m:
T_b: 68.4°C (341.5 K)

T_c:
P_c:

μ: 2.68 D
IP: 9.86 eV

TRANSITION PROPERTIES

$\Delta_{fus}H$ (T_m):
$\Delta_{vap}H$ (T_b):
$\Delta_{vap}H$ (25°C):

Vapor pressure (0°C):
Vapor pressure (25°C):
Vapor pressure (100°C):

PROPERTIES AT 25°C AND 100 kPa

	Solid	Liquid	Gas		Liquid
$\Delta_f H°$/kJ mol^{-1}:				d:	0.840 g/mL
$S°$/J mol^{-1}K^{-1}:				η:	
C_p/J mol^{-1}K^{-1}:				k:	

COMMENTS: Highly toxic

Name: Methyl propyl ether
Synonyms: 1-Methoxypropane
        Propyl methyl ether
Mol. Form.: $C_4H_{10}O$

CAS RN: 557-17-5
Merck No.: 6031
Mol. Wt.: 74.123

PHYSICAL CONSTANTS

T_m:
T_b: 39.1°C (312.2 K)

T_c: 203.10°C (476.25 K)
P_c: 3.801 MPa

μ: 1.107 D
IP: 9.42 eV

TRANSITION PROPERTIES

$\Delta_{fus}H$ (T_m):
$\Delta_{vap}H$ (T_b): 26.75 kJ/mol
$\Delta_{vap}H$ (25°C): 27.60 kJ/mol

Vapor pressure (0°C): 19.4 kPa
Vapor pressure (25°C): 60.9 kPa
Vapor pressure (100°C):

PROPERTIES AT 25°C AND 100 kPa

	Solid	Liquid	Gas		Liquid
$\Delta_f H°$/kJ mol^{-1}:		-266.0	-238.2	d:	0.727 g/mL
$S°$/J mol^{-1}K^{-1}:		262.9		η:	
C_p/J mol^{-1}K^{-1}:		165.4		k:	

COMMENTS: Flammable

Name: 2-Methylpyridine
Synonym: 2-Picoline

Mol. Form.: C_6H_7N

CAS RN: 109-06-8
Merck No.: 7372
Mol. Wt.: 93.128

PHYSICAL CONSTANTS

T_m: -66.68°C (206.47 K)
T_b: 129.38°C (402.53 K)

T_c: 347.9°C (621.0 K)
P_c: 4.60 MPa

μ: 1.85 D
IP: 9.02 eV

TRANSITION PROPERTIES

$\Delta_{fus}H$ (T_m): 9.72 kJ/mol
$\Delta_{vap}H$ (T_b): 36.17 kJ/mol
$\Delta_{vap}H$ (25°C): 42.48 kJ/mol

Vapor pressure (0°C):
Vapor pressure (25°C): 1.50 kPa
Vapor pressure (100°C): 41.1 kPa

PROPERTIES AT 25°C AND 100 kPa

	Solid	Liquid	Gas	Liquid
$\Delta_f H°$/kJ mol^{-1}:		56.7	99.2	d: 0.9398 g/mL
$S°$/J mol^{-1}K^{-1}:				η: 0.753 mPa s
C_p/J mol^{-1}K^{-1}:		158.6		k:
COMMENTS:				

Name: 3-Methylpyridine
Synonym: 3-Picoline

Mol. Form.: C_6H_7N

CAS RN: 108-99-6
Merck No.: 7373
Mol. Wt.: 93.128

PHYSICAL CONSTANTS

T_m: -18.14°C (255.01 K)
T_b: 144.14°C (417.29 K)

T_c: 371.9°C (645.0 K)
P_c: 4.48 MPa

μ:
IP: 9.04 eV

TRANSITION PROPERTIES

$\Delta_{fus}H$ (T_m): 14.18 kJ/mol
$\Delta_{vap}H$ (T_b): 37.35 kJ/mol
$\Delta_{vap}H$ (25°C): 44.44 kJ/mol

Vapor pressure (0°C):
Vapor pressure (25°C): 0.795 kPa
Vapor pressure (100°C): 26.0 kPa

PROPERTIES AT 25°C AND 100 kPa

	Solid	Liquid	Gas	Liquid
$\Delta_f H°$/kJ mol^{-1}:		61.9	106.4	d: 0.9520 g/mL
$S°$/J mol^{-1}K^{-1}:		216.3		η: 0.872 mPa s
C_p/J mol^{-1}K^{-1}:		158.7		k:
COMMENTS:				

Name: 4-Methylpyridine
Synonym: 4-Picoline

Mol. Form.: C_6H_7N

CAS RN: 108-89-4
Merck No.: 7374
Mol. Wt.: 93.128

PHYSICAL CONSTANTS

T_m: 3.66°C (276.81 K)
T_b: 145.36°C (418.51 K)

T_c: 372.6°C (645.7 K)
P_c: 4.70 MPa

μ: 2.70 D
IP: 9.04 eV

TRANSITION PROPERTIES

$\Delta_{fus}H$ (T_m): 11.57 kJ/mol
$\Delta_{vap}H$ (T_b): 37.51 kJ/mol
$\Delta_{vap}H$ (25°C): 44.56 kJ/mol

Vapor pressure (0°C):
Vapor pressure (25°C): 0.759 kPa
Vapor pressure (100°C): 25.0 kPa

PROPERTIES AT 25°C AND 100 kPa

	Solid	Liquid	Gas	Liquid
$\Delta_f H°$/kJ mol^{-1}:		59.2	103.8	d: 0.9502 g/mL
$S°$/J mol^{-1}K^{-1}:		209.1		η:
C_p/J mol^{-1}K^{-1}:		159.0		k:
COMMENTS:				

Name: *N*-Methyl-2-pyrrolidone
Synonym: 1-Methyl-2-pyrrolidinone

Mol. Form.: C_5H_9NO

CAS RN: 872-50-4
Merck No.:
Mol. Wt.: 99.133

PHYSICAL CONSTANTS

T_m: -24°C (249 K) T_c: 448.7°C (721.8 K) μ:
T_b: 202°C (475 K) P_c: IP: 9.17 eV

TRANSITION PROPERTIES

$\Delta_{fus}H$ (T_m): Vapor pressure (0°C):
$\Delta_{vap}H$ (T_b): Vapor pressure (25°C): 0.040 kPa
$\Delta_{vap}H$ (25°C): Vapor pressure (100°C): 3.28 kPa

PROPERTIES AT 25°C AND 100 kPa

	Solid	Liquid	Gas		Liquid
$\Delta_f H°$/kJ mol^{-1}:		-262.2		d:	1.0251 g/mL
$S°$/J mol^{-1}K^{-1}:				η:	
C_p/J mol^{-1}K^{-1}:		307.8		k:	
COMMENTS:					

Name: Methyl salicylate
Synonym: Methyl 2-hydroxybenzoate

Mol. Form.: $C_8H_8O_3$

CAS RN: 119-36-8
Merck No.: 6038
Mol. Wt.: 152.150

PHYSICAL CONSTANTS

T_m: -8°C (265 K) T_c: 436°C (709 K) μ:
T_b: 222.95°C (496.10 K) P_c: IP:

TRANSITION PROPERTIES

$\Delta_{fus}H$ (T_m): Vapor pressure (0°C):
$\Delta_{vap}H$ (T_b): 46.67 kJ/mol Vapor pressure (25°C): 0.015 kPa
$\Delta_{vap}H$ (25°C): Vapor pressure (100°C): 1.74 kPa

PROPERTIES AT 25°C AND 100 kPa

	Solid	Liquid	Gas		Liquid
$\Delta_f H°$/kJ mol^{-1}:				d:	1.179 g/mL
$S°$/J mol^{-1}K^{-1}:				η:	
C_p/J mol^{-1}K^{-1}:		249.0		k:	
COMMENTS:					

Name: Methyl vinyl ether
Synonyms: Methoxyethene
      Vinyl methyl ether
Mol. Form.: C_3H_6O

CAS RN: 107-25-5
Merck No.:
Mol. Wt.: 58.080

PHYSICAL CONSTANTS

T_m: -122°C (151 K) T_c: μ: 0.96 D
T_b: 5.5°C (278.6 K) P_c: IP: 8.93 eV

TRANSITION PROPERTIES

$\Delta_{fus}H$ (T_m): Vapor pressure (0°C): 86.5 kPa
$\Delta_{vap}H$ (T_b): Vapor pressure (25°C): 180 kPa
$\Delta_{vap}H$ (25°C): Vapor pressure (100°C):

PROPERTIES AT 25°C AND 100 kPa

	Solid	Liquid	Gas		Gas
$\Delta_f H°$/kJ mol^{-1}:				d:	2.374 g/L
$S°$/J mol^{-1}K^{-1}:				η:	
C_p/J mol^{-1}K^{-1}:				k:	
COMMENTS: Very flammable					

Name: Molybdenum

CAS RN: 7439-98-7
Merck No.: 6144
Mol. Wt.: 95.940

Mol. Form.: Mo

PHYSICAL CONSTANTS

T_m: 2623°C (2896 K) T_c: μ:
T_b: 4639°C (4912 K) P_c: IP: 7.09 eV

TRANSITION PROPERTIES

$\Delta_{fus}H$ (T_m): 37.48 kJ/mol Vapor pressure (0°C): N/A
$\Delta_{vap}H$ (T_b): Vapor pressure (25°C): N/A
$\Delta_{vap}H$ (25°C): Vapor pressure (100°C): N/A

PROPERTIES AT 25°C AND 100 kPa

	Solid	Liquid	Gas		Solid
$\Delta_f H°$/kJ mol^{-1}:	0.0		658.1		d: 10.2 g/mL
$S°$/J mol^{-1}K^{-1}:	28.7		182.0		η: N/A
C_p/J mol^{-1}K^{-1}:	24.1		20.8		k: 138 W/m K
COMMENTS:					

Name: Molybdenum fluoride (MoF$_6$)
Synonyms: Molybdenum hexafluoride
      Molybdenum(VI) fluoride
Mol. Form.: F$_6$Mo

CAS RN: 7783-77-9
Merck No.: 6146
Mol. Wt.: 209.930

PHYSICAL CONSTANTS

T_m: 17.5°C (290.6 K) T_c: 200°C (473 K) μ:
T_b: 34°C (307 K) P_c: 4.75 MPa IP: 14.50 eV

TRANSITION PROPERTIES

$\Delta_{fus}H$ (T_m): 4.33 kJ/mol Vapor pressure (0°C):
$\Delta_{vap}H$ (T_b): 27.20 kJ/mol Vapor pressure (25°C):
$\Delta_{vap}H$ (25°C): 28.00 kJ/mol Vapor pressure (100°C):

PROPERTIES AT 25°C AND 100 kPa

	Solid	Liquid	Gas		Liquid
$\Delta_f H°$/kJ mol^{-1}:		-1585.5	-1557.7		d: 2.54 g/mL
$S°$/J mol^{-1}K^{-1}:		259.7	350.5		η:
C_p/J mol^{-1}K^{-1}:		169.8	120.6		k:
COMMENTS:					

Name: Molybdenum oxide (MoO$_3$)
Synonyms: Molybdenum trioxide
      Molybdenum(VI) oxide
Mol. Form.: MoO$_3$

CAS RN: 1313-27-5
Merck No.: 6148
Mol. Wt.: 143.938

PHYSICAL CONSTANTS

T_m: 801°C (1074 K) T_c: μ:
T_b: 1155°C (1428 K) P_c: IP:

TRANSITION PROPERTIES

$\Delta_{fus}H$ (T_m): 48.00 kJ/mol Vapor pressure (0°C): N/A
$\Delta_{vap}H$ (T_b): 138.00 kJ/mol Vapor pressure (25°C): N/A
$\Delta_{vap}H$ (25°C): Vapor pressure (100°C): N/A

PROPERTIES AT 25°C AND 100 kPa

	Solid	Liquid	Gas		Solid
$\Delta_f H°$/kJ mol^{-1}:	-745.1				d: 4.70 g/mL
$S°$/J mol^{-1}K^{-1}:	77.7				η: N/A
C_p/J mol^{-1}K^{-1}:	75.0				k:
COMMENTS: Highly toxic					

Name: Morpholine
Synonyms: Tetrahydro-1,4-oxazine
            Diethyleneimine oxide
Mol. Form.: C_4H_9NO

CAS RN: 110-91-8
Merck No.: 6194
Mol. Wt.: 87.122

PHYSICAL CONSTANTS

T_m: -4.9°C (268.2 K)
T_b: 128°C (401 K)

T_c:
P_c:

μ: 1.55 D
IP: 8.20 eV

TRANSITION PROPERTIES

$\Delta_{fus}H$ (T_m):
$\Delta_{vap}H$ (T_b): 37.05 kJ/mol
$\Delta_{vap}H$ (25°C): 43.96 kJ/mol

Vapor pressure (0°C):
Vapor pressure (25°C): 1.34 kPa
Vapor pressure (100°C): 40.4 kPa

PROPERTIES AT 25°C AND 100 kPa

	Solid	Liquid	Gas	Liquid
$\Delta_f H°$/kJ mol^{-1}:				d: 0.9959 g/mL
$S°$/J mol^{-1}K^{-1}:				η: 2.02 mPa s
C_p/J mol^{-1}K^{-1}:		164.8		k:

COMMENTS: TLV=20 ppm; flammable

Name: Naphthalene

CAS RN: 91-20-3
Merck No.: 6289
Mol. Wt.: 128.174

Mol. Form.: $C_{10}H_8$

PHYSICAL CONSTANTS

T_m: 80.2°C (353.3 K)
T_b: 217.99°C (491.14 K)

T_c: 475.3°C (748.4 K)
P_c: 4.051 MPa

μ: 0 D
IP: 8.14 eV

TRANSITION PROPERTIES

$\Delta_{fus}H$ (T_m): 17.87 kJ/mol
$\Delta_{vap}H$ (T_b): 43.18 kJ/mol
$\Delta_{vap}H$ (25°C):

Vapor pressure (0°C):
Vapor pressure (25°C): 0.011 kPa
Vapor pressure (100°C): 2.50 kPa

PROPERTIES AT 25°C AND 100 kPa

	Solid	Liquid	Gas	Solid
$\Delta_f H°$/kJ mol^{-1}:	77.9		150.6	d: 1.1536 g/mL
$S°$/J mol^{-1}K^{-1}:	167.4		333.1	η: N/A
C_p/J mol^{-1}K^{-1}:	165.7		131.9	k:

COMMENTS: TLV=10 ppm; highly toxic

Name: 1-Naphthol
Synonyms: 1-Hydroxynaphthalene
            α-Naphthyl alcohol
Mol. Form.: $C_{10}H_8O$

CAS RN: 90-15-3
Merck No.: 6303
Mol. Wt.: 144.173

PHYSICAL CONSTANTS

T_m: 96°C (369 K)
T_b: 288°C (561 K)

T_c:
P_c:

μ:
IP: 7.76 eV

TRANSITION PROPERTIES

$\Delta_{fus}H$ (T_m): 23.33 kJ/mol
$\Delta_{vap}H$ (T_b):
$\Delta_{vap}H$ (25°C):

Vapor pressure (0°C):
Vapor pressure (25°C):
Vapor pressure (100°C):

PROPERTIES AT 25°C AND 100 kPa

	Solid	Liquid	Gas	Solid
$\Delta_f H°$/kJ mol^{-1}:	-121.0		-29.9	d: 1.292 g/mL
$S°$/J mol^{-1}K^{-1}:				η: N/A
C_p/J mol^{-1}K^{-1}:	166.9			k:

COMMENTS:

Name: 2-Naphthol   CAS RN: 135-19-3
Synonyms: 2-Hydroxynaphthalene   Merck No.: 6304
          β-Naphthyl alcohol   Mol. Wt.: 144.173
Mol. Form.: $C_{10}H_8O$

PHYSICAL CONSTANTS

T_m: 123°C (396 K) T_c: μ:
T_b: 285°C (558 K) P_c: IP: 7.85 eV

TRANSITION PROPERTIES

$\Delta_{fus}H$ (T_m): 17.51 kJ/mol   Vapor pressure (0°C):
$\Delta_{vap}H$ (T_b):   Vapor pressure (25°C):
$\Delta_{vap}H$ (25°C): 94.2 kJ/mol   Vapor pressure (100°C):

PROPERTIES AT 25°C AND 100 kPa

	Solid	Liquid	Gas		Solid
$\Delta_f H°$/kJ mol^{-1}:		-124.2	-30.0		d: 1.252 g/mL
$S°$/J mol^{-1}K^{-1}:					η: N/A
C_p/J mol^{-1}K^{-1}:					k:
COMMENTS:					

Name: Neodymium   CAS RN: 7440-00-8
                                                                                  Merck No.: 6365
                                                                                  Mol. Wt.: 144.240
Mol. Form.: Nd

PHYSICAL CONSTANTS

T_m: 1016°C (1289 K) T_c: μ:
T_b: 3066°C (3339 K) P_c: IP: 5.53 eV

TRANSITION PROPERTIES

$\Delta_{fus}H$ (T_m): 7.14 kJ/mol   Vapor pressure (0°C): N/A
$\Delta_{vap}H$ (T_b):   Vapor pressure (25°C): N/A
$\Delta_{vap}H$ (25°C):   Vapor pressure (100°C): N/A

PROPERTIES AT 25°C AND 100 kPa

	Solid	Liquid	Gas		Solid
$\Delta_f H°$/kJ mol^{-1}:	0.0		327.6		d: 7.01 g/mL
$S°$/J mol^{-1}K^{-1}:	71.5		189.4		η: N/A
C_p/J mol^{-1}K^{-1}:	27.5		22.1		k: 16.5 W/m K
COMMENTS:					

Name: Neon   CAS RN: 7440-01-9
                                                                                  Merck No.: 6371
                                                                                  Mol. Wt.: 20.180
Mol. Form.: Ne

PHYSICAL CONSTANTS

T_m: -248.59°C (24.56 K) T_c: -228.7°C (44.4 K)   μ: 0 D
T_b: -246.08°C (27.07 K) P_c: 2.76 MPa   IP: 21.56 eV

TRANSITION PROPERTIES

$\Delta_{fus}H$ (T_m): 0.34 kJ/mol   Vapor pressure (0°C): N/A
$\Delta_{vap}H$ (T_b): 1.71 kJ/mol   Vapor pressure (25°C): N/A
$\Delta_{vap}H$ (25°C):   Vapor pressure (100°C): N/A

PROPERTIES AT 25°C AND 100 kPa

	Solid	Liquid	Gas		Gas
$\Delta_f H°$/kJ mol^{-1}:			0.0		d: 0.825 g/L
$S°$/J mol^{-1}K^{-1}:			146.3		η: 32.0 μPa s
C_p/J mol^{-1}K^{-1}:			20.8		k: 0.0498 W/m K
COMMENTS:					

Name: Neopentane

Synonyms: 2,2-Dimethylpropane

           Tetramethylmethane

Mol. Form.: C_5H_{12}

CAS RN: 463-82-1

Merck No.: 6372

Mol. Wt.: 72.150

PHYSICAL CONSTANTS

T_m: -16.6°C (256.5 K) T_c: 160.7°C (433.8 K) μ: 0 D

T_b: 9.48°C (282.63 K) P_c: 3.197 MPa   IP: 10.21 eV

TRANSITION PROPERTIES

$\Delta_{fus}H$ (T_m): 3.10 kJ/mol

$\Delta_{vap}H$ (T_b): 22.74 kJ/mol

$\Delta_{vap}H$ (25°C): 21.84 kJ/mol

Vapor pressure (0°C): 71.0 kPa

Vapor pressure (25°C): 171 kPa

Vapor pressure (100°C):

PROPERTIES AT 25°C AND 100 kPa

	Solid	Liquid	Gas		Gas
$\Delta_f H°$/kJ mol^{-1}:		-190.2	-168.1	d:	2.949 g/L
$S°$/J mol^{-1}K^{-1}:				η:	
C_p/J mol^{-1}K^{-1}:				k:	

COMMENTS: Very flammable

Name: Nickel

CAS RN: 7440-02-0

Merck No.: 8124

Mol. Wt.: 58.693

Mol. Form.: Ni

PHYSICAL CONSTANTS

T_m: 1455°C (1728.3 K) T_c: μ:

T_b: 2913°C (3186 K) P_c: IP: 7.64 eV

TRANSITION PROPERTIES

$\Delta_{fus}H$ (T_m): 17.48 kJ/mol

$\Delta_{vap}H$ (T_b):

$\Delta_{vap}H$ (25°C):

Vapor pressure (0°C): N/A

Vapor pressure (25°C): N/A

Vapor pressure (100°C): N/A

PROPERTIES AT 25°C AND 100 kPa

	Solid	Liquid	Gas		Solid
$\Delta_f H°$/kJ mol^{-1}:	0.0		429.7	d:	8.90 g/mL
$S°$/J mol^{-1}K^{-1}:	29.9		182.2	η:	N/A
C_p/J mol^{-1}K^{-1}:	26.1		23.4	k:	90.9 W/m K

COMMENTS: Carcinogen; highly toxic

Name: Nickel carbonyl (Ni(CO)$_4$)

CAS RN: 13463-39-3

Merck No.: 6411

Mol. Wt.: 170.735

Mol. Form.: C_4NiO_4

PHYSICAL CONSTANTS

T_m: -19.3°C (253.8 K) T_c: μ:

T_b: 43°C (316 K) P_c: IP: 8.27 eV

TRANSITION PROPERTIES

$\Delta_{fus}H$ (T_m):

$\Delta_{vap}H$ (T_b):

$\Delta_{vap}H$ (25°C): 30.1 kJ/mol

Vapor pressure (0°C):

Vapor pressure (25°C):

Vapor pressure (100°C):

PROPERTIES AT 25°C AND 100 kPa

	Solid	Liquid	Gas		Liquid
$\Delta_f H°$/kJ mol^{-1}:		-633.0	-602.9	d:	1.31 g/mL
$S°$/J mol^{-1}K^{-1}:		313.4	410.6	η:	
C_p/J mol^{-1}K^{-1}:		204.6	145.2	k:	

COMMENTS: TLV=0.05 ppm; carcinogen; highly toxic; flammable

Name: Niobium

CAS RN: 7440-03-1
Merck No.: 6472
Mol. Wt.: 92.906

Mol. Form.: Nb

PHYSICAL CONSTANTS

T_m: 2477°C (2750 K) T_c: μ:
T_b: 4744°C (5017 K) P_c: IP: 6.76 eV

TRANSITION PROPERTIES

$\Delta_{fus}H$ (T_m): 30.00 kJ/mol   Vapor pressure (0°C): N/A
$\Delta_{vap}H$ (T_b):   Vapor pressure (25°C): N/A
$\Delta_{vap}H$ (25°C):   Vapor pressure (100°C): N/A

PROPERTIES AT 25°C AND 100 kPa

	Solid	Liquid	Gas		Solid
$\Delta_f H°$/kJ mol^{-1}:	0.0		725.9		d: 8.57 g/mL
$S°$/J mol^{-1}K^{-1}:	36.4		186.3		η: N/A
C_p/J mol^{-1}K^{-1}:	24.6		30.2		k: 53.7 W/m K
COMMENTS:					

Name: Niobium chloride (NbCl$_5$)
Synonyms: Niobium pentachloride
         Niobium(V) chloride
Mol. Form.: Cl$_5$Nb

CAS RN: 10026-12-7
Merck No.: 6473
Mol. Wt.: 270.170

PHYSICAL CONSTANTS

T_m: 204.7°C (477.8 K) T_c: 530.4°C (803.5 K) μ:
T_b: 254.05°C (527.20 K) P_c: 4.88 MPa   IP: 10.97 eV

TRANSITION PROPERTIES

$\Delta_{fus}H$ (T_m): 33.90 kJ/mol   Vapor pressure (0°C):
$\Delta_{vap}H$ (T_b): 52.70 kJ/mol   Vapor pressure (25°C):
$\Delta_{vap}H$ (25°C):   Vapor pressure (100°C):

PROPERTIES AT 25°C AND 100 kPa

	Solid	Liquid	Gas		Solid
$\Delta_f H°$/kJ mol^{-1}:	-797.5		-703.7		d: 2.73 g/mL
$S°$/J mol^{-1}K^{-1}:	210.5		400.6		η: N/A
C_p/J mol^{-1}K^{-1}:	148.1		120.8		k:
COMMENTS:					

Name: Niobium fluoride (NbF$_5$)
Synonyms: Niobium pentafluoride
         Niobium(V) fluoride
Mol. Form.: F$_5$Nb

CAS RN: 7783-68-8
Merck No.: 6474
Mol. Wt.: 187.898

PHYSICAL CONSTANTS

T_m: 80°C (353 K) T_c: 464°C (737 K) μ:
T_b: 229°C (502 K) P_c: 6.28 MPa   IP:

TRANSITION PROPERTIES

$\Delta_{fus}H$ (T_m): 36.00 kJ/mol   Vapor pressure (0°C):
$\Delta_{vap}H$ (T_b): 52.30 kJ/mol   Vapor pressure (25°C):
$\Delta_{vap}H$ (25°C):   Vapor pressure (100°C):

PROPERTIES AT 25°C AND 100 kPa

	Solid	Liquid	Gas		Solid
$\Delta_f H°$/kJ mol^{-1}:	-1813.8		-1739.7		d: 2.7 g/mL
$S°$/J mol^{-1}K^{-1}:	160.2		321.9		η: N/A
C_p/J mol^{-1}K^{-1}:	134.7		97.1		k:
COMMENTS:					

Name: Nitric acid

Mol. Form.: HNO_3

CAS RN: 7697-37-2
Merck No.: 6496
Mol. Wt.: 63.013

PHYSICAL CONSTANTS

T_m: -41.6°C (231.5 K)	T_c:	μ: 2.17 D
T_b: 83°C (356 K)	P_c:	IP: 11.95 eV

TRANSITION PROPERTIES

$\Delta_{fus}H$ (T_m): 10.50 kJ/mol	Vapor pressure (0°C):
$\Delta_{vap}H$ (T_b):	Vapor pressure (25°C):
$\Delta_{vap}H$ (25°C): 39.10 kJ/mol	Vapor pressure (100°C):

PROPERTIES AT 25°C AND 100 kPa

	Solid	Liquid	Gas		Liquid
$\Delta_f H°$/kJ mol^{-1}:		-174.1	-135.1		d: 1.55 g/mL
$S°$/J mol^{-1}K^{-1}:		155.6	266.4		η:
C_p/J mol^{-1}K^{-1}:		109.9	53.4		k:

COMMENTS: TLV=2 ppm

Name: Nitric oxide
Synonym: Nitrogen monoxide

Mol. Form.: NO

CAS RN: 10102-43-9
Merck No.: 6498
Mol. Wt.: 30.006

PHYSICAL CONSTANTS

T_m: -163.6°C (109.5 K)	T_c: -93°C (180 K)	μ: 0.159 D
T_b: -151.74°C (121.41 K)	P_c: 6.48 MPa	IP: 9.26 eV

TRANSITION PROPERTIES

$\Delta_{fus}H$ (T_m): 2.30 kJ/mol	Vapor pressure (0°C): N/A
$\Delta_{vap}H$ (T_b): 13.83 kJ/mol	Vapor pressure (25°C): N/A
$\Delta_{vap}H$ (25°C):	Vapor pressure (100°C): N/A

PROPERTIES AT 25°C AND 100 kPa

	Solid	Liquid	Gas		Gas
$\Delta_f H°$/kJ mol^{-1}:			90.3		d: 1.226 g/L
$S°$/J mol^{-1}K^{-1}:			210.8		η: 19.1 μPa s
C_p/J mol^{-1}K^{-1}:			29.8		k: 0.0259 W/m K

COMMENTS: TLV=25 ppm; highly toxic

Name: *o*-Nitroaniline
Synonym: 2-Nitrobenzenamine

Mol. Form.: $C_6H_6N_2O_2$

CAS RN: 88-74-4
Merck No.: 6504
Mol. Wt.: 138.126

PHYSICAL CONSTANTS

T_m: 71.2°C (344.3 K)	T_c:	μ:
T_b: 284°C (557 K)	P_c:	IP: 8.27 eV

TRANSITION PROPERTIES

$\Delta_{fus}H$ (T_m): 16.11 kJ/mol	Vapor pressure (0°C):
$\Delta_{vap}H$ (T_b):	Vapor pressure (25°C):
$\Delta_{vap}H$ (25°C): 73.2 kJ/mol	Vapor pressure (100°C):

PROPERTIES AT 25°C AND 100 kPa

	Solid	Liquid	Gas		Solid
$\Delta_f H°$/kJ mol^{-1}:	-26.1	-9.4	63.8		d: 0.9015 g/mL
$S°$/J mol^{-1}K^{-1}:					η: N/A
C_p/J mol^{-1}K^{-1}:	166.0				k:

COMMENTS:

Name: *m*-Nitroaniline

Synonym: 3-Nitrobenzenamine

Mol. Form.: $C_6H_6N_2O_2$

CAS RN: 99-09-2

Merck No.: 6503

Mol. Wt.: 138.126

PHYSICAL CONSTANTS

| T_m: 114°C (387 K) | T_c: | μ: |
| T_b: | P_c: | IP: 8.31 eV |

TRANSITION PROPERTIES

$\Delta_{fus}H$ (T_m): 23.68 kJ/mol

$\Delta_{vap}H$ (T_b):

$\Delta_{vap}H$ (25°C): 72.8 kJ/mol

Vapor pressure (0°C):

Vapor pressure (25°C):

Vapor pressure (100°C):

PROPERTIES AT 25°C AND 100 kPa

	Solid	Liquid	Gas		Solid
$\Delta_f H°$/kJ mol^{-1}:	-38.3	-14.4	58.4		*d*: 0.9011 g/mL
$S°$/J mol^{-1}K^{-1}:					η: N/A
C_p/J mol^{-1}K^{-1}:	158.8				*k*:
COMMENTS:					

Name: *p*-Nitroaniline

Synonym: 4-Nitrobenzenamine

Mol. Form.: $C_6H_6N_2O_2$

CAS RN: 100-01-6

Merck No.: 6505

Mol. Wt.: 138.126

PHYSICAL CONSTANTS

| T_m: 147°C (420 K) | T_c: | μ: |
| T_b: 332°C (605 K) | P_c: | IP: 8.34 eV |

TRANSITION PROPERTIES

$\Delta_{fus}H$ (T_m): 21.10 kJ/mol

$\Delta_{vap}H$ (T_b):

$\Delta_{vap}H$ (25°C): 79.5 kJ/mol

Vapor pressure (0°C):

Vapor pressure (25°C):

Vapor pressure (100°C):

PROPERTIES AT 25°C AND 100 kPa

	Solid	Liquid	Gas		Solid
$\Delta_f H°$/kJ mol^{-1}:	-42.0	-20.7	58.8		*d*: 1.42 g/mL
$S°$/J mol^{-1}K^{-1}:					η: N/A
C_p/J mol^{-1}K^{-1}:	167.0				*k*:
COMMENTS:					

Name: Nitrobenzene

Mol. Form.: $C_6H_5NO_2$

CAS RN: 98-95-3

Merck No.: 6509

Mol. Wt.: 123.111

PHYSICAL CONSTANTS

| T_m: 5.7°C (278.8 K) | T_c: | μ: 4.22 D |
| T_b: 210.8°C (483.9 K) | P_c: | IP: 9.86 eV |

TRANSITION PROPERTIES

$\Delta_{fus}H$ (T_m): 11.59 kJ/mol

$\Delta_{vap}H$ (T_b):

$\Delta_{vap}H$ (25°C): 55.01 kJ/mol

Vapor pressure (0°C):

Vapor pressure (25°C): 0.030 kPa

Vapor pressure (100°C):

PROPERTIES AT 25°C AND 100 kPa

	Solid	Liquid	Gas		Liquid
$\Delta_f H°$/kJ mol^{-1}:		12.5	67.5		*d*: 1.1985 g/mL
$S°$/J mol^{-1}K^{-1}:					η: 1.86 mPa s
C_p/J mol^{-1}K^{-1}:		185.8			*k*:
COMMENTS: TLV=1 ppm; highly toxic					

Name: Nitroethane

CAS RN: 79-24-3
Merck No.: 6518
Mol. Wt.: 75.067

Mol. Form.: $C_2H_5NO_2$

PHYSICAL CONSTANTS

T_m: -89.52°C (183.63 K) T_c: μ: 3.23 D
T_b: 114.07°C (387.22 K) P_c: IP: 10.88 eV

TRANSITION PROPERTIES

$\Delta_{fus}H$ (T_m): 9.85 kJ/mol Vapor pressure (0°C):
$\Delta_{vap}H$ (T_b): 37.97 kJ/mol Vapor pressure (25°C): 2.79 kPa
$\Delta_{vap}H$ (25°C): 41.59 kJ/mol Vapor pressure (100°C): 66.6 kPa

PROPERTIES AT 25°C AND 100 kPa

	Solid	Liquid	Gas		Liquid
$\Delta_f H°$/kJ mol^{-1}:		-143.9	-102.3		d: 1.0427 g/mL
$S°$/J mol^{-1}K^{-1}:					η: 0.688 mPa s
C_p/J mol^{-1}K^{-1}:		134.4			k:

COMMENTS: TLV=100 ppm; flammable

Name: Nitrogen (N_2)
Synonym: Dinitrogen

CAS RN: 7727-37-9
Merck No.: 6522
Mol. Wt.: 28.014

Mol. Form.: N_2

PHYSICAL CONSTANTS

T_m: -210.00°C (63.15 K) T_c: -146.94°C (126.21 K) μ: 0 D
T_b: -195.79°C (77.36 K) P_c: 3.39 MPa IP: 15.58 eV

TRANSITION PROPERTIES

$\Delta_{fus}H$ (T_m): 0.71 kJ/mol Vapor pressure (0°C): N/A
$\Delta_{vap}H$ (T_b): 5.57 kJ/mol Vapor pressure (25°C): N/A
$\Delta_{vap}H$ (25°C): Vapor pressure (100°C): N/A

PROPERTIES AT 25°C AND 100 kPa

	Solid	Liquid	Gas		Gas
$\Delta_f H°$/kJ mol^{-1}:			0.0		d: 1.145 g/L
$S°$/J mol^{-1}K^{-1}:			191.6		η: 17.8 μPa s
C_p/J mol^{-1}K^{-1}:			29.1		k: 0.0260 W/m K

COMMENTS:

Name: Nitrogen dioxide

CAS RN: 10102-44-0
Merck No.: 6524
Mol. Wt.: 46.006

Mol. Form.: NO_2

PHYSICAL CONSTANTS

T_m: T_c: μ: 0.316 D
T_b: P_c: IP: 9.75 eV

TRANSITION PROPERTIES

$\Delta_{fus}H$ (T_m): Vapor pressure (0°C):
$\Delta_{vap}H$ (T_b): Vapor pressure (25°C):
$\Delta_{vap}H$ (25°C): Vapor pressure (100°C):

PROPERTIES AT 25°C AND 100 kPa

	Solid	Liquid	Gas		Gas
$\Delta_f H°$/kJ mol^{-1}:			33.2		d: 1.880 g/L
$S°$/J mol^{-1}K^{-1}:			240.1		η:
C_p/J mol^{-1}K^{-1}:			37.2		k:

COMMENTS: TLV=3 ppm; highly toxic; see entry for nitrogen tetroxide

---

Name: Nitrogen pentoxide
Synonym: Dinitrogen pentoxide

Mol. Form.: N_2O_5

CAS RN: 10102-03-1
Merck No.: 6526
Mol. Wt.: 108.011

PHYSICAL CONSTANTS

T_m: 30°C (303 K)	T_c:	μ:
T_b: 47°C (320 K)	P_c:	IP: 11.90 eV

TRANSITION PROPERTIES

$\Delta_{fus}H$ (T_m):	Vapor pressure (0°C):
$\Delta_{vap}H$ (T_b):	Vapor pressure (25°C):
$\Delta_{vap}H$ (25°C):	Vapor pressure (100°C):

PROPERTIES AT 25°C AND 100 kPa

	Solid	Liquid	Gas		Solid
$\Delta_f H°$/kJ mol^{-1}:	-43.1		11.3	d:	2.0 g/mL
$S°$/J mol^{-1}K^{-1}:	178.2		355.7	η:	N/A
C_p/J mol^{-1}K^{-1}:	143.1		84.5	k:	
COMMENTS:					

Name: Nitrogen tetroxide
Synonym: Dinitrogen tetraoxide

Mol. Form.: N_2O_4

CAS RN: 10544-72-6
Merck No.: 6524
Mol. Wt.: 92.011

PHYSICAL CONSTANTS

T_m: -9.3°C (263.8 K)	T_c: 158°C (431 K)	μ:
T_b: 21.15°C (294.30 K)	P_c: 10.1 MPa	IP: 10.80 eV

TRANSITION PROPERTIES

$\Delta_{fus}H$ (T_m): 14.65 kJ/mol	Vapor pressure (0°C): 28.0 kPa
$\Delta_{vap}H$ (T_b): 38.12 kJ/mol	Vapor pressure (25°C):
$\Delta_{vap}H$ (25°C): 28.7 kJ/mol	Vapor pressure (100°C):

PROPERTIES AT 25°C AND 100 kPa

	Solid	Liquid	Gas		Gas
$\Delta_f H°$/kJ mol^{-1}:		-19.5	9.2	d:	3.761 g/L
$S°$/J mol^{-1}K^{-1}:		209.2	304.3	η:	
C_p/J mol^{-1}K^{-1}:		142.7	77.3	k:	
COMMENTS: Equilibrium mixture of N_2O_4 and NO_2					

Name: Nitrogen trifluoride

Mol. Form.: F_3N

CAS RN: 7783-54-2
Merck No.: 6525
Mol. Wt.: 71.002

PHYSICAL CONSTANTS

T_m: -206.79°C (66.36 K)	T_c: -39.1°C (234.0 K)	μ: 0.235 D
T_b: -128.75°C (144.40 K)	P_c: 4.46 MPa	IP: 13.00 eV

TRANSITION PROPERTIES

$\Delta_{fus}H$ (T_m):	Vapor pressure (0°C): N/A
$\Delta_{vap}H$ (T_b): 11.56 kJ/mol	Vapor pressure (25°C): N/A
$\Delta_{vap}H$ (25°C):	Vapor pressure (100°C): N/A

PROPERTIES AT 25°C AND 100 kPa

	Solid	Liquid	Gas		Gas
$\Delta_f H°$/kJ mol^{-1}:			-132.1	d:	2.902 g/L
$S°$/J mol^{-1}K^{-1}:			260.8	η:	
C_p/J mol^{-1}K^{-1}:			53.4	k:	
COMMENTS: TLV=10 ppm					

Name: Nitromethane

CAS RN: 75-52-5
Merck No.: 6532
Mol. Wt.: 61.040

Mol. Form.: CH_3NO_2

PHYSICAL CONSTANTS

T_m: -28.55°C (244.60 K) T_c: 315°C (588 K)   μ: 3.46 D
T_b: 101.19°C (374.34 K) P_c: 5.87 MPa   IP: 11.02 eV

TRANSITION PROPERTIES

$\Delta_{fus}H$ (T_m): 9.70 kJ/mol   Vapor pressure (0°C):
$\Delta_{vap}H$ (T_b): 33.99 kJ/mol   Vapor pressure (25°C): 4.79 kPa
$\Delta_{vap}H$ (25°C): 38.27 kJ/mol   Vapor pressure (100°C): 97.6 kPa

PROPERTIES AT 25°C AND 100 kPa

	Solid	Liquid	Gas		Liquid
$\Delta_f H°$/kJ mol⁻¹:		-113.1	-74.7		d: 1.1313 g/mL
$S°$/J mol⁻¹K⁻¹:		171.8	275.0		η: 0.630 mPa s
C_p/J mol⁻¹K⁻¹:		106.6	57.3		k:

COMMENTS: TLV=20 ppm; flammable

Name: *o*-Nitrophenol
Synonym: 2-Nitrophenol

CAS RN: 88-75-5
Merck No.: 6541
Mol. Wt.: 139.111

Mol. Form.: $C_6H_5NO_3$

PHYSICAL CONSTANTS

T_m: 44.8°C (317.9 K) T_c: μ:
T_b: 216°C (489 K) P_c: IP: 9.10 eV

TRANSITION PROPERTIES

$\Delta_{fus}H$ (T_m): 17.44 kJ/mol   Vapor pressure (0°C):
$\Delta_{vap}H$ (T_b):   Vapor pressure (25°C):
$\Delta_{vap}H$ (25°C):   Vapor pressure (100°C):

PROPERTIES AT 25°C AND 100 kPa

	Solid	Liquid	Gas		Solid
$\Delta_f H°$/kJ mol⁻¹:					d:
$S°$/J mol⁻¹K⁻¹:					η: N/A
C_p/J mol⁻¹K⁻¹:					k:

COMMENTS: Highly toxic

Name: 1-Nitropropane

CAS RN: 108-03-2
Merck No.: 6548
Mol. Wt.: 89.094

Mol. Form.: $C_3H_7NO_2$

PHYSICAL CONSTANTS

T_m: -108°C (165 K) T_c: μ: 3.66 D
T_b: 131.18°C (404.33 K) P_c: IP: 10.81 eV

TRANSITION PROPERTIES

$\Delta_{fus}H$ (T_m):   Vapor pressure (0°C):
$\Delta_{vap}H$ (T_b): 38.47 kJ/mol   Vapor pressure (25°C): 1.36 kPa
$\Delta_{vap}H$ (25°C): 43.39 kJ/mol   Vapor pressure (100°C): 38.4 kPa

PROPERTIES AT 25°C AND 100 kPa

	Solid	Liquid	Gas		Liquid
$\Delta_f H°$/kJ mol⁻¹:		-167.2	-123.8		d: 0.9961 g/mL
$S°$/J mol⁻¹K⁻¹:					η: 0.798 mPa s
C_p/J mol⁻¹K⁻¹:		175.3			k:

COMMENTS:

Name: 2-Nitropropane
Synonym: Isonitropropane

Mol. Form.: $C_3H_7NO_2$

CAS RN: 79-46-9
Merck No.: 6549
Mol. Wt.: 89.094

PHYSICAL CONSTANTS

T_m: -91.32°C (181.83 K)
T_b: 120.25°C (393.40 K)

T_c: μ: 3.73 D
P_c:

IP: 10.71 eV

TRANSITION PROPERTIES

$\Delta_{fus}H$ (T_m):
$\Delta_{vap}H$ (T_b): 36.79 kJ/mol
$\Delta_{vap}H$ (25°C): 41.34 kJ/mol

Vapor pressure (0°C):
Vapor pressure (25°C): 2.30 kPa
Vapor pressure (100°C): 55.4 kPa

PROPERTIES AT 25°C AND 100 kPa

	Solid	Liquid	Gas	Liquid
$\Delta_f H°$/kJ mol^{-1}:		-180.3	-139.0	d: 0.9835 g/mL
$S°$/J mol^{-1}K^{-1}:				η: 0.721 mPa s
C_p/J mol^{-1}K^{-1}:		170.3		k:

COMMENTS: TLV=10 ppm; carcinogen; highly toxic

Name: Nitrosyl chloride
Synonym: Nitrogen oxychloride

Mol. Form.: ClNO

CAS RN: 2696-92-6
Merck No.: 6568
Mol. Wt.: 65.459

PHYSICAL CONSTANTS

T_m: -61.5°C (211.6 K)
T_b: -5.55°C (267.60 K)

T_c: 167°C (440 K)
P_c:

μ:
IP: 10.87 eV

TRANSITION PROPERTIES

$\Delta_{fus}H$ (T_m):
$\Delta_{vap}H$ (T_b): 25.78 kJ/mol
$\Delta_{vap}H$ (25°C):

Vapor pressure (0°C):
Vapor pressure (25°C):
Vapor pressure (100°C):

PROPERTIES AT 25°C AND 100 kPa

	Solid	Liquid	Gas	Gas
$\Delta_f H°$/kJ mol^{-1}:			51.7	d: 2.676 g/L
$S°$/J mol^{-1}K^{-1}:			261.7	η:
C_p/J mol^{-1}K^{-1}:			44.7	k:

COMMENTS:

Name: Nitrosyl fluoride
Synonym: Nitrogen oxyfluoride

Mol. Form.: FNO

CAS RN: 7789-25-5
Merck No.: 6569
Mol. Wt.: 49.005

PHYSICAL CONSTANTS

T_m: -132.5°C (140.6 K)
T_b: -59.9°C (213.2 K)

T_c:
P_c:

μ: 1.730 D
IP: 12.63 eV

TRANSITION PROPERTIES

$\Delta_{fus}H$ (T_m):
$\Delta_{vap}H$ (T_b): 19.28 kJ/mol
$\Delta_{vap}H$ (25°C):

Vapor pressure (0°C):
Vapor pressure (25°C):
Vapor pressure (100°C):

PROPERTIES AT 25°C AND 100 kPa

	Solid	Liquid	Gas	Gas
$\Delta_f H°$/kJ mol^{-1}:			-66.5	d: 2.003 g/L
$S°$/J mol^{-1}K^{-1}:			248.1	η:
C_p/J mol^{-1}K^{-1}:			41.3	k:

COMMENTS:

Name: *o*-Nitrotoluene
Synonym: 1-Methyl-2-nitrobenzene

Mol. Form.: $C_7H_7NO_2$

CAS RN: 88-72-2
Merck No.: 6572
Mol. Wt.: 137.138

PHYSICAL CONSTANTS

T_m: -10°C (263 K)
T_b: 222°C (495 K)

T_c:
P_c:

μ:
IP: 9.45 eV

TRANSITION PROPERTIES

$\Delta_{fus}H\ (T_m)$:
$\Delta_{vap}H\ (T_b)$:
$\Delta_{vap}H\ (25°C)$:

Vapor pressure (0°C):
Vapor pressure (25°C):
Vapor pressure (100°C):

PROPERTIES AT 25°C AND 100 kPa

	Solid	Liquid	Gas	Liquid
$\Delta_f H°$/kJ mol^{-1}:		-9.7		d: 1.154 g/mL
$S°$/J mol^{-1}K^{-1}:				η: 2.17 mPa s
C_p/J mol^{-1}K^{-1}:		202.0		k:
COMMENTS: TLV=2 ppm				

Name: *m*-Nitrotoluene
Synonym: 1-Methyl-3-nitrobenzene

Mol. Form.: $C_7H_7NO_2$

CAS RN: 99-08-1
Merck No.: 6572
Mol. Wt.: 137.138

PHYSICAL CONSTANTS

T_m: 15.5°C (288.6 K)
T_b: 232°C (505 K)

T_c:
P_c:

μ:
IP: 9.48 eV

TRANSITION PROPERTIES

$\Delta_{fus}H\ (T_m)$:
$\Delta_{vap}H\ (T_b)$:
$\Delta_{vap}H\ (25°C)$:

Vapor pressure (0°C):
Vapor pressure (25°C):
Vapor pressure (100°C): 1.58 kPa

PROPERTIES AT 25°C AND 100 kPa

	Solid	Liquid	Gas	Liquid
$\Delta_f H°$/kJ mol^{-1}:		-31.5		d: 1.152 g/mL
$S°$/J mol^{-1}K^{-1}:				η: 2.07 mPa s
C_p/J mol^{-1}K^{-1}:		202.0		k:
COMMENTS: TLV=2 ppm				

Name: *p*-Nitrotoluene
Synonym: 1-Methyl-4-nitrobenzene

Mol. Form.: $C_7H_7NO_2$

CAS RN: 99-99-0
Merck No.: 6572
Mol. Wt.: 137.138

PHYSICAL CONSTANTS

T_m: 51.6°C (324.7 K)
T_b: 238.3°C (511.4 K)

T_c:
P_c:

μ:
IP: 9.40 eV

TRANSITION PROPERTIES

$\Delta_{fus}H\ (T_m)$: 16.81 kJ/mol
$\Delta_{vap}H\ (T_b)$:
$\Delta_{vap}H\ (25°C)$:

Vapor pressure (0°C):
Vapor pressure (25°C):
Vapor pressure (100°C):

PROPERTIES AT 25°C AND 100 kPa

	Solid	Liquid	Gas	Solid
$\Delta_f H°$/kJ mol^{-1}:	-48.1		31.0	d: 1.286 g/mL
$S°$/J mol^{-1}K^{-1}:				η: N/A
C_p/J mol^{-1}K^{-1}:	172.3			k:
COMMENTS: TLV=2 ppm				

Name: Nitrous oxide   CAS RN: 10024-97-2
Synonym: Dinitrogen oxide   Merck No.: 6575
                                                                 Mol. Wt.: 44.013
Mol. Form.: N_2O

PHYSICAL CONSTANTS

T_m: -90.8°C (182.3 K) T_c: 36.42°C (309.57 K)   μ: 0.161 D
T_b: -88.48°C (184.67 K) P_c: 7.255 MPa   IP: 12.89 eV

TRANSITION PROPERTIES

$\Delta_{fus}H$ (T_m): 6.54 kJ/mol   Vapor pressure (0°C): 3205 kPa
$\Delta_{vap}H$ (T_b): 16.53 kJ/mol   Vapor pressure (25°C): 5717 kPa
$\Delta_{vap}H$ (25°C):   Vapor pressure (100°C): N/A

PROPERTIES AT 25°C AND 100 kPa

	Solid	Liquid	Gas	Gas
$\Delta_f H°$/kJ mol^{-1}:			82.1	d: 1.799 g/L
$S°$/J mol^{-1}K^{-1}:			219.9	η: 14.8 μPa s
C_p/J mol^{-1}K^{-1}:			38.5	k: 0.0174 W/m K
COMMENTS:				

Name: Nitryl fluoride   CAS RN: 10022-50-1
                                                                 Merck No.: 6580
                                                                 Mol. Wt.: 65.004
Mol. Form.: FNO_2

PHYSICAL CONSTANTS

T_m: -166°C (107 K) T_c: 76.4°C (349.5 K)   μ: 0.466 D
T_b: -72.4°C (200.7 K) P_c: IP: 13.09 eV

TRANSITION PROPERTIES

$\Delta_{fus}H$ (T_m):   Vapor pressure (0°C):
$\Delta_{vap}H$ (T_b): 18.05 kJ/mol   Vapor pressure (25°C):
$\Delta_{vap}H$ (25°C):   Vapor pressure (100°C): N/A

PROPERTIES AT 25°C AND 100 kPa

	Solid	Liquid	Gas	Gas
$\Delta_f H°$/kJ mol^{-1}:				d: 2.657 g/L
$S°$/J mol^{-1}K^{-1}:			260.4	η:
C_p/J mol^{-1}K^{-1}:			49.8	k:
COMMENTS:				

Name: Nonadecane   CAS RN: 629-92-5
                                                                 Merck No.:
                                                                 Mol. Wt.: 268.527
Mol. Form.: $C_{19}H_{40}$

PHYSICAL CONSTANTS

T_m: 32.1°C (305.2 K) T_c: 485°C (758 K)   μ:
T_b: 329.9°C (603.0 K) P_c: 1.23 MPa   IP:

TRANSITION PROPERTIES

$\Delta_{fus}H$ (T_m): 45.82 kJ/mol   Vapor pressure (0°C):
$\Delta_{vap}H$ (T_b): 56 kJ/mol   Vapor pressure (25°C):
$\Delta_{vap}H$ (25°C):   Vapor pressure (100°C):

PROPERTIES AT 25°C AND 100 kPa

	Solid	Liquid	Gas	Solid
$\Delta_f H°$/kJ mol^{-1}:				d: 0.776 g/mL
$S°$/J mol^{-1}K^{-1}:				η: N/A
C_p/J mol^{-1}K^{-1}:				k:
COMMENTS:				

Name: Nonanal	CAS RN: 124-19-6
Synonyms: Pelargonaldehyde	Merck No.:
Nonaldehyde	Mol. Wt.: 142.241
Mol. Form.: $C_9H_{18}O$	

PHYSICAL CONSTANTS

T_m:	T_c:	μ:
T_b: 191°C (464 K)	P_c:	IP:

TRANSITION PROPERTIES

$\Delta_{fus}H$ (T_m):	Vapor pressure (0°C):
$\Delta_{vap}H$ (T_b):	Vapor pressure (25°C): 0.084 kPa
$\Delta_{vap}H$ (25°C):	Vapor pressure (100°C): 5.25 kPa

PROPERTIES AT 25°C AND 100 kPa

	Solid	Liquid	Gas	Liquid
$\Delta_f H°$/kJ mol^{-1}:				d: 0.824 g/mL
$S°$/J mol^{-1}K^{-1}:				η:
C_p/J mol^{-1}K^{-1}:				k:
COMMENTS:				

Name: Nonane	CAS RN: 111-84-2
	Merck No.:
	Mol. Wt.: 128.258
Mol. Form.: C_9H_{20}	

PHYSICAL CONSTANTS

T_m: -53.5°C (219.6 K)	T_c: 321.8°C (594.9 K)	μ:
T_b: 150.82°C (423.97 K)	P_c: 2.288 MPa	IP: 9.72 eV

TRANSITION PROPERTIES

$\Delta_{fus}H$ (T_m): 15.47 kJ/mol	Vapor pressure (0°C):
$\Delta_{vap}H$ (T_b): 36.91 kJ/mol	Vapor pressure (25°C): 0.570 kPa
$\Delta_{vap}H$ (25°C): 46.41 kJ/mol	Vapor pressure (100°C): 21.0 kPa

PROPERTIES AT 25°C AND 100 kPa

	Solid	Liquid	Gas	Liquid
$\Delta_f H°$/kJ mol^{-1}:		-274.7	-228.2	d: 0.7138 g/mL
$S°$/J mol^{-1}K^{-1}:				η: 0.665 mPa s
C_p/J mol^{-1}K^{-1}:		284.4		k: 0.131 W/m K
COMMENTS: TLV=200 ppm; flammable				

Name: Nonanoic acid	CAS RN: 112-05-0
Synonyms: Pelargonic acid	Merck No.: 7013
Nonylic acid	Mol. Wt.: 158.241
Mol. Form.: $C_9H_{18}O_2$	

PHYSICAL CONSTANTS

T_m: 12.35°C (285.50 K)	T_c: 438°C (711 K)	μ:
T_b: 254.55°C (527.70 K)	P_c: 2.40 MPa	IP:

TRANSITION PROPERTIES

$\Delta_{fus}H$ (T_m): 20.28 kJ/mol	Vapor pressure (0°C):
$\Delta_{vap}H$ (T_b):	Vapor pressure (25°C):
$\Delta_{vap}H$ (25°C): 82.4 kJ/mol	Vapor pressure (100°C): 0.126 kPa

PROPERTIES AT 25°C AND 100 kPa

	Solid	Liquid	Gas	Liquid
$\Delta_f H°$/kJ mol^{-1}:		-659.7	-577.3	d: 0.9013 g/mL
$S°$/J mol^{-1}K^{-1}:				η: 7.01 mPa s
C_p/J mol^{-1}K^{-1}:		362.4		k:
COMMENTS:				

Name: 1-Nonanol   CAS RN: 143-08-8
Synonyms: Pelargonic alcohol   Merck No.: 6598
        Nonyl alcohol   Mol. Wt.: 144.257
Mol. Form.: $C_9H_{20}O$

PHYSICAL CONSTANTS

T_m: -5°C (268 K) T_c: 398.4°C (671.5 K)   μ:
T_b: 213.37°C (486.52 K) P_c: 2.63 MPa   IP:

TRANSITION PROPERTIES

$\Delta_{fus}H$ (T_m):   Vapor pressure (0°C):
$\Delta_{vap}H$ (T_b):   Vapor pressure (25°C):
$\Delta_{vap}H$ (25°C): 76.86 kJ/mol   Vapor pressure (100°C): 1.20 kPa

PROPERTIES AT 25°C AND 100 kPa

	Solid	Liquid	Gas		Liquid
$\Delta_f H°$/kJ mol^{-1}:		-453.4	-376.3		d: 0.8247 g/mL
$S°$/J mol^{-1}K^{-1}:					η: 9.12 mPa s
C_p/J mol^{-1}K^{-1}:					k: 0.161 W/m K
COMMENTS:					

Name: Octadecane   CAS RN: 593-45-3
                                                                     Merck No.:
                                                                     Mol. Wt.: 254.500
Mol. Form.: $C_{18}H_{38}$

PHYSICAL CONSTANTS

T_m: 28.2°C (301.3 K) T_c: 473°C (746 K)   μ:
T_b: 316.35°C (589.50 K) P_c: 1.30 MPa   IP:

TRANSITION PROPERTIES

$\Delta_{fus}H$ (T_m): 61.39 kJ/mol   Vapor pressure (0°C):
$\Delta_{vap}H$ (T_b): 52.8 kJ/mol   Vapor pressure (25°C):
$\Delta_{vap}H$ (25°C):   Vapor pressure (100°C):

PROPERTIES AT 25°C AND 100 kPa

	Solid	Liquid	Gas		Solid
$\Delta_f H°$/kJ mol^{-1}:	-567.4		-414.6		d: 0.7791 g/mL
$S°$/J mol^{-1}K^{-1}:	480.2				η: N/A
C_p/J mol^{-1}K^{-1}:	485.6				k:
COMMENTS:					

Name: 1-Octadecanol   CAS RN: 112-92-5
Synonyms: Stearyl alcohol   Merck No.:
        Octadecyl alcohol   Mol. Wt.: 270.499
Mol. Form.: $C_{18}H_{38}O$

PHYSICAL CONSTANTS

T_m: 59.5°C (332.6 K) T_c: 517°C (790 K)   μ:
T_b: P_c: 1.44 MPa   IP:

TRANSITION PROPERTIES

$\Delta_{fus}H$ (T_m):   Vapor pressure (0°C):
$\Delta_{vap}H$ (T_b):   Vapor pressure (25°C):
$\Delta_{vap}H$ (25°C):   Vapor pressure (100°C):

PROPERTIES AT 25°C AND 100 kPa

	Solid	Liquid	Gas		Solid
$\Delta_f H°$/kJ mol^{-1}:					d:
$S°$/J mol^{-1}K^{-1}:					η: N/A
C_p/J mol^{-1}K^{-1}:					k:
COMMENTS:					

Name: Octane

CAS RN: 111-65-9
Merck No.: 6672
Mol. Wt.: 114.231

Mol. Form.: C_8H_{18}

PHYSICAL CONSTANTS

T_m: -56.8°C (216.3 K) T_c: 295.8°C (568.9 K) μ:
T_b: 125.67°C (398.82 K) P_c: 2.493 MPa IP: 9.82 eV

TRANSITION PROPERTIES

$\Delta_{fus}H$ (T_m): 20.65 kJ/mol Vapor pressure (0°C): 0.386 kPa
$\Delta_{vap}H$ (T_b): 34.41 kJ/mol Vapor pressure (25°C): 1.86 kPa
$\Delta_{vap}H$ (25°C): 41.49 kJ/mol Vapor pressure (100°C): 46.8 kPa

PROPERTIES AT 25°C AND 100 kPa

	Solid	Liquid	Gas		Liquid
$\Delta_f H°$/kJ mol⁻¹:		-250.1	-208.6		d: 0.6986 g/mL
$S°$/J mol⁻¹K⁻¹:					η: 0.508 mPa s
C_p/J mol⁻¹K⁻¹:		254.6			k: 0.128 W/m K

COMMENTS: TLV=300 ppm; flammable

Name: Octanoic acid
Synonym: Caprylic acid

CAS RN: 124-07-2
Merck No.: 1765
Mol. Wt.: 144.214

Mol. Form.: $C_8H_{16}O_2$

PHYSICAL CONSTANTS

T_m: 16.3°C (289.4 K) T_c: 422°C (695 K) μ:
T_b: 239°C (512 K) P_c: 2.64 MPa IP:

TRANSITION PROPERTIES

$\Delta_{fus}H$ (T_m): 21.36 kJ/mol Vapor pressure (0°C):
$\Delta_{vap}H$ (T_b): 58.45 kJ/mol Vapor pressure (25°C):
$\Delta_{vap}H$ (25°C): 81.7 kJ/mol Vapor pressure (100°C):

PROPERTIES AT 25°C AND 100 kPa

	Solid	Liquid	Gas		Liquid
$\Delta_f H°$/kJ mol⁻¹:		-636.0	-554.3		d: 0.9066 g/mL
$S°$/J mol⁻¹K⁻¹:					η: 5.02 mPa s
C_p/J mol⁻¹K⁻¹:		297.9			k:

COMMENTS:

Name: 1-Octanol
Synonyms: Caprylic alcohol
        Octyl alcohol
Mol. Form.: $C_8H_{18}O$

CAS RN: 111-87-5
Merck No.: 6674
Mol. Wt.: 130.230

PHYSICAL CONSTANTS

T_m: -15.5°C (257.6 K) T_c: 379.4°C (652.5 K) μ:
T_b: 195.16°C (468.31 K) P_c: 2.86 MPa IP:

TRANSITION PROPERTIES

$\Delta_{fus}H$ (T_m): 42.30 kJ/mol Vapor pressure (0°C):
$\Delta_{vap}H$ (T_b): 46.90 kJ/mol Vapor pressure (25°C): 0.010 kPa
$\Delta_{vap}H$ (25°C): 70.98 kJ/mol Vapor pressure (100°C):

PROPERTIES AT 25°C AND 100 kPa

	Solid	Liquid	Gas		Liquid
$\Delta_f H°$/kJ mol⁻¹:		-426.5	-355.5		d: 0.8223 g/mL
$S°$/J mol⁻¹K⁻¹:					η: 7.29 mPa s
C_p/J mol⁻¹K⁻¹:		305.2			k: 0.161 W/m K

COMMENTS:

Name: 2-Octanol   CAS RN: 123-96-6
Synonym: *sec*-Caprylic alcohol   Merck No.: 6675
                                                                                   Mol. Wt.: 130.230
Mol. Form.: $C_8H_{18}O$

PHYSICAL CONSTANTS

T_m: -32°C (241 K) T_c: 365°C (638 K)   μ:
T_b: 179.88°C (453.03 K) P_c: 2.9 MPa   IP:

TRANSITION PROPERTIES

$\Delta_{fus}H$ (T_m):   Vapor pressure (0°C):
$\Delta_{vap}H$ (T_b): 44.35 kJ/mol   Vapor pressure (25°C):
$\Delta_{vap}H$ (25°C):   Vapor pressure (100°C): 5.59 kPa

PROPERTIES AT 25°C AND 100 kPa

	Solid	Liquid	Gas		Liquid
$\Delta_f H°$/kJ mol^{-1}:					d: 0.8171 g/mL
$S°$/J mol^{-1}K^{-1}:					η:
C_p/J mol^{-1}K^{-1}:		330.1			k:
COMMENTS:					

Name: 1-Octene   CAS RN: 111-66-0
Synonym: Caprylene   Merck No.: 1764
                                                                                   Mol. Wt.: 112.215
Mol. Form.: C_8H_{16}

PHYSICAL CONSTANTS

T_m: -101.7°C (171.4 K) T_c: 293.6°C (566.7 K)   μ:
T_b: 121.29°C (394.44 K) P_c: 2.675 MPa   IP: 9.43 eV

TRANSITION PROPERTIES

$\Delta_{fus}H$ (T_m): 15.57 kJ/mol   Vapor pressure (0°C):
$\Delta_{vap}H$ (T_b): 34.07 kJ/mol   Vapor pressure (25°C): 2.30 kPa
$\Delta_{vap}H$ (25°C): 40.39 kJ/mol   Vapor pressure (100°C): 53.7 kPa

PROPERTIES AT 25°C AND 100 kPa

	Solid	Liquid	Gas		Liquid
$\Delta_f H°$/kJ mol^{-1}:		-121.8	-81.4		d: 0.711 g/mL
$S°$/J mol^{-1}K^{-1}:					η: 0.45 mPa s
C_p/J mol^{-1}K^{-1}:		241.0			k:
COMMENTS: Flammable					

Name: Oleic acid   CAS RN: 112-80-1
Synonym: *cis*-9-Octadecenoic acid   Merck No.: 6788
                                                                                   Mol. Wt.: 282.467
Mol. Form.: $C_{18}H_{34}O_2$

PHYSICAL CONSTANTS

T_m: 13.4°C (286.5 K) T_c: μ:
T_b: P_c: IP:

TRANSITION PROPERTIES

$\Delta_{fus}H$ (T_m):   Vapor pressure (0°C):
$\Delta_{vap}H$ (T_b): 67.36 kJ/mol   Vapor pressure (25°C):
$\Delta_{vap}H$ (25°C):   Vapor pressure (100°C):

PROPERTIES AT 25°C AND 100 kPa

	Solid	Liquid	Gas		Liquid
$\Delta_f H°$/kJ mol^{-1}:					d: 0.898 g/mL
$S°$/J mol^{-1}K^{-1}:					η: 23 mPa s
C_p/J mol^{-1}K^{-1}:					k:
COMMENTS:					

Name: Osmium

Mol. Form.: Os

CAS RN: 7440-04-2
Merck No.: 6845
Mol. Wt.: 190.230

PHYSICAL CONSTANTS

T_m: 3033°C (3306 K) T_c: μ:
T_b: 5012°C (5285 K) P_c: IP: 8.70 eV

TRANSITION PROPERTIES

$\Delta_{fus}H$ (T_m): 57.85 kJ/mol   Vapor pressure (0°C): N/A
$\Delta_{vap}H$ (T_b):   Vapor pressure (25°C): N/A
$\Delta_{vap}H$ (25°C):   Vapor pressure (100°C): N/A

PROPERTIES AT 25°C AND 100 kPa

	Solid	Liquid	Gas	Solid
$\Delta_f H°$/kJ mol^{-1}:	0.0		791.0	d: 22.5 g/mL
$S°$/J mol^{-1}K^{-1}:	32.6		192.6	η: N/A
C_p/J mol^{-1}K^{-1}:	24.7		20.8	k: 87.6 W/m K
COMMENTS:				

Name: Osmium oxide (OsO$_4$)
Synonyms: Osmium tetroxide
        Osmium(VIII) oxide
Mol. Form.: O$_4$Os

CAS RN: 20816-12-0
Merck No.: 6848
Mol. Wt.: 254.228

PHYSICAL CONSTANTS

T_m: 41°C (314 K) T_c: 405°C (678 K)   μ:
T_b: 135°C (408 K) P_c: IP: 12.32 eV

TRANSITION PROPERTIES

$\Delta_{fus}H$ (T_m): 9.80 kJ/mol   Vapor pressure (0°C):
$\Delta_{vap}H$ (T_b):   Vapor pressure (25°C):
$\Delta_{vap}H$ (25°C):   Vapor pressure (100°C):

PROPERTIES AT 25°C AND 100 kPa

	Solid	Liquid	Gas	Solid
$\Delta_f H°$/kJ mol^{-1}:	-394.1		-337.2	d: 5.0 g/mL
$S°$/J mol^{-1}K^{-1}:	143.9		293.8	η: N/A
C_p/J mol^{-1}K^{-1}:			74.1	k:
COMMENTS: Highly toxic				

Name: Oxetane
Synonyms: Trimethylene oxide
        1,3-Propylene oxide
Mol. Form.: C$_3$H$_6$O

CAS RN: 503-30-0
Merck No.: 9630
Mol. Wt.: 58.080

PHYSICAL CONSTANTS

T_m: T_c: μ: 1.94 D
T_b: 47.6°C (320.7 K) P_c: IP: 9.67 eV

TRANSITION PROPERTIES

$\Delta_{fus}H$ (T_m):   Vapor pressure (0°C):
$\Delta_{vap}H$ (T_b): 28.67 kJ/mol   Vapor pressure (25°C):
$\Delta_{vap}H$ (25°C): 29.85 kJ/mol   Vapor pressure (100°C):

PROPERTIES AT 25°C AND 100 kPa

	Solid	Liquid	Gas	Liquid
$\Delta_f H°$/kJ mol^{-1}:			-80.5	d: 0.8930 g/mL
$S°$/J mol^{-1}K^{-1}:				η:
C_p/J mol^{-1}K^{-1}:				k:
COMMENTS:				

Name: 2-Oxetanone   CAS RN: 57-57-8
Synonym: β-Propiolactone   Merck No.: 7832
                                                            Mol. Wt.: 72.064
Mol. Form.: $C_3H_4O_2$

PHYSICAL CONSTANTS

T_m: -33.4°C (239.7 K)	T_c:	μ: 4.18 D
T_b: 162°C (435 K)	P_c:	IP: 9.70 eV

TRANSITION PROPERTIES

$\Delta_{fus}H$ (T_m):   Vapor pressure (0°C):
$\Delta_{vap}H$ (T_b):   Vapor pressure (25°C):
$\Delta_{vap}H$ (25°C): 47.03 kJ/mol   Vapor pressure (100°C): 12.9 kPa

PROPERTIES AT 25°C AND 100 kPa

	Solid	Liquid	Gas		Liquid
$\Delta_f H°$/kJ mol^{-1}:		-329.9	-282.9		d: 1.1420 g/mL
$S°$/J mol^{-1}K^{-1}:		175.3			η:
C_p/J mol^{-1}K^{-1}:		122.1			k:

COMMENTS: TLV=0.05 ppm; carcinogen; highly toxic

Name: Oxygen (O_2)   CAS RN: 7782-44-7
Synonym: Dioxygen   Merck No.: 6917
                                                            Mol. Wt.: 31.999
Mol. Form.: O_2

PHYSICAL CONSTANTS

T_m: -218.79°C (54.36 K)	T_c: -118.56°C (154.59 K)	μ: 0 D
T_b: -182.95°C (90.20 K)	P_c: 5.043 MPa	IP: 12.07 eV

TRANSITION PROPERTIES

$\Delta_{fus}H$ (T_m): 0.44 kJ/mol   Vapor pressure (0°C): N/A
$\Delta_{vap}H$ (T_b): 6.82 kJ/mol   Vapor pressure (25°C): N/A
$\Delta_{vap}H$ (25°C):   Vapor pressure (100°C): N/A

PROPERTIES AT 25°C AND 100 kPa

	Solid	Liquid	Gas		Gas
$\Delta_f H°$/kJ mol^{-1}:			0.0		d: 1.308 g/L
$S°$/J mol^{-1}K^{-1}:			205.2		η: 20.7 μPa s
C_p/J mol^{-1}K^{-1}:			29.4		k: 0.0263 W/m K

COMMENTS:

Name: Oxygen dichloride   CAS RN: 7791-21-1
Synonyms: Chlorine monoxide   Merck No.: 2099
          Dichlorine oxide   Mol. Wt.: 86.905
Mol. Form.: Cl_2O

PHYSICAL CONSTANTS

T_m: -120.6°C (152.5 K)	T_c:	μ:
T_b: 2.2°C (275.3 K)	P_c:	IP: 10.94 eV

TRANSITION PROPERTIES

$\Delta_{fus}H$ (T_m):   Vapor pressure (0°C):
$\Delta_{vap}H$ (T_b): 25.90 kJ/mol   Vapor pressure (25°C):
$\Delta_{vap}H$ (25°C):   Vapor pressure (100°C):

PROPERTIES AT 25°C AND 100 kPa

	Solid	Liquid	Gas		Gas
$\Delta_f H°$/kJ mol^{-1}:			80.3		d: 3.552 g/L
$S°$/J mol^{-1}K^{-1}:			266.2		η:
C_p/J mol^{-1}K^{-1}:			45.4		k:

COMMENTS: Highly toxic; very flammable

Name: Oxygen difluoride
Synonym: Fluorine oxide (F_2O)

Mol. Form.: F_2O

CAS RN: 7783-41-7
Merck No.: 4092
Mol. Wt.: 53.996

PHYSICAL CONSTANTS

T_m: -223.8°C (49.3 K)
T_b: -144.75°C (128.40 K)

T_c: -58°C (215 K)
P_c:

μ: 0.297 D
IP: 13.11 eV

TRANSITION PROPERTIES

$\Delta_{fus}H$ (T_m):
$\Delta_{vap}H$ (T_b): 11.09 kJ/mol
$\Delta_{vap}H$ (25°C):

Vapor pressure (0°C): N/A
Vapor pressure (25°C): N/A
Vapor pressure (100°C): N/A

PROPERTIES AT 25°C AND 100 kPa

	Solid	Liquid	Gas		Gas
$\Delta_f H°$/kJ mol^{-1}:			24.7		d: 2.207 g/L
$S°$/J mol^{-1}K^{-1}:			247.4		η:
C_p/J mol^{-1}K^{-1}:			43.3		k:
COMMENTS: TLV=0.05 ppm					

Name: Ozone

Mol. Form.: O_3

CAS RN: 10028-15-6
Merck No.: 6936
Mol. Wt.: 47.998

PHYSICAL CONSTANTS

T_m: -193°C (80 K)
T_b: -111.35°C (161.80 K)

T_c: -12.0°C (261.1 K)
P_c: 5.57 MPa

μ: 0.534 D
IP: 12.43 eV

TRANSITION PROPERTIES

$\Delta_{fus}H$ (T_m):
$\Delta_{vap}H$ (T_b):
$\Delta_{vap}H$ (25°C):

Vapor pressure (0°C): N/A
Vapor pressure (25°C): N/A
Vapor pressure (100°C): N/A

PROPERTIES AT 25°C AND 100 kPa

	Solid	Liquid	Gas		Gas
$\Delta_f H°$/kJ mol^{-1}:			142.7		d: 1.962 g/L
$S°$/J mol^{-1}K^{-1}:			238.9		η:
C_p/J mol^{-1}K^{-1}:			39.2		k:
COMMENTS: TLV=0.1 ppm					

Name: Palladium

Mol. Form.: Pd

CAS RN: 7440-05-3
Merck No.: 6940
Mol. Wt.: 106.420

PHYSICAL CONSTANTS

T_m: 1554.9°C (1828.0 K)
T_b: 2963°C (3236 K)

T_c:
P_c:

μ:
IP: 8.34 eV

TRANSITION PROPERTIES

$\Delta_{fus}H$ (T_m): 16.74 kJ/mol
$\Delta_{vap}H$ (T_b):
$\Delta_{vap}H$ (25°C):

Vapor pressure (0°C): N/A
Vapor pressure (25°C): N/A
Vapor pressure (100°C): N/A

PROPERTIES AT 25°C AND 100 kPa

	Solid	Liquid	Gas		Solid
$\Delta_f H°$/kJ mol^{-1}:	0.0		378.2		d: 12.0 g/mL
$S°$/J mol^{-1}K^{-1}:	37.6		167.1		η: N/A
C_p/J mol^{-1}K^{-1}:	26.0		20.8		k: 71.8 W/m K
COMMENTS:					

Name: Paraldehyde
Synonyms: 2,4,6-Trimethyl-1,3,5-trioxane
            Acetaldehyde trimer
Mol. Form.: $C_6H_{12}O_3$

CAS RN: 123-63-7
Merck No.: 6975
Mol. Wt.: 132.160

PHYSICAL CONSTANTS

T_m: 12.6°C (285.7 K)
T_b: 124.35°C (397.50 K)

T_c: 290°C (563 K)
P_c:

μ: 1.43 D
IP:

TRANSITION PROPERTIES

$\Delta_{fus}H$ (T_m):
$\Delta_{vap}H$ (T_b):
$\Delta_{vap}H$ (25°C): 41.4 kJ/mol

Vapor pressure (0°C):
Vapor pressure (25°C):
Vapor pressure (100°C): 45.2 kPa

PROPERTIES AT 25°C AND 100 kPa

	Solid	Liquid	Gas		Liquid
$\Delta_f H°$/kJ mol^{-1}:		-673.2	-631.8		d: 0.991 g/mL
$S°$/J mol^{-1}K^{-1}:					η: 1.08 mPa s
C_p/J mol^{-1}K^{-1}:		250.0			k:
COMMENTS: Flammable					

Name: Pentaborane
Synonym: Pentaboron nonahydride

Mol. Form.: B_5H_9

CAS RN: 19624-22-7
Merck No.: 7054
Mol. Wt.: 63.127

PHYSICAL CONSTANTS

T_m: -46.6°C (226.5 K)
T_b: 60°C (333 K)

T_c:
P_c:

μ: 2.13 D
IP: 9.90 eV

TRANSITION PROPERTIES

$\Delta_{fus}H$ (T_m):
$\Delta_{vap}H$ (T_b):
$\Delta_{vap}H$ (25°C): 30.5 kJ/mol

Vapor pressure (0°C):
Vapor pressure (25°C):
Vapor pressure (100°C):

PROPERTIES AT 25°C AND 100 kPa

	Solid	Liquid	Gas		Liquid
$\Delta_f H°$/kJ mol^{-1}:		42.7	73.2		d: 0.60 g/mL
$S°$/J mol^{-1}K^{-1}:		184.2	275.9		η:
C_p/J mol^{-1}K^{-1}:		151.1	96.8		k:
COMMENTS: TLV=0.005 ppm; highly toxic; flammable					

Name: Pentachloroethane
Synonym: Pentalin

Mol. Form.: C_2HCl_5

CAS RN: 76-01-7
Merck No.: 7058
Mol. Wt.: 202.293

PHYSICAL CONSTANTS

T_m: -29°C (244 K)
T_b: 159.88°C (433.03 K)

T_c:
P_c:

μ: 0.92 D
IP: 11.00 eV

TRANSITION PROPERTIES

$\Delta_{fus}H$ (T_m): 11.34 kJ/mol
$\Delta_{vap}H$ (T_b): 36.94 kJ/mol
$\Delta_{vap}H$ (25°C): 45.6 kJ/mol

Vapor pressure (0°C):
Vapor pressure (25°C): 0.478 kPa
Vapor pressure (100°C): 16.9 kPa

PROPERTIES AT 25°C AND 100 kPa

	Solid	Liquid	Gas		Liquid
$\Delta_f H°$/kJ mol^{-1}:		-187.6	-142.0		d: 1.6749 g/mL
$S°$/J mol^{-1}K^{-1}:					η: 2.25 mPa s
C_p/J mol^{-1}K^{-1}:		173.8			k:
COMMENTS:					

Name: Pentadecane

CAS RN: 629-62-9
Merck No.:
Mol. Wt.: 212.419

Mol. Form.: $C_{15}H_{32}$

PHYSICAL CONSTANTS

T_m: 9.9°C (283.0 K) T_c: 435°C (708 K)   μ:
T_b: 270.6°C (543.7 K) P_c: 1.515 MPa   IP:

TRANSITION PROPERTIES

$\Delta_{fus}H$ (T_m):   Vapor pressure (0°C):
$\Delta_{vap}H$ (T_b):   Vapor pressure (25°C):
$\Delta_{vap}H$ (25°C): 76.11 kJ/mol   Vapor pressure (100°C):

PROPERTIES AT 25°C AND 100 kPa

	Solid	Liquid	Gas		Liquid
$\Delta_f H°$/kJ mol^{-1}:					d: 0.765 g/mL
$S°$/J mol^{-1}K^{-1}:					η:
C_p/J mol^{-1}K^{-1}:		469.9			k:
COMMENTS:					

Name: *cis*-1,3-Pentadiene
Synonym: *cis*-Piperylene

CAS RN: 1574-41-0
Merck No.:
Mol. Wt.: 68.119

Mol. Form.: C_5H_8

PHYSICAL CONSTANTS

T_m: -140.8°C (132.3 K) T_c: μ: 0.500 D
T_b: 44.1°C (317.2 K) P_c: IP: 8.63 eV

TRANSITION PROPERTIES

$\Delta_{fus}H$ (T_m):   Vapor pressure (0°C): 17.1 kPa
$\Delta_{vap}H$ (T_b):   Vapor pressure (25°C): 50.6 kPa
$\Delta_{vap}H$ (25°C):   Vapor pressure (100°C):

PROPERTIES AT 25°C AND 100 kPa

	Solid	Liquid	Gas		Liquid
$\Delta_f H°$/kJ mol^{-1}:			81.4		d: 0.688 g/mL
$S°$/J mol^{-1}K^{-1}:					η:
C_p/J mol^{-1}K^{-1}:					k:
COMMENTS:					

Name: *trans*-1,3-Pentadiene
Synonym: *trans*-Piperylene

CAS RN: 2004-70-8
Merck No.:
Mol. Wt.: 68.119

Mol. Form.: C_5H_8

PHYSICAL CONSTANTS

T_m: -87°C (186 K) T_c: μ: 0.585 D
T_b: 42°C (315 K) P_c: IP: 8.59 eV

TRANSITION PROPERTIES

$\Delta_{fus}H$ (T_m):   Vapor pressure (0°C): 18.8 kPa
$\Delta_{vap}H$ (T_b):   Vapor pressure (25°C): 54.8 kPa
$\Delta_{vap}H$ (25°C):   Vapor pressure (100°C):

PROPERTIES AT 25°C AND 100 kPa

	Solid	Liquid	Gas		Liquid
$\Delta_f H°$/kJ mol^{-1}:			76.1		d: 0.671 g/mL
$S°$/J mol^{-1}K^{-1}:					η:
C_p/J mol^{-1}K^{-1}:					k:
COMMENTS:					

Name: 1,4-Pentadiene

CAS RN: 591-93-5
Merck No.:
Mol. Wt.: 68.119

Mol. Form.: C_5H_8

PHYSICAL CONSTANTS

T_m: -148.8°C (124.3 K)	T_c:	μ:
T_b: 26°C (299 K)	P_c:	IP: 9.62 eV

TRANSITION PROPERTIES

$\Delta_{fus}H$ (T_m): 6.14 kJ/mol
$\Delta_{vap}H$ (T_b):
$\Delta_{vap}H$ (25°C):

Vapor pressure (0°C): 36.5 kPa
Vapor pressure (25°C): 97.9 kPa
Vapor pressure (100°C):

PROPERTIES AT 25°C AND 100 kPa

	Solid	Liquid	Gas	Liquid
$\Delta_f H°$/kJ mol^{-1}:		105.6		d: 0.658 g/mL
$S°$/J mol^{-1}K^{-1}:				η:
C_p/J mol^{-1}K^{-1}:				k:
COMMENTS:				

Name: Pentaerythritol
Synonyms: 2,2-Bis(hydroxymethyl)-1,3-propanediol
            Tetra(hydroxymethyl)methane
Mol. Form.: $C_5H_{12}O_4$

CAS RN: 115-77-5
Merck No.: 7062
Mol. Wt.: 136.148

PHYSICAL CONSTANTS

T_m: 260°C (533 K)	T_c:	μ:
T_b:	P_c:	IP:

TRANSITION PROPERTIES

$\Delta_{fus}H$ (T_m):
$\Delta_{vap}H$ (T_b):
$\Delta_{vap}H$ (25°C):

Vapor pressure (0°C): N/A
Vapor pressure (25°C): N/A
Vapor pressure (100°C): N/A

PROPERTIES AT 25°C AND 100 kPa

	Solid	Liquid	Gas	Solid
$\Delta_f H°$/kJ mol^{-1}:	-920.6		-776.7	d: 0.886 g/mL
$S°$/J mol^{-1}K^{-1}:				η: N/A
C_p/J mol^{-1}K^{-1}:				k:
COMMENTS:				

Name: Pentanal
Synonyms: Valeraldehyde
            Butanecarbaldehyde
Mol. Form.: $C_5H_{10}O$

CAS RN: 110-62-3
Merck No.: 9813
Mol. Wt.: 86.134

PHYSICAL CONSTANTS

T_m: -91.5°C (181.6 K)	T_c: 293.0°C (566.1 K)	μ:
T_b: 103°C (376 K)	P_c: 3.97 MPa	IP: 9.74 eV

TRANSITION PROPERTIES

$\Delta_{fus}H$ (T_m):
$\Delta_{vap}H$ (T_b):
$\Delta_{vap}H$ (25°C): -21.2 kJ/mol

Vapor pressure (0°C):
Vapor pressure (25°C): 4.58 kPa
Vapor pressure (100°C): 92.5 kPa

PROPERTIES AT 25°C AND 100 kPa

	Solid	Liquid	Gas	Liquid
$\Delta_f H°$/kJ mol^{-1}:		-267.3	-288.5	d: 0.805 g/mL
$S°$/J mol^{-1}K^{-1}:				η:
C_p/J mol^{-1}K^{-1}:				k:
COMMENTS: TLV=50 ppm; flammable				

Name: Pentane

CAS RN: 109-66-0
Merck No.: 7072
Mol. Wt.: 72.150

Mol. Form.: C_5H_{12}

PHYSICAL CONSTANTS

T_m: -129.7°C (143.4 K) T_c: 196.54°C (469.69 K)   μ:
T_b: 36.06°C (309.21 K) P_c: 3.364 MPa   IP: 10.35 eV

TRANSITION PROPERTIES

$\Delta_{fus}H$ (T_m): 8.42 kJ/mol   Vapor pressure (0°C): 24.5 kPa
$\Delta_{vap}H$ (T_b): 25.79 kJ/mol   Vapor pressure (25°C): 68.3 kPa
$\Delta_{vap}H$ (25°C): 26.43 kJ/mol   Vapor pressure (100°C):

PROPERTIES AT 25°C AND 100 kPa

	Solid	Liquid	Gas		Liquid
$\Delta_f H°$/kJ mol^{-1}:		-173.5	-146.9		d: 0.6214 g/mL
$S°$/J mol^{-1}K^{-1}:		263.1			η: 0.224 mPa s
C_p/J mol^{-1}K^{-1}:		167.2			k: 0.113 W/m K

COMMENTS: TLV=600 ppm; very flammable

Name: 1,5-Pentanediol
Synonym: Pentamethylene glycol

CAS RN: 111-29-5
Merck No.: 7073
Mol. Wt.: 104.149

Mol. Form.: $C_5H_{12}O_2$

PHYSICAL CONSTANTS

T_m: -18°C (255 K) T_c: μ:
T_b: 239°C (512 K) P_c: IP:

TRANSITION PROPERTIES

$\Delta_{fus}H$ (T_m):   Vapor pressure (0°C):
$\Delta_{vap}H$ (T_b): 60.70 kJ/mol   Vapor pressure (25°C):
$\Delta_{vap}H$ (25°C): 82.40 kJ/mol   Vapor pressure (100°C):

PROPERTIES AT 25°C AND 100 kPa

	Solid	Liquid	Gas		Liquid
$\Delta_f H°$/kJ mol^{-1}:		-531.5	-449.1		d: 0.9858 g/mL
$S°$/J mol^{-1}K^{-1}:					η:
C_p/J mol^{-1}K^{-1}:					k:

COMMENTS:

Name: 2,4-Pentanedione
Synonyms: Acetylacetone
          Diacetylmethane
Mol. Form.: $C_5H_8O_2$

CAS RN: 123-54-6
Merck No.: 75
Mol. Wt.: 100.117

PHYSICAL CONSTANTS

T_m: -23°C (250 K) T_c: μ:
T_b: 138°C (411 K) P_c: IP: 8.85 eV

TRANSITION PROPERTIES

$\Delta_{fus}H$ (T_m):   Vapor pressure (0°C):
$\Delta_{vap}H$ (T_b): 34.30 kJ/mol   Vapor pressure (25°C): 1.02 kPa
$\Delta_{vap}H$ (25°C): 41.77 kJ/mol   Vapor pressure (100°C): 33.9 kPa

PROPERTIES AT 25°C AND 100 kPa

	Solid	Liquid	Gas		Liquid
$\Delta_f H°$/kJ mol^{-1}:		-423.8	-380.6		d: 0.972 g/mL
$S°$/J mol^{-1}K^{-1}:					η:
C_p/J mol^{-1}K^{-1}:					k:

COMMENTS:

Name: Pentanenitrile
Synonyms: Valeronitrile
         Butyl cyanide
Mol. Form.: C_5H_9N

CAS RN: 110-59-8
Merck No.:
Mol. Wt.: 83.133

PHYSICAL CONSTANTS

T_m: -96.2°C (176.9 K) T_c: 337.2°C (610.3 K) μ: 4.12 D
T_b: 141.3°C (414.4 K) P_c: 3.58 MPa   IP:

TRANSITION PROPERTIES

$\Delta_{fus}H$ (T_m): 4.73 kJ/mol   Vapor pressure (0°C):
$\Delta_{vap}H$ (T_b): 36.09 kJ/mol   Vapor pressure (25°C): 0.943 kPa
$\Delta_{vap}H$ (25°C): 43.60 kJ/mol   Vapor pressure (100°C): 28.7 kPa

PROPERTIES AT 25°C AND 100 kPa

	Solid	Liquid	Gas		Liquid
$\Delta_fH°$/kJ mol^{-1}:		-33.1	10.5		d: 0.797 g/mL
$S°$/J mol^{-1}K^{-1}:					η: 0.693 mPa s
C_p/J mol^{-1}K^{-1}:					k:
COMMENTS: Highly toxic					

Name: 1-Pentanethiol
Synonyms: Amyl mercaptan
         Pentyl mercaptan
Mol. Form.: $C_5H_{12}S$

CAS RN: 110-66-7
Merck No.: 648
Mol. Wt.: 104.216

PHYSICAL CONSTANTS

T_m: -75.7°C (197.4 K) T_c: μ:
T_b: 126.6°C (399.7 K) P_c: IP:

TRANSITION PROPERTIES

$\Delta_{fus}H$ (T_m):   Vapor pressure (0°C):
$\Delta_{vap}H$ (T_b): 34.88 kJ/mol   Vapor pressure (25°C):
$\Delta_{vap}H$ (25°C): 41.24 kJ/mol   Vapor pressure (100°C): 45.6 kPa

PROPERTIES AT 25°C AND 100 kPa

	Solid	Liquid	Gas		Liquid
$\Delta_fH°$/kJ mol^{-1}:		-151.3	-110.1		d: 0.846 g/mL
$S°$/J mol^{-1}K^{-1}:					η:
C_p/J mol^{-1}K^{-1}:		201.0			k:
COMMENTS: Flammable					

Name: Pentanoic acid
Synonym: Valeric acid

Mol. Form.: $C_5H_{10}O_2$

CAS RN: 109-52-4
Merck No.: 9815
Mol. Wt.: 102.133

PHYSICAL CONSTANTS

T_m: -34°C (239 K) T_c: 370°C (643 K) μ:
T_b: 186.15°C (459.30 K) P_c: 3.58 MPa   IP: 10.53 eV

TRANSITION PROPERTIES

$\Delta_{fus}H$ (T_m): 14.16 kJ/mol   Vapor pressure (0°C):
$\Delta_{vap}H$ (T_b): 44.06 kJ/mol   Vapor pressure (25°C): 0.019 kPa
$\Delta_{vap}H$ (25°C): 69.29 kJ/mol   Vapor pressure (100°C): 22.1 kPa

PROPERTIES AT 25°C AND 100 kPa

	Solid	Liquid	Gas		Liquid
$\Delta_fH°$/kJ mol^{-1}:		-559.4	-491.9		d: 0.9345 g/mL
$S°$/J mol^{-1}K^{-1}:		259.8			η: 1.97 mPa s
C_p/J mol^{-1}K^{-1}:		210.3			k:
COMMENTS:					

Name: 1-Pentanol
Synonyms: Amyl alcohol
          Pentyl alcohol
Mol. Form.: $C_5H_{12}O$

CAS RN: 71-41-0
Merck No.: 7074
Mol. Wt.: 88.150

PHYSICAL CONSTANTS

T_m: -78.9°C (194.2 K)
T_b: 137.98°C (411.13 K)

T_c: 315.00°C (588.15 K)
P_c: 3.909 MPa

μ:
IP: 10.00 eV

TRANSITION PROPERTIES

$\Delta_{fus}H$ (T_m): 9.83 kJ/mol
$\Delta_{vap}H$ (T_b): 44.36 kJ/mol
$\Delta_{vap}H$ (25°C): 57.02 kJ/mol

Vapor pressure (0°C):
Vapor pressure (25°C): 0.259 kPa
Vapor pressure (100°C): 24.0 kPa

PROPERTIES AT 25°C AND 100 kPa

	Solid	Liquid	Gas	Liquid
$\Delta_f H°$/kJ mol^{-1}:		-351.6	-294.7	*d*: 0.8108 g/mL
$S°$/J mol^{-1}K^{-1}:				η: 3.62 mPa s
C_p/J mol^{-1}K^{-1}:		208.1		*k*: 0.153 W/m K
COMMENTS: Flammable				

Name: 2-Pentanol
Synonyms: *sec*-Amyl alcohol
          sec-Pentyl alcohol
Mol. Form.: $C_5H_{12}O$

CAS RN: 6032-29-7
Merck No.: 7075
Mol. Wt.: 88.150

PHYSICAL CONSTANTS

T_m: -73°C (200 K)
T_b: 119.3°C (392.4 K)

T_c: 287.3°C (560.4 K)
P_c:

μ:
IP: 9.78 eV

TRANSITION PROPERTIES

$\Delta_{fus}H$ (T_m):
$\Delta_{vap}H$ (T_b): 41.40 kJ/mol
$\Delta_{vap}H$ (25°C): 54.21 kJ/mol

Vapor pressure (0°C):
Vapor pressure (25°C): 0.804 kPa
Vapor pressure (100°C): 49.9 kPa

PROPERTIES AT 25°C AND 100 kPa

	Solid	Liquid	Gas	Liquid
$\Delta_f H°$/kJ mol^{-1}:		-365.2	-312.7	*d*: 0.8054 g/mL
$S°$/J mol^{-1}K^{-1}:				η: 3.47 mPa s
C_p/J mol^{-1}K^{-1}:		239.4		*k*:
COMMENTS:				

Name: 3-Pentanol

Mol. Form.: $C_5H_{12}O$

CAS RN: 584-02-1
Merck No.: 7076
Mol. Wt.: 88.150

PHYSICAL CONSTANTS

T_m: -69°C (204 K)
T_b: 116.25°C (389.40 K)

T_c: 286.5°C (559.6 K)
P_c:

μ:
IP: 9.78 eV

TRANSITION PROPERTIES

$\Delta_{fus}H$ (T_m):
$\Delta_{vap}H$ (T_b): 43.50 kJ/mol
$\Delta_{vap}H$ (25°C): 54.00 kJ/mol

Vapor pressure (0°C):
Vapor pressure (25°C): 1.10 kPa
Vapor pressure (100°C): 57.7 kPa

PROPERTIES AT 25°C AND 100 kPa

	Solid	Liquid	Gas	Liquid
$\Delta_f H°$/kJ mol^{-1}:		-368.9	-317.2	*d*: 0.8160 g/mL
$S°$/J mol^{-1}K^{-1}:				η: 4.15 mPa s
C_p/J mol^{-1}K^{-1}:		239.7		*k*:
COMMENTS: Flammable				

Name: 2-Pentanone   CAS RN: 107-87-9
Synonyms: Methyl propyl ketone   Merck No.: 6032
            Propyl methyl ketone   Mol. Wt.: 86.134
Mol. Form.: $C_5H_{10}O$

PHYSICAL CONSTANTS

T_m: -76.9°C (196.2 K) T_c: 287.93°C (561.08 K)   μ:
T_b: 102.26°C (375.41 K) P_c: 3.694 MPa   IP: 9.38 eV

TRANSITION PROPERTIES

$\Delta_{fus}H$ (T_m): 10.63 kJ/mol   Vapor pressure (0°C):
$\Delta_{vap}H$ (T_b): 33.44 kJ/mol   Vapor pressure (25°C): 4.97 kPa
$\Delta_{vap}H$ (25°C): 38.40 kJ/mol   Vapor pressure (100°C): 94.6 kPa

PROPERTIES AT 25°C AND 100 kPa

	Solid	Liquid	Gas	Liquid
$\Delta_f H°$/kJ mol^{-1}:		-297.3	-259.0	d: 0.8020 g/mL
$S°$/J mol^{-1}K^{-1}:				η: 0.470 mPa s
C_p/J mol^{-1}K^{-1}:		184.1		k:

COMMENTS: TLV=200 ppm; flammable

Name: 3-Pentanone   CAS RN: 96-22-0
Synonym: Diethyl ketone   Merck No.: 3111
                                                                             Mol. Wt.: 86.134

Mol. Form.: $C_5H_{10}O$

PHYSICAL CONSTANTS

T_m: -39°C (234 K) T_c: 288.31°C (561.46 K)   μ:
T_b: 101.96°C (375.11 K) P_c: 3.729 MPa   IP: 9.31 eV

TRANSITION PROPERTIES

$\Delta_{fus}H$ (T_m): 11.59 kJ/mol   Vapor pressure (0°C):
$\Delta_{vap}H$ (T_b): 33.45 kJ/mol   Vapor pressure (25°C): 4.72 kPa
$\Delta_{vap}H$ (25°C): 38.52 kJ/mol   Vapor pressure (100°C): 95.5 kPa

PROPERTIES AT 25°C AND 100 kPa

	Solid	Liquid	Gas	Liquid
$\Delta_f H°$/kJ mol^{-1}:		-296.5	-257.9	d: 0.811 g/mL
$S°$/J mol^{-1}K^{-1}:		266.0		η: 0.444 mPa s
C_p/J mol^{-1}K^{-1}:		190.9		k:

COMMENTS: TLV=200 ppm; flammable

Name: 1-Pentene   CAS RN: 109-67-1
Synonym: α-Amylene   Merck No.: 7079
                                                                             Mol. Wt.: 70.134

Mol. Form.: C_5H_{10}

PHYSICAL CONSTANTS

T_m: -165.2°C (107.9 K) T_c: 191.63°C (464.78 K)   μ: 0.5 D
T_b: 29.96°C (303.11 K) P_c: 3.527 MPa   IP: 9.52 eV

TRANSITION PROPERTIES

$\Delta_{fus}H$ (T_m): 5.81 kJ/mol   Vapor pressure (0°C): 31.4 kPa
$\Delta_{vap}H$ (T_b): 25.20 kJ/mol   Vapor pressure (25°C): 85.0 kPa
$\Delta_{vap}H$ (25°C): 25.47 kJ/mol   Vapor pressure (100°C): 695 kPa

PROPERTIES AT 25°C AND 100 kPa

	Solid	Liquid	Gas	Liquid
$\Delta_f H°$/kJ mol^{-1}:		-46.9	-21.3	d: 0.6353 g/mL
$S°$/J mol^{-1}K^{-1}:		262.6		η: 0.195 mPa s
C_p/J mol^{-1}K^{-1}:		154.0		k: 0.116 W/m K

COMMENTS: Very flammable

Name: *cis*-2-Pentene
Synonym: *cis*-β-Amylene

Mol. Form.: C_5H_{10}

CAS RN: 627-20-3
Merck No.: 7080
Mol. Wt.: 70.134

PHYSICAL CONSTANTS

T_m: -151.4°C (121.7 K)
T_b: 36.93°C (310.08 K)

T_c: 202°C (475 K)
P_c: 3.69 MPa

μ:
IP: 9.04 eV

TRANSITION PROPERTIES

$\Delta_{fus}H$ (T_m): 7.12 kJ/mol
$\Delta_{vap}H$ (T_b): 26.11 kJ/mol
$\Delta_{vap}H$ (25°C): 26.86 kJ/mol

Vapor pressure (0°C): 23.2 kPa
Vapor pressure (25°C): 66.0 kPa
Vapor pressure (100°C): 586 kPa

PROPERTIES AT 25°C AND 100 kPa

	Solid	Liquid	Gas		Liquid
$\Delta_f H°$/kJ mol^{-1}:		-53.7	-27.6		d: 0.6508 g/mL
$S°$/J mol^{-1}K^{-1}:		258.6			η:
C_p/J mol^{-1}K^{-1}:		151.7			k:
COMMENTS: Very flammable					

Name: *trans*-2-Pentene
Synonym: *trans*-β-Amylene

Mol. Form.: C_5H_{10}

CAS RN: 646-04-8
Merck No.: 7080
Mol. Wt.: 70.134

PHYSICAL CONSTANTS

T_m: -140.2°C (132.9 K)
T_b: 36.34°C (309.49 K)

T_c: 198°C (471 K)
P_c: 3.52 MPa

μ:
IP: 9.04 eV

TRANSITION PROPERTIES

$\Delta_{fus}H$ (T_m): 8.36 kJ/mol
$\Delta_{vap}H$ (T_b): 26.07 kJ/mol
$\Delta_{vap}H$ (25°C): 26.76 kJ/mol

Vapor pressure (0°C): 23.9 kPa
Vapor pressure (25°C): 67.4 kPa
Vapor pressure (100°C): 597 kPa

PROPERTIES AT 25°C AND 100 kPa

	Solid	Liquid	Gas		Liquid
$\Delta_f H°$/kJ mol^{-1}:		-58.2	-31.9		d: 0.6431 g/mL
$S°$/J mol^{-1}K^{-1}:		256.5			η:
C_p/J mol^{-1}K^{-1}:		157.0			k:
COMMENTS: Very flammable					

Name: Pentyl acetate
Synonyms: Amyl acetate
         Pentyl ethanoate
Mol. Form.: $C_7H_{14}O_2$

CAS RN: 628-63-7
Merck No.:
Mol. Wt.: 130.187

PHYSICAL CONSTANTS

T_m: -70.8°C (202.3 K)
T_b: 149.25°C (422.40 K)

T_c:
P_c:

μ: 1.75 D
IP:

TRANSITION PROPERTIES

$\Delta_{fus}H$ (T_m):
$\Delta_{vap}H$ (T_b): 41.00 kJ/mol
$\Delta_{vap}H$ (25°C):

Vapor pressure (0°C):
Vapor pressure (25°C): 0.600 kPa
Vapor pressure (100°C): 20.1 kPa

PROPERTIES AT 25°C AND 100 kPa

	Solid	Liquid	Gas		Liquid
$\Delta_f H°$/kJ mol^{-1}:					d: 0.8721 g/mL
$S°$/J mol^{-1}K^{-1}:					η: 0.862 mPa s
C_p/J mol^{-1}K^{-1}:		261.0			k:
COMMENTS: TLV=100 ppm; flammable					

Name: Pentylamine
Synonyms: Amylamine
        1-Pentanamine
Mol. Form.: $C_5H_{13}N$

CAS RN: 110-58-7
Merck No.: 630
Mol. Wt.: 87.165

PHYSICAL CONSTANTS

T_m: -55°C (218 K)	T_c:	μ:
T_b: 104.3°C (377.4 K)	P_c:	IP: 8.67 eV

TRANSITION PROPERTIES

$\Delta_{fus}H$ (T_m):
$\Delta_{vap}H$ (T_b): 34.01 kJ/mol
$\Delta_{vap}H$ (25°C): 40.08 kJ/mol

Vapor pressure (0°C):
Vapor pressure (25°C): 4.00 kPa
Vapor pressure (100°C): 88.3 kPa

PROPERTIES AT 25°C AND 100 kPa

	Solid	Liquid	Gas	Liquid
$\Delta_f H°$/kJ mol^{-1}:				d: 0.750 g/mL
$S°$/J mol^{-1}K^{-1}:				η: 0.702 mPa s
C_p/J mol^{-1}K^{-1}:		218.0		k:

COMMENTS: Highly toxic; flammable

Name: Perchloryl fluoride

Mol. Form.: $ClFO_3$

CAS RN: 7616-94-6
Merck No.: 7111
Mol. Wt.: 102.449

PHYSICAL CONSTANTS

T_m: -147°C (126 K)	T_c: 95.3°C (368.4 K)	μ: 0.023 D
T_b: -46.75°C (226.40 K)	P_c: 5.37 MPa	IP: 12.95 eV

TRANSITION PROPERTIES

$\Delta_{fus}H$ (T_m):
$\Delta_{vap}H$ (T_b): 19.33 kJ/mol
$\Delta_{vap}H$ (25°C):

Vapor pressure (0°C):
Vapor pressure (25°C):
Vapor pressure (100°C): N/A

PROPERTIES AT 25°C AND 100 kPa

	Solid	Liquid	Gas	Gas
$\Delta_f H°$/kJ mol^{-1}:			-23.8	d: 4.187 g/L
$S°$/J mol^{-1}K^{-1}:			279.0	η:
C_p/J mol^{-1}K^{-1}:			64.9	k:

COMMENTS: TLV=3 ppm

Name: Perfluoroacetone
Synonym: Hexafluoro-2-propanone

Mol. Form.: C_3F_6O

CAS RN: 684-16-2
Merck No.:
Mol. Wt.: 166.023

PHYSICAL CONSTANTS

T_m: -125°C (148 K)	T_c: 83.99°C (357.14 K)	μ:
T_b: -27.45°C (245.70 K)	P_c: 2.84 MPa	IP: 11.44 eV

TRANSITION PROPERTIES

$\Delta_{fus}H$ (T_m):
$\Delta_{vap}H$ (T_b):
$\Delta_{vap}H$ (25°C):

Vapor pressure (0°C):
Vapor pressure (25°C):
Vapor pressure (100°C): N/A

PROPERTIES AT 25°C AND 100 kPa

	Solid	Liquid	Gas	Gas
$\Delta_f H°$/kJ mol^{-1}:				d: 6.786 g/L
$S°$/J mol^{-1}K^{-1}:				η:
C_p/J mol^{-1}K^{-1}:				k:

COMMENTS: TLV=0.1 ppm

Name: Perfluorobutane
Synonym: Decafluorobutane

Mol. Form.: C_4F_{10}

CAS RN: 355-25-9
Merck No.:
Mol. Wt.: 238.028

PHYSICAL CONSTANTS

T_m: -128.2°C (144.9 K) T_c: 113.3°C (386.4 K) μ:
T_b: -1.95°C (271.20 K) P_c: 2.323 MPa   IP:

TRANSITION PROPERTIES

$\Delta_{fus}H$ (T_m):
$\Delta_{vap}H$ (T_b): 22.90 kJ/mol
$\Delta_{vap}H$ (25°C):

Vapor pressure (0°C): 111 kPa
Vapor pressure (25°C):
Vapor pressure (100°C):

PROPERTIES AT 25°C AND 100 kPa

	Solid	Liquid	Gas	Gas
$\Delta_f H°$/kJ mol^{-1}:				d: 9.729 g/L
$S°$/J mol^{-1}K^{-1}:				η:
C_p/J mol^{-1}K^{-1}:				k:
COMMENTS:				

Name: Perfluorocyclobutane
Synonyms: Octafluorocyclobutane
            Refrigerant C318
Mol. Form.: C_4F_8

CAS RN: 115-25-3
Merck No.: 6666
Mol. Wt.: 200.031

PHYSICAL CONSTANTS

T_m: -40.19°C (232.96 K) T_c: 115.31°C (388.46 K) μ: 0 D
T_b: -5.99°C (267.16 K) P_c: 2.784 MPa   IP:

TRANSITION PROPERTIES

$\Delta_{fus}H$ (T_m): 2.77 kJ/mol
$\Delta_{vap}H$ (T_b): 23.24 kJ/mol
$\Delta_{vap}H$ (25°C):

Vapor pressure (0°C): 129 kPa
Vapor pressure (25°C):
Vapor pressure (100°C):

PROPERTIES AT 25°C AND 100 kPa

	Solid	Liquid	Gas	Gas
$\Delta_f H°$/kJ mol^{-1}:			-1542.6	d: 8.176 g/L
$S°$/J mol^{-1}K^{-1}:				η:
C_p/J mol^{-1}K^{-1}:				k: 0.0125 W/m K
COMMENTS:				

Name: Perfluoropropane
Synonyms: Octafluoropropane
            Refrigerant 218
Mol. Form.: C_3F_8

CAS RN: 76-19-7
Merck No.:
Mol. Wt.: 188.020

PHYSICAL CONSTANTS

T_m: -147.69°C (125.46 K) T_c: 72.0°C (345.1 K) μ:
T_b: -36.65°C (236.50 K) P_c: 2.680 MPa   IP: 13.38 eV

TRANSITION PROPERTIES

$\Delta_{fus}H$ (T_m):
$\Delta_{vap}H$ (T_b):
$\Delta_{vap}H$ (25°C):

Vapor pressure (0°C):
Vapor pressure (25°C):
Vapor pressure (100°C): N/A

PROPERTIES AT 25°C AND 100 kPa

	Solid	Liquid	Gas	Gas
$\Delta_f H°$/kJ mol^{-1}:			-1783.2	d: 7.685 g/L
$S°$/J mol^{-1}K^{-1}:				η:
C_p/J mol^{-1}K^{-1}:				k:
COMMENTS:				

Name: Perfluoropropene
Synonym: Hexafluoropropene

Mol. Form.: C_3F_6

CAS RN: 116-15-4
Merck No.:
Mol. Wt.: 150.023

PHYSICAL CONSTANTS

T_m: -156.5°C (116.6 K)	T_c:	μ:
T_b: -29.6°C (243.5 K)	P_c:	IP: 10.60 eV

TRANSITION PROPERTIES

$\Delta_{fus}H$ (T_m):
$\Delta_{vap}H$ (T_b):
$\Delta_{vap}H$ (25°C):

Vapor pressure (0°C): 342 kPa
Vapor pressure (25°C): 768 kPa
Vapor pressure (100°C):

PROPERTIES AT 25°C AND 100 kPa

	Solid	Liquid	Gas		Gas
$\Delta_fH°$/kJ mol^{-1}:					d: 6.132 g/L
$S°$/J mol^{-1}K^{-1}:					η:
C_p/J mol^{-1}K^{-1}:					k:
COMMENTS:					

Name: Peroxyacetic acid
Synonyms: Peracetic acid
        Ethaneperoxoic acid
Mol. Form.: $C_2H_4O_3$

CAS RN: 79-21-0
Merck No.: 7107
Mol. Wt.: 76.052

PHYSICAL CONSTANTS

T_m: -0.2°C (272.9 K)	T_c:	μ:
T_b: 110°C (383 K)	P_c:	IP:

TRANSITION PROPERTIES

$\Delta_{fus}H$ (T_m):
$\Delta_{vap}H$ (T_b):
$\Delta_{vap}H$ (25°C):

Vapor pressure (0°C): 0.376 kPa
Vapor pressure (25°C): 1.93 kPa
Vapor pressure (100°C): 69.6 kPa

PROPERTIES AT 25°C AND 100 kPa

	Solid	Liquid	Gas		Liquid
$\Delta_fH°$/kJ mol^{-1}:					d:
$S°$/J mol^{-1}K^{-1}:					η:
C_p/J mol^{-1}K^{-1}:					k:
COMMENTS: Highly toxic					

Name: Perylene
Synonym: Dibenz[de,kl]anthracene

Mol. Form.: $C_{20}H_{12}$

CAS RN: 198-55-0
Merck No.: 7137
Mol. Wt.: 252.315

PHYSICAL CONSTANTS

T_m: 273.5°C (546.6 K)	T_c:	μ:
T_b:	P_c:	IP: 6.90 eV

TRANSITION PROPERTIES

$\Delta_{fus}H$ (T_m): 31.75 kJ/mol
$\Delta_{vap}H$ (T_b):
$\Delta_{vap}H$ (25°C):

Vapor pressure (0°C): N/A
Vapor pressure (25°C): N/A
Vapor pressure (100°C): N/A

PROPERTIES AT 25°C AND 100 kPa

	Solid	Liquid	Gas		Solid
$\Delta_fH°$/kJ mol^{-1}:	182.8				d: 1.35 g/mL
$S°$/J mol^{-1}K^{-1}:	264.6				η: N/A
C_p/J mol^{-1}K^{-1}:	274.9				k:
COMMENTS:					

Name: Phenanthrene

Mol. Form.: $C_{14}H_{10}$

CAS RN: 85-01-8
Merck No.: 7167
Mol. Wt.: 178.233

PHYSICAL CONSTANTS

T_m: 99.24°C (372.39 K)
T_b: 340°C (613 K)

T_c: 600°C (873 K)
P_c:

μ:
IP: 7.86 eV

TRANSITION PROPERTIES

$\Delta_{fus}H$ (T_m): 16.46 kJ/mol
$\Delta_{vap}H$ (T_b):
$\Delta_{vap}H$ (25°C): 75.50 kJ/mol

Vapor pressure (0°C):
Vapor pressure (25°C):
Vapor pressure (100°C):

PROPERTIES AT 25°C AND 100 kPa

	Solid	Liquid	Gas		Solid
$\Delta_f H°$/kJ mol^{-1}:	109.8		201.2	d:	1.179 g/mL
$S°$/J mol^{-1}K^{-1}:	215.1			η:	N/A
C_p/J mol^{-1}K^{-1}:	220.6			k:	
COMMENTS: Highly toxic					

Name: Phenetole
Synonyms: Ethyl phenyl ether
         Ethoxybenzene
Mol. Form.: $C_8H_{10}O$

CAS RN: 103-73-1
Merck No.: 7189
Mol. Wt.: 122.167

PHYSICAL CONSTANTS

T_m: -29.52°C (243.63 K)
T_b: 169.81°C (442.96 K)

T_c: 374°C (647 K)
P_c: 3.42 MPa

μ: 1.45 D
IP: 8.13 eV

TRANSITION PROPERTIES

$\Delta_{fus}H$ (T_m):
$\Delta_{vap}H$ (T_b): 40.70 kJ/mol
$\Delta_{vap}H$ (25°C): 51.04 kJ/mol

Vapor pressure (0°C):
Vapor pressure (25°C): 0.204 kPa
Vapor pressure (100°C):

PROPERTIES AT 25°C AND 100 kPa

	Solid	Liquid	Gas		Liquid
$\Delta_f H°$/kJ mol^{-1}:		-152.6	-101.6	d:	0.9605 g/mL
$S°$/J mol^{-1}K^{-1}:				η:	1.20 mPa s
C_p/J mol^{-1}K^{-1}:		228.5		k:	
COMMENTS:					

Name: Phenol
Synonyms: Hydroxybenzene
         Carbolic acid
Mol. Form.: C_6H_6O

CAS RN: 108-95-2
Merck No.: 7206
Mol. Wt.: 94.113

PHYSICAL CONSTANTS

T_m: 40.9°C (314.0 K)
T_b: 181.87°C (455.02 K)

T_c: 421.1°C (694.2 K)
P_c: 6.13 MPa

μ: 1.224 D
IP: 8.47 eV

TRANSITION PROPERTIES

$\Delta_{fus}H$ (T_m): 11.29 kJ/mol
$\Delta_{vap}H$ (T_b): 45.69 kJ/mol
$\Delta_{vap}H$ (25°C): 57.82 kJ/mol

Vapor pressure (0°C):
Vapor pressure (25°C): 0.055 kPa
Vapor pressure (100°C): 5.32 kPa

PROPERTIES AT 25°C AND 100 kPa

	Solid	Liquid	Gas		Solid
$\Delta_f H°$/kJ mol^{-1}:	-165.1		-96.4	d:	1.132 g/mL
$S°$/J mol^{-1}K^{-1}:	144.0		314.8	η:	N/A
C_p/J mol^{-1}K^{-1}:	127.4		103.2	k:	
COMMENTS: TLV=5 ppm; highly toxic					

Name: *o*-Phenylenediamine   CAS RN: 95-54-5
Synonyms: 1,2-Benzenediamine   Merck No.: 7255
         1,2-Diaminobenzene   Mol. Wt.: 108.143
Mol. Form.: $C_6H_8N_2$

PHYSICAL CONSTANTS

T_m: 102.5°C (375.6 K) T_c: μ:
T_b: 257°C (530 K) P_c: IP: 7.20 eV

TRANSITION PROPERTIES

$\Delta_{fus}H$ (T_m):   Vapor pressure (0°C):
$\Delta_{vap}H$ (T_b):   Vapor pressure (25°C):
$\Delta_{vap}H$ (25°C):   Vapor pressure (100°C):

PROPERTIES AT 25°C AND 100 kPa

	Solid	Liquid	Gas		Solid
$\Delta_f H°$/kJ mol^{-1}:	-0.3			d:	
$S°$/J mol^{-1}K^{-1}:				η: N/A	
C_p/J mol^{-1}K^{-1}:				k:	
COMMENTS:					

Name: *m*-Phenylenediamine   CAS RN: 108-45-2
Synonyms: 1,3-Benzenediamine   Merck No.: 7254
         1,3-Diaminobenzene   Mol. Wt.: 108.143
Mol. Form.: $C_6H_8N_2$

PHYSICAL CONSTANTS

T_m: 63.5°C (336.6 K) T_c: μ:
T_b: 285°C (558 K) P_c: IP: 7.14 eV

TRANSITION PROPERTIES

$\Delta_{fus}H$ (T_m):   Vapor pressure (0°C):
$\Delta_{vap}H$ (T_b):   Vapor pressure (25°C):
$\Delta_{vap}H$ (25°C):   Vapor pressure (100°C):

PROPERTIES AT 25°C AND 100 kPa

	Solid	Liquid	Gas		Solid
$\Delta_f H°$/kJ mol^{-1}:	-7.8			d: 1.23 g/mL	
$S°$/J mol^{-1}K^{-1}:	154.5			η: N/A	
C_p/J mol^{-1}K^{-1}:	159.6			k:	
COMMENTS:					

Name: *p*-Phenylenediamine   CAS RN: 106-50-3
Synonyms: 1,4-Benzenediamine   Merck No.: 7256
         1,4-Diaminobenzene   Mol. Wt.: 108.143
Mol. Form.: $C_6H_8N_2$

PHYSICAL CONSTANTS

T_m: 146°C (419 K) T_c: μ:
T_b: 267°C (540 K) P_c: IP: 6.87 eV

TRANSITION PROPERTIES

$\Delta_{fus}H$ (T_m):   Vapor pressure (0°C):
$\Delta_{vap}H$ (T_b):   Vapor pressure (25°C):
$\Delta_{vap}H$ (25°C):   Vapor pressure (100°C):

PROPERTIES AT 25°C AND 100 kPa

	Solid	Liquid	Gas		Solid
$\Delta_f H°$/kJ mol^{-1}:	3.1			d: 1.2 g/mL	
$S°$/J mol^{-1}K^{-1}:				η: N/A	
C_p/J mol^{-1}K^{-1}:				k:	
COMMENTS: Highly toxic					

Name: Phenylhydrazine

CAS RN: 100-63-0
Merck No.: 7264
Mol. Wt.: 108.143

Mol. Form.: $C_6H_8N_2$

PHYSICAL CONSTANTS

T_m: 19.6°C (292.7 K)	T_c:	μ:
T_b: 243.55°C (516.70 K)	P_c:	IP:

TRANSITION PROPERTIES

$\Delta_{fus}H$ (T_m): 16.43 kJ/mol

$\Delta_{vap}H$ (T_b):

$\Delta_{vap}H$ (25°C): 61.9 kJ/mol

Vapor pressure (0°C):
Vapor pressure (25°C):
Vapor pressure (100°C):

PROPERTIES AT 25°C AND 100 kPa

	Solid	Liquid	Gas		Liquid
$\Delta_f H°$/kJ mol^{-1}:		141.0	202.9		d: 1.095 g/mL
$S°$/J mol^{-1}K^{-1}:					η: 13.0 mPa s
C_p/J mol^{-1}K^{-1}:		217.0			k:

COMMENTS: TLV=0.1 ppm; carcinogen; highly toxic

Name: Phosphine

CAS RN: 7803-51-2
Merck No.: 7313
Mol. Wt.: 33.998

Mol. Form.: H_3P

PHYSICAL CONSTANTS

T_m: -133°C (140 K)	T_c: 51.4°C (324.5 K)	μ: 0.574 D
T_b: -87.75°C (185.40 K)	P_c: 6.54 MPa	IP: 9.87 eV

TRANSITION PROPERTIES

$\Delta_{fus}H$ (T_m):

$\Delta_{vap}H$ (T_b): 14.60 kJ/mol

$\Delta_{vap}H$ (25°C):

Vapor pressure (0°C):
Vapor pressure (25°C):
Vapor pressure (100°C): N/A

PROPERTIES AT 25°C AND 100 kPa

	Solid	Liquid	Gas		Gas
$\Delta_f H°$/kJ mol^{-1}:			5.4		d: 1.390 g/L
$S°$/J mol^{-1}K^{-1}:			210.2		η:
C_p/J mol^{-1}K^{-1}:			37.1		k:

COMMENTS: TLV=0.3 ppm; highly toxic; very flammable

Name: Phosphoric acid
Synonym: Orthophosphoric acid

CAS RN: 7664-38-2
Merck No.: 7318
Mol. Wt.: 97.995

Mol. Form.: H_3O_4P

PHYSICAL CONSTANTS

T_m: 42.4°C (315.5 K)	T_c:	μ:
T_b: 407°C (680 K)	P_c:	IP:

TRANSITION PROPERTIES

$\Delta_{fus}H$ (T_m): 13.40 kJ/mol

$\Delta_{vap}H$ (T_b):

$\Delta_{vap}H$ (25°C):

Vapor pressure (0°C):
Vapor pressure (25°C):
Vapor pressure (100°C):

PROPERTIES AT 25°C AND 100 kPa

	Solid	Liquid	Gas		Solid
$\Delta_f H°$/kJ mol^{-1}:	-1284.4	-1271.7			d:
$S°$/J mol^{-1}K^{-1}:	110.5	150.8			η: N/A
C_p/J mol^{-1}K^{-1}:	106.1	145.0			k:

COMMENTS: Highly toxic

Name: Phosphorous acid   CAS RN: 13598-36-2
Synonym: Phosphonic acid   Merck No.: 7320
                                                                          Mol. Wt.: 81.996
Mol. Form.: H_3O_3P

PHYSICAL CONSTANTS

T_m: 74.4°C (347.5 K) T_c: μ:
T_b: 200°C (473 K) P_c: IP:

TRANSITION PROPERTIES

$\Delta_{fus}H$ (T_m): 12.80 kJ/mol   Vapor pressure (0°C):
$\Delta_{vap}H$ (T_b):   Vapor pressure (25°C):
$\Delta_{vap}H$ (25°C):   Vapor pressure (100°C):

PROPERTIES AT 25°C AND 100 kPa

	Solid	Liquid	Gas		Solid
$\Delta_fH°$/kJ mol^{-1}:	-964.4				d: 1.65 g/mL
$S°$/J mol^{-1}K^{-1}:					η: N/A
C_p/J mol^{-1}K^{-1}:					k:
COMMENTS:					

Name: Phosphorus   CAS RN: 7723-14-0
                                                                          Merck No.: 7321
                                                                          Mol. Wt.: 30.974
Mol. Form.: P

PHYSICAL CONSTANTS

T_m: 44.15°C (317.30 K) T_c: 721°C (994 K)   μ:
T_b: 277°C (550 K) P_c: IP: 10.49 eV

TRANSITION PROPERTIES

$\Delta_{fus}H$ (T_m): 0.66 kJ/mol   Vapor pressure (0°C):
$\Delta_{vap}H$ (T_b): 12.40 kJ/mol   Vapor pressure (25°C):
$\Delta_{vap}H$ (25°C): 14.20 kJ/mol   Vapor pressure (100°C):

PROPERTIES AT 25°C AND 100 kPa

	Solid	Liquid	Gas		Solid
$\Delta_fH°$/kJ mol^{-1}:	0.0		316.5		d: 2.69 g/mL
$S°$/J mol^{-1}K^{-1}:	41.1		163.2		η: N/A
C_p/J mol^{-1}K^{-1}:	23.8		20.8		k: 0.236 W/m K
COMMENTS: Highly toxic; data refer to white phosphorus					

Name: Phosphorus pentachloride   CAS RN: 10026-13-8
Synonym: Phosphorus (V) chloride   Merck No.: 7326
                                                                          Mol. Wt.: 208.237
Mol. Form.: Cl_5P

PHYSICAL CONSTANTS

T_m: T_c: 373°C (646 K)   μ:
T_b: 160°C (433 K) P_c: IP: 10.70 eV

TRANSITION PROPERTIES

$\Delta_{fus}H$ (T_m):   Vapor pressure (0°C):
$\Delta_{vap}H$ (T_b):   Vapor pressure (25°C):
$\Delta_{vap}H$ (25°C):   Vapor pressure (100°C):

PROPERTIES AT 25°C AND 100 kPa

	Solid	Liquid	Gas		Solid
$\Delta_fH°$/kJ mol^{-1}:	-443.5		-374.9		d:
$S°$/J mol^{-1}K^{-1}:			364.6		η: N/A
C_p/J mol^{-1}K^{-1}:			112.8		k:
COMMENTS: TLV=0.1 ppm; highly toxic					

Name: Phosphorus pentafluoride

Synonym: Phosphorus (V) fluoride

CAS RN: 7647-19-0

Merck No.: 7327

Mol. Wt.: 125.966

Mol. Form.: F_5P

PHYSICAL CONSTANTS

T_m:

T_b: -84.6°C (188.5 K)

T_c:

P_c:

μ:

IP: 15.10 eV

TRANSITION PROPERTIES

$\Delta_{fus}H$ (T_m):

$\Delta_{vap}H$ (T_b): 17.20 kJ/mol

$\Delta_{vap}H$ (25°C):

Vapor pressure (0°C):

Vapor pressure (25°C):

Vapor pressure (100°C):

PROPERTIES AT 25°C AND 100 kPa

	Solid	Liquid	Gas		Gas
$\Delta_fH°$/kJ mol⁻¹:			-1594.4	d:	5.149 g/L
$S°$/J mol⁻¹K⁻¹:			300.8	η:	
C_p/J mol⁻¹K⁻¹:			84.8	k:	
COMMENTS:					

Name: Phosphorus pentoxide

Synonyms: Diphosphorus pentoxide

Phosphorus (V) oxide

Mol. Form.: O_5P_2

CAS RN: 1314-56-3

Merck No.: 7330

Mol. Wt.: 141.945

PHYSICAL CONSTANTS

T_m: 420°C (693 K)

T_b:

T_c:

P_c:

μ:

IP:

TRANSITION PROPERTIES

$\Delta_{fus}H$ (T_m): 27.20 kJ/mol

$\Delta_{vap}H$ (T_b):

$\Delta_{vap}H$ (25°C):

Vapor pressure (0°C): N/A

Vapor pressure (25°C): N/A

Vapor pressure (100°C): N/A

PROPERTIES AT 25°C AND 100 kPa

	Solid	Liquid	Gas		Solid
$\Delta_fH°$/kJ mol⁻¹:				d:	2.30 g/mL
$S°$/J mol⁻¹K⁻¹:				η:	N/A
C_p/J mol⁻¹K⁻¹:				k:	
COMMENTS: Highly toxic					

Name: Phosphorus thiochloride

Synonym: Thiophosphoryl chloride

CAS RN: 3982-91-0

Merck No.: 7331

Mol. Wt.: 169.398

Mol. Form.: Cl_3PS

PHYSICAL CONSTANTS

T_m: -36.2°C (236.9 K)

T_b: 125°C (398 K)

T_c:

P_c:

μ:

IP: 9.71 eV

TRANSITION PROPERTIES

$\Delta_{fus}H$ (T_m):

$\Delta_{vap}H$ (T_b):

$\Delta_{vap}H$ (25°C):

Vapor pressure (0°C):

Vapor pressure (25°C):

Vapor pressure (100°C):

PROPERTIES AT 25°C AND 100 kPa

	Solid	Liquid	Gas		Liquid
$\Delta_fH°$/kJ mol⁻¹:				d:	
$S°$/J mol⁻¹K⁻¹:				η:	
C_p/J mol⁻¹K⁻¹:				k:	
COMMENTS:					

Name: Phosphorus tribromide
Synonym: Phosphorus(III) bromide

CAS RN: 7789-60-8
Merck No.: 7332
Mol. Wt.: 270.686

Mol. Form.: Br_3P

PHYSICAL CONSTANTS

T_m: -40°C (233 K)
T_b: 172.95°C (446.10 K)

T_c: 438°C (711 K)
P_c:

μ:
IP: 9.70 eV

TRANSITION PROPERTIES

$\Delta_{fus}H$ (T_m):
$\Delta_{vap}H$ (T_b): 38.80 kJ/mol
$\Delta_{vap}H$ (25°C): 45.2 kJ/mol

Vapor pressure (0°C):
Vapor pressure (25°C):
Vapor pressure (100°C):

PROPERTIES AT 25°C AND 100 kPa

	Solid	Liquid	Gas	Liquid
$\Delta_f H°$/kJ mol^{-1}:		-184.5	-139.3	d: 2.8 g/mL
$S°$/J mol^{-1}K^{-1}:		240.2	348.1	η:
C_p/J mol^{-1}K^{-1}:			76.0	k:
COMMENTS:				

Name: Phosphorus trichloride
Synonym: Phosphorus (III) chloride

CAS RN: 7719-12-2
Merck No.: 7333
Mol. Wt.: 137.332

Mol. Form.: Cl_3P

PHYSICAL CONSTANTS

T_m: -112°C (161 K)
T_b: 75.95°C (349.10 K)

T_c: 290°C (563 K)
P_c:

μ: 0.56 D
IP: 9.91 eV

TRANSITION PROPERTIES

$\Delta_{fus}H$ (T_m): 7.10 kJ/mol
$\Delta_{vap}H$ (T_b): 30.50 kJ/mol
$\Delta_{vap}H$ (25°C): 32.10 kJ/mol

Vapor pressure (0°C):
Vapor pressure (25°C):
Vapor pressure (100°C):

PROPERTIES AT 25°C AND 100 kPa

	Solid	Liquid	Gas	Liquid
$\Delta_f H°$/kJ mol^{-1}:		-319.7	-287.0	d: 1.574 g/mL
$S°$/J mol^{-1}K^{-1}:		217.1	311.8	η: 0.529 mPa s
C_p/J mol^{-1}K^{-1}:			71.8	k:
COMMENTS: TLV=0.2 ppm; highly toxic				

Name: Phosphorus trifluoride
Synonym: Phosphorus (III) fluoride

CAS RN: 7783-55-3
Merck No.: 7334
Mol. Wt.: 87.969

Mol. Form.: F_3P

PHYSICAL CONSTANTS

T_m: -151.5°C (121.6 K)
T_b: -101.5°C (171.6 K)

T_c: -1.9°C (271.2 K)
P_c: 4.33 MPa

μ: 1.03 D
IP: 11.44 eV

TRANSITION PROPERTIES

$\Delta_{fus}H$ (T_m):
$\Delta_{vap}H$ (T_b): 16.50 kJ/mol
$\Delta_{vap}H$ (25°C):

Vapor pressure (0°C): N/A
Vapor pressure (25°C): N/A
Vapor pressure (100°C): N/A

PROPERTIES AT 25°C AND 100 kPa

	Solid	Liquid	Gas	Gas
$\Delta_f H°$/kJ mol^{-1}:			-958.4	d: 3.596 g/L
$S°$/J mol^{-1}K^{-1}:			273.1	η:
C_p/J mol^{-1}K^{-1}:			58.7	k:
COMMENTS:				

Name: Phosphoryl chloride
Synonym: Phosphorus oxychloride

Mol. Form.: Cl_3OP

CAS RN: 10025-87-3
Merck No.: 7324
Mol. Wt.: 153.331

PHYSICAL CONSTANTS

T_m: 1°C (274 K)
T_b: 105.5°C (378.6 K)

T_c:
P_c:

μ: 2.54 D
IP: 11.36 eV

TRANSITION PROPERTIES

$\Delta_{fus}H$ (T_m): 13.10 kJ/mol
$\Delta_{vap}H$ (T_b): 34.35 kJ/mol
$\Delta_{vap}H$ (25°C): 38.60 kJ/mol

Vapor pressure (0°C):
Vapor pressure (25°C):
Vapor pressure (100°C):

PROPERTIES AT 25°C AND 100 kPa

	Solid	Liquid	Gas		Liquid
$\Delta_f H°$/kJ mol^{-1}:		-597.1	-558.5		d: 1.645 g/mL
$S°$/J mol^{-1}K^{-1}:		222.5	325.5		η:
C_p/J mol^{-1}K^{-1}:		138.8	84.9		k:

COMMENTS: TLV=0.1 ppm; highly toxic

Name: Phthalic acid
Synonym: 1,2-Benzenedicarboxylic acid

Mol. Form.: $C_8H_6O_4$

CAS RN: 88-99-3
Merck No.: 7345
Mol. Wt.: 166.133

PHYSICAL CONSTANTS

T_m: 230°C (503 K)
T_b:

T_c:
P_c:

μ:
IP:

TRANSITION PROPERTIES

$\Delta_{fus}H$ (T_m):
$\Delta_{vap}H$ (T_b):
$\Delta_{vap}H$ (25°C):

Vapor pressure (0°C): N/A
Vapor pressure (25°C): N/A
Vapor pressure (100°C): N/A

PROPERTIES AT 25°C AND 100 kPa

	Solid	Liquid	Gas		Solid
$\Delta_f H°$/kJ mol^{-1}:	-782.0				d:
$S°$/J mol^{-1}K^{-1}:	207.9				η: N/A
C_p/J mol^{-1}K^{-1}:	188.1				k:

COMMENTS:

Name: Phthalic anhydride
Synonyms: 1,2-Benzenedicarboxylic anhydride
           1,3-Isobenzofurandione
Mol. Form.: $C_8H_4O_3$

CAS RN: 85-44-9
Merck No.: 7346
Mol. Wt.: 148.118

PHYSICAL CONSTANTS

T_m: 130.8°C (403.9 K)
T_b: 295°C (568 K)

T_c:
P_c:

μ:
IP: 10.00 eV

TRANSITION PROPERTIES

$\Delta_{fus}H$ (T_m):
$\Delta_{vap}H$ (T_b):
$\Delta_{vap}H$ (25°C):

Vapor pressure (0°C):
Vapor pressure (25°C):
Vapor pressure (100°C):

PROPERTIES AT 25°C AND 100 kPa

	Solid	Liquid	Gas		Solid
$\Delta_f H°$/kJ mol^{-1}:	-460.1		-371.4		d: 1.494 g/mL
$S°$/J mol^{-1}K^{-1}:	180.0				η: N/A
C_p/J mol^{-1}K^{-1}:	160.0				k:

COMMENTS: TLV=1 ppm; highly toxic

Name: α-Pinene
Synonym: 2,7,7-Trimethylbicyclo[3.1.1]hept-2-ene

Mol. Form.: $C_{10}H_{16}$

CAS RN: 80-56-8
Merck No.: 7414
Mol. Wt.: 136.237

PHYSICAL CONSTANTS

T_m: -64°C (209 K)
T_b: 156°C (429 K)

T_c:
P_c:

μ:
IP: 8.07 eV

TRANSITION PROPERTIES

$\Delta_{fus}H$ (T_m):
$\Delta_{vap}H$ (T_b):
$\Delta_{vap}H$ (25°C): 44.7 kJ/mol

Vapor pressure (0°C):
Vapor pressure (25°C): 0.650 kPa
Vapor pressure (100°C):

PROPERTIES AT 25°C AND 100 kPa

	Solid	Liquid	Gas		Liquid
$\Delta_f H°$/kJ mol^{-1}:		-16.4	28.3		d: 0.855 g/mL
$S°$/J mol^{-1}K^{-1}:					η:
C_p/J mol^{-1}K^{-1}:					k:
COMMENTS:					

Name: β-Pinene
Synonym: 7,7-Dimethyl-2-methylenebicyclo[3.1.1]heptane

Mol. Form.: $C_{10}H_{16}$

CAS RN: 127-91-3
Merck No.: 7415
Mol. Wt.: 136.237

PHYSICAL CONSTANTS

T_m: -61.5°C (211.6 K)
T_b: 166°C (439 K)

T_c:
P_c:

μ:
IP:

TRANSITION PROPERTIES

$\Delta_{fus}H$ (T_m):
$\Delta_{vap}H$ (T_b):
$\Delta_{vap}H$ (25°C): 46.4 kJ/mol

Vapor pressure (0°C):
Vapor pressure (25°C): 0.610 kPa
Vapor pressure (100°C):

PROPERTIES AT 25°C AND 100 kPa

	Solid	Liquid	Gas		Liquid
$\Delta_f H°$/kJ mol^{-1}:		-7.7	38.7		d: 0.860 g/mL
$S°$/J mol^{-1}K^{-1}:					η:
C_p/J mol^{-1}K^{-1}:					k:
COMMENTS:					

Name: Piperidine
Synonyms: Azacyclohexane
          Hexahydropyridine
Mol. Form.: $C_5H_{11}N$

CAS RN: 110-89-4
Merck No.: 7438
Mol. Wt.: 85.149

PHYSICAL CONSTANTS

T_m: -11.03°C (262.12 K)
T_b: 106.22°C (379.37 K)

T_c: 321.0°C (594.1 K)
P_c: 4.94 MPa

μ:
IP: 8.05 eV

TRANSITION PROPERTIES

$\Delta_{fus}H$ (T_m): 14.85 kJ/mol
$\Delta_{vap}H$ (T_b):
$\Delta_{vap}H$ (25°C): 39.29 kJ/mol

Vapor pressure (0°C):
Vapor pressure (25°C): 4.28 kPa
Vapor pressure (100°C): 84.1 kPa

PROPERTIES AT 25°C AND 100 kPa

	Solid	Liquid	Gas		Liquid
$\Delta_f H°$/kJ mol^{-1}:		-86.4	-47.2		d: 0.8578 g/mL
$S°$/J mol^{-1}K^{-1}:		210.0			η: 1.57 mPa s
C_p/J mol^{-1}K^{-1}:		179.9			k:
COMMENTS: Highly toxic; flammable					

Name: Platinum

CAS RN: 7440-06-4
Merck No.: 7503
Mol. Wt.: 195.080

Mol. Form.: Pt

PHYSICAL CONSTANTS

T_m: 1768.4°C (2041.5 K) T_c: μ:
T_b: 3825°C (4098 K) P_c: IP: 9.00 eV

TRANSITION PROPERTIES

$\Delta_{fus}H$ (T_m): 22.17 kJ/mol   Vapor pressure (0°C): N/A
$\Delta_{vap}H$ (T_b):   Vapor pressure (25°C): N/A
$\Delta_{vap}H$ (25°C):   Vapor pressure (100°C): N/A

PROPERTIES AT 25°C AND 100 kPa

	Solid	Liquid	Gas		Solid
$\Delta_f H°$/kJ mol^{-1}:	0.0		565.3	d:	21.5 g/mL
$S°$/J mol^{-1}K^{-1}:	41.6		192.4	η:	N/A
C_p/J mol^{-1}K^{-1}:	25.9		25.5	k:	71.6 W/m K
COMMENTS:					

Name: Plutonium

CAS RN: 7440-07-5
Merck No.: 7514
Mol. Wt.: 244.064

Mol. Form.: Pu

PHYSICAL CONSTANTS

T_m: 640°C (913 K) T_c: μ:
T_b: 3228°C (3501 K) P_c: IP: 6.06 eV

TRANSITION PROPERTIES

$\Delta_{fus}H$ (T_m): 2.82 kJ/mol   Vapor pressure (0°C): N/A
$\Delta_{vap}H$ (T_b):   Vapor pressure (25°C): N/A
$\Delta_{vap}H$ (25°C):   Vapor pressure (100°C): N/A

PROPERTIES AT 25°C AND 100 kPa

	Solid	Liquid	Gas		Solid
$\Delta_f H°$/kJ mol^{-1}:	0.0			d:	19.7 g/mL
$S°$/J mol^{-1}K^{-1}:				η:	N/A
C_p/J mol^{-1}K^{-1}:				k:	6.70 W/m K
COMMENTS: Highly toxic					

Name: Potassium

CAS RN: 7440-09-7
Merck No.: 7579
Mol. Wt.: 39.098

Mol. Form.: K

PHYSICAL CONSTANTS

T_m: 63.38°C (336.53 K) T_c: μ:
T_b: 759°C (1032 K) P_c: IP: 4.34 eV

TRANSITION PROPERTIES

$\Delta_{fus}H$ (T_m): 2.32 kJ/mol   Vapor pressure (0°C):
$\Delta_{vap}H$ (T_b):   Vapor pressure (25°C):
$\Delta_{vap}H$ (25°C):   Vapor pressure (100°C):

PROPERTIES AT 25°C AND 100 kPa

	Solid	Liquid	Gas		Solid
$\Delta_f H°$/kJ mol^{-1}:	0.0		89.0	d:	0.89 g/mL
$S°$/J mol^{-1}K^{-1}:	64.7		160.3	η:	N/A
C_p/J mol^{-1}K^{-1}:	29.6		20.8	k:	102.5 W/m K
COMMENTS: Flammable					

Name: Potassium bromide

CAS RN: 7758-02-3
Merck No.: 7597
Mol. Wt.: 119.002

Mol. Form.: BrK

PHYSICAL CONSTANTS

T_m: 734°C (1007 K) T_c: μ: 10.628 D
T_b: P_c: IP: 7.85 eV

TRANSITION PROPERTIES

$\Delta_{fus}H$ (T_m): 25.50 kJ/mol   Vapor pressure (0°C): N/A
$\Delta_{vap}H$ (T_b):   Vapor pressure (25°C): N/A
$\Delta_{vap}H$ (25°C):   Vapor pressure (100°C): N/A

PROPERTIES AT 25°C AND 100 kPa

	Solid	Liquid	Gas		Solid
$\Delta_f H°$/kJ mol^{-1}:	-393.8			d:	2.74 g/mL
$S°$/J mol^{-1}K^{-1}:	95.9			η:	N/A
C_p/J mol^{-1}K^{-1}:	52.3			k:	4.9 W/m K

COMMENTS:

Name: Potassium chloride
Synonym: Sylvite

CAS RN: 7447-40-7
Merck No.: 7601
Mol. Wt.: 74.551

Mol. Form.: ClK

PHYSICAL CONSTANTS

T_m: 771°C (1044 K) T_c: μ: 10.269 D
T_b: P_c: IP: 8.00 eV

TRANSITION PROPERTIES

$\Delta_{fus}H$ (T_m): 26.53 kJ/mol   Vapor pressure (0°C): N/A
$\Delta_{vap}H$ (T_b):   Vapor pressure (25°C): N/A
$\Delta_{vap}H$ (25°C):   Vapor pressure (100°C): N/A

PROPERTIES AT 25°C AND 100 kPa

	Solid	Liquid	Gas		Solid
$\Delta_f H°$/kJ mol^{-1}:	-436.5			d:	1.988 g/mL
$S°$/J mol^{-1}K^{-1}:	82.6			η:	N/A
C_p/J mol^{-1}K^{-1}:	51.3			k:	6.7 W/m K

COMMENTS:

Name: Potassium fluoride

CAS RN: 7789-23-3
Merck No.: 7613
Mol. Wt.: 58.097

Mol. Form.: FK

PHYSICAL CONSTANTS

T_m: 858°C (1131 K) T_c: μ: 8.585 D
T_b: 1502°C (1775 K) P_c: IP:

TRANSITION PROPERTIES

$\Delta_{fus}H$ (T_m): 27.20 kJ/mol   Vapor pressure (0°C): N/A
$\Delta_{vap}H$ (T_b):   Vapor pressure (25°C): N/A
$\Delta_{vap}H$ (25°C):   Vapor pressure (100°C): N/A

PROPERTIES AT 25°C AND 100 kPa

	Solid	Liquid	Gas		Solid
$\Delta_f H°$/kJ mol^{-1}:	-567.3			d:	2.48 g/mL
$S°$/J mol^{-1}K^{-1}:	66.6			η:	N/A
C_p/J mol^{-1}K^{-1}:	49.0			k:	

COMMENTS: Highly toxic

Name: Potassium hydroxide
Synonym: Caustic potash

Mol. Form.: HKO

CAS RN: 1310-58-3
Merck No.: 7625
Mol. Wt.: 56.106

PHYSICAL CONSTANTS

T_m: 406°C (679 K)
T_b: 1327°C (1600 K)

T_c:
P_c:

μ:
IP:

TRANSITION PROPERTIES

$\Delta_{fus}H$ (T_m): 8.60 kJ/mol
$\Delta_{vap}H$ (T_b):
$\Delta_{vap}H$ (25°C):

Vapor pressure (0°C): N/A
Vapor pressure (25°C): N/A
Vapor pressure (100°C): N/A

PROPERTIES AT 25°C AND 100 kPa

	Solid	Liquid	Gas		Solid
$\Delta_f H°$/kJ mol^{-1}:	-424.8			d:	
$S°$/J mol^{-1}K^{-1}:	78.9			η: N/A	
C_p/J mol^{-1}K^{-1}:	64.9			k:	
COMMENTS:					

Name: Potassium iodide

Mol. Form.: IK

CAS RN: 7681-11-0
Merck No.: 7628
Mol. Wt.: 166.003

PHYSICAL CONSTANTS

T_m: 681°C (954 K)
T_b: 1323°C (1596 K)

T_c:
P_c:

μ:
IP: 7.21 eV

TRANSITION PROPERTIES

$\Delta_{fus}H$ (T_m): 24.00 kJ/mol
$\Delta_{vap}H$ (T_b):
$\Delta_{vap}H$ (25°C):

Vapor pressure (0°C): N/A
Vapor pressure (25°C): N/A
Vapor pressure (100°C): N/A

PROPERTIES AT 25°C AND 100 kPa

	Solid	Liquid	Gas		Solid
$\Delta_f H°$/kJ mol^{-1}:	-327.9			d: 3.12 g/mL	
$S°$/J mol^{-1}K^{-1}:	106.3			η: N/A	
C_p/J mol^{-1}K^{-1}:	52.9			k:	
COMMENTS:					

Name: Potassium nitrate
Synonym: Saltpeter

Mol. Form.: KNO$_3$

CAS RN: 7757-79-1
Merck No.: 7634
Mol. Wt.: 101.103

PHYSICAL CONSTANTS

T_m: 337°C (610 K)
T_b:

T_c:
P_c:

μ:
IP:

TRANSITION PROPERTIES

$\Delta_{fus}H$ (T_m): 10.10 kJ/mol
$\Delta_{vap}H$ (T_b):
$\Delta_{vap}H$ (25°C):

Vapor pressure (0°C): N/A
Vapor pressure (25°C): N/A
Vapor pressure (100°C): N/A

PROPERTIES AT 25°C AND 100 kPa

	Solid	Liquid	Gas		Solid
$\Delta_f H°$/kJ mol^{-1}:	-494.6			d: 2.11 g/mL	
$S°$/J mol^{-1}K^{-1}:	133.1			η: N/A	
C_p/J mol^{-1}K^{-1}:	96.4			k:	
COMMENTS:					

Name: Potassium permanganate

CAS RN: 7722-64-7
Merck No.: 7643
Mol. Wt.: 158.034

Mol. Form.: $KMnO_4$

PHYSICAL CONSTANTS

| T_m: | T_c: | μ: |
| T_b: | P_c: | IP: |

TRANSITION PROPERTIES

$\Delta_{fus}H$ (T_m):   Vapor pressure (0°C):
$\Delta_{vap}H$ (T_b):   Vapor pressure (25°C):
$\Delta_{vap}H$ (25°C):   Vapor pressure (100°C):

PROPERTIES AT 25°C AND 100 kPa

	Solid	Liquid	Gas		Solid
$\Delta_f H°$/kJ mol^{-1}:	-837.2			d:	2.7 g/mL
$S°$/J mol^{-1}K^{-1}:	171.7			η:	N/A
C_p/J mol^{-1}K^{-1}:	117.6			k:	
COMMENTS:					

Name: Propanal
Synonym: Propionaldehyde

CAS RN: 123-38-6
Merck No.: 7835
Mol. Wt.: 58.080

Mol. Form.: C_3H_6O

PHYSICAL CONSTANTS

| T_m: -80°C (193 K) | T_c: 231.3°C (504.4 K) | μ: 2.52 D |
| T_b: 48°C (321 K) | P_c: 5.27 MPa | IP: 9.95 eV |

TRANSITION PROPERTIES

$\Delta_{fus}H$ (T_m):   Vapor pressure (0°C): 13.5 kPa
$\Delta_{vap}H$ (T_b): 28.31 kJ/mol   Vapor pressure (25°C): 42.4 kPa
$\Delta_{vap}H$ (25°C): 29.62 kJ/mol   Vapor pressure (100°C):

PROPERTIES AT 25°C AND 100 kPa

	Solid	Liquid	Gas		Liquid
$\Delta_f H°$/kJ mol^{-1}:		-215.3	-185.6	d:	0.797 g/mL
$S°$/J mol^{-1}K^{-1}:			304.5	η:	0.321 mPa s
C_p/J mol^{-1}K^{-1}:		137.2	80.7	k:	
COMMENTS: Highly toxic; flammable					

Name: Propane

CAS RN: 74-98-6
Merck No.: 7809
Mol. Wt.: 44.097

Mol. Form.: C_3H_8

PHYSICAL CONSTANTS

| T_m: -187.69°C (85.46 K) | T_c: 96.67°C (369.82 K) | μ: 0.084 D |
| T_b: -42.1°C (231.0 K) | P_c: 4.250 MPa | IP: 10.95 eV |

TRANSITION PROPERTIES

$\Delta_{fus}H$ (T_m): 3.53 kJ/mol   Vapor pressure (0°C): 472 kPa
$\Delta_{vap}H$ (T_b): 19.04 kJ/mol   Vapor pressure (25°C): 939 kPa
$\Delta_{vap}H$ (25°C): 14.79 kJ/mol   Vapor pressure (100°C): N/A

PROPERTIES AT 25°C AND 100 kPa

	Solid	Liquid	Gas		Gas
$\Delta_f H°$/kJ mol^{-1}:			-103.8	d:	1.802 g/L
$S°$/J mol^{-1}K^{-1}:			270.3	η:	8.2 μPa s
C_p/J mol^{-1}K^{-1}:			73.6	k:	0.0180 W/m K
COMMENTS: Very flammable					

Name: 1,2-Propanediol
Synonyms: Propylene glycol
        2-Hydroxypropanol
Mol. Form.: $C_3H_8O_2$

CAS RN: 57-55-6
Merck No.: 7868
Mol. Wt.: 76.095

PHYSICAL CONSTANTS

T_m: -60°C (213 K)
T_b: 187.6°C (460.7 K)

T_c:
P_c:

μ:
IP:

TRANSITION PROPERTIES

$\Delta_{fus}H$ (T_m):
$\Delta_{vap}H$ (T_b): 52.35 kJ/mol
$\Delta_{vap}H$ (25°C): 64.40 kJ/mol

Vapor pressure (0°C):
Vapor pressure (25°C): 0.020 kPa
Vapor pressure (100°C): 3.15 kPa

PROPERTIES AT 25°C AND 100 kPa

	Solid	Liquid	Gas		Liquid
$\Delta_f H°$/kJ mol⁻¹:		-485.7	-421.3		d: 1.0327 g/mL
$S°$/J mol⁻¹K⁻¹:					η: 40.4 mPa s
C_p/J mol⁻¹K⁻¹:		190.8			k: 0.200 W/m K
COMMENTS:					

Name: 1,3-Propanediol
Synonyms: Trimethylene glycol
        1,3-Dihydroxypropane
Mol. Form.: $C_3H_8O_2$

CAS RN: 504-63-2
Merck No.: 9629
Mol. Wt.: 76.095

PHYSICAL CONSTANTS

T_m: -26.7°C (246.4 K)
T_b: 214.4°C (487.5 K)

T_c:
P_c:

μ:
IP:

TRANSITION PROPERTIES

$\Delta_{fus}H$ (T_m):
$\Delta_{vap}H$ (T_b): 57.86 kJ/mol
$\Delta_{vap}H$ (25°C): 72.80 kJ/mol

Vapor pressure (0°C):
Vapor pressure (25°C):
Vapor pressure (100°C):

PROPERTIES AT 25°C AND 100 kPa

	Solid	Liquid	Gas		Liquid
$\Delta_f H°$/kJ mol⁻¹:		-464.9	-392.1		d: 1.050 g/mL
$S°$/J mol⁻¹K⁻¹:					η:
C_p/J mol⁻¹K⁻¹:					k:
COMMENTS:					

Name: 1,3-Propanedithiol
Synonym: Trimethylene dimercaptan

Mol. Form.: $C_3H_8S_2$

CAS RN: 109-80-8
Merck No.: 7811
Mol. Wt.: 108.229

PHYSICAL CONSTANTS

T_m: -79°C (194 K)
T_b: 172.9°C (446.0 K)

T_c:
P_c:

μ:
IP:

TRANSITION PROPERTIES

$\Delta_{fus}H$ (T_m):
$\Delta_{vap}H$ (T_b):
$\Delta_{vap}H$ (25°C): 49.66 kJ/mol

Vapor pressure (0°C):
Vapor pressure (25°C):
Vapor pressure (100°C): 11.3 kPa

PROPERTIES AT 25°C AND 100 kPa

	Solid	Liquid	Gas		Liquid
$\Delta_f H°$/kJ mol⁻¹:					d: 1.072 g/mL
$S°$/J mol⁻¹K⁻¹:					η:
C_p/J mol⁻¹K⁻¹:					k:
COMMENTS:					

Name: Propanenitrile   CAS RN: 107-12-0
Synonym: Ethyl cyanide   Merck No.: 7839
                                                                            Mol. Wt.: 55.079
Mol. Form.: C_3H_5N

PHYSICAL CONSTANTS

T_m: -92.89°C (180.26 K) T_c: 288.2°C (561.3 K)   μ: 4.05 D
T_b: 97.14°C (370.29 K) P_c: 4.26 MPa   IP: 11.84 eV

TRANSITION PROPERTIES

$\Delta_{fus}H$ (T_m): 5.05 kJ/mol   Vapor pressure (0°C): 1.66 kPa
$\Delta_{vap}H$ (T_b): 31.81 kJ/mol   Vapor pressure (25°C): 6.14 kPa
$\Delta_{vap}H$ (25°C): 36.03 kJ/mol   Vapor pressure (100°C):

PROPERTIES AT 25°C AND 100 kPa

	Solid	Liquid	Gas		Liquid
$\Delta_f H°$/kJ mol^{-1}:		15.5	51.5	d: 0.7768 g/mL	
$S°$/J mol^{-1}K^{-1}:				η: 0.294 mPa s	
C_p/J mol^{-1}K^{-1}:		119.3		k:	

COMMENTS: Highly toxic; flammable

Name: 2-Propanethiol   CAS RN: 75-33-2
Synonym: Isopropyl mercaptan   Merck No.:
                                                                            Mol. Wt.: 76.162
Mol. Form.: C_3H_8S

PHYSICAL CONSTANTS

T_m: -130.5°C (142.6 K) T_c: μ: 1.61 D
T_b: 52.6°C (325.7 K) P_c: IP: 9.14 eV

TRANSITION PROPERTIES

$\Delta_{fus}H$ (T_m):   Vapor pressure (0°C): 11.9 kPa
$\Delta_{vap}H$ (T_b): 27.91 kJ/mol   Vapor pressure (25°C): 36.9 kPa
$\Delta_{vap}H$ (25°C): 29.45 kJ/mol   Vapor pressure (100°C): 386 kPa

PROPERTIES AT 25°C AND 100 kPa

	Solid	Liquid	Gas		Liquid
$\Delta_f H°$/kJ mol^{-1}:		-105.9	-76.2	d: 0.810 g/mL	
$S°$/J mol^{-1}K^{-1}:		233.5		η: 0.357 mPa s	
C_p/J mol^{-1}K^{-1}:		145.3		k:	

COMMENTS:

Name: Propanoic acid   CAS RN: 79-09-4
Synonym: Propionic acid   Merck No.: 7837
                                                                            Mol. Wt.: 74.079
Mol. Form.: $C_3H_6O_2$

PHYSICAL CONSTANTS

T_m: -20.7°C (252.4 K) T_c: 331°C (604 K)   μ: 1.75 D
T_b: 141.15°C (414.30 K) P_c: 4.53 MPa   IP: 10.53 eV

TRANSITION PROPERTIES

$\Delta_{fus}H$ (T_m): 10.66 kJ/mol   Vapor pressure (0°C):
$\Delta_{vap}H$ (T_b): 32.28 kJ/mol   Vapor pressure (25°C): 0.451 kPa
$\Delta_{vap}H$ (25°C): 32.14 kJ/mol   Vapor pressure (100°C): 23.7 kPa

PROPERTIES AT 25°C AND 100 kPa

	Solid	Liquid	Gas		Liquid
$\Delta_f H°$/kJ mol^{-1}:		-510.7	-453.5	d: 0.9881 g/mL	
$S°$/J mol^{-1}K^{-1}:		191.0		η: 1.03 mPa s	
C_p/J mol^{-1}K^{-1}:		152.8		k:	

COMMENTS: TLV=10 ppm

Name: Propanoic anhydride
Synonym: Propionic anhydride

Mol. Form.: $C_6H_{10}O_3$

CAS RN: 123-62-6
Merck No.: 7838
Mol. Wt.: 130.144

PHYSICAL CONSTANTS

T_m: -45°C (228 K)	T_c:	μ:
T_b: 167.0°C (440.1 K)	P_c:	IP:

TRANSITION PROPERTIES

$\Delta_{fus}H$ (T_m):
$\Delta_{vap}H$ (T_b): 41.70 kJ/mol
$\Delta_{vap}H$ (25°C): 52.6 kJ/mol

Vapor pressure (0°C):
Vapor pressure (25°C):
Vapor pressure (100°C): 25.1 kPa

PROPERTIES AT 25°C AND 100 kPa

	Solid	Liquid	Gas		Liquid
$\Delta_fH°$/kJ mol^{-1}:		-679.1	-626.5		d: 1.007 g/mL
$S°$/J mol^{-1}K^{-1}:					η: 1.06 mPa s
C_p/J mol^{-1}K^{-1}:		235.0			k:
COMMENTS:					

Name: 1-Propanol
Synonym: Propyl alcohol

Mol. Form.: C_3H_8O

CAS RN: 71-23-8
Merck No.: 7854
Mol. Wt.: 60.096

PHYSICAL CONSTANTS

T_m: -126.1°C (147.0 K)	T_c: 263.63°C (536.78 K)	μ: 1.55 D
T_b: 97.2°C (370.3 K)	P_c: 5.168 MPa	IP: 10.22 eV

TRANSITION PROPERTIES

$\Delta_{fus}H$ (T_m): 5.20 kJ/mol
$\Delta_{vap}H$ (T_b): 41.44 kJ/mol
$\Delta_{vap}H$ (25°C): 47.45 kJ/mol

Vapor pressure (0°C):
Vapor pressure (25°C): 2.76 kPa
Vapor pressure (100°C): 113 kPa

PROPERTIES AT 25°C AND 100 kPa

	Solid	Liquid	Gas		Liquid
$\Delta_fH°$/kJ mol^{-1}:		-302.6	-255.1		d: 0.7996 g/mL
$S°$/J mol^{-1}K^{-1}:		193.6	322.6		η: 1.95 mPa s
C_p/J mol^{-1}K^{-1}:		143.9	85.6		k: 0.154 W/m K
COMMENTS: TLV=200 ppm; flammable					

Name: 2-Propanol
Synonym: Isopropyl alcohol

Mol. Form.: C_3H_8O

CAS RN: 67-63-0
Merck No.: 5096
Mol. Wt.: 60.096

PHYSICAL CONSTANTS

T_m: -89.5°C (183.6 K)	T_c: 235.2°C (508.3 K)	μ: 1.56 D
T_b: 82.3°C (355.4 K)	P_c: 4.762 MPa	IP: 10.12 eV

TRANSITION PROPERTIES

$\Delta_{fus}H$ (T_m): 5.37 kJ/mol
$\Delta_{vap}H$ (T_b): 39.85 kJ/mol
$\Delta_{vap}H$ (25°C): 45.39 kJ/mol

Vapor pressure (0°C): 1.11 kPa
Vapor pressure (25°C): 6.02 kPa
Vapor pressure (100°C): 198 kPa

PROPERTIES AT 25°C AND 100 kPa

	Solid	Liquid	Gas		Liquid
$\Delta_fH°$/kJ mol^{-1}:		-318.1	-272.8		d: 0.7813 g/mL
$S°$/J mol^{-1}K^{-1}:		181.1	309.2		η: 2.04 mPa s
C_p/J mol^{-1}K^{-1}:		156.5	89.3		k: 0.135 W/m K
COMMENTS: TLV=400 ppm; highly toxic; flammable					

Name: Propargyl alcohol
Synonyms: 2-Propyn-1-ol
       3-Hydroxy-1-propyne
Mol. Form.: C_3H_4O

CAS RN: 107-19-7
Merck No.: 7819
Mol. Wt.: 56.064

PHYSICAL CONSTANTS

T_m: -51.8°C (221.3 K)
T_b: 113.6°C (386.7 K)

T_c:
P_c:

μ: 1.13 D
IP: 10.51 eV

TRANSITION PROPERTIES

$\Delta_{fus}H\ (T_m)$:
$\Delta_{vap}H\ (T_b)$:
$\Delta_{vap}H\ (25°C)$:

Vapor pressure (0°C):
Vapor pressure (25°C): 2.07 kPa
Vapor pressure (100°C):

PROPERTIES AT 25°C AND 100 kPa

	Solid	Liquid	Gas		Liquid
$\Delta_f H°$/kJ mol^{-1}:				d:	0.9450 g/mL
$S°$/J mol^{-1}K^{-1}:				η:	
C_p/J mol^{-1}K^{-1}:				k:	

COMMENTS: TLV=1 ppm; flammable

Name: 2-Propenal
Synonyms: Acrolein
      Acrylaldehyde
Mol. Form.: C_3H_4O

CAS RN: 107-02-8
Merck No.: 122
Mol. Wt.: 56.064

PHYSICAL CONSTANTS

T_m: -87.7°C (185.4 K)
T_b: 52.69°C (325.84 K)

T_c:
P_c:

μ: 3.12 D
IP: 10.10 eV

TRANSITION PROPERTIES

$\Delta_{fus}H\ (T_m)$:
$\Delta_{vap}H\ (T_b)$: 28.33 kJ/mol
$\Delta_{vap}H\ (25°C)$:

Vapor pressure (0°C): 11.7 kPa
Vapor pressure (25°C): 36.2 kPa
Vapor pressure (100°C):

PROPERTIES AT 25°C AND 100 kPa

	Solid	Liquid	Gas		Liquid
$\Delta_f H°$/kJ mol^{-1}:				d:	0.837 g/mL
$S°$/J mol^{-1}K^{-1}:				η:	
C_p/J mol^{-1}K^{-1}:				k:	

COMMENTS: TLV=0.1 ppm; highly toxic; flammable

Name: Propenamide
Synonym: Acrylamide

Mol. Form.: C_3H_5NO

CAS RN: 79-06-1
Merck No.: 123
Mol. Wt.: 71.079

PHYSICAL CONSTANTS

T_m: 84.5°C (357.6 K)
T_b: 192.6°C (465.7 K)

T_c:
P_c:

μ:
IP: 9.50 eV

TRANSITION PROPERTIES

$\Delta_{fus}H\ (T_m)$:
$\Delta_{vap}H\ (T_b)$:
$\Delta_{vap}H\ (25°C)$:

Vapor pressure (0°C):
Vapor pressure (25°C):
Vapor pressure (100°C): 0.060 kPa

PROPERTIES AT 25°C AND 100 kPa

	Solid	Liquid	Gas		Solid
$\Delta_f H°$/kJ mol^{-1}:				d:	1.12 g/mL
$S°$/J mol^{-1}K^{-1}:				η:	N/A
C_p/J mol^{-1}K^{-1}:				k:	

COMMENTS: Carcinogen; highly toxic

Name: Propene
Synonyms: Propylene
            Refrigerant 1270
Mol. Form.: C_3H_6

CAS RN: 115-07-1
Merck No.: 7862
Mol. Wt.: 42.081

PHYSICAL CONSTANTS

T_m: -185.25°C (87.90 K) T_c: 91.70°C (364.85 K) μ: 0.366 D
T_b: -47.69°C (225.46 K) P_c: 4.601 MPa   IP: 9.73 eV

TRANSITION PROPERTIES

$\Delta_{fus}H$ (T_m): 3.00 kJ/mol   Vapor pressure (0°C):
$\Delta_{vap}H$ (T_b): 18.42 kJ/mol   Vapor pressure (25°C):
$\Delta_{vap}H$ (25°C): 14.24 kJ/mol   Vapor pressure (100°C): N/A

PROPERTIES AT 25°C AND 100 kPa

	Solid	Liquid	Gas		Gas
$\Delta_f H°$/kJ mol^{-1}:		1.7	20.0		d: 1.720 g/L
$S°$/J mol^{-1}K^{-1}:					η:
C_p/J mol^{-1}K^{-1}:					k:

COMMENTS: Very flammable

Name: Propenenitrile
Synonyms: Acrylonitrile
            Vinyl cyanide
Mol. Form.: C_3H_3N

CAS RN: 107-13-1
Merck No.: 125
Mol. Wt.: 53.064

PHYSICAL CONSTANTS

T_m: -83.5°C (189.6 K) T_c: μ: 3.87 D
T_b: 77.35°C (350.50 K) P_c: IP: 10.91 eV

TRANSITION PROPERTIES

$\Delta_{fus}H$ (T_m): 6.23 kJ/mol   Vapor pressure (0°C): 3.97 kPa
$\Delta_{vap}H$ (T_b): 32.55 kJ/mol   Vapor pressure (25°C): 14.1 kPa
$\Delta_{vap}H$ (25°C): 33.5 kJ/mol   Vapor pressure (100°C): 194 kPa

PROPERTIES AT 25°C AND 100 kPa

	Solid	Liquid	Gas		Liquid
$\Delta_f H°$/kJ mol^{-1}:		147.1	180.6		d: 0.8002 g/mL
$S°$/J mol^{-1}K^{-1}:					η: 0.342 mPa s
C_p/J mol^{-1}K^{-1}:		108.8			k:

COMMENTS: TLV=2 ppm; carcinogen; highly toxic; flammable

Name: 2-Propenoic acid
Synonym: Acrylic acid

Mol. Form.: $C_3H_4O_2$

CAS RN: 79-10-7
Merck No.: 124
Mol. Wt.: 72.064

PHYSICAL CONSTANTS

T_m: 12.3°C (285.4 K) T_c: μ:
T_b: 141°C (414 K) P_c: IP: 10.60 eV

TRANSITION PROPERTIES

$\Delta_{fus}H$ (T_m): 11.16 kJ/mol   Vapor pressure (0°C):
$\Delta_{vap}H$ (T_b):   Vapor pressure (25°C):
$\Delta_{vap}H$ (25°C):   Vapor pressure (100°C): 25.2 kPa

PROPERTIES AT 25°C AND 100 kPa

	Solid	Liquid	Gas		Liquid
$\Delta_f H°$/kJ mol^{-1}:		-383.8			d: 1.046 g/mL
$S°$/J mol^{-1}K^{-1}:					η:
C_p/J mol^{-1}K^{-1}:		145.7			k:

COMMENTS: TLV=2 ppm; highly toxic

Name: Propyl acetate   CAS RN: 109-60-4
Synonym: Propyl ethanoate   Merck No.: 7853
                                                                              Mol. Wt.: 102.133
Mol. Form.: $C_5H_{10}O_2$

PHYSICAL CONSTANTS

T_m: -93°C (180 K) T_c: 276.6°C (549.7 K)   μ:
T_b: 101.54°C (374.69 K) P_c: 3.36 MPa   IP: 10.04 eV

TRANSITION PROPERTIES

$\Delta_{fus}H$ (T_m):   Vapor pressure (0°C):
$\Delta_{vap}H$ (T_b): 33.92 kJ/mol   Vapor pressure (25°C): 4.49 kPa
$\Delta_{vap}H$ (25°C): 39.72 kJ/mol   Vapor pressure (100°C): 96.6 kPa

PROPERTIES AT 25°C AND 100 kPa

	Solid	Liquid	Gas	Liquid
$\Delta_f H°$/kJ mol^{-1}:				d: 0.8830 g/mL
$S°$/J mol^{-1}K^{-1}:				η: 0.544 mPa s
C_p/J mol^{-1}K^{-1}:		196.2		k:

COMMENTS: TLV=200 ppm; flammable

Name: Propylamine   CAS RN: 107-10-8
Synonyms: 1-Propanamine   Merck No.: 7855
          1-Aminopropane   Mol. Wt.: 59.111
Mol. Form.: C_3H_9N

PHYSICAL CONSTANTS

T_m: -83°C (190 K) T_c: 223.9°C (497.0 K)   μ: 1.17 D
T_b: 47.22°C (320.37 K) P_c: 4.72 MPa   IP: 8.78 eV

TRANSITION PROPERTIES

$\Delta_{fus}H$ (T_m): 10.97 kJ/mol   Vapor pressure (0°C): 12.4 kPa
$\Delta_{vap}H$ (T_b): 29.55 kJ/mol   Vapor pressure (25°C): 42.1 kPa
$\Delta_{vap}H$ (25°C): 31.27 kJ/mol   Vapor pressure (100°C):

PROPERTIES AT 25°C AND 100 kPa

	Solid	Liquid	Gas	Liquid
$\Delta_f H°$/kJ mol^{-1}:		-101.5	-70.1	d: 0.7121 g/mL
$S°$/J mol^{-1}K^{-1}:		227.4	325.4	η: 0.376 mPa s
C_p/J mol^{-1}K^{-1}:		162.5	91.2	k:

COMMENTS: Highly toxic; flammable

Name: Propylbenzene   CAS RN: 103-65-1
Synonym: Isocumene   Merck No.: 7856
                                                                              Mol. Wt.: 120.194
Mol. Form.: C_9H_{12}

PHYSICAL CONSTANTS

T_m: -99.56°C (173.59 K) T_c: 365.17°C (638.32 K)   μ:
T_b: 159.24°C (432.39 K) P_c: 3.200 MPa   IP: 8.72 eV

TRANSITION PROPERTIES

$\Delta_{fus}H$ (T_m): 9.27 kJ/mol   Vapor pressure (0°C):
$\Delta_{vap}H$ (T_b):   Vapor pressure (25°C):
$\Delta_{vap}H$ (25°C): 46.22 kJ/mol   Vapor pressure (100°C): 16.6 kPa

PROPERTIES AT 25°C AND 100 kPa

	Solid	Liquid	Gas	Liquid
$\Delta_f H°$/kJ mol^{-1}:		-38.3	7.9	d: 0.8579 g/mL
$S°$/J mol^{-1}K^{-1}:		287.8		η:
C_p/J mol^{-1}K^{-1}:		214.7		k:

COMMENTS: Flammable

Name: Propyl butanoate
Synonym: Propyl butyrate

Mol. Form.: $C_7H_{14}O_2$

CAS RN: 105-66-8
Merck No.: 7858
Mol. Wt.: 130.187

PHYSICAL CONSTANTS

T_m: -95.2°C (177.9 K)
T_b: 143.05°C (416.20 K)

T_c: 327°C (600 K)
P_c:

μ:
IP:

TRANSITION PROPERTIES

$\Delta_{fus}H$ (T_m):
$\Delta_{vap}H$ (T_b):
$\Delta_{vap}H$ (25°C):

Vapor pressure (0°C):
Vapor pressure (25°C): 0.617 kPa
Vapor pressure (100°C): 27.0 kPa

PROPERTIES AT 25°C AND 100 kPa

	Solid	Liquid	Gas		Liquid
$\Delta_f H°$/kJ mol^{-1}:					d: 0.869 g/mL
$S°$/J mol^{-1}K^{-1}:					η:
C_p/J mol^{-1}K^{-1}:					k:
COMMENTS:					

Name: Propylcyclohexane

CAS RN: 1678-92-8
Merck No.:
Mol. Wt.: 126.242

Mol. Form.: C_9H_{18}

PHYSICAL CONSTANTS

T_m: -94.9°C (178.2 K)
T_b: 156.7°C (429.8 K)

T_c:
P_c:

μ:
IP: 9.46 eV

TRANSITION PROPERTIES

$\Delta_{fus}H$ (T_m): 10.37 kJ/mol
$\Delta_{vap}H$ (T_b):
$\Delta_{vap}H$ (25°C): 45.08 kJ/mol

Vapor pressure (0°C):
Vapor pressure (25°C):
Vapor pressure (100°C): 18.5 kPa

PROPERTIES AT 25°C AND 100 kPa

	Solid	Liquid	Gas		Liquid
$\Delta_f H°$/kJ mol^{-1}:		-237.4	-192.5		d: 0.790 g/mL
$S°$/J mol^{-1}K^{-1}:		311.9			η:
C_p/J mol^{-1}K^{-1}:		242.0			k:
COMMENTS:					

Name: Propyl formate
Synonym: Propyl methanoate

CAS RN: 110-74-7
Merck No.: 7871
Mol. Wt.: 88.106

Mol. Form.: $C_4H_8O_2$

PHYSICAL CONSTANTS

T_m: -92.9°C (180.2 K)
T_b: 80.9°C (354.0 K)

T_c: 264.9°C (538.0 K)
P_c: 4.06 MPa

μ:
IP: 10.52 eV

TRANSITION PROPERTIES

$\Delta_{fus}H$ (T_m):
$\Delta_{vap}H$ (T_b): 33.61 kJ/mol
$\Delta_{vap}H$ (25°C): 37.53 kJ/mol

Vapor pressure (0°C): 2.76 kPa
Vapor pressure (25°C): 11.0 kPa
Vapor pressure (100°C): 180 kPa

PROPERTIES AT 25°C AND 100 kPa

	Solid	Liquid	Gas		Liquid
$\Delta_f H°$/kJ mol^{-1}:		-500.3			d: 0.8996 g/mL
$S°$/J mol^{-1}K^{-1}:					η: 0.485 mPa s
C_p/J mol^{-1}K^{-1}:					k:
COMMENTS: Flammable					

Name: Propyl propanoate
Synonym: Propyl propionate

Mol. Form.: $C_6H_{12}O_2$

CAS RN: 106-36-5
Merck No.: 7880
Mol. Wt.: 116.160

PHYSICAL CONSTANTS

T_m: -75.9°C (197.2 K)
T_b: 122.5°C (395.6 K)

T_c: 305°C (578 K)
P_c:

μ:
IP:

TRANSITION PROPERTIES

$\Delta_{fus}H$ (T_m):
$\Delta_{vap}H$ (T_b): 35.54 kJ/mol
$\Delta_{vap}H$ (25°C): 43.45 kJ/mol

Vapor pressure (0°C):
Vapor pressure (25°C): 1.88 kPa
Vapor pressure (100°C): 49.8 kPa

PROPERTIES AT 25°C AND 100 kPa

	Solid	Liquid	Gas		Liquid
$\Delta_fH°$/kJ mol^{-1}:					d: 0.878 g/mL
$S°$/J mol^{-1}K^{-1}:					η:
C_p/J mol^{-1}K^{-1}:					k:
COMMENTS:					

Name: Propyne
Synonym: Methylacetylene

Mol. Form.: C_3H_4

CAS RN: 74-99-7
Merck No.:
Mol. Wt.: 40.065

PHYSICAL CONSTANTS

T_m: -102.7°C (170.4 K)
T_b: -23.2°C (249.9 K)

T_c: 129.23°C (402.38 K)
P_c: 5.628 MPa

μ: 0.784 D
IP: 10.36 eV

TRANSITION PROPERTIES

$\Delta_{fus}H$ (T_m):
$\Delta_{vap}H$ (T_b):
$\Delta_{vap}H$ (25°C):

Vapor pressure (0°C):
Vapor pressure (25°C):
Vapor pressure (100°C):

PROPERTIES AT 25°C AND 100 kPa

	Solid	Liquid	Gas		Gas
$\Delta_fH°$/kJ mol^{-1}:			184.9		d: 1.638 g/L
$S°$/J mol^{-1}K^{-1}:					η:
C_p/J mol^{-1}K^{-1}:					k:
COMMENTS: TLV=1000 ppm; very flammable					

Name: Pyrazine
Synonyms: 1,4-Diazine
            Paradiazine
Mol. Form.: $C_4H_4N_2$

CAS RN: 290-37-9
Merck No.: 7971
Mol. Wt.: 80.089

PHYSICAL CONSTANTS

T_m: 55°C (328 K)
T_b: 115°C (388 K)

T_c:
P_c:

μ:
IP:

TRANSITION PROPERTIES

$\Delta_{fus}H$ (T_m):
$\Delta_{vap}H$ (T_b):
$\Delta_{vap}H$ (25°C):

Vapor pressure (0°C):
Vapor pressure (25°C):
Vapor pressure (100°C):

PROPERTIES AT 25°C AND 100 kPa

	Solid	Liquid	Gas		Solid
$\Delta_fH°$/kJ mol^{-1}:	139.8		196.1		d: 1.27 g/mL
$S°$/J mol^{-1}K^{-1}:					η: N/A
C_p/J mol^{-1}K^{-1}:					k:
COMMENTS:					

Name: Pyrazole

Mol. Form.: $C_3H_4N_2$

CAS RN: 288-13-1
Merck No.: 7974
Mol. Wt.: 68.078

PHYSICAL CONSTANTS

T_m: 68°C (341 K)
T_b: 187°C (460 K)

T_c:
P_c:

μ:
IP:

TRANSITION PROPERTIES

$\Delta_{fus}H$ (T_m):
$\Delta_{vap}H$ (T_b):
$\Delta_{vap}H$ (25°C):

Vapor pressure (0°C):
Vapor pressure (25°C):
Vapor pressure (100°C):

PROPERTIES AT 25°C AND 100 kPa

	Solid	Liquid	Gas		Solid
$\Delta_fH°$/kJ mol^{-1}:	116.0			d:	
$S°$/J mol^{-1}K^{-1}:				η: N/A	
C_p/J mol^{-1}K^{-1}:				k:	
COMMENTS:					

Name: Pyrene
Synonym: Benzo[def]phenanthrene

Mol. Form.: $C_{16}H_{10}$

CAS RN: 129-00-0
Merck No.: 7977
Mol. Wt.: 202.255

PHYSICAL CONSTANTS

T_m: 151.2°C (424.3 K)
T_b: 404°C (677 K)

T_c:
P_c:

μ:
IP: 7.41 eV

TRANSITION PROPERTIES

$\Delta_{fus}H$ (T_m): 17.11 kJ/mol
$\Delta_{vap}H$ (T_b):
$\Delta_{vap}H$ (25°C):

Vapor pressure (0°C):
Vapor pressure (25°C):
Vapor pressure (100°C):

PROPERTIES AT 25°C AND 100 kPa

	Solid	Liquid	Gas		Solid
$\Delta_fH°$/kJ mol^{-1}:	125.5		225.7	d: 1.268 g/mL	
$S°$/J mol^{-1}K^{-1}:	224.9			η: N/A	
C_p/J mol^{-1}K^{-1}:	229.7			k:	
COMMENTS: Highly toxic					

Name: Pyridazine
Synonyms: 1,2-Diazabenzene
          1,2-Diazine
Mol. Form.: $C_4H_4N_2$

CAS RN: 289-80-5
Merck No.: 7982
Mol. Wt.: 80.089

PHYSICAL CONSTANTS

T_m: -8°C (265 K)
T_b: 208°C (481 K)

T_c:
P_c:

μ: 4.22 D
IP: 8.64 eV

TRANSITION PROPERTIES

$\Delta_{fus}H$ (T_m):
$\Delta_{vap}H$ (T_b):
$\Delta_{vap}H$ (25°C): 53.47 kJ/mol

Vapor pressure (0°C):
Vapor pressure (25°C):
Vapor pressure (100°C):

PROPERTIES AT 25°C AND 100 kPa

	Solid	Liquid	Gas		Liquid
$\Delta_fH°$/kJ mol^{-1}:		224.8	278.3	d: 1.102 g/mL	
$S°$/J mol^{-1}K^{-1}:				η:	
C_p/J mol^{-1}K^{-1}:				k:	
COMMENTS:					

Name: Pyridine
Synonyms: Azine
       Azabenzene
Mol. Form.: C_5H_5N

CAS RN: 110-86-1
Merck No.: 7983
Mol. Wt.: 79.101

PHYSICAL CONSTANTS

T_m: -41.66°C (231.49 K)
T_b: 115.23°C (388.38 K)

T_c: 346.9°C (620.0 K)
P_c: 5.67 MPa

μ: 2.215 D
IP: 9.25 eV

TRANSITION PROPERTIES

$\Delta_{fus}H$ (T_m): 8.28 kJ/mol
$\Delta_{vap}H$ (T_b): 35.09 kJ/mol
$\Delta_{vap}H$ (25°C): 40.21 kJ/mol

Vapor pressure (0°C):
Vapor pressure (25°C): 2.76 kPa
Vapor pressure (100°C): 63.7 kPa

PROPERTIES AT 25°C AND 100 kPa

	Solid	Liquid	Gas		Liquid
$\Delta_f H°$/kJ mol^{-1}:		100.2	140.4		d: 0.9786 g/mL
$S°$/J mol^{-1}K^{-1}:					η: 0.879 mPa s
C_p/J mol^{-1}K^{-1}:		132.7			k: 0.165 W/m K

COMMENTS: TLV=5 ppm; highly toxic; flammable

Name: Pyrimidine

CAS RN: 289-95-2
Merck No.: 7998
Mol. Wt.: 80.089

Mol. Form.: $C_4H_4N_2$

PHYSICAL CONSTANTS

T_m: 22°C (295 K)
T_b: 123.8°C (396.9 K)

T_c:
P_c:

μ: 2.334 D
IP: 9.23 eV

TRANSITION PROPERTIES

$\Delta_{fus}H$ (T_m):
$\Delta_{vap}H$ (T_b): 43.09 kJ/mol
$\Delta_{vap}H$ (25°C): 49.79 kJ/mol

Vapor pressure (0°C):
Vapor pressure (25°C):
Vapor pressure (100°C):

PROPERTIES AT 25°C AND 100 kPa

	Solid	Liquid	Gas		Liquid
$\Delta_f H°$/kJ mol^{-1}:		145.9	195.9		d:
$S°$/J mol^{-1}K^{-1}:					η:
C_p/J mol^{-1}K^{-1}:					k:

COMMENTS:

Name: 1H-Pyrrole
Synonyms: Azole
      Imidole
Mol. Form.: C_4H_5N

CAS RN: 109-97-7
Merck No.: 8025
Mol. Wt.: 67.090

PHYSICAL CONSTANTS

T_m: -23.42°C (249.73 K)
T_b: 129.79°C (402.94 K)

T_c: 366.6°C (639.7 K)
P_c: 6.34 MPa

μ: 1.74 D
IP: 8.21 eV

TRANSITION PROPERTIES

$\Delta_{fus}H$ (T_m): 7.91 kJ/mol
$\Delta_{vap}H$ (T_b): 38.75 kJ/mol
$\Delta_{vap}H$ (25°C): 45.09 kJ/mol

Vapor pressure (0°C):
Vapor pressure (25°C): 1.10 kPa
Vapor pressure (100°C): 38.2 kPa

PROPERTIES AT 25°C AND 100 kPa

	Solid	Liquid	Gas		Liquid
$\Delta_f H°$/kJ mol^{-1}:		63.1	108.2		d: 0.9656 g/mL
$S°$/J mol^{-1}K^{-1}:		156.4			η: 1.23 mPa s
C_p/J mol^{-1}K^{-1}:		127.7			k:

COMMENTS:

Name: Pyrrolidine

Synonyms: Azacyclopentane

Tetrahydropyrrole

Mol. Form.: C_4H_9N

CAS RN: 123-75-1

Merck No.: 8026

Mol. Wt.: 71.122

PHYSICAL CONSTANTS

T_m: -57.84°C (215.31 K)

T_b: 86.56°C (359.71 K)

T_c: 295.1°C (568.2 K)

P_c: 5.59 MPa

μ:

IP: 8.00 eV

TRANSITION PROPERTIES

$\Delta_{fus}H$ (T_m): 8.58 kJ/mol

$\Delta_{vap}H$ (T_b): 33.01 kJ/mol

$\Delta_{vap}H$ (25°C): 37.52 kJ/mol

Vapor pressure (0°C):

Vapor pressure (25°C): 8.40 kPa

Vapor pressure (100°C): 153 kPa

PROPERTIES AT 25° AND 100 kPa

	Solid	Liquid	Gas		Liquid
$\Delta_f H°$/kJ mol^{-1}:		-41.0	-3.6		d: 0.8538 g/mL
$S°$/J mol^{-1}K^{-1}:		204.1			η: 0.704 mPa s
C_p/J mol^{-1}K^{-1}:		156.6			k:
COMMENTS: Flammable					

Name: 2-Pyrrolidone

Synonyms: 2-Pyrrolidinone

γ-Butyrolactam

Mol. Form.: C_4H_7NO

CAS RN: 616-45-5

Merck No.: 8027

Mol. Wt.: 85.106

PHYSICAL CONSTANTS

T_m: 25°C (298 K)

T_b: 245°C (518 K)

T_c:

P_c:

μ:

IP: 9.20 eV

TRANSITION PROPERTIES

$\Delta_{fus}H$ (T_m):

$\Delta_{vap}H$ (T_b):

$\Delta_{vap}H$ (25°C):

Vapor pressure (0°C):

Vapor pressure (25°C):

Vapor pressure (100°C):

PROPERTIES AT 25° AND 100 kPa

	Solid	Liquid	Gas		Liquid
$\Delta_f H°$/kJ mol^{-1}:		-286.2			d: 1.107 g/mL
$S°$/J mol^{-1}K^{-1}:					η: 13.3 mPa s
C_p/J mol^{-1}K^{-1}:					k:
COMMENTS:					

Name: Quinoline

Synonyms: 1-Azanaphthalene

Benzopyridine

Mol. Form.: C_9H_7N

CAS RN: 91-22-5

Merck No.: 8097

Mol. Wt.: 129.161

PHYSICAL CONSTANTS

T_m: -14.78°C (258.37 K)

T_b: 237.16°C (510.31 K)

T_c: 509°C (782 K)

P_c: 4.86 MPa

μ: 2.29 D

IP: 8.62 eV

TRANSITION PROPERTIES

$\Delta_{fus}H$ (T_m): 10.66 kJ/mol

$\Delta_{vap}H$ (T_b): 49.71 kJ/mol

$\Delta_{vap}H$ (25°C): 53.90 kJ/mol

Vapor pressure (0°C):

Vapor pressure (25°C): 0.011 kPa

Vapor pressure (100°C):

PROPERTIES AT 25° AND 100 kPa

	Solid	Liquid	Gas		Liquid
$\Delta_f H°$/kJ mol^{-1}:		141.2	200.5		d: 1.090 g/mL
$S°$/J mol^{-1}K^{-1}:					η: 3.34 mPa s
C_p/J mol^{-1}K^{-1}:		200.0			k:
COMMENTS: Highly toxic					

Name: Radon

CAS RN: 10043-92-2
Merck No.: 8119
Mol. Wt.: 222.017

Mol. Form.: Rn

PHYSICAL CONSTANTS

T_m: -71°C (202 K) T_c: 104°C (377 K)   μ: 0 D
T_b: -61.7°C (211.4 K) P_c: 6.28 MPa   IP: 10.75 eV

TRANSITION PROPERTIES

$\Delta_{fus}H$ (T_m):   Vapor pressure (0°C):
$\Delta_{vap}H$ (T_b):   Vapor pressure (25°C):
$\Delta_{vap}H$ (25°C):   Vapor pressure (100°C):

PROPERTIES AT 25° AND 100 kPa

	Solid	Liquid	Gas	Gas
$\Delta_f H°$/kJ mol^{-1}:			0.0	d: 9.075 g/L
$S°$/J mol^{-1}K^{-1}:			176.2	η:
C_p/J mol^{-1}K^{-1}:			20.8	k: 0.0036 W/m K
COMMENTS:				

Name: Rhenium

CAS RN: 7440-15-5
Merck No.: 8176
Mol. Wt.: 186.207

Mol. Form.: Re

PHYSICAL CONSTANTS

T_m: 3186°C (3459 K) T_c: μ:
T_b: 5596°C (5869 K) P_c: IP: 7.88 eV

TRANSITION PROPERTIES

$\Delta_{fus}H$ (T_m): 60.43 kJ/mol   Vapor pressure (0°C): N/A
$\Delta_{vap}H$ (T_b):   Vapor pressure (25°C): N/A
$\Delta_{vap}H$ (25°C):   Vapor pressure (100°C): N/A

PROPERTIES AT 25° AND 100 kPa

	Solid	Liquid	Gas	Solid
$\Delta_f H°$/kJ mol^{-1}:	0.0		769.9	d: 20.8 g/mL
$S°$/J mol^{-1}K^{-1}:	36.9		188.9	η: N/A
C_p/J mol^{-1}K^{-1}:	25.5		20.8	k: 48.0 W/m K
COMMENTS:				

Name: Rhenium oxide (Re_2O_7)
Synonyms: Dirhenium heptoxide
          Rhenium(VII) oxide
Mol. Form.: O_7Re_2

CAS RN: 1314-68-7
Merck No.: 8177
Mol. Wt.: 484.410

PHYSICAL CONSTANTS

T_m: 297°C (570 K) T_c: 669°C (942 K)   μ:
T_b: 450°C (723 K) P_c: IP: 12.70 eV

TRANSITION PROPERTIES

$\Delta_{fus}H$ (T_m): 64.20 kJ/mol   Vapor pressure (0°C): N/A
$\Delta_{vap}H$ (T_b):   Vapor pressure (25°C): N/A
$\Delta_{vap}H$ (25°C):   Vapor pressure (100°C): N/A

PROPERTIES AT 25° AND 100 kPa

	Solid	Liquid	Gas	Solid
$\Delta_f H°$/kJ mol^{-1}:	-1240.1		-1100.0	d: 6.103 g/mL
$S°$/J mol^{-1}K^{-1}:	207.1		452.0	η: N/A
C_p/J mol^{-1}K^{-1}:	166.1			k:
COMMENTS:				

Name: Rhodium

CAS RN: 7440-16-6
Merck No.: 8187
Mol. Wt.: 102.906

Mol. Form.: Rh

PHYSICAL CONSTANTS

T_m: 1964°C (2237 K)	T_c:	μ:
T_b: 3695°C (3968 K)	P_c:	IP: 7.46 eV

TRANSITION PROPERTIES

$\Delta_{fus}H$ (T_m): 26.59 kJ/mol

$\Delta_{vap}H$ (T_b):

$\Delta_{vap}H$ (25°C):

Vapor pressure (0°C): N/A
Vapor pressure (25°C): N/A
Vapor pressure (100°C): N/A

PROPERTIES AT 25° AND 100 kPa

	Solid	Liquid	Gas		Solid
$\Delta_f H°$/kJ mol⁻¹:	0.0		556.9	d:	12.4 g/mL
$S°$/J mol⁻¹K⁻¹:	31.5		185.8	η:	N/A
C_p/J mol⁻¹K⁻¹:	25.0		21.0	k:	150 W/m K
COMMENTS:					

Name: Rubidium

CAS RN: 7440-17-7
Merck No.: 8259
Mol. Wt.: 85.468

Mol. Form.: Rb

PHYSICAL CONSTANTS

T_m: 39.31°C (312.46 K)	T_c:	μ:
T_b: 688°C (961 K)	P_c:	IP: 4.18 eV

TRANSITION PROPERTIES

$\Delta_{fus}H$ (T_m): 2.19 kJ/mol

$\Delta_{vap}H$ (T_b):

$\Delta_{vap}H$ (25°C):

Vapor pressure (0°C):
Vapor pressure (25°C):
Vapor pressure (100°C):

PROPERTIES AT 25° AND 100 kPa

	Solid	Liquid	Gas		Solid
$\Delta_f H°$/kJ mol⁻¹:	0.0		80.9	d:	1.53 g/mL
$S°$/J mol⁻¹K⁻¹:	76.8		170.1	η:	N/A
C_p/J mol⁻¹K⁻¹:	31.1		20.8	k:	58.2 W/m K
COMMENTS: Flammable					

Name: Rubidium chloride

CAS RN: 7791-11-9
Merck No.: 8261
Mol. Wt.: 120.921

Mol. Form.: ClRb

PHYSICAL CONSTANTS

T_m: 715°C (988 K)	T_c:	μ: 10.510 D
T_b: 1390°C (1663 K)	P_c:	IP: 8.50 eV

TRANSITION PROPERTIES

$\Delta_{fus}H$ (T_m): 18.40 kJ/mol

$\Delta_{vap}H$ (T_b):

$\Delta_{vap}H$ (25°C):

Vapor pressure (0°C): N/A
Vapor pressure (25°C): N/A
Vapor pressure (100°C): N/A

PROPERTIES AT 25° AND 100 kPa

	Solid	Liquid	Gas		Solid
$\Delta_f H°$/kJ mol⁻¹:	-435.4			d:	2.76 g/mL
$S°$/J mol⁻¹K⁻¹:	95.9			η:	N/A
C_p/J mol⁻¹K⁻¹:	52.4			k:	
COMMENTS:					

Name: Ruthenium

CAS RN: 7440-18-8
Merck No.:
Mol. Wt.: 101.070

Mol. Form.: Ru

PHYSICAL CONSTANTS

T_m: 2334°C (2607 K)	T_c:	μ:
T_b: 4150°C (4423 K)	P_c:	IP: 7.36 eV

TRANSITION PROPERTIES

$\Delta_{fus}H$ (T_m): 38.59 kJ/mol   Vapor pressure (0°C): N/A
$\Delta_{vap}H$ (T_b):   Vapor pressure (25°C): N/A
$\Delta_{vap}H$ (25°C):   Vapor pressure (100°C): N/A

PROPERTIES AT 25° AND 100 kPa

	Solid	Liquid	Gas		Solid
$\Delta_f H°$/kJ mol⁻¹:	0.0		642.7		d: 12.1 g/mL
$S°$/J mol⁻¹K⁻¹:	28.5		186.5		η: N/A
C_p/J mol⁻¹K⁻¹:	24.1		21.5		k: 117 W/m K
COMMENTS:					

Name: Salicylaldehyde
Synonym: 2-Hydroxybenzaldehyde

CAS RN: 90-02-8
Merck No.: 8295
Mol. Wt.: 122.123

Mol. Form.: $C_7H_6O_2$

PHYSICAL CONSTANTS

T_m: -7°C (266 K)	T_c:	μ:
T_b: 197°C (470 K)	P_c:	IP:

TRANSITION PROPERTIES

$\Delta_{fus}H$ (T_m):   Vapor pressure (0°C):
$\Delta_{vap}H$ (T_b): 38.24 kJ/mol   Vapor pressure (25°C):
$\Delta_{vap}H$ (25°C):   Vapor pressure (100°C): 4.39 kPa

PROPERTIES AT 25° AND 100 kPa

	Solid	Liquid	Gas		Liquid
$\Delta_f H°$/kJ mol⁻¹:					d: 1.162 g/mL
$S°$/J mol⁻¹K⁻¹:					η:
C_p/J mol⁻¹K⁻¹:					k:
COMMENTS:					

Name: Salicylic acid
Synonym: 2-Hydroxybenzoic acid

CAS RN: 69-72-7
Merck No.: 8301
Mol. Wt.: 138.123

Mol. Form.: $C_7H_6O_3$

PHYSICAL CONSTANTS

T_m: 158°C (431 K)	T_c:	μ:
T_b:	P_c:	IP:

TRANSITION PROPERTIES

$\Delta_{fus}H$ (T_m):   Vapor pressure (0°C):
$\Delta_{vap}H$ (T_b):   Vapor pressure (25°C):
$\Delta_{vap}H$ (25°C):   Vapor pressure (100°C):

PROPERTIES AT 25° AND 100 kPa

	Solid	Liquid	Gas		Solid
$\Delta_f H°$/kJ mol⁻¹:	-589.9		-494.8		d:
$S°$/J mol⁻¹K⁻¹:					η: N/A
C_p/J mol⁻¹K⁻¹:					k:
COMMENTS:					

Name: Scandium

CAS RN: 7440-20-2
Merck No.: 8348
Mol. Wt.: 44.956

Mol. Form.: Sc

PHYSICAL CONSTANTS

T_m: 1541°C (1814 K) T_c: μ:
T_b: 2830°C (3103 K) P_c: IP: 6.56 eV

TRANSITION PROPERTIES

$\Delta_{fus}H$ (T_m): 14.10 kJ/mol   Vapor pressure (0°C): N/A
$\Delta_{vap}H$ (T_b):   Vapor pressure (25°C): N/A
$\Delta_{vap}H$ (25°C):   Vapor pressure (100°C): N/A

PROPERTIES AT 25° AND 100 kPa

	Solid	Liquid	Gas		Solid
$\Delta_f H°$/kJ mol^{-1}:	0.0		377.8	d: 2.99 g/mL
$S°$/J mol^{-1}K^{-1}:	34.6		174.8	η: N/A
C_p/J mol^{-1}K^{-1}:	25.5		22.1	k: 15.8 W/m K
COMMENTS:				

Name: Selenium

CAS RN: 7782-49-2
Merck No.: 8382
Mol. Wt.: 78.960

Mol. Form.: Se

PHYSICAL CONSTANTS

T_m: 221°C (494 K) T_c: 1493°C (1766 K) μ:
T_b: 685°C (958 K) P_c: 27.2 MPa   IP: 9.75 eV

TRANSITION PROPERTIES

$\Delta_{fus}H$ (T_m): 6.69 kJ/mol   Vapor pressure (0°C):
$\Delta_{vap}H$ (T_b): 95.48 kJ/mol   Vapor pressure (25°C):
$\Delta_{vap}H$ (25°C):   Vapor pressure (100°C):

PROPERTIES AT 25° AND 100 kPa

	Solid	Liquid	Gas		Solid
$\Delta_f H°$/kJ mol^{-1}:	0.0		227.1	d: 4.79 g/mL
$S°$/J mol^{-1}K^{-1}:	42.4		176.7	η: N/A
C_p/J mol^{-1}K^{-1}:	25.4		20.8	k: 4.52 W/m K
COMMENTS: Highly toxic				

Name: Selenium dioxide
Synonym: Selenium(IV) oxide

CAS RN: 7446-08-4
Merck No.: 8386
Mol. Wt.: 110.959

Mol. Form.: O$_2$Se

PHYSICAL CONSTANTS

T_m: 340°C (613 K) T_c: μ: 2.62 D
T_b: P_c: IP:

TRANSITION PROPERTIES

$\Delta_{fus}H$ (T_m):   Vapor pressure (0°C): N/A
$\Delta_{vap}H$ (T_b):   Vapor pressure (25°C): N/A
$\Delta_{vap}H$ (25°C):   Vapor pressure (100°C): N/A

PROPERTIES AT 25° AND 100 kPa

	Solid	Liquid	Gas		Solid
$\Delta_f H°$/kJ mol^{-1}:	-225.4			d: 3.9 g/mL
$S°$/J mol^{-1}K^{-1}:				η: N/A
C_p/J mol^{-1}K^{-1}:				k:
COMMENTS: Highly toxic				

Name: Selenium fluoride (SeF$_6$)   CAS RN: 7783-79-1
Synonym: Selenium hexafluoride   Merck No.: 8385
                                                                               Mol. Wt.: 192.950

Mol. Form.: F$_6$Se

PHYSICAL CONSTANTS

T_m: T_c: 72.4°C (345.5 K)   μ:
T_b: -46°C (227 K) P_c: IP:

TRANSITION PROPERTIES

$\Delta_{fus}H$ (T_m):   Vapor pressure (0°C):
$\Delta_{vap}H$ (T_b):   Vapor pressure (25°C):
$\Delta_{vap}H$ (25°C):   Vapor pressure (100°C): N/A

PROPERTIES AT 25° AND 100 kPa

	Solid	Liquid	Gas		Gas
$\Delta_f H°$/kJ mol^{-1}:			-1117.0	d:	7.887 g/L
$S°$/J mol^{-1}K^{-1}:			313.9	η:	
C_p/J mol^{-1}K^{-1}:			110.5	k:	

COMMENTS: TLV=0.05 ppm

Name: Silane   CAS RN: 7803-62-5
Synonym: Silicon tetrahydride   Merck No.: 8435
                                                                               Mol. Wt.: 32.117

Mol. Form.: H$_4$Si

PHYSICAL CONSTANTS

T_m: -185°C (88 K) T_c: μ: 0 D
T_b: -112°C (161 K) P_c: IP: 11.65 eV

TRANSITION PROPERTIES

$\Delta_{fus}H$ (T_m):   Vapor pressure (0°C):
$\Delta_{vap}H$ (T_b): 12.10 kJ/mol   Vapor pressure (25°C):
$\Delta_{vap}H$ (25°C):   Vapor pressure (100°C):

PROPERTIES AT 25° AND 100 kPa

	Solid	Liquid	Gas		Gas
$\Delta_f H°$/kJ mol^{-1}:			34.3	d:	1.313 g/L
$S°$/J mol^{-1}K^{-1}:			204.6	η:	
C_p/J mol^{-1}K^{-1}:			42.8	k:	

COMMENTS: TLV=5 ppm; flammable

Name: Silicon   CAS RN: 7440-21-3
                                                                               Merck No.: 8438
                                                                               Mol. Wt.: 28.086

Mol. Form.: Si

PHYSICAL CONSTANTS

T_m: 1414°C (1687 K) T_c: μ:
T_b: 3265°C (3538 K) P_c: IP: 8.15 eV

TRANSITION PROPERTIES

$\Delta_{fus}H$ (T_m): 50.21 kJ/mol   Vapor pressure (0°C): N/A
$\Delta_{vap}H$ (T_b):   Vapor pressure (25°C): N/A
$\Delta_{vap}H$ (25°C):   Vapor pressure (100°C): N/A

PROPERTIES AT 25° AND 100 kPa

	Solid	Liquid	Gas		Solid
$\Delta_f H°$/kJ mol^{-1}:	0.0		450.0	d:	2.3290 g/mL
$S°$/J mol^{-1}K^{-1}:	18.8		168.0	η:	N/A
C_p/J mol^{-1}K^{-1}:	20.0		22.3	k:	149 W/m K

COMMENTS:

Name: Silicon dioxide
Synonym: Silicon(IV) oxide

Mol. Form.: O_2Si

CAS RN: 14808-60-7
Merck No.:
Mol. Wt.: 60.084

PHYSICAL CONSTANTS

T_m: 1610°C (1883 K) T_c: μ:
T_b: 2230°C (2503 K) P_c: IP:

TRANSITION PROPERTIES

$\Delta_{fus}H$ (T_m): 8.51 kJ/mol
$\Delta_{vap}H$ (T_b):
$\Delta_{vap}H$ (25°C):

Vapor pressure (0°C): N/A
Vapor pressure (25°C): N/A
Vapor pressure (100°C): N/A

PROPERTIES AT 25° AND 100 kPa

	Solid	Liquid	Gas		Solid
$\Delta_f H°$/kJ mol^{-1}:	-910.7		-322.0		d: 2.6481 g/mL
$S°$/J mol^{-1}K^{-1}:	41.5				η: N/A
C_p/J mol^{-1}K^{-1}:	44.4				k: 1.4 W/m K

COMMENTS: Data refer to α-quartz

Name: Silicon tetrabromide
Synonym: Tetrabromosilane

Mol. Form.: Br_4Si

CAS RN: 7789-66-4
Merck No.: 8446
Mol. Wt.: 347.702

PHYSICAL CONSTANTS

T_m: 5.2°C (278.3 K) T_c: 390°C (663 K)   μ: 0 D
T_b: 154°C (427 K) P_c: IP:

TRANSITION PROPERTIES

$\Delta_{fus}H$ (T_m):
$\Delta_{vap}H$ (T_b): 37.90 kJ/mol
$\Delta_{vap}H$ (25°C): 41.8 kJ/mol

Vapor pressure (0°C):
Vapor pressure (25°C):
Vapor pressure (100°C):

PROPERTIES AT 25° AND 100 kPa

	Solid	Liquid	Gas		Liquid
$\Delta_f H°$/kJ mol^{-1}:		-457.3	-415.5		d: 2.8 g/mL
$S°$/J mol^{-1}K^{-1}:		277.8	377.9		η:
C_p/J mol^{-1}K^{-1}:			97.1		k:

COMMENTS:

Name: Silicon tetrachloride
Synonym: Tetrachlorosilane

Mol. Form.: Cl_4Si

CAS RN: 10026-04-7
Merck No.: 8447
Mol. Wt.: 169.896

PHYSICAL CONSTANTS

T_m: -68.85°C (204.30 K) T_c: 235.0°C (508.1 K)   μ: 0 D
T_b: 57.65°C (330.80 K) P_c: 3.593 MPa   IP: 11.79 eV

TRANSITION PROPERTIES

$\Delta_{fus}H$ (T_m): 7.60 kJ/mol
$\Delta_{vap}H$ (T_b): 28.70 kJ/mol
$\Delta_{vap}H$ (25°C): 29.70 kJ/mol

Vapor pressure (0°C):
Vapor pressure (25°C): 31.3 kPa
Vapor pressure (100°C):

PROPERTIES AT 25° AND 100 kPa

	Solid	Liquid	Gas		Liquid
$\Delta_f H°$/kJ mol^{-1}:		-687.0	-657.0		d: 1.5 g/mL
$S°$/J mol^{-1}K^{-1}:		239.7	330.7		η: 99 mPa s
C_p/J mol^{-1}K^{-1}:		145.3	90.3		k: 0.099 W/m K

COMMENTS:

Name: Silicon tetrafluoride
Synonym: Tetrafluorosilane

CAS RN: 7783-61-1
Merck No.: 8448
Mol. Wt.: 104.079

Mol. Form.: F_4Si

PHYSICAL CONSTANTS

T_m: -90.2°C (182.9 K)
T_b: -86°C (187 K)

T_c: -14.1°C (259.0 K)
P_c: 3.72 MPa

μ: 0 D
IP: 15.70 eV

TRANSITION PROPERTIES

$\Delta_{fus}H$ (T_m):
$\Delta_{vap}H$ (T_b):
$\Delta_{vap}H$ (25°C):

Vapor pressure (0°C): N/A
Vapor pressure (25°C): N/A
Vapor pressure (100°C): N/A

PROPERTIES AT 25° AND 100 kPa

	Solid	Liquid	Gas	Gas
$\Delta_f H°$/kJ mol^{-1}:			-1615.0	d: 4.254 g/L
$S°$/J mol^{-1}K^{-1}:			282.8	η:
C_p/J mol^{-1}K^{-1}:			73.6	k:
COMMENTS:				

Name: Silver

CAS RN: 7440-22-4
Merck No.: 8450
Mol. Wt.: 107.868

Mol. Form.: Ag

PHYSICAL CONSTANTS

T_m: 961.78°C (1234.93 K)
T_b: 2162°C (2435 K)

T_c:
P_c:

μ:
IP: 7.58 eV

TRANSITION PROPERTIES

$\Delta_{fus}H$ (T_m): 11.30 kJ/mol
$\Delta_{vap}H$ (T_b):
$\Delta_{vap}H$ (25°C):

Vapor pressure (0°C): N/A
Vapor pressure (25°C): N/A
Vapor pressure (100°C): N/A

PROPERTIES AT 25° AND 100 kPa

	Solid	Liquid	Gas	Solid
$\Delta_f H°$/kJ mol^{-1}:	0.0		284.9	d: 10.5 g/mL
$S°$/J mol^{-1}K^{-1}:	42.6		173.0	η: N/A
C_p/J mol^{-1}K^{-1}:	25.4		20.8	k: 429 W/m K
COMMENTS:				

Name: Silver bromide

CAS RN: 7785-23-1
Merck No.: 8452
Mol. Wt.: 187.772

Mol. Form.: AgBr

PHYSICAL CONSTANTS

T_m: 432°C (705 K)
T_b: 1502°C (1775 K)

T_c:
P_c:

μ:
IP:

TRANSITION PROPERTIES

$\Delta_{fus}H$ (T_m): 9.12 kJ/mol
$\Delta_{vap}H$ (T_b): 198.00 kJ/mol
$\Delta_{vap}H$ (25°C):

Vapor pressure (0°C): N/A
Vapor pressure (25°C): N/A
Vapor pressure (100°C): N/A

PROPERTIES AT 25° AND 100 kPa

	Solid	Liquid	Gas	Solid
$\Delta_f H°$/kJ mol^{-1}:	-100.4			d: 6.47 g/mL
$S°$/J mol^{-1}K^{-1}:	107.1			η: N/A
C_p/J mol^{-1}K^{-1}:	52.4			k: 0.9 W/m K
COMMENTS: Highly toxic				

Name: Silver chloride

CAS RN: 7783-90-6
Merck No.: 8455
Mol. Wt.: 143.321

Mol. Form.: AgCl

PHYSICAL CONSTANTS

T_m: 455°C (728 K) T_c: μ: 5.70 D
T_b: 1547°C (1820 K) P_c: IP: 10.08 eV

TRANSITION PROPERTIES

$\Delta_{fus}H$ (T_m): 13.20 kJ/mol   Vapor pressure (0°C): N/A
$\Delta_{vap}H$ (T_b): 199.00 kJ/mol   Vapor pressure (25°C): N/A
$\Delta_{vap}H$ (25°C):   Vapor pressure (100°C): N/A

PROPERTIES AT 25° AND 100 kPa

	Solid	Liquid	Gas		Solid
$\Delta_f H°$/kJ mol^{-1}:	-127.0			d:	5.56 g/mL
$S°$/J mol^{-1}K^{-1}:	96.3			η:	N/A
C_p/J mol^{-1}K^{-1}:	50.8			k:	1.1 W/m K

COMMENTS: Highly toxic

Name: Silver iodide

CAS RN: 7783-96-2
Merck No.: 8462
Mol. Wt.: 234.773

Mol. Form.: AgI

PHYSICAL CONSTANTS

T_m: 558°C (831 K) T_c: μ: 5.10 D
T_b: 1506°C (1779 K) P_c: IP:

TRANSITION PROPERTIES

$\Delta_{fus}H$ (T_m): 9.41 kJ/mol   Vapor pressure (0°C): N/A
$\Delta_{vap}H$ (T_b): 143.90 kJ/mol   Vapor pressure (25°C): N/A
$\Delta_{vap}H$ (25°C):   Vapor pressure (100°C): N/A

PROPERTIES AT 25° AND 100 kPa

	Solid	Liquid	Gas		Solid
$\Delta_f H°$/kJ mol^{-1}:	-61.8			d:	5.68 g/mL
$S°$/J mol^{-1}K^{-1}:	115.5			η:	N/A
C_p/J mol^{-1}K^{-1}:	56.8			k:	

COMMENTS: Highly toxic

Name: Silver nitrate

CAS RN: 7761-88-8
Merck No.: 8464
Mol. Wt.: 169.873

Mol. Form.: AgNO$_3$

PHYSICAL CONSTANTS

T_m: 212°C (485 K) T_c: μ:
T_b: P_c: IP:

TRANSITION PROPERTIES

$\Delta_{fus}H$ (T_m): 11.50 kJ/mol   Vapor pressure (0°C):
$\Delta_{vap}H$ (T_b):   Vapor pressure (25°C):
$\Delta_{vap}H$ (25°C):   Vapor pressure (100°C):

PROPERTIES AT 25° AND 100 kPa

	Solid	Liquid	Gas		Solid
$\Delta_f H°$/kJ mol^{-1}:	-124.4			d:	4.35 g/mL
$S°$/J mol^{-1}K^{-1}:	140.9			η:	N/A
C_p/J mol^{-1}K^{-1}:	93.1			k:	

COMMENTS: Highly toxic

Name: Sodium   CAS RN: 7440-23-5
                                                                Merck No.: 8512
                                                                Mol. Wt.: 22.990

Mol. Form.: Na

PHYSICAL CONSTANTS

T_m: 97.72°C (370.87 K) T_c: μ:
T_b: 883°C (1156 K) P_c: IP: 5.14 eV

TRANSITION PROPERTIES

$\Delta_{fus}H$ (T_m): 2.60 kJ/mol   Vapor pressure (0°C):
$\Delta_{vap}H$ (T_b):   Vapor pressure (25°C):
$\Delta_{vap}H$ (25°C):   Vapor pressure (100°C):

PROPERTIES AT 25° AND 100 kPa

	Solid	Liquid	Gas		Solid
$\Delta_f H°$/kJ mol^{-1}:	0.0		107.5	d: 0.97 g/mL	
$S°$/J mol^{-1}K^{-1}:	51.3		153.7	η: N/A	
C_p/J mol^{-1}K^{-1}:	28.2		20.8	k: 142 W/m K	
COMMENTS: Flammable					

Name: Sodium bromide   CAS RN: 7647-15-6
                                                                Merck No.: 8539
                                                                Mol. Wt.: 102.894

Mol. Form.: BrNa

PHYSICAL CONSTANTS

T_m: 747°C (1020 K) T_c: μ: 9.118 D
T_b: P_c: IP: 8.31 eV

TRANSITION PROPERTIES

$\Delta_{fus}H$ (T_m): 26.11 kJ/mol   Vapor pressure (0°C): N/A
$\Delta_{vap}H$ (T_b):   Vapor pressure (25°C): N/A
$\Delta_{vap}H$ (25°C):   Vapor pressure (100°C): N/A

PROPERTIES AT 25° AND 100 kPa

	Solid	Liquid	Gas		Solid
$\Delta_f H°$/kJ mol^{-1}:	-361.1		-143.1	d: 3.200 g/mL	
$S°$/J mol^{-1}K^{-1}:	86.8		241.2	η: N/A	
C_p/J mol^{-1}K^{-1}:	51.4		36.3	k:	
COMMENTS:					

Name: Sodium carbonate   CAS RN: 497-19-8
                                                                Merck No.: 8541
                                                                Mol. Wt.: 105.989

Mol. Form.: CNa$_2$O$_3$

PHYSICAL CONSTANTS

T_m: 858.1°C (1131.2 K) T_c: μ:
T_b: P_c: IP:

TRANSITION PROPERTIES

$\Delta_{fus}H$ (T_m): 29.70 kJ/mol   Vapor pressure (0°C): N/A
$\Delta_{vap}H$ (T_b):   Vapor pressure (25°C): N/A
$\Delta_{vap}H$ (25°C):   Vapor pressure (100°C): N/A

PROPERTIES AT 25° AND 100 kPa

	Solid	Liquid	Gas		Solid
$\Delta_f H°$/kJ mol^{-1}:	-1130.7			d: 2.54 g/mL	
$S°$/J mol^{-1}K^{-1}:	135.0			η: N/A	
C_p/J mol^{-1}K^{-1}:	112.3			k:	
COMMENTS:					

Name: Sodium chlorate

CAS RN: 7775-09-9
Merck No.: 8543
Mol. Wt.: 106.441

Mol. Form.: $ClNaO_3$

PHYSICAL CONSTANTS

T_m: 248°C (521 K)	T_c:	μ:
T_b:	P_c:	IP:

TRANSITION PROPERTIES

$\Delta_{fus}H$ (T_m): 22.10 kJ/mol
$\Delta_{vap}H$ (T_b):
$\Delta_{vap}H$ (25°C):

Vapor pressure (0°C): N/A
Vapor pressure (25°C): N/A
Vapor pressure (100°C): N/A

PROPERTIES AT 25° AND 100 kPa

	Solid	Liquid	Gas		Solid
$\Delta_f H°$/kJ mol^{-1}:	-365.8			d:	2.5 g/mL
$S°$/J mol^{-1}K^{-1}:	123.4			η:	N/A
C_p/J mol^{-1}K^{-1}:				k:	
COMMENTS:					

Name: Sodium chloride
Synonyms: Halite
         Rock salt
Mol. Form.: ClNa

CAS RN: 7647-14-5
Merck No.: 8544
Mol. Wt.: 58.442

PHYSICAL CONSTANTS

T_m: 800.7°C (1073.8 K)	T_c:	μ: 9.001 D
T_b: 1465°C (1738 K)	P_c:	IP: 8.92 eV

TRANSITION PROPERTIES

$\Delta_{fus}H$ (T_m): 28.16 kJ/mol
$\Delta_{vap}H$ (T_b):
$\Delta_{vap}H$ (25°C):

Vapor pressure (0°C): N/A
Vapor pressure (25°C): N/A
Vapor pressure (100°C): N/A

PROPERTIES AT 25° AND 100 kPa

	Solid	Liquid	Gas		Solid
$\Delta_f H°$/kJ mol^{-1}:	-411.2			d:	2.17 g/mL
$S°$/J mol^{-1}K^{-1}:	72.1			η:	N/A
C_p/J mol^{-1}K^{-1}:	50.2			k:	6.0 W/m K
COMMENTS:					

Name: Sodium fluoride

CAS RN: 7681-49-4
Merck No.: 8565
Mol. Wt.: 41.988

Mol. Form.: FNa

PHYSICAL CONSTANTS

T_m: 996°C (1269 K)	T_c:	μ: 8.156 D
T_b: 1704°C (1977 K)	P_c:	IP:

TRANSITION PROPERTIES

$\Delta_{fus}H$ (T_m): 33.35 kJ/mol
$\Delta_{vap}H$ (T_b):
$\Delta_{vap}H$ (25°C):

Vapor pressure (0°C): N/A
Vapor pressure (25°C): N/A
Vapor pressure (100°C): N/A

PROPERTIES AT 25° AND 100 kPa

	Solid	Liquid	Gas		Solid
$\Delta_f H°$/kJ mol^{-1}:	-576.6			d:	2.78 g/mL
$S°$/J mol^{-1}K^{-1}:	51.1			η:	N/A
C_p/J mol^{-1}K^{-1}:	46.9			k:	
COMMENTS: Highly toxic					

Name: Sodium hydroxide   CAS RN: 1310-73-2
Synonym: Caustic soda   Merck No.: 8575
                                                                Mol. Wt.: 39.997
Mol. Form.: HNaO

PHYSICAL CONSTANTS

T_m: 323°C (596 K) T_c: μ:
T_b: 1388°C (1661 K) P_c: IP:

TRANSITION PROPERTIES

$\Delta_{fus}H$ (T_m): 6.60 kJ/mol   Vapor pressure (0°C): N/A
$\Delta_{vap}H$ (T_b): 175.00 kJ/mol   Vapor pressure (25°C): N/A
$\Delta_{vap}H$ (25°C):   Vapor pressure (100°C): N/A

PROPERTIES AT 25° AND 100 kPa

	Solid	Liquid	Gas		Solid
$\Delta_f H°$/kJ mol^{-1}:	-425.6			d:	2.13 g/mL
$S°$/J mol^{-1}K^{-1}:	64.5			η:	N/A
C_p/J mol^{-1}K^{-1}:	59.5			k:	
COMMENTS:					

Name: Sodium iodide   CAS RN: 7681-82-5
                                                                Merck No.: 8582
                                                                Mol. Wt.: 149.894
Mol. Form.: INa

PHYSICAL CONSTANTS

T_m: 660°C (933 K) T_c: μ: 9.236 D
T_b: 1304°C (1577 K) P_c: IP: 7.64 eV

TRANSITION PROPERTIES

$\Delta_{fus}H$ (T_m): 23.60 kJ/mol   Vapor pressure (0°C): N/A
$\Delta_{vap}H$ (T_b):   Vapor pressure (25°C): N/A
$\Delta_{vap}H$ (25°C):   Vapor pressure (100°C): N/A

PROPERTIES AT 25° AND 100 kPa

	Solid	Liquid	Gas		Solid
$\Delta_f H°$/kJ mol^{-1}:	-287.8			d:	3.67 g/mL
$S°$/J mol^{-1}K^{-1}:	98.5			η:	N/A
C_p/J mol^{-1}K^{-1}:	52.1			k:	
COMMENTS:					

Name: Sodium nitrate   CAS RN: 7631-99-4
Synonym: Chile saltpeter   Merck No.: 8598
                                                                Mol. Wt.: 84.995
Mol. Form.: NNaO$_3$

PHYSICAL CONSTANTS

T_m: 307°C (580 K) T_c: μ:
T_b: P_c: IP:

TRANSITION PROPERTIES

$\Delta_{fus}H$ (T_m): 15.00 kJ/mol   Vapor pressure (0°C): N/A
$\Delta_{vap}H$ (T_b):   Vapor pressure (25°C): N/A
$\Delta_{vap}H$ (25°C):   Vapor pressure (100°C): N/A

PROPERTIES AT 25° AND 100 kPa

	Solid	Liquid	Gas		Solid
$\Delta_f H°$/kJ mol^{-1}:	-467.9			d:	2.26 g/mL
$S°$/J mol^{-1}K^{-1}:	116.5			η:	N/A
C_p/J mol^{-1}K^{-1}:	92.9			k:	
COMMENTS:					

Name: Sodium sulfate
Synonym: Disodium sulfate

Mol. Form.: Na_2O_4S

CAS RN: 7757-82-6
Merck No.:
Mol. Wt.: 142.043

PHYSICAL CONSTANTS

T_m: 884°C (1157 K) T_c: μ:
T_b: P_c: IP:

TRANSITION PROPERTIES

$\Delta_{fus}H$ (T_m): 23.60 kJ/mol   Vapor pressure (0°C): N/A
$\Delta_{vap}H$ (T_b):   Vapor pressure (25°C): N/A
$\Delta_{vap}H$ (25°C):   Vapor pressure (100°C): N/A

PROPERTIES AT 25° AND 100 kPa

	Solid	Liquid	Gas		Solid
$\Delta_f H°$/kJ mol^{-1}:	-1387.1			d:	2.7 g/mL
$S°$/J mol^{-1}K^{-1}:	149.6			η:	N/A
C_p/J mol^{-1}K^{-1}:	128.2			k:	
COMMENTS:					

Name: Sorbitol
Synonym: *D*-Glucitol

Mol. Form.: $C_6H_{14}O_6$

CAS RN: 50-70-4
Merck No.: 8680
Mol. Wt.: 182.174

PHYSICAL CONSTANTS

T_m: 111°C (384 K) T_c: μ:
T_b: P_c: IP:

TRANSITION PROPERTIES

$\Delta_{fus}H$ (T_m):   Vapor pressure (0°C):
$\Delta_{vap}H$ (T_b):   Vapor pressure (25°C):
$\Delta_{vap}H$ (25°C):   Vapor pressure (100°C):

PROPERTIES AT 25° AND 100 kPa

	Solid	Liquid	Gas		Solid
$\Delta_f H°$/kJ mol^{-1}:				d:	
$S°$/J mol^{-1}K^{-1}:				η:	N/A
C_p/J mol^{-1}K^{-1}:				k:	
COMMENTS:					

Name: Stannane
Synonym: Tin tetrahydride

Mol. Form.: H_4Sn

CAS RN: 2406-52-2
Merck No.:
Mol. Wt.: 122.742

PHYSICAL CONSTANTS

T_m: T_c: μ: 0 D
T_b: -51.8°C (221.3 K) P_c: IP: 10.75 eV

TRANSITION PROPERTIES

$\Delta_{fus}H$ (T_m):   Vapor pressure (0°C):
$\Delta_{vap}H$ (T_b): 19.05 kJ/mol   Vapor pressure (25°C):
$\Delta_{vap}H$ (25°C):   Vapor pressure (100°C):

PROPERTIES AT 25° AND 100 kPa

	Solid	Liquid	Gas		Gas
$\Delta_f H°$/kJ mol^{-1}:			162.8	d:	5.017 g/L
$S°$/J mol^{-1}K^{-1}:			227.7	η:	
C_p/J mol^{-1}K^{-1}:			49.0	k:	
COMMENTS:					

Name: Stearic acid

Synonym: Octadecanoic acid

CAS RN: 57-11-4

Merck No.: 8761

Mol. Wt.: 284.483

Mol. Form.: $C_{18}H_{36}O_2$

PHYSICAL CONSTANTS

T_m: 68.82°C (341.97 K) T_c: μ:

T_b: P_c: IP:

TRANSITION PROPERTIES

$\Delta_{fus}H$ (T_m): 63.00 kJ/mol   Vapor pressure (0°C):

$\Delta_{vap}H$ (T_b):   Vapor pressure (25°C):

$\Delta_{vap}H$ (25°C): 103.5 kJ/mol   Vapor pressure (100°C):

PROPERTIES AT 25° AND 100 kPa

	Solid	Liquid	Gas		Solid
$\Delta_f H°$/kJ mol^{-1}:	-947.7	-884.7	-781.2	d:	
$S°$/J mol^{-1}K^{-1}:				η: N/A	
C_p/J mol^{-1}K^{-1}:	501.5			k:	

COMMENTS:

Name: Stibine

Synonym: Antimony trihydride

CAS RN: 7803-52-3

Merck No.: 8767

Mol. Wt.: 124.781

Mol. Form.: H_3Sb

PHYSICAL CONSTANTS

T_m: -88°C (185 K) T_c: μ: 0.12 D

T_b: -17°C (256 K) P_c: IP: 9.54 eV

TRANSITION PROPERTIES

$\Delta_{fus}H$ (T_m):   Vapor pressure (0°C):

$\Delta_{vap}H$ (T_b): 21.30 kJ/mol   Vapor pressure (25°C):

$\Delta_{vap}H$ (25°C):   Vapor pressure (100°C):

PROPERTIES AT 25° AND 100 kPa

	Solid	Liquid	Gas		Gas
$\Delta_f H°$/kJ mol^{-1}:			145.1	d:	5.100 g/L
$S°$/J mol^{-1}K^{-1}:			232.8	η:	
C_p/J mol^{-1}K^{-1}:			41.1	k:	

COMMENTS: TLV=0.1 ppm

Name: *cis*-Stilbene

Synonym: *cis*-1,2-Diphenylethene

CAS RN: 645-49-8

Merck No.: 8774

Mol. Wt.: 180.249

Mol. Form.: $C_{14}H_{12}$

PHYSICAL CONSTANTS

T_m: -5°C (268 K) T_c: μ:

T_b: P_c: IP: 7.80 eV

TRANSITION PROPERTIES

$\Delta_{fus}H$ (T_m):   Vapor pressure (0°C):

$\Delta_{vap}H$ (T_b):   Vapor pressure (25°C):

$\Delta_{vap}H$ (25°C): 69.0 kJ/mol   Vapor pressure (100°C):

PROPERTIES AT 25° AND 100 kPa

	Solid	Liquid	Gas		Liquid
$\Delta_f H°$/kJ mol^{-1}:		183.3	252.3	d:	
$S°$/J mol^{-1}K^{-1}:				η:	
C_p/J mol^{-1}K^{-1}:				k:	

COMMENTS:

Name: *trans*-Stilbene
Synonym: *trans*-1,2-Diphenylethene

Mol. Form.: $C_{14}H_{12}$

CAS RN: 103-30-0
Merck No.: 8774
Mol. Wt.: 180.249

PHYSICAL CONSTANTS

T_m: 123°C (396 K)
T_b: 307°C (580 K)

T_c:
P_c:

μ: 0 D
IP: 7.70 eV

TRANSITION PROPERTIES

$\Delta_{fus}H$ (T_m): 27.40 kJ/mol
$\Delta_{vap}H$ (T_b):
$\Delta_{vap}H$ (25°C):

Vapor pressure (0°C):
Vapor pressure (25°C):
Vapor pressure (100°C):

PROPERTIES AT 25° AND 100 kPa

	Solid	Liquid	Gas		Solid
$\Delta_f H°$/kJ mol^{-1}:	136.9		236.1		d: 1.046 g/mL
$S°$/J mol^{-1}K^{-1}:					η: N/A
C_p/J mol^{-1}K^{-1}:					k:
COMMENTS:					

Name: Strontium

Mol. Form.: Sr

CAS RN: 7440-24-6
Merck No.: 8797
Mol. Wt.: 87.620

PHYSICAL CONSTANTS

T_m: 777°C (1050 K)
T_b: 1382°C (1655 K)

T_c:
P_c:

μ:
IP: 5.69 eV

TRANSITION PROPERTIES

$\Delta_{fus}H$ (T_m): 7.43 kJ/mol
$\Delta_{vap}H$ (T_b):
$\Delta_{vap}H$ (25°C):

Vapor pressure (0°C): N/A
Vapor pressure (25°C): N/A
Vapor pressure (100°C): N/A

PROPERTIES AT 25° AND 100 kPa

	Solid	Liquid	Gas		Solid
$\Delta_f H°$/kJ mol^{-1}:	0.0		164.4		d: 2.64 g/mL
$S°$/J mol^{-1}K^{-1}:	52.3		164.6		η: N/A
C_p/J mol^{-1}K^{-1}:	26.4		20.8		k: 35.4 W/m K
COMMENTS:					

Name: Strontium iodide
Synonym: Strontium diiodide

Mol. Form.: I_2Sr

CAS RN: 10476-86-5
Merck No.: 8808
Mol. Wt.: 341.429

PHYSICAL CONSTANTS

T_m: 538°C (811 K)
T_b: 1773°C (2046 K)

T_c:
P_c:

μ:
IP:

TRANSITION PROPERTIES

$\Delta_{fus}H$ (T_m): 19.70 kJ/mol
$\Delta_{vap}H$ (T_b):
$\Delta_{vap}H$ (25°C):

Vapor pressure (0°C): N/A
Vapor pressure (25°C): N/A
Vapor pressure (100°C): N/A

PROPERTIES AT 25° AND 100 kPa

	Solid	Liquid	Gas		Solid
$\Delta_f H°$/kJ mol^{-1}:	-558.1				d: 4.4 g/mL
$S°$/J mol^{-1}K^{-1}:					η: N/A
C_p/J mol^{-1}K^{-1}:	81.6				k:
COMMENTS:					

Name: Styrene
Synonyms: Vinylbenzene
        Ethenylbenzene
Mol. Form.: C_8H_8

CAS RN: 100-42-5
Merck No.: 8830
Mol. Wt.: 104.152

PHYSICAL CONSTANTS

T_m: -31°C (242 K)	T_c:	μ:
T_b: 145°C (418 K)	P_c:	IP: 8.43 eV

TRANSITION PROPERTIES

$\Delta_{fus}H$ (T_m): 10.95 kJ/mol
$\Delta_{vap}H$ (T_b): 38.70 kJ/mol
$\Delta_{vap}H$ (25°C): 43.93 kJ/mol

Vapor pressure (0°C):
Vapor pressure (25°C): 0.810 kPa
Vapor pressure (100°C): 25.5 kPa

PROPERTIES AT 25° AND 100 kPa

	Solid	Liquid	Gas		Liquid
$\Delta_fH°$/kJ mol^{-1}:		103.8	147.9	d:	0.9001 g/mL
$S°$/J mol^{-1}K^{-1}:				η:	0.701 mPa s
C_p/J mol^{-1}K^{-1}:		182.0		k:	0.137 W/m K

COMMENTS: TLV=50 ppm; carcinogen; highly toxic; flammable

Name: Succinic acid
Synonyms: Butanedioic acid
        1,2-Ethanedicarboxylic acid
Mol. Form.: $C_4H_6O_4$

CAS RN: 110-15-6
Merck No.: 8840
Mol. Wt.: 118.089

PHYSICAL CONSTANTS

T_m: 188°C (461 K)	T_c:	μ:
T_b:	P_c:	IP:

TRANSITION PROPERTIES

$\Delta_{fus}H$ (T_m): 32.95 kJ/mol
$\Delta_{vap}H$ (T_b):
$\Delta_{vap}H$ (25°C):

Vapor pressure (0°C):
Vapor pressure (25°C):
Vapor pressure (100°C):

PROPERTIES AT 25° AND 100 kPa

	Solid	Liquid	Gas		Solid
$\Delta_fH°$/kJ mol^{-1}:	-940.5		-823.0	d:	1.572 g/mL
$S°$/J mol^{-1}K^{-1}:	167.3			η:	N/A
C_p/J mol^{-1}K^{-1}:	153.1			k:	

COMMENTS:

Name: Succinic anhydride
Synonym: Butanedioic anhydride

Mol. Form.: $C_4H_4O_3$

CAS RN: 108-30-5
Merck No.: 8841
Mol. Wt.: 100.074

PHYSICAL CONSTANTS

T_m: 119°C (392 K)	T_c:	μ:
T_b: 261°C (534 K)	P_c:	IP: 10.60 eV

TRANSITION PROPERTIES

$\Delta_{fus}H$ (T_m): 20.41 kJ/mol
$\Delta_{vap}H$ (T_b):
$\Delta_{vap}H$ (25°C):

Vapor pressure (0°C):
Vapor pressure (25°C):
Vapor pressure (100°C):

PROPERTIES AT 25° AND 100 kPa

	Solid	Liquid	Gas		Solid
$\Delta_fH°$/kJ mol^{-1}:	-607.8			d:	1.2 g/mL
$S°$/J mol^{-1}K^{-1}:				η:	N/A
C_p/J mol^{-1}K^{-1}:				k:	

COMMENTS:

Name: Succinonitrile
Synonyms: Butanedinitrile
        1,2-Dicyanoethane
Mol. Form.: $C_4H_4N_2$

CAS RN: 110-61-2
Merck No.: 8843
Mol. Wt.: 80.089

PHYSICAL CONSTANTS

T_m: 54.5°C (327.6 K)
T_b: 266°C (539 K)

T_c:
P_c:

μ:
IP: 12.10 eV

TRANSITION PROPERTIES

$\Delta_{fus}H$ (T_m): 3.70 kJ/mol
$\Delta_{vap}H$ (T_b): 48.50 kJ/mol
$\Delta_{vap}H$ (25°C): 63.97 kJ/mol

Vapor pressure (0°C):
Vapor pressure (25°C): 0.001 kPa
Vapor pressure (100°C):

PROPERTIES AT 25° AND 100 kPa

	Solid	Liquid	Gas		Solid
$\Delta_fH°$/kJ mol^{-1}:		139.7	209.7		d: 1.023 g/mL
$S°$/J mol^{-1}K^{-1}:		191.6			η: N/A
C_p/J mol^{-1}K^{-1}:		145.6			k:

COMMENTS: Highly toxic

Name: Sulfolane
Synonyms: Thiolane 1,1-dioxide
        Tetrahydrothiophene 1,1-dioxide
Mol. Form.: $C_4H_8O_2S$

CAS RN: 126-33-0
Merck No.: 8934
Mol. Wt.: 120.172

PHYSICAL CONSTANTS

T_m: 27.6°C (300.7 K)
T_b: 285°C (558 K)

T_c:
P_c:

μ:
IP: 9.80 eV

TRANSITION PROPERTIES

$\Delta_{fus}H$ (T_m): 1.43 kJ/mol
$\Delta_{vap}H$ (T_b):
$\Delta_{vap}H$ (25°C):

Vapor pressure (0°C):
Vapor pressure (25°C):
Vapor pressure (100°C):

PROPERTIES AT 25° AND 100 kPa

	Solid	Liquid	Gas		Solid
$\Delta_fH°$/kJ mol^{-1}:					d: 1.2660 g/mL
$S°$/J mol^{-1}K^{-1}:					η: N/A
C_p/J mol^{-1}K^{-1}:					k:

COMMENTS:

Name: Sulfur

Mol. Form.: S

CAS RN: 7704-34-9
Merck No.: 8956
Mol. Wt.: 32.066

PHYSICAL CONSTANTS

T_m: 115.21°C (388.36 K)
T_b: 444.60°C (717.75 K)

T_c: 1041°C (1314 K)
P_c: 20.7 MPa

μ:
IP: 10.36 eV

TRANSITION PROPERTIES

$\Delta_{fus}H$ (T_m): 1.72 kJ/mol
$\Delta_{vap}H$ (T_b): 45.00 kJ/mol
$\Delta_{vap}H$ (25°C):

Vapor pressure (0°C):
Vapor pressure (25°C):
Vapor pressure (100°C):

PROPERTIES AT 25° AND 100 kPa

	Solid	Liquid	Gas		Solid
$\Delta_fH°$/kJ mol^{-1}:	0.0		277.2		d: 2.07 g/mL
$S°$/J mol^{-1}K^{-1}:	32.1		167.8		η: N/A
C_p/J mol^{-1}K^{-1}:	22.6		23.7		k: 0.27 W/m K

COMMENTS: Data refer to rhombic form

Name: Sulfur dioxide

Synonym: Sulfur(IV) oxide

CAS RN: 7446-09-5

Merck No.: 8950

Mol. Wt.: 64.065

Mol. Form.: O_2S

PHYSICAL CONSTANTS

T_m: -75.5°C (197.6 K)

T_b: -10.05°C (263.10 K)

T_c: 157.7°C (430.8 K)

P_c: 7.884 MPa

μ: 1.633 D

IP: 12.32 eV

TRANSITION PROPERTIES

$\Delta_{fus}H$ (T_m):

$\Delta_{vap}H$ (T_b): 24.94 kJ/mol

$\Delta_{vap}H$ (25°C): 22.92 kJ/mol

Vapor pressure (0°C): 155 kPa

Vapor pressure (25°C):

Vapor pressure (100°C):

PROPERTIES AT 25° AND 100 kPa

	Solid	Liquid	Gas		Gas
$\Delta_fH°$/kJ mol^{-1}:		-320.5	-296.8		d: 2.619 g/L
$S°$/J mol^{-1}K^{-1}:			248.2		η: 12.8 μPa s
C_p/J mol^{-1}K^{-1}:			39.9		k: 0.0096 W/m K

COMMENTS: TLV=2 ppm; highly toxic

Name: Sulfur fluoride (SF_4)

Synonym: Sulfur tetrafluoride

CAS RN: 7783-60-0

Merck No.: 8957

Mol. Wt.: 108.060

Mol. Form.: F_4S

PHYSICAL CONSTANTS

T_m: -125°C (148 K)

T_b: -40.45°C (232.70 K)

T_c: 91°C (364 K)

P_c:

μ: 0.632 D

IP: 12.03 eV

TRANSITION PROPERTIES

$\Delta_{fus}H$ (T_m):

$\Delta_{vap}H$ (T_b): 26.44 kJ/mol

$\Delta_{vap}H$ (25°C):

Vapor pressure (0°C):

Vapor pressure (25°C):

Vapor pressure (100°C): N/A

PROPERTIES AT 25° AND 100 kPa

	Solid	Liquid	Gas		Gas
$\Delta_fH°$/kJ mol^{-1}:			-763.2		d: 4.417 g/L
$S°$/J mol^{-1}K^{-1}:			299.6		η:
C_p/J mol^{-1}K^{-1}:			77.6		k:

COMMENTS: TLV=0.1 ppm; highly toxic

Name: Sulfur fluoride (SF_6)

Synonym: Sulfur hexafluoride

CAS RN: 2551-62-4

Merck No.: 8952

Mol. Wt.: 146.056

Mol. Form.: F_6S

PHYSICAL CONSTANTS

T_t: -50.7°C (222.4 K)*

T_s: -63.8°C (209.3 K)†

T_c: 45.54°C (318.69 K)

P_c: 3.77 MPa

μ:

IP: 15.33 eV

TRANSITION PROPERTIES

$\Delta_{fus}H$ (T_m): 5.02 kJ/mol

$\Delta_{vap}H$ (T_b):

$\Delta_{vap}H$ (25°C): 8.99 kJ/mol

Vapor pressure (0°C):

Vapor pressure (25°C):

Vapor pressure (100°C): N/A

PROPERTIES AT 25° AND 100 kPa

	Solid	Liquid	Gas		Gas
$\Delta_fH°$/kJ mol^{-1}:			-1220.5		d: 5.970 g/L
$S°$/J mol^{-1}K^{-1}:			291.5		η: 15.3 μPa s
C_p/J mol^{-1}K^{-1}:			97.0		k: 0.0130 W/m K

COMMENTS: TLV=1000 ppm. *Triple point. †Sublimation point.

Name: Sulfuric acid

Mol. Form.: H_2O_4S

CAS RN: 7664-93-9
Merck No.: 8953
Mol. Wt.: 98.080

PHYSICAL CONSTANTS

T_m: 10.31°C (283.46 K) T_c: μ:
T_b: 337°C (610 K) P_c: IP:

TRANSITION PROPERTIES

$\Delta_{fus}H$ (T_m): 10.71 kJ/mol   Vapor pressure (0°C):
$\Delta_{vap}H$ (T_b):   Vapor pressure (25°C):
$\Delta_{vap}H$ (25°C):   Vapor pressure (100°C):

PROPERTIES AT 25° AND 100 kPa

	Solid	Liquid	Gas		Liquid
$\Delta_f H°$/kJ mol^{-1}:		-814.0		d:	1.8 g/mL
$S°$/J mol^{-1}K^{-1}:		156.9		η:	
C_p/J mol^{-1}K^{-1}:		138.9		k:	
COMMENTS:					

Name: Sulfur trioxide
Synonym: Sulfur(VI) oxide

Mol. Form.: O_3S

CAS RN: 7446-11-9
Merck No.: 8958
Mol. Wt.: 80.064

PHYSICAL CONSTANTS

T_m: 16.8°C (289.9 K) T_c: 217.9°C (491.0 K) μ:
T_b: 45°C (318 K) P_c: 8.2 MPa   IP: 12.80 eV

TRANSITION PROPERTIES

$\Delta_{fus}H$ (T_m): 13.5 kJ/mol   Vapor pressure (0°C):
$\Delta_{vap}H$ (T_b): 40.69 kJ/mol   Vapor pressure (25°C):
$\Delta_{vap}H$ (25°C): 43.14 kJ/mol   Vapor pressure (100°C):

PROPERTIES AT 25° AND 100 kPa

	Solid	Liquid	Gas		Liquid
$\Delta_f H°$/kJ mol^{-1}:	-454.5	-441.0	-395.7	d:	1.92 g/mL
$S°$/J mol^{-1}K^{-1}:	70.7	113.8	256.8	η:	
C_p/J mol^{-1}K^{-1}:			50.7	k:	
COMMENTS: Highly toxic; T_m refers to γ-form					

Name: Sulfuryl chloride
Synonym: Sulfonyl dichloride

Mol. Form.: Cl_2O_2S

CAS RN: 7791-25-5
Merck No.: 8959
Mol. Wt.: 134.970

PHYSICAL CONSTANTS

T_m: -51°C (222 K) T_c: μ: 1.81 D
T_b: 69.4°C (342.5 K) P_c: IP: 12.05 eV

TRANSITION PROPERTIES

$\Delta_{fus}H$ (T_m):   Vapor pressure (0°C):
$\Delta_{vap}H$ (T_b): 31.40 kJ/mol   Vapor pressure (25°C):
$\Delta_{vap}H$ (25°C): 30.10 kJ/mol   Vapor pressure (100°C):

PROPERTIES AT 25° AND 100 kPa

	Solid	Liquid	Gas		Liquid
$\Delta_f H°$/kJ mol^{-1}:		-394.1	-364.0	d:	1.680 g/mL
$S°$/J mol^{-1}K^{-1}:			311.9	η:	
C_p/J mol^{-1}K^{-1}:		134.0	77.0	k:	
COMMENTS:					

Name: Tantalum

CAS RN: 7440-25-7
Merck No.: 9025
Mol. Wt.: 180.948

Mol. Form.: Ta

PHYSICAL CONSTANTS

T_m: 3017°C (3290 K) T_c: μ:
T_b: 5458°C (5731 K) P_c: IP: 7.89 eV

TRANSITION PROPERTIES

$\Delta_{fus}H$ (T_m): 36.57 kJ/mol   Vapor pressure (0°C): N/A
$\Delta_{vap}H$ (T_b):   Vapor pressure (25°C): N/A
$\Delta_{vap}H$ (25°C):   Vapor pressure (100°C): N/A

PROPERTIES AT 25° AND 100 kPa

	Solid	Liquid	Gas		Solid
$\Delta_f H°$/kJ mol^{-1}:	0.0		782.0		d: 16.4 g/mL
$S°$/J mol^{-1}K^{-1}:	41.5		185.2		η: N/A
C_p/J mol^{-1}K^{-1}:	25.4		20.9		k: 57.5 W/m K
COMMENTS:					

Name: Tantalum chloride (TaCl$_5$)
Synonyms: Tantalum pentachloride
          Tantalum(V) chloride
Mol. Form.: Cl$_5$Ta

CAS RN: 7721-01-9
Merck No.: 9027
Mol. Wt.: 358.211

PHYSICAL CONSTANTS

T_m: 216°C (489 K) T_c: 494°C (767 K)   μ:
T_b: 239.35°C (512.50 K) P_c: IP: 11.08 eV

TRANSITION PROPERTIES

$\Delta_{fus}H$ (T_m): 35.10 kJ/mol   Vapor pressure (0°C):
$\Delta_{vap}H$ (T_b): 54.80 kJ/mol   Vapor pressure (25°C):
$\Delta_{vap}H$ (25°C):   Vapor pressure (100°C):

PROPERTIES AT 25° AND 100 kPa

	Solid	Liquid	Gas		Solid
$\Delta_f H°$/kJ mol^{-1}:	-859.0				d: 3.68 g/mL
$S°$/J mol^{-1}K^{-1}:					η: N/A
C_p/J mol^{-1}K^{-1}:					k:
COMMENTS:					

Name: Tellurium

CAS RN: 13494-80-9
Merck No.: 9064
Mol. Wt.: 127.600

Mol. Form.: Te

PHYSICAL CONSTANTS

T_m: 449.51°C (722.66 K) T_c: μ:
T_b: 988°C (1261 K) P_c: IP: 9.01 eV

TRANSITION PROPERTIES

$\Delta_{fus}H$ (T_m): 17.49 kJ/mol   Vapor pressure (0°C): N/A
$\Delta_{vap}H$ (T_b): 114.10 kJ/mol   Vapor pressure (25°C): N/A
$\Delta_{vap}H$ (25°C):   Vapor pressure (100°C): N/A

PROPERTIES AT 25° AND 100 kPa

	Solid	Liquid	Gas		Solid
$\Delta_f H°$/kJ mol^{-1}:	0.0		196.7		d: 6.24 g/mL
$S°$/J mol^{-1}K^{-1}:	49.7		182.7		η: N/A
C_p/J mol^{-1}K^{-1}:	25.7		20.8		k: 3.38 W/m K
COMMENTS: Highly toxic					

Name: Tellurium chloride (TeCl$_4$)
Synonyms: Tellurium tetrachloride
             Tellurium(IV) chloride
Mol. Form.: Cl$_4$Te

CAS RN: 10026-07-0
Merck No.: 9070
Mol. Wt.: 269.411

PHYSICAL CONSTANTS

T_m: 224°C (497 K)
T_b: 387°C (660 K)

T_c: 729°C (1002 K)
P_c: 8.56 MPa

μ:
IP:

TRANSITION PROPERTIES

$\Delta_{fus}H$ (T_m):
$\Delta_{vap}H$ (T_b): 77.00 kJ/mol
$\Delta_{vap}H$ (25°C):

Vapor pressure (0°C):
Vapor pressure (25°C):
Vapor pressure (100°C):

PROPERTIES AT 25° AND 100 kPa

	Solid	Liquid	Gas		Solid
$\Delta_f H°$/kJ mol^{-1}:	-326.4			d: 3.0 g/mL	
$S°$/J mol^{-1}K^{-1}:				η: N/A	
C_p/J mol^{-1}K^{-1}:	138.5			k:	
COMMENTS:					

Name: Tellurium fluoride (TeF$_6$)
Synonyms: Tellurium hexafluoride
             Tellurium(VI) fluoride
Mol. Form.: F$_6$Te

CAS RN: 7783-80-4
Merck No.: 9068
Mol. Wt.: 241.590

PHYSICAL CONSTANTS

T_m:
T_b: -39°C (234 K)

T_c: 83°C (356 K)
P_c:

μ:
IP:

TRANSITION PROPERTIES

$\Delta_{fus}H$ (T_m):
$\Delta_{vap}H$ (T_b):
$\Delta_{vap}H$ (25°C):

Vapor pressure (0°C):
Vapor pressure (25°C):
Vapor pressure (100°C): N/A

PROPERTIES AT 25° AND 100 kPa

	Solid	Liquid	Gas		Gas
$\Delta_f H°$/kJ mol^{-1}:			-1318.0	d: 9.875 g/L	
$S°$/J mol^{-1}K^{-1}:				η:	
C_p/J mol^{-1}K^{-1}:				k:	
COMMENTS: Highly toxic					

Name: Terephthalic acid
Synonym: 1,4-Benzenedicarboxylic acid

Mol. Form.: C$_8$H$_6$O$_4$

CAS RN: 100-21-0
Merck No.: 9093
Mol. Wt.: 166.133

PHYSICAL CONSTANTS

T_m:
T_b:

T_c:
P_c:

μ:
IP: 9.86 eV

TRANSITION PROPERTIES

$\Delta_{fus}H$ (T_m):
$\Delta_{vap}H$ (T_b):
$\Delta_{vap}H$ (25°C):

Vapor pressure (0°C):
Vapor pressure (25°C):
Vapor pressure (100°C):

PROPERTIES AT 25° AND 100 kPa

	Solid	Liquid	Gas		Solid
$\Delta_f H°$/kJ mol^{-1}:	-816.1		-717.9	d:	
$S°$/J mol^{-1}K^{-1}:				η: N/A	
C_p/J mol^{-1}K^{-1}:				k:	
COMMENTS:					

Name: *o*-Terphenyl
Synonym: 1,2-Diphenylbenzene

Mol. Form.: $C_{18}H_{14}$

CAS RN: 84-15-1
Merck No.:
Mol. Wt.: 230.309

PHYSICAL CONSTANTS

T_m: 56.2°C (329.3 K)
T_b: 332°C (605 K)

T_c: 617.9°C (891.0 K)
P_c: 3.90 MPa

μ:
IP: 8.00 eV

TRANSITION PROPERTIES

$\Delta_{fus}H$ (T_m):
$\Delta_{vap}H$ (T_b):
$\Delta_{vap}H$ (25°C):

Vapor pressure (0°C):
Vapor pressure (25°C):
Vapor pressure (100°C):

PROPERTIES AT 25° AND 100 kPa

	Solid	Liquid	Gas		Solid
$\Delta_f H°$/kJ mol⁻¹:				*d*:	1.16 g/mL
$S°$/J mol⁻¹K⁻¹:				η:	N/A
C_p/J mol⁻¹K⁻¹:				*k*:	
COMMENTS:					

Name: *m*-Terphenyl
Synonym: 1,3-Diphenylbenzene

Mol. Form.: $C_{18}H_{14}$

CAS RN: 92-06-8
Merck No.:
Mol. Wt.: 230.309

PHYSICAL CONSTANTS

T_m: 87°C (360 K)
T_b: 363°C (636 K)

T_c: 651.8°C (924.9 K)
P_c: 3.51 MPa

μ:
IP: 8.01 eV

TRANSITION PROPERTIES

$\Delta_{fus}H$ (T_m):
$\Delta_{vap}H$ (T_b):
$\Delta_{vap}H$ (25°C):

Vapor pressure (0°C):
Vapor pressure (25°C):
Vapor pressure (100°C):

PROPERTIES AT 25° AND 100 kPa

	Solid	Liquid	Gas		Solid
$\Delta_f H°$/kJ mol⁻¹:				*d*:	1.195 g/mL
$S°$/J mol⁻¹K⁻¹:				η:	N/A
C_p/J mol⁻¹K⁻¹:				*k*:	
COMMENTS:					

Name: *p*-Terphenyl
Synonym: 1,4-Diphenylbenzene

Mol. Form.: $C_{18}H_{14}$

CAS RN: 92-94-4
Merck No.:
Mol. Wt.: 230.309

PHYSICAL CONSTANTS

T_m: 210.1°C (483.2 K)
T_b: 376°C (649 K)

T_c: 652.9°C (926.0 K)
P_c: 3.32 MPa

μ: 0 D
IP: 7.78 eV

TRANSITION PROPERTIES

$\Delta_{fus}H$ (T_m): 35.50 kJ/mol
$\Delta_{vap}H$ (T_b):
$\Delta_{vap}H$ (25°C):

Vapor pressure (0°C):
Vapor pressure (25°C):
Vapor pressure (100°C):

PROPERTIES AT 25° AND 100 kPa

	Solid	Liquid	Gas		Solid
$\Delta_f H°$/kJ mol⁻¹:				*d*:	1.213 g/mL
$S°$/J mol⁻¹K⁻¹:				η:	N/A
C_p/J mol⁻¹K⁻¹:	278.7			*k*:	
COMMENTS:					

Name: Tetraborane

Synonyms: Tetraboron decahydride
           Tetraborane (10)

Mol. Form.: B_4H_{10}

CAS RN: 18283-93-7

Merck No.: 9119

Mol. Wt.: 53.323

PHYSICAL CONSTANTS

T_m: -120°C (153 K)	T_c:	μ: 0.486 D
T_b: 18°C (291 K)	P_c:	IP: 10.76 eV

TRANSITION PROPERTIES

$\Delta_{fus}H$ (T_m):

$\Delta_{vap}H$ (T_b): 27.10 kJ/mol

$\Delta_{vap}H$ (25°C):

Vapor pressure (0°C):

Vapor pressure (25°C):

Vapor pressure (100°C):

PROPERTIES AT 25° AND 100 kPa

	Solid	Liquid	Gas		Gas
$\Delta_f H°$/kJ mol^{-1}:			66.1		d: 2.180 g/L
$S°$/J mol^{-1}K^{-1}:					η:
C_p/J mol^{-1}K^{-1}:					k:
COMMENTS: Highly toxic					

Name: 1,1,2,2-Tetrabromoethane

Synonym: Acetylene tetrabromide

Mol. Form.: $C_2H_2Br_4$

CAS RN: 79-27-6

Merck No.: 9121

Mol. Wt.: 345.654

PHYSICAL CONSTANTS

T_m: 0.0°C (273 K)	T_c:	μ:
T_b: 243.55°C (516.70 K)	P_c:	IP:

TRANSITION PROPERTIES

$\Delta_{fus}H$ (T_m):

$\Delta_{vap}H$ (T_b): 48.65 kJ/mol

$\Delta_{vap}H$ (25°C): 70.00 kJ/mol

Vapor pressure (0°C):

Vapor pressure (25°C): 0.003 kPa

Vapor pressure (100°C):

PROPERTIES AT 25° AND 100 kPa

	Solid	Liquid	Gas		Liquid
$\Delta_f H°$/kJ mol^{-1}:					d: 2.9529 g/mL
$S°$/J mol^{-1}K^{-1}:					η: 9.68 mPa s
C_p/J mol^{-1}K^{-1}:		165.7			k:
COMMENTS: Highly toxic					

Name: Tetrabromomethane

Synonym: Carbon tetrabromide

Mol. Form.: CBr_4

CAS RN: 558-13-4

Merck No.:

Mol. Wt.: 331.627

PHYSICAL CONSTANTS

T_m: 90.1°C (363.2 K)	T_c:	μ: 0 D
T_b: 189.5°C (462.6 K)	P_c:	IP: 10.31 eV

TRANSITION PROPERTIES

$\Delta_{fus}H$ (T_m):

$\Delta_{vap}H$ (T_b): 43.5 kJ/mol

$\Delta_{vap}H$ (25°C):

Vapor pressure (0°C):

Vapor pressure (25°C): 0.096 kPa

Vapor pressure (100°C):

PROPERTIES AT 25° AND 100 kPa

	Solid	Liquid	Gas		Solid
$\Delta_f H°$/kJ mol^{-1}:	18.8		79.0		d: 3.42 g/mL
$S°$/J mol^{-1}K^{-1}:	212.5		358.1		η: N/A
C_p/J mol^{-1}K^{-1}:	144.3		91.2		k:
COMMENTS: TLV=0.1 ppm					

Name: Tetrachlorodiborane(4)
Synonym: Diboron tetrachloride

Mol. Form.: B_2Cl_4

CAS RN: 13701-67-2
Merck No.: 2998
Mol. Wt.: 163.433

PHYSICAL CONSTANTS

T_m: -92.6°C (180.5 K) T_c: μ:
T_b: 65°C (338 K) P_c: IP:

TRANSITION PROPERTIES

$\Delta_{fus}H$ (T_m):   Vapor pressure (0°C):
$\Delta_{vap}H$ (T_b):   Vapor pressure (25°C):
$\Delta_{vap}H$ (25°C): 32.6 kJ/mol   Vapor pressure (100°C):

PROPERTIES AT 25° AND 100 kPa

	Solid	Liquid	Gas		Liquid
$\Delta_f H°$/kJ mol^{-1}:		-523.0	-490.4	d:	
$S°$/J mol^{-1}K^{-1}:		262.3	357.4	η:	
C_p/J mol^{-1}K^{-1}:		137.7	95.4	k:	
COMMENTS:					

Name: Tetrachloro-1,2-difluoroethane
Synonyms: Refrigerant 112
         CFC-112
Mol. Form.: $C_2Cl_4F_2$

CAS RN: 76-12-0
Merck No.:
Mol. Wt.: 203.830

PHYSICAL CONSTANTS

T_m: 26°C (299 K) T_c: 278°C (551 K)   μ:
T_b: 93°C (366 K) P_c: IP: 11.30 eV

TRANSITION PROPERTIES

$\Delta_{fus}H$ (T_m): 3.70 kJ/mol   Vapor pressure (0°C): 1.73 kPa
$\Delta_{vap}H$ (T_b):   Vapor pressure (25°C): 7.51 kPa
$\Delta_{vap}H$ (25°C):   Vapor pressure (100°C): 124 kPa

PROPERTIES AT 25° AND 100 kPa

	Solid	Liquid	Gas		Solid
$\Delta_f H°$/kJ mol^{-1}:				d:	1.6447 g/mL
$S°$/J mol^{-1}K^{-1}:				η:	N/A
C_p/J mol^{-1}K^{-1}:	175.5	173.6	132.3	k:	
COMMENTS: TLV=500 ppm					

Name: 1,1,1,2-Tetrachloroethane

Mol. Form.: $C_2H_2Cl_4$

CAS RN: 630-20-6
Merck No.:
Mol. Wt.: 167.849

PHYSICAL CONSTANTS

T_m: -70.21°C (202.94 K) T_c: μ:
T_b: 130.5°C (403.6 K) P_c: IP: 11.10 eV

TRANSITION PROPERTIES

$\Delta_{fus}H$ (T_m):   Vapor pressure (0°C):
$\Delta_{vap}H$ (T_b):   Vapor pressure (25°C): 1.60 kPa
$\Delta_{vap}H$ (25°C):   Vapor pressure (100°C): 41.2 kPa

PROPERTIES AT 25° AND 100 kPa

	Solid	Liquid	Gas		Liquid
$\Delta_f H°$/kJ mol^{-1}:				d:	1.5346 g/mL
$S°$/J mol^{-1}K^{-1}:			356.0	η:	1.44 mPa s
C_p/J mol^{-1}K^{-1}:		153.8	102.7	k:	
COMMENTS:					

Name: 1,1,2,2-Tetrachloroethane
Synonym: Acetylene tetrachloride

Mol. Form.: $C_2H_2Cl_4$

CAS RN: 79-34-5
Merck No.: 9125
Mol. Wt.: 167.849

PHYSICAL CONSTANTS

T_m: -43.8°C (229.3 K)
T_b: 146.5°C (419.6 K)

T_c: 388.00°C (661.15 K)
P_c:

μ: 1.32 D
IP: 11.62 eV

TRANSITION PROPERTIES

$\Delta_{fus}H$ (T_m):
$\Delta_{vap}H$ (T_b): 37.64 kJ/mol
$\Delta_{vap}H$ (25°C): 45.71 kJ/mol

Vapor pressure (0°C):
Vapor pressure (25°C): 0.622 kPa
Vapor pressure (100°C): 24.9 kPa

PROPERTIES AT 25° AND 100 kPa

	Solid	Liquid	Gas		Liquid
$\Delta_f H°$/kJ mol^{-1}:		-195.0	-149.2		d: 1.5872 g/mL
$S°$/J mol^{-1}K^{-1}:		246.9	362.8		η: 1.6 mPa s
C_p/J mol^{-1}K^{-1}:		162.3	100.8		k:

COMMENTS: TLV=1 ppm; carcinogen; highly toxic

Name: Tetrachloroethylene
Synonym: Tetrachloroethene

Mol. Form.: C_2Cl_4

CAS RN: 127-18-4
Merck No.: 9126
Mol. Wt.: 165.833

PHYSICAL CONSTANTS

T_m: -22.35°C (250.80 K)
T_b: 121.3°C (394.4 K)

T_c: 347.1°C (620.2 K)
P_c:

μ: 0 D
IP: 9.32 eV

TRANSITION PROPERTIES

$\Delta_{fus}H$ (T_m): 10.56 kJ/mol
$\Delta_{vap}H$ (T_b): 34.68 kJ/mol
$\Delta_{vap}H$ (25°C): 39.68 kJ/mol

Vapor pressure (0°C):
Vapor pressure (25°C): 2.42 kPa
Vapor pressure (100°C): 54.2 kPa

PROPERTIES AT 25° AND 100 kPa

	Solid	Liquid	Gas		Liquid
$\Delta_f H°$/kJ mol^{-1}:		-50.6	-10.8		d: 1.6130 g/mL
$S°$/J mol^{-1}K^{-1}:		266.9			η: 0.845 mPa s
C_p/J mol^{-1}K^{-1}:		143.4			k: 0.110 W/m K

COMMENTS: TLV=50 ppm; carcinogen; highly toxic

Name: Tetrachloromethane
Synonyms: Carbon tetrachloride
         Refrigerant 10
Mol. Form.: CCl_4

CAS RN: 56-23-5
Merck No.: 1822
Mol. Wt.: 153.822

PHYSICAL CONSTANTS

T_m: -23°C (250 K)
T_b: 76.8°C (349.9 K)

T_c: 283.5°C (556.6 K)
P_c: 4.516 MPa

μ: 0 D
IP: 11.47 eV

TRANSITION PROPERTIES

$\Delta_{fus}H$ (T_m): 3.28 kJ/mol
$\Delta_{vap}H$ (T_b): 29.82 kJ/mol
$\Delta_{vap}H$ (25°C): 32.43 kJ/mol

Vapor pressure (0°C): 4.49 kPa
Vapor pressure (25°C): 15.2 kPa
Vapor pressure (100°C): 197 kPa

PROPERTIES AT 25° AND 100 kPa

	Solid	Liquid	Gas		Liquid
$\Delta_f H°$/kJ mol^{-1}:		-128.2	-95.8		d: 1.5844 g/mL
$S°$/J mol^{-1}K^{-1}:					η: 0.908 mPa s
C_p/J mol^{-1}K^{-1}:		130.7	83.3		k: 0.099 W/m K

COMMENTS: TLV=5 ppm; carcinogen; highly toxic

Name: Tetradecane

CAS RN: 629-59-4
Merck No.:
Mol. Wt.: 198.392

Mol. Form.: $C_{14}H_{30}$

PHYSICAL CONSTANTS

T_m: 5.86°C (279.01 K) T_c: 420°C (693 K)   μ:
T_b: 253.58°C (526.73 K) P_c: 1.61 MPa   IP:

TRANSITION PROPERTIES

$\Delta_{fus}H$ (T_m):   Vapor pressure (0°C):
$\Delta_{vap}H$ (T_b): 47.5 kJ/mol   Vapor pressure (25°C):
$\Delta_{vap}H$ (25°C): 71.30 kJ/mol   Vapor pressure (100°C):

PROPERTIES AT 25° AND 100 kPa

	Solid	Liquid	Gas		Liquid
$\Delta_f H°$/kJ mol^{-1}:				d: 0.7592 g/mL	
$S°$/J mol^{-1}K^{-1}:				η: 2.13 mPa s	
C_p/J mol^{-1}K^{-1}:		438.3		k: 0.136 W/m K	
COMMENTS:					

Name: Tetradecanoic acid
Synonym: Myristic acid

CAS RN: 544-63-8
Merck No.: 6246
Mol. Wt.: 228.375

Mol. Form.: $C_{14}H_{28}O_2$

PHYSICAL CONSTANTS

T_m: 53.96°C (327.11 K) T_c: μ:
T_b: P_c: IP:

TRANSITION PROPERTIES

$\Delta_{fus}H$ (T_m): 45.38 kJ/mol   Vapor pressure (0°C):
$\Delta_{vap}H$ (T_b):   Vapor pressure (25°C):
$\Delta_{vap}H$ (25°C): 95.1 kJ/mol   Vapor pressure (100°C):

PROPERTIES AT 25° AND 100 kPa

	Solid	Liquid	Gas		Solid
$\Delta_f H°$/kJ mol^{-1}:	-833.5	-788.8	-693.7	d:	
$S°$/J mol^{-1}K^{-1}:				η: N/A	
C_p/J mol^{-1}K^{-1}:	432.0			k:	
COMMENTS:					

Name: 1-Tetradecanol
Synonyms: Myristyl alcohol
          Tetradecyl alcohol
Mol. Form.: $C_{14}H_{30}O$

CAS RN: 112-72-1
Merck No.: 6248
Mol. Wt.: 214.392

PHYSICAL CONSTANTS

T_m: 39.5°C (312.6 K) T_c: 474°C (747 K)   μ:
T_b: 289°C (562 K) P_c: 1.81 MPa   IP:

TRANSITION PROPERTIES

$\Delta_{fus}H$ (T_m): 49.0 kJ/mol   Vapor pressure (0°C):
$\Delta_{vap}H$ (T_b):   Vapor pressure (25°C):
$\Delta_{vap}H$ (25°C): 102.20 kJ/mol   Vapor pressure (100°C):

PROPERTIES AT 25° AND 100 kPa

	Solid	Liquid	Gas		Solid
$\Delta_f H°$/kJ mol^{-1}:	-629.6	-580.6		d:	
$S°$/J mol^{-1}K^{-1}:				η: N/A	
C_p/J mol^{-1}K^{-1}:	388.0			k:	
COMMENTS:					

Name: Tetrafluoroethylene
Synonym: Tetrafluoroethene

Mol. Form.: C_2F_4

CAS RN: 116-14-3
Merck No.:
Mol. Wt.: 100.016

PHYSICAL CONSTANTS

T_m: -142.5°C (130.6 K)
T_b: -75.95°C (197.20 K)

T_c: 33.4°C (306.5 K)
P_c: 3.94 MPa

μ: 0 D
IP: 10.12 eV

TRANSITION PROPERTIES

$\Delta_{fus}H$ (T_m):
$\Delta_{vap}H$ (T_b): 16.8 kJ/mol
$\Delta_{vap}H$ (25°C):

Vapor pressure (0°C):
Vapor pressure (25°C):
Vapor pressure (100°C): N/A

PROPERTIES AT 25° AND 100 kPa

	Solid	Liquid	Gas		Gas
$\Delta_f H°$/kJ mol^{-1}:	-820.5		-658.9	d:	4.088 g/L
$S°$/J mol^{-1}K^{-1}:			300.1	η:	
C_p/J mol^{-1}K^{-1}:			80.5	k:	

COMMENTS: Highly toxic; very flammable

Name: Tetrafluoromethane
Synonyms: Carbon tetrafluoride
         Refrigerant 14
Mol. Form.: CF_4

CAS RN: 75-73-0
Merck No.: 1823
Mol. Wt.: 88.005

PHYSICAL CONSTANTS

T_m: -183.59°C (89.56 K)
T_b: -128.02°C (145.13 K)

T_c: -45.5°C (227.6 K)
P_c: 3.74 MPa

μ: 0 D
IP:

TRANSITION PROPERTIES

$\Delta_{fus}H$ (T_m):
$\Delta_{vap}H$ (T_b): 12.3 kJ/mol
$\Delta_{vap}H$ (25°C):

Vapor pressure (0°C): N/A
Vapor pressure (25°C): N/A
Vapor pressure (100°C): N/A

PROPERTIES AT 25° AND 100 kPa

	Solid	Liquid	Gas		Gas
$\Delta_f H°$/kJ mol^{-1}:			-933.6	d:	3.597 g/L
$S°$/J mol^{-1}K^{-1}:			261.6	η:	
C_p/J mol^{-1}K^{-1}:			61.1	k:	0.0160 W/m K

COMMENTS:

Name: Tetrahydrofuran
Synonyms: Oxolane
         1,4-Epoxybutane
Mol. Form.: C_4H_8O

CAS RN: 109-99-9
Merck No.: 9144
Mol. Wt.: 72.107

PHYSICAL CONSTANTS

T_m: -108.39°C (164.76 K)
T_b: 65°C (338 K)

T_c: 267.0°C (540.1 K)
P_c: 5.19 MPa

μ: 1.75 D
IP: 9.41 eV

TRANSITION PROPERTIES

$\Delta_{fus}H$ (T_m): 8.54 kJ/mol
$\Delta_{vap}H$ (T_b): 29.81 kJ/mol
$\Delta_{vap}H$ (25°C): 31.99 kJ/mol

Vapor pressure (0°C):
Vapor pressure (25°C): 21.6 kPa
Vapor pressure (100°C): 272 kPa

PROPERTIES AT 25° AND 100 kPa

	Solid	Liquid	Gas		Liquid
$\Delta_f H°$/kJ mol^{-1}:		-216.2	-184.2	d:	0.8800 g/mL
$S°$/J mol^{-1}K^{-1}:		204.3	302.4	η:	0.456 mPa s
C_p/J mol^{-1}K^{-1}:		124.0	76.3	k:	0.120 W/m K

COMMENTS: TLV=200 ppm; flammable

Name: Tetrahydrofurfuryl alcohol
Synonym: Tetrahydro-2-furanmethanol

Mol. Form.: $C_5H_{10}O_2$

CAS RN: 97-99-4
Merck No.: 9146
Mol. Wt.: 102.133

PHYSICAL CONSTANTS

T_m: <-80°C (<193 K)
T_b: 178°C (451 K)

T_c:
P_c:

μ:
IP:

TRANSITION PROPERTIES

$\Delta_{fus}H$ (T_m):
$\Delta_{vap}H$ (T_b): 45.19 kJ/mol
$\Delta_{vap}H$ (25°C): 51.55 kJ/mol

Vapor pressure (0°C):
Vapor pressure (25°C): 0.100 kPa
Vapor pressure (100°C):

PROPERTIES AT 25° AND 100 kPa

	Solid	Liquid	Gas	Liquid
$\Delta_f H°$/kJ mol^{-1}:		-435.7	-369.2	d: 1.048 g/mL
$S°$/J mol^{-1}K^{-1}:				η:
C_p/J mol^{-1}K^{-1}:		181.2		k:
COMMENTS:				

Name: 1,2,3,4-Tetrahydronaphthalene
Synonyms: Tetralin
          Benzocyclohexane
Mol. Form.: $C_{10}H_{12}$

CAS RN: 119-64-2
Merck No.: 9152
Mol. Wt.: 132.205

PHYSICAL CONSTANTS

T_m: -35.75°C (237.40 K)
T_b: 207.62°C (480.77 K)

T_c: 446.8°C (719.9 K)
P_c:

μ:
IP: 8.47 eV

TRANSITION PROPERTIES

$\Delta_{fus}H$ (T_m): 12.45 kJ/mol
$\Delta_{vap}H$ (T_b): 43.85 kJ/mol
$\Delta_{vap}H$ (25°C): 55.23 kJ/mol

Vapor pressure (0°C):
Vapor pressure (25°C): 0.050 kPa
Vapor pressure (100°C): 3.48 kPa

PROPERTIES AT 25° AND 100 kPa

	Solid	Liquid	Gas	Liquid
$\Delta_f H°$/kJ mol^{-1}:		-29.2	26.0	d: 0.9671 g/mL
$S°$/J mol^{-1}K^{-1}:				η: 2.14 mPa s
C_p/J mol^{-1}K^{-1}:		217.5		k:
COMMENTS:				

Name: Tetrahydropyran
Synonyms: Oxane
          Pentamethylene oxide
Mol. Form.: $C_5H_{10}O$

CAS RN: 142-68-7
Merck No.: 9149
Mol. Wt.: 86.134

PHYSICAL CONSTANTS

T_m: -45°C (228 K)
T_b: 88°C (361 K)

T_c: 299.1°C (572.2 K)
P_c: 4.77 MPa

μ: 1.74 D
IP: 9.25 eV

TRANSITION PROPERTIES

$\Delta_{fus}H$ (T_m):
$\Delta_{vap}H$ (T_b): 31.17 kJ/mol
$\Delta_{vap}H$ (25°C): 34.58 kJ/mol

Vapor pressure (0°C): 2.61 kPa
Vapor pressure (25°C): 9.54 kPa
Vapor pressure (100°C):

PROPERTIES AT 25° AND 100 kPa

	Solid	Liquid	Gas	Liquid
$\Delta_f H°$/kJ mol^{-1}:		-258.3	-223.4	d: 0.8772 g/mL
$S°$/J mol^{-1}K^{-1}:				η: 0.764 mPa s
C_p/J mol^{-1}K^{-1}:		156.5		k:
COMMENTS:				

Name: Tetrahydrothiophene
Synonyms: Thiacyclopentane
        Thiolane
Mol. Form.: C_4H_8S

CAS RN: 110-01-0
Merck No.:
Mol. Wt.: 88.173

PHYSICAL CONSTANTS

T_m: -96.16°C (176.99 K)
T_b: 121.0°C (394.1 K)

T_c: 358.9°C (632.0 K)
P_c:

μ:
IP: 8.47 eV

TRANSITION PROPERTIES

$\Delta_{fus}H$ (T_m):
$\Delta_{vap}H$ (T_b): 34.66 kJ/mol
$\Delta_{vap}H$ (25°C): 39.43 kJ/mol

Vapor pressure (0°C):
Vapor pressure (25°C): 2.45 kPa
Vapor pressure (100°C): 54.5 kPa

PROPERTIES AT 25° AND 100 kPa

	Solid	Liquid	Gas		Liquid
$\Delta_f H°$/kJ mol^{-1}:		-72.9	-34.1		d: 0.9938 g/mL
$S°$/J mol^{-1}K^{-1}:					η: 0.973 mPa s
C_p/J mol^{-1}K^{-1}:					k:
COMMENTS:					

Name: Tetraiodomethane
Synonym: Carbon tetraiodide

Mol. Form.: CI_4

CAS RN: 507-25-5
Merck No.: 1824
Mol. Wt.: 519.629

PHYSICAL CONSTANTS

T_m: 171°C (444 K)
T_b:

T_c:
P_c:

μ: 0 D
IP:

TRANSITION PROPERTIES

$\Delta_{fus}H$ (T_m):
$\Delta_{vap}H$ (T_b):
$\Delta_{vap}H$ (25°C):

Vapor pressure (0°C):
Vapor pressure (25°C):
Vapor pressure (100°C):

PROPERTIES AT 25° AND 100 kPa

	Solid	Liquid	Gas		Solid
$\Delta_f H°$/kJ mol^{-1}:					d: 4.3 g/mL
$S°$/J mol^{-1}K^{-1}:			391.9		η: N/A
C_p/J mol^{-1}K^{-1}:			95.9		k:
COMMENTS:					

Name: 1,2,4,5-Tetramethylbenzene
Synonym: Durene

Mol. Form.: $C_{10}H_{14}$

CAS RN: 95-93-2
Merck No.: 3450
Mol. Wt.: 134.221

PHYSICAL CONSTANTS

T_m: 79.3°C (352.4 K)
T_b: 196.84°C (469.99 K)

T_c: 402°C (675 K)
P_c: 2.9 MPa

μ:
IP: 8.04 eV

TRANSITION PROPERTIES

$\Delta_{fus}H$ (T_m): 21.00 kJ/mol
$\Delta_{vap}H$ (T_b):
$\Delta_{vap}H$ (25°C):

Vapor pressure (0°C):
Vapor pressure (25°C):
Vapor pressure (100°C): 4.48 kPa

PROPERTIES AT 25° AND 100 kPa

	Solid	Liquid	Gas		Solid
$\Delta_f H°$/kJ mol^{-1}:	-119.9				d: 1.03 g/mL
$S°$/J mol^{-1}K^{-1}:	245.6				η: N/A
C_p/J mol^{-1}K^{-1}:	215.1				k:
COMMENTS:					

Name: 2,2,3,3-Tetramethylpentane   CAS RN: 7154-79-2
                                                                                     Merck No.:
                                                                                     Mol. Wt.: 128.258
Mol. Form.: C_9H_{20}

PHYSICAL CONSTANTS

T_m: -9.8°C (263.3 K) T_c: 334.6°C (607.7 K)   μ:
T_b: 140.29°C (413.44 K) P_c: 2.741 MPa   IP:

TRANSITION PROPERTIES

$\Delta_{fus}H$ (T_m): 2.33 kJ/mol   Vapor pressure (0°C):
$\Delta_{vap}H$ (T_b):   Vapor pressure (25°C):
$\Delta_{vap}H$ (25°C): 41.2 kJ/mol   Vapor pressure (100°C): 31.5 kPa

PROPERTIES AT 25° AND 100 kPa

	Solid	Liquid	Gas		Liquid
$\Delta_f H°$/kJ mol^{-1}:		-278.3	-237.1	d: 0.753 g/mL	
$S°$/J mol^{-1}K^{-1}:				η:	
C_p/J mol^{-1}K^{-1}:		271.5		k:	

COMMENTS: Flammable

Name: 2,2,3,4-Tetramethylpentane   CAS RN: 1186-53-4
                                                                                     Merck No.:
                                                                                     Mol. Wt.: 128.258
Mol. Form.: C_9H_{20}

PHYSICAL CONSTANTS

T_m: -121.09°C (152.06 K) T_c: 319.6°C (592.7 K)   μ:
T_b: 133.03°C (406.18 K) P_c: 2.602 MPa   IP:

TRANSITION PROPERTIES

$\Delta_{fus}H$ (T_m):   Vapor pressure (0°C):
$\Delta_{vap}H$ (T_b):   Vapor pressure (25°C):
$\Delta_{vap}H$ (25°C): 40.8 kJ/mol   Vapor pressure (100°C): 39.0 kPa

PROPERTIES AT 25° AND 100 kPa

	Solid	Liquid	Gas		Liquid
$\Delta_f H°$/kJ mol^{-1}:		-277.7	-236.9	d: 0.735 g/mL	
$S°$/J mol^{-1}K^{-1}:				η:	
C_p/J mol^{-1}K^{-1}:				k:	

COMMENTS: Flammable

Name: 2,2,4,4-Tetramethylpentane   CAS RN: 1070-87-7
Synonym: Di-*tert*-butylmethane   Merck No.:
                                                                                     Mol. Wt.: 128.258
Mol. Form.: C_9H_{20}

PHYSICAL CONSTANTS

T_m: -66.54°C (206.61 K) T_c: 301.6°C (574.7 K)   μ:
T_b: 122.29°C (395.44 K) P_c: 2.485 MPa   IP:

TRANSITION PROPERTIES

$\Delta_{fus}H$ (T_m): 9.75 kJ/mol   Vapor pressure (0°C):
$\Delta_{vap}H$ (T_b): 32.51 kJ/mol   Vapor pressure (25°C):
$\Delta_{vap}H$ (25°C): 38.49 kJ/mol   Vapor pressure (100°C): 53.7 kPa

PROPERTIES AT 25° AND 100 kPa

	Solid	Liquid	Gas		Liquid
$\Delta_f H°$/kJ mol^{-1}:		-280.0	-241.6	d: 0.716 g/mL	
$S°$/J mol^{-1}K^{-1}:				η:	
C_p/J mol^{-1}K^{-1}:		266.3		k:	

COMMENTS: Flammable

Name: Tetranitromethane

CAS RN: 509-14-8
Merck No.: 9164
Mol. Wt.: 196.033

Mol. Form.: CN_4O_8

PHYSICAL CONSTANTS

T_m: 13.8°C (286.9 K)
T_b: 126.1°C (399.2 K)

T_c:
P_c:

μ: 0 D
IP:

TRANSITION PROPERTIES

$\Delta_{fus}H$ (T_m):
$\Delta_{vap}H$ (T_b): 40.74 kJ/mol
$\Delta_{vap}H$ (25°C): 49.93 kJ/mol

Vapor pressure (0°C):
Vapor pressure (25°C):
Vapor pressure (100°C):

PROPERTIES AT 25° AND 100 kPa

	Solid	Liquid	Gas		Liquid
$\Delta_f H°$/kJ mol^{-1}:		38.4	82.0		d: 1.6229 g/mL
$S°$/J mol^{-1}K^{-1}:					η:
C_p/J mol^{-1}K^{-1}:					k:

COMMENTS: Highly toxic

Name: Thallium

CAS RN: 7440-28-0
Merck No.: 9183
Mol. Wt.: 204.383

Mol. Form.: Tl

PHYSICAL CONSTANTS

T_m: 304°C (577 K)
T_b: 1473°C (1746 K)

T_c:
P_c:

μ:
IP: 6.11 eV

TRANSITION PROPERTIES

$\Delta_{fus}H$ (T_m): 4.14 kJ/mol
$\Delta_{vap}H$ (T_b):
$\Delta_{vap}H$ (25°C):

Vapor pressure (0°C): N/A
Vapor pressure (25°C): N/A
Vapor pressure (100°C): N/A

PROPERTIES AT 25° AND 100 kPa

	Solid	Liquid	Gas		Solid
$\Delta_f H°$/kJ mol^{-1}:	0.0		182.2		d: 11.8 g/mL
$S°$/J mol^{-1}K^{-1}:	64.2		181.0		η: N/A
C_p/J mol^{-1}K^{-1}:	26.3		20.8		k: 46.1 W/m K

COMMENTS: Highly toxic

Name: Thallium chloride (TlCl)
Synonyms: Thallous chloride
Thallium(I) chloride
Mol. Form.: ClTl

CAS RN: 7791-12-0
Merck No.: 9187
Mol. Wt.: 239.836

PHYSICAL CONSTANTS

T_m: 430°C (703 K)
T_b: 807°C (1080 K)

T_c:
P_c:

μ: 4.543 D
IP: 9.70 eV

TRANSITION PROPERTIES

$\Delta_{fus}H$ (T_m): 17.80 kJ/mol
$\Delta_{vap}H$ (T_b): 102.20 kJ/mol
$\Delta_{vap}H$ (25°C):

Vapor pressure (0°C): N/A
Vapor pressure (25°C): N/A
Vapor pressure (100°C): N/A

PROPERTIES AT 25° AND 100 kPa

	Solid	Liquid	Gas		Solid
$\Delta_f H°$/kJ mol^{-1}:	-204.1		-67.8		d: 7.0 g/mL
$S°$/J mol^{-1}K^{-1}:	111.3				η: N/A
C_p/J mol^{-1}K^{-1}:	50.9				k:

COMMENTS: Highly toxic

Name: Thallium fluoride (TlF)
Synonyms: Thallium(I) fluoride
         Thallous fluoride
Mol. Form.: FTl

CAS RN: 7789-27-7
Merck No.: 9189
Mol. Wt.: 223.382

PHYSICAL CONSTANTS

T_m: 322°C (595 K)
T_b: 655°C (928 K)

T_c:
P_c:

μ: 4.228 D
IP: 10.52 eV

TRANSITION PROPERTIES

$\Delta_{fus}H$ (T_m): 14.00 kJ/mol
$\Delta_{vap}H$ (T_b):
$\Delta_{vap}H$ (25°C):

Vapor pressure (0°C): N/A
Vapor pressure (25°C): N/A
Vapor pressure (100°C): N/A

PROPERTIES AT 25° AND 100 kPa

	Solid	Liquid	Gas		Solid
$\Delta_f H°$/kJ mol^{-1}:	-324.7		-182.4	d:	8.36 g/mL
$S°$/J mol^{-1}K^{-1}:				η:	N/A
C_p/J mol^{-1}K^{-1}:				k:	

COMMENTS: Highly toxic

Name: Thallium iodide (TlI)
Synonyms: Thallous iodide
         Thallium(I) iodide
Mol. Form.: ITl

CAS RN: 7790-30-9
Merck No.: 9191
Mol. Wt.: 331.288

PHYSICAL CONSTANTS

T_m: 440°C (713 K)
T_b: 824°C (1097 K)

T_c:
P_c:

μ: 4.61 D
IP: 8.47 eV

TRANSITION PROPERTIES

$\Delta_{fus}H$ (T_m): 13.10 kJ/mol
$\Delta_{vap}H$ (T_b): 104.70 kJ/mol
$\Delta_{vap}H$ (25°C):

Vapor pressure (0°C): N/A
Vapor pressure (25°C): N/A
Vapor pressure (100°C): N/A

PROPERTIES AT 25° AND 100 kPa

	Solid	Liquid	Gas		Solid
$\Delta_f H°$/kJ mol^{-1}:	-123.8		7.1	d:	7.1 g/mL
$S°$/J mol^{-1}K^{-1}:	127.6			η:	N/A
C_p/J mol^{-1}K^{-1}:				k:	

COMMENTS: Highly toxic

Name: Thallium oxide (Tl$_2$O)
Synonyms: Dithallium oxide
         Thallium(I) oxide
Mol. Form.: OTl$_2$

CAS RN: 1314-12-1
Merck No.: 9193
Mol. Wt.: 424.766

PHYSICAL CONSTANTS

T_m: 300°C (573 K)
T_b: 500°C (773 K)

T_c:
P_c:

μ:
IP:

TRANSITION PROPERTIES

$\Delta_{fus}H$ (T_m):
$\Delta_{vap}H$ (T_b):
$\Delta_{vap}H$ (25°C):

Vapor pressure (0°C): N/A
Vapor pressure (25°C): N/A
Vapor pressure (100°C): N/A

PROPERTIES AT 25° AND 100 kPa

	Solid	Liquid	Gas		Solid
$\Delta_f H°$/kJ mol^{-1}:	-178.7			d:	
$S°$/J mol^{-1}K^{-1}:	126.0			η:	N/A
C_p/J mol^{-1}K^{-1}:				k:	

COMMENTS: Highly toxic

Name: Thallium sulfide (Tl$_2$S)
Synonyms: Dithallium sulfide
        Thallium(I) sulfide
Mol. Form.: STl$_2$

CAS RN: 1314-97-2
Merck No.: 9198
Mol. Wt.: 440.833

PHYSICAL CONSTANTS

T_m: 448°C (721 K)
T_b: 1367°C (1640 K)

T_c:
P_c:

μ:
IP:

TRANSITION PROPERTIES

$\Delta_{fus}H$ (T_m): 12.00 kJ/mol
$\Delta_{vap}H$ (T_b): 154.00 kJ/mol
$\Delta_{vap}H$ (25°C):

Vapor pressure (0°C): N/A
Vapor pressure (25°C): N/A
Vapor pressure (100°C): N/A

PROPERTIES AT 25° AND 100 kPa

	Solid	Liquid	Gas		Solid
$\Delta_f H°$/kJ mol^{-1}:	-97.1			d:	8.39 g/mL
$S°$/J mol^{-1}K^{-1}:	151.0			η:	N/A
C_p/J mol^{-1}K^{-1}:				k:	

COMMENTS: Highly toxic

Name: Thionyl chloride
Synonym: Sulfinyl dichloride

Mol. Form.: Cl$_2$OS

CAS RN: 7719-09-7
Merck No.: 9278
Mol. Wt.: 118.971

PHYSICAL CONSTANTS

T_m: -101°C (172 K)
T_b: 75.6°C (348.7 K)

T_c:
P_c:

μ: 1.45 D
IP: 10.96 eV

TRANSITION PROPERTIES

$\Delta_{fus}H$ (T_m):
$\Delta_{vap}H$ (T_b): 31.70 kJ/mol
$\Delta_{vap}H$ (25°C): 31.00 kJ/mol

Vapor pressure (0°C):
Vapor pressure (25°C):
Vapor pressure (100°C):

PROPERTIES AT 25° AND 100 kPa

	Solid	Liquid	Gas		Liquid
$\Delta_f H°$/kJ mol^{-1}:		-245.6	-212.5	d:	1.631 g/mL
$S°$/J mol^{-1}K^{-1}:			309.8	η:	
C_p/J mol^{-1}K^{-1}:		121.0	66.5	k:	

COMMENTS: TLV=1 ppm

Name: Thionyl fluoride
Synonym: Sulfinyl difluoride

Mol. Form.: F$_2$OS

CAS RN: 7783-42-8
Merck No.: 9279
Mol. Wt.: 86.062

PHYSICAL CONSTANTS

T_m: -129.5°C (143.6 K)
T_b: -43.8°C (229.3 K)

T_c:
P_c:

μ: 1.63 D
IP: 12.25 eV

TRANSITION PROPERTIES

$\Delta_{fus}H$ (T_m):
$\Delta_{vap}H$ (T_b): 21.80 kJ/mol
$\Delta_{vap}H$ (25°C):

Vapor pressure (0°C):
Vapor pressure (25°C):
Vapor pressure (100°C):

PROPERTIES AT 25° AND 100 kPa

	Solid	Liquid	Gas		Gas
$\Delta_f H°$/kJ mol^{-1}:				d:	3.518 g/L
$S°$/J mol^{-1}K^{-1}:			278.7	η:	
C_p/J mol^{-1}K^{-1}:			56.8	k:	

COMMENTS:

Name: Thiophene

CAS RN: 110-02-1
Merck No.: 9283
Mol. Wt.: 84.142

Synonym: Thiofuran

Mol. Form.: C_4H_4S

PHYSICAL CONSTANTS

T_m: -39.4°C (233.7 K) T_c: 306.3°C (579.4 K)   μ: 0.55 D
T_b: 84.0°C (357.1 K) P_c: 5.69 MPa   IP: 8.87 eV

TRANSITION PROPERTIES

$\Delta_{fus}H$ (T_m): 5.09 kJ/mol   Vapor pressure (0°C): 2.85 kPa
$\Delta_{vap}H$ (T_b): 31.48 kJ/mol   Vapor pressure (25°C): 10.6 kPa
$\Delta_{vap}H$ (25°C): 34.70 kJ/mol   Vapor pressure (100°C): 161 kPa

PROPERTIES AT 25° AND 100 kPa

	Solid	Liquid	Gas		Liquid
$\Delta_f H°$/kJ mol^{-1}:		80.2	114.9		d: 1.0588 g/mL
$S°$/J mol^{-1}K^{-1}:		181.2			η: 0.613 mPa s
C_p/J mol^{-1}K^{-1}:		123.8			k: 0.199 W/m K

COMMENTS: Flammable

Name: Thorium

CAS RN: 7440-29-1
Merck No.: 9308
Mol. Wt.: 232.038

Mol. Form.: Th

PHYSICAL CONSTANTS

T_m: 1750°C (2023 K) T_c: μ:
T_b: 4788°C (5061 K) P_c: IP: 6.08 eV

TRANSITION PROPERTIES

$\Delta_{fus}H$ (T_m): 13.81 kJ/mol   Vapor pressure (0°C): N/A
$\Delta_{vap}H$ (T_b):   Vapor pressure (25°C): N/A
$\Delta_{vap}H$ (25°C):   Vapor pressure (100°C): N/A

PROPERTIES AT 25° AND 100 kPa

	Solid	Liquid	Gas		Solid
$\Delta_f H°$/kJ mol^{-1}:	0.0		602.0		d: 11.7 g/mL
$S°$/J mol^{-1}K^{-1}:	51.8		190.2		η: N/A
C_p/J mol^{-1}K^{-1}:	27.3		20.8		k: 54.0 W/m K

COMMENTS:

Name: Thorium chloride (ThCl$_4$)

CAS RN: 10026-08-1
Merck No.: 9309
Mol. Wt.: 373.849

Synonyms: Thorium tetrachloride
                Thorium(IV) chloride

Mol. Form.: Cl$_4$Th

PHYSICAL CONSTANTS

T_m: 770°C (1043 K) T_c: μ:
T_b: 921°C (1194 K) P_c: IP:

TRANSITION PROPERTIES

$\Delta_{fus}H$ (T_m): 40.20 kJ/mol   Vapor pressure (0°C): N/A
$\Delta_{vap}H$ (T_b): 146.40 kJ/mol   Vapor pressure (25°C): N/A
$\Delta_{vap}H$ (25°C):   Vapor pressure (100°C): N/A

PROPERTIES AT 25° AND 100 kPa

	Solid	Liquid	Gas		Solid
$\Delta_f H°$/kJ mol^{-1}:	-1186.6				d: 4.59 g/mL
$S°$/J mol^{-1}K^{-1}:	190.4				η: N/A
C_p/J mol^{-1}K^{-1}:					k:

COMMENTS:

Name: Thymol
Synonyms: 2-Isopropyl-5-methylphenol
     p-Cymen-3-ol
Mol. Form.: $C_{10}H_{14}O$

CAS RN: 89-83-8
Merck No.: 9333
Mol. Wt.: 150.221

PHYSICAL CONSTANTS

T_m: 51.5°C (324.6 K)
T_b: 232.55°C (505.70 K)

T_c: 425°C (698 K)
P_c:

μ:
IP:

TRANSITION PROPERTIES

$\Delta_{fus}H$ (T_m): 17.27 kJ/mol
$\Delta_{vap}H$ (T_b):
$\Delta_{vap}H$ (25°C):

Vapor pressure (0°C):
Vapor pressure (25°C):
Vapor pressure (100°C):

PROPERTIES AT 25° AND 100 kPa

	Solid	Liquid	Gas		Solid
$\Delta_f H°$/kJ mol^{-1}:	-309.7		-218.5		*d*: 0.970 g/mL
$S°$/J mol^{-1}K^{-1}:					η: N/A
C_p/J mol^{-1}K^{-1}:					*k*:
COMMENTS:					

Name: Tin

Mol. Form.: Sn

CAS RN: 7440-31-5
Merck No.: 9376
Mol. Wt.: 118.710

PHYSICAL CONSTANTS

T_m: 231.93°C (505.08 K)
T_b: 2602°C (2875 K)

T_c:
P_c:

μ:
IP: 7.34 eV

TRANSITION PROPERTIES

$\Delta_{fus}H$ (T_m): 7.03 kJ/mol
$\Delta_{vap}H$ (T_b):
$\Delta_{vap}H$ (25°C):

Vapor pressure (0°C): N/A
Vapor pressure (25°C): N/A
Vapor pressure (100°C): N/A

PROPERTIES AT 25° AND 100 kPa

	Solid	Liquid	Gas		Solid
$\Delta_f H°$/kJ mol^{-1}:	0.0		301.2		*d*: 7.28 g/mL
$S°$/J mol^{-1}K^{-1}:	51.2		168.5		η: N/A
C_p/J mol^{-1}K^{-1}:	27.0		21.3		*k*: 66.8 W/m K
COMMENTS: Data refer to white tin					

Name: Tin bromide (SnBr$_2$)
Synonyms: Stannous bromide
     Tin(II) bromide
Mol. Form.: Br$_2$Sn

CAS RN: 10031-24-0
Merck No.: 8741
Mol. Wt.: 278.518

PHYSICAL CONSTANTS

T_m: 215°C (488 K)
T_b: 639°C (912 K)

T_c:
P_c:

μ:
IP: 9.00 eV

TRANSITION PROPERTIES

$\Delta_{fus}H$ (T_m):
$\Delta_{vap}H$ (T_b): 102.00 kJ/mol
$\Delta_{vap}H$ (25°C):

Vapor pressure (0°C):
Vapor pressure (25°C):
Vapor pressure (100°C):

PROPERTIES AT 25° AND 100 kPa

	Solid	Liquid	Gas		Solid
$\Delta_f H°$/kJ mol^{-1}:	-243.5				*d*: 5.12 g/mL
$S°$/J mol^{-1}K^{-1}:					η: N/A
C_p/J mol^{-1}K^{-1}:					*k*:
COMMENTS:					

Name: Tin bromide (SnBr$_4$)   CAS RN: 7789-67-5
Synonyms: Stannic bromide   Merck No.: 8731
        Tin(IV) bromide   Mol. Wt.: 438.326
Mol. Form.: Br$_4$Sn

PHYSICAL CONSTANTS

T_m: 31°C (304 K)	T_c: 471°C (744 K)	μ:
T_b: 205°C (478 K)	P_c:	IP: 10.60 eV

TRANSITION PROPERTIES

$\Delta_{fus}H$ (T_m): 12.00 kJ/mol   Vapor pressure (0°C):
$\Delta_{vap}H$ (T_b): 43.50 kJ/mol   Vapor pressure (25°C):
$\Delta_{vap}H$ (25°C):   Vapor pressure (100°C):

PROPERTIES AT 25° AND 100 kPa

	Solid	Liquid	Gas		Solid
$\Delta_f H°$/kJ mol^{-1}:	-377.4		-314.6	d:	3.7 g/mL
$S°$/J mol^{-1}K^{-1}:	264.4		411.9	η:	N/A
C_p/J mol^{-1}K^{-1}:			103.4	k:	
COMMENTS:					

Name: Tin chloride (SnCl$_2$)   CAS RN: 7772-99-8
Synonyms: Stannous chloride   Merck No.: 8742
        Tin(II) chloride   Mol. Wt.: 189.615
Mol. Form.: Cl$_2$Sn

PHYSICAL CONSTANTS

T_m: 247°C (520 K)	T_c:	μ:
T_b: 623°C (896 K)	P_c:	IP: 10.00 eV

TRANSITION PROPERTIES

$\Delta_{fus}H$ (T_m): 12.80 kJ/mol   Vapor pressure (0°C): N/A
$\Delta_{vap}H$ (T_b): 86.80 kJ/mol   Vapor pressure (25°C): N/A
$\Delta_{vap}H$ (25°C):   Vapor pressure (100°C): N/A

PROPERTIES AT 25° AND 100 kPa

	Solid	Liquid	Gas		Solid
$\Delta_f H°$/kJ mol^{-1}:	-325.1			d:	3.90 g/mL
$S°$/J mol^{-1}K^{-1}:				η:	N/A
C_p/J mol^{-1}K^{-1}:				k:	
COMMENTS:					

Name: Tin chloride (SnCl$_4$)   CAS RN: 7646-78-8
Synonyms: Stannic chloride   Merck No.: 8732
        Tin(IV) chloride   Mol. Wt.: 260.521
Mol. Form.: Cl$_4$Sn

PHYSICAL CONSTANTS

T_m: -33°C (240 K)	T_c: 318.8°C (591.9 K)	μ: 0 D
T_b: 114.15°C (387.30 K)	P_c: 3.75 MPa	IP: 11.88 eV

TRANSITION PROPERTIES

$\Delta_{fus}H$ (T_m): 9.20 kJ/mol   Vapor pressure (0°C):
$\Delta_{vap}H$ (T_b): 34.90 kJ/mol   Vapor pressure (25°C):
$\Delta_{vap}H$ (25°C): 39.8 kJ/mol   Vapor pressure (100°C):

PROPERTIES AT 25° AND 100 kPa

	Solid	Liquid	Gas		Liquid
$\Delta_f H°$/kJ mol^{-1}:		-511.3	-471.5	d:	2.3 g/mL
$S°$/J mol^{-1}K^{-1}:		258.6	365.8	η:	
C_p/J mol^{-1}K^{-1}:		165.3	98.3	k:	
COMMENTS:					

Name: Tin iodide (SnI_4)
Synonyms: Stannic iodide
           Tin(IV) iodide
Mol. Form.: I_4Sn

CAS RN: 7790-47-8
Merck No.: 8735
Mol. Wt.: 626.328

PHYSICAL CONSTANTS

T_m: 143°C (416 K)
T_b: 364.35°C (637.50 K)

T_c: 695°C (968 K)
P_c:

μ:
IP:

TRANSITION PROPERTIES

$\Delta_{fus}H$ (T_m):
$\Delta_{vap}H$ (T_b): 56.90 kJ/mol
$\Delta_{vap}H$ (25°C):

Vapor pressure (0°C):
Vapor pressure (25°C):
Vapor pressure (100°C):

PROPERTIES AT 25° AND 100 kPa

	Solid	Liquid	Gas		Solid
$\Delta_f H°$/kJ mol^{-1}:					d: 4.46 g/mL
$S°$/J mol^{-1}K^{-1}:			446.1		η: N/A
C_p/J mol^{-1}K^{-1}:	84.9		105.4		k:
COMMENTS:					

Name: Titanium

Mol. Form.: Ti

CAS RN: 7440-32-6
Merck No.: 9396
Mol. Wt.: 47.880

PHYSICAL CONSTANTS

T_m: 1668°C (1941 K)
T_b: 3287°C (3560 K)

T_c:
P_c:

μ:
IP: 6.83 eV

TRANSITION PROPERTIES

$\Delta_{fus}H$ (T_m): 14.15 kJ/mol
$\Delta_{vap}H$ (T_b):
$\Delta_{vap}H$ (25°C):

Vapor pressure (0°C): N/A
Vapor pressure (25°C): N/A
Vapor pressure (100°C): N/A

PROPERTIES AT 25° AND 100 kPa

	Solid	Liquid	Gas		Solid
$\Delta_f H°$/kJ mol^{-1}:	0.0		473.0		d: 4.5 g/mL
$S°$/J mol^{-1}K^{-1}:	30.7		180.3		η: N/A
C_p/J mol^{-1}K^{-1}:	25.0		24.4		k: 21.9 W/m K
COMMENTS:					

Name: Titanium bromide ($TiBr_4$)
Synonyms: Titanium tetrabromide
           Titanium(IV) bromide
Mol. Form.: Br_4Ti

CAS RN: 7789-68-6
Merck No.: 9403
Mol. Wt.: 367.496

PHYSICAL CONSTANTS

T_m: 39°C (312 K)
T_b: 230°C (503 K)

T_c: 522.6°C (795.7 K)
P_c:

μ:
IP: 10.30 eV

TRANSITION PROPERTIES

$\Delta_{fus}H$ (T_m): 12.90 kJ/mol
$\Delta_{vap}H$ (T_b): 44.37 kJ/mol
$\Delta_{vap}H$ (25°C):

Vapor pressure (0°C):
Vapor pressure (25°C):
Vapor pressure (100°C):

PROPERTIES AT 25° AND 100 kPa

	Solid	Liquid	Gas		Solid
$\Delta_f H°$/kJ mol^{-1}:	-616.7		-549.4		d: 3.37 g/mL
$S°$/J mol^{-1}K^{-1}:	243.5		398.4		η: N/A
C_p/J mol^{-1}K^{-1}:	131.5		100.8		k:
COMMENTS:					

Name: Titanium chloride (TiCl$_4$)
Synonyms: Titanium tetrachloride
            Titanium(IV) chloride
Mol. Form.: Cl$_4$Ti

CAS RN: 7550-45-0
Merck No.: 9404
Mol. Wt.: 189.691

PHYSICAL CONSTANTS

T_m: -25°C (248 K)
T_b: 136.45°C (409.60 K)

T_c: 365°C (638 K)
P_c: 4.66 MPa

μ:
IP: 11.65 eV

TRANSITION PROPERTIES

$\Delta_{fus}H$ (T_m): 9.97 kJ/mol
$\Delta_{vap}H$ (T_b): 36.20 kJ/mol
$\Delta_{vap}H$ (25°C): 41.0 kJ/mol

Vapor pressure (0°C):
Vapor pressure (25°C):
Vapor pressure (100°C):

PROPERTIES AT 25° AND 100 kPa

	Solid	Liquid	Gas		Liquid
$\Delta_f H°$/kJ mol^{-1}:		-804.2	-763.2	d:	1.73 g/mL
$S°$/J mol^{-1}K^{-1}:		252.3	353.2	η:	
C_p/J mol^{-1}K^{-1}:		145.2	95.4	k:	

COMMENTS: Highly toxic

Name: Titanium oxide (TiO$_2$)
Synonyms: Titanium dioxide
            Titanium(IV) oxide
Mol. Form.: O$_2$Ti

CAS RN: 13463-67-7
Merck No.: 9398
Mol. Wt.: 79.879

PHYSICAL CONSTANTS

T_m: 1843°C (2116 K)
T_b:

T_c:
P_c:

μ:
IP: 9.54 eV

TRANSITION PROPERTIES

$\Delta_{fus}H$ (T_m):
$\Delta_{vap}H$ (T_b):
$\Delta_{vap}H$ (25°C):

Vapor pressure (0°C): N/A
Vapor pressure (25°C): N/A
Vapor pressure (100°C): N/A

PROPERTIES AT 25° AND 100 kPa

	Solid	Liquid	Gas		Solid
$\Delta_f H°$/kJ mol^{-1}:	-944.0			d:	
$S°$/J mol^{-1}K^{-1}:	50.6			η:	N/A
C_p/J mol^{-1}K^{-1}:	55.0			k:	

COMMENTS: Data refer to rutile

Name: Toluene
Synonyms: Methylbenzene
            Phenylmethane
Mol. Form.: C$_7$H$_8$

CAS RN: 108-88-3
Merck No.: 9455
Mol. Wt.: 92.141

PHYSICAL CONSTANTS

T_m: -94.99°C (178.16 K)
T_b: 110.63°C (383.78 K)

T_c: 318.64°C (591.79 K)
P_c: 4.104 MPa

μ: 0.375 D
IP: 8.82 eV

TRANSITION PROPERTIES

$\Delta_{fus}H$ (T_m): 6.85 kJ/mol
$\Delta_{vap}H$ (T_b): 33.18 kJ/mol
$\Delta_{vap}H$ (25°C): 38.01 kJ/mol

Vapor pressure (0°C):
Vapor pressure (25°C): 3.79 kPa
Vapor pressure (100°C): 74.6 kPa

PROPERTIES AT 25° AND 100 kPa

	Solid	Liquid	Gas		Liquid
$\Delta_f H°$/kJ mol^{-1}:		12.4	50.4	d:	0.8622 g/mL
$S°$/J mol^{-1}K^{-1}:				η:	0.555 mPa s
C_p/J mol^{-1}K^{-1}:		157.3		k:	0.1311 W/m K

COMMENTS: TLV=50 ppm; flammable

Name: Toluene-2,4-diamine
Synonyms: 4-Methyl-1,3-benzenediamine
           2,4-Diaminotoluene
Mol. Form.: $C_7H_{10}N_2$

CAS RN: 95-80-7
Merck No.:
Mol. Wt.: 122.170

PHYSICAL CONSTANTS

T_m: 99°C (372 K)
T_b: 292°C (565 K)

T_c:
P_c:

μ:
IP:

TRANSITION PROPERTIES

$\Delta_{fus}H$ (T_m):
$\Delta_{vap}H$ (T_b):
$\Delta_{vap}H$ (25°C):

Vapor pressure (0°C):
Vapor pressure (25°C):
Vapor pressure (100°C):

PROPERTIES AT 25° AND 100 kPa

	Solid	Liquid	Gas		Solid
$\Delta_f H°$/kJ mol^{-1}:				d:	
$S°$/J mol^{-1}K^{-1}:				η: N/A	
C_p/J mol^{-1}K^{-1}:				k:	

COMMENTS: Highly toxic

Name: Toluene-2,4-diisocyanate
Synonyms: 2,4-Diisocyanato-1-methylbenzene
           2,4-Diisocyanatotoluene
Mol. Form.: $C_9H_6N_2O_2$

CAS RN: 584-84-9
Merck No.: 9456
Mol. Wt.: 174.159

PHYSICAL CONSTANTS

T_m: 20.5°C (293.6 K)
T_b: 251°C (524 K)

T_c:
P_c:

μ:
IP:

TRANSITION PROPERTIES

$\Delta_{fus}H$ (T_m):
$\Delta_{vap}H$ (T_b):
$\Delta_{vap}H$ (25°C):

Vapor pressure (0°C):
Vapor pressure (25°C):
Vapor pressure (100°C):

PROPERTIES AT 25° AND 100 kPa

	Solid	Liquid	Gas		Liquid
$\Delta_f H°$/kJ mol^{-1}:				d: 1.221 g/mL	
$S°$/J mol^{-1}K^{-1}:				η:	
C_p/J mol^{-1}K^{-1}:				k:	

COMMENTS: TLV=0.005 ppm; carcinogen; highly toxic

Name: *o*-Toluic acid
Synonym: 2-Methylbenzoic acid

Mol. Form.: $C_8H_8O_2$

CAS RN: 118-90-1
Merck No.: 9461
Mol. Wt.: 136.150

PHYSICAL CONSTANTS

T_m: 103.7°C (376.8 K)
T_b:

T_c:
P_c:

μ:
IP: 9.10 eV

TRANSITION PROPERTIES

$\Delta_{fus}H$ (T_m): 20.17 kJ/mol
$\Delta_{vap}H$ (T_b):
$\Delta_{vap}H$ (25°C):

Vapor pressure (0°C):
Vapor pressure (25°C):
Vapor pressure (100°C):

PROPERTIES AT 25° AND 100 kPa

	Solid	Liquid	Gas		Solid
$\Delta_f H°$/kJ mol^{-1}:	-416.5			d: 1.073 g/mL	
$S°$/J mol^{-1}K^{-1}:				η: N/A	
C_p/J mol^{-1}K^{-1}:	174.9			k:	

COMMENTS:

Name: *p*-Toluic acid   CAS RN: 99-94-5
Synonyms: 4-Methylbenzoic acid   Merck No.: 9461
          Crithminic acid   Mol. Wt.: 136.150
Mol. Form.: $C_8H_8O_2$

PHYSICAL CONSTANTS

T_m: 179.6°C (452.7 K) T_c: μ:
T_b: P_c: IP: 9.23 eV

TRANSITION PROPERTIES

$\Delta_{fus}H$ (T_m): 22.73 kJ/mol   Vapor pressure (0°C):
$\Delta_{vap}H$ (T_b):   Vapor pressure (25°C):
$\Delta_{vap}H$ (25°C):   Vapor pressure (100°C):

PROPERTIES AT 25° AND 100 kPa

	Solid	Liquid	Gas		Solid
$\Delta_f H°$/kJ mol^{-1}:	-429.2			*d*:	1.049 g/mL
$S°$/J mol^{-1}K^{-1}:				η:	N/A
C_p/J mol^{-1}K^{-1}:	169.0			*k*:	

COMMENTS:

Name: *o*-Tolunitrile   CAS RN: 529-19-1
Synonyms: 2-Methylbenzonitrile   Merck No.: 9463
          2-Cyanotoluene   Mol. Wt.: 117.150
Mol. Form.: C_8H_7N

PHYSICAL CONSTANTS

T_m: -13.5°C (259.6 K) T_c: μ:
T_b: 205°C (478 K) P_c: IP: 9.38 eV

TRANSITION PROPERTIES

$\Delta_{fus}H$ (T_m):   Vapor pressure (0°C):
$\Delta_{vap}H$ (T_b):   Vapor pressure (25°C):
$\Delta_{vap}H$ (25°C):   Vapor pressure (100°C):

PROPERTIES AT 25° AND 100 kPa

	Solid	Liquid	Gas		Liquid
$\Delta_f H°$/kJ mol^{-1}:				*d*:	0.991 g/mL
$S°$/J mol^{-1}K^{-1}:				η:	
C_p/J mol^{-1}K^{-1}:				*k*:	

COMMENTS: Highly toxic

Name: *p*-Tolunitrile   CAS RN: 104-85-8
Synonyms: 4-Methylbenzonitrile   Merck No.: 9464
          4-Cyanotoluene   Mol. Wt.: 117.150
Mol. Form.: C_8H_7N

PHYSICAL CONSTANTS

T_m: 29.5°C (302.6 K) T_c: 450°C (723 K) μ:
T_b: 217.05°C (490.20 K) P_c: IP: 9.32 eV

TRANSITION PROPERTIES

$\Delta_{fus}H$ (T_m):   Vapor pressure (0°C):
$\Delta_{vap}H$ (T_b):   Vapor pressure (25°C):
$\Delta_{vap}H$ (25°C):   Vapor pressure (100°C):

PROPERTIES AT 25° AND 100 kPa

	Solid	Liquid	Gas		Solid
$\Delta_f H°$/kJ mol^{-1}:				*d*:	
$S°$/J mol^{-1}K^{-1}:				η:	N/A
C_p/J mol^{-1}K^{-1}:				*k*:	

COMMENTS: Highly toxic

Name: Triacetin
Synonyms: 1,2,3-Propanetriol triacetate
            Glycerol triacetate
Mol. Form.: $C_9H_{14}O_6$

CAS RN: 102-76-1
Merck No.: 9504
Mol. Wt.: 218.207

PHYSICAL CONSTANTS

T_m: -78°C (195 K) T_c: μ:
T_b: 259°C (532 K) P_c: IP:

TRANSITION PROPERTIES

$\Delta_{fus}H$ (T_m):   Vapor pressure (0°C):
$\Delta_{vap}H$ (T_b): 57.80 kJ/mol   Vapor pressure (25°C):
$\Delta_{vap}H$ (25°C): 85.74 kJ/mol   Vapor pressure (100°C):

PROPERTIES AT 25° AND 100 kPa

	Solid	Liquid	Gas		Liquid
$\Delta_f H°$/kJ mol^{-1}:		-1330.8	-1248.8		d: 1.154 g/mL
$S°$/J mol^{-1}K^{-1}:		458.3			η: 16 mPa s
C_p/J mol^{-1}K^{-1}:		384.7			k:
COMMENTS:					

Name: s-Triazaborane
Synonyms: Cyclotriborazane
            Borazine
Mol. Form.: $B_3H_6N_3$

CAS RN: 6569-51-3
Merck No.: 9516
Mol. Wt.: 80.501

PHYSICAL CONSTANTS

T_m: -58°C (215 K) T_c: μ:
T_b: 53°C (326 K) P_c: IP:

TRANSITION PROPERTIES

$\Delta_{fus}H$ (T_m):   Vapor pressure (0°C):
$\Delta_{vap}H$ (T_b):   Vapor pressure (25°C):
$\Delta_{vap}H$ (25°C):   Vapor pressure (100°C):

PROPERTIES AT 25° AND 100 kPa

	Solid	Liquid	Gas		Liquid
$\Delta_f H°$/kJ mol^{-1}:		-541.0			d: 0.80 g/mL
$S°$/J mol^{-1}K^{-1}:		199.6			η:
C_p/J mol^{-1}K^{-1}:					k:
COMMENTS:					

Name: Tribromomethane
Synonym: Bromoform

Mol. Form.: $CHBr_3$

CAS RN: 75-25-2
Merck No.: 1407
Mol. Wt.: 252.731

PHYSICAL CONSTANTS

T_m: 8.05°C (281.20 K) T_c: μ: 0.99 D
T_b: 149.1°C (422.2 K) P_c: IP: 10.48 eV

TRANSITION PROPERTIES

$\Delta_{fus}H$ (T_m):   Vapor pressure (0°C):
$\Delta_{vap}H$ (T_b): 39.66 kJ/mol   Vapor pressure (25°C): 0.726 kPa
$\Delta_{vap}H$ (25°C): 46.05 kJ/mol   Vapor pressure (100°C): 22.7 kPa

PROPERTIES AT 25° AND 100 kPa

	Solid	Liquid	Gas		Liquid
$\Delta_f H°$/kJ mol^{-1}:		-28.5	17.0		d: 2.8761 g/mL
$S°$/J mol^{-1}K^{-1}:		220.9	330.9		η: 1.86 mPa s
C_p/J mol^{-1}K^{-1}:		130.7	71.2		k:
COMMENTS: TLV=0.5 ppm; highly toxic					

Name: Tribromosilane   CAS RN: 7789-57-3
Synonym: Silicobromoform   Merck No.: 9528
                                                                Mol. Wt.: 268.805
Mol. Form.: Br$_3$HSi

PHYSICAL CONSTANTS

| T_m: -73°C (200 K) | T_c: 336.9°C (610.0 K) | μ: |
| T_b: 109°C (382 K) | P_c: | IP: |

TRANSITION PROPERTIES

$\Delta_{fus}H$ (T_m):   Vapor pressure (0°C):
$\Delta_{vap}H$ (T_b): 34.80 kJ/mol   Vapor pressure (25°C):
$\Delta_{vap}H$ (25°C): 38.0 kJ/mol   Vapor pressure (100°C):

PROPERTIES AT 25° AND 100 kPa

	Solid	Liquid	Gas		Liquid
$\Delta_f H°$/kJ mol^{-1}:		-355.6	-317.6	d:	2.7 g/mL
$S°$/J mol^{-1}K^{-1}:		248.1	348.6	η:	
C_p/J mol^{-1}K^{-1}:			80.8	k:	
COMMENTS:					

Name: Tributylamine   CAS RN: 102-82-9
Synonym: *N,N*-Dibutyl-1-butanamine   Merck No.: 9530
                                                                Mol. Wt.: 185.353
Mol. Form.: C$_{12}$H$_{27}$N

PHYSICAL CONSTANTS

| T_m: -70°C (203 K) | T_c: | μ: |
| T_b: 216.5°C (489.6 K) | P_c: | IP: 7.40 eV |

TRANSITION PROPERTIES

$\Delta_{fus}H$ (T_m):   Vapor pressure (0°C):
$\Delta_{vap}H$ (T_b): 46.90 kJ/mol   Vapor pressure (25°C): 0.010 kPa
$\Delta_{vap}H$ (25°C):   Vapor pressure (100°C): 2.69 kPa

PROPERTIES AT 25° AND 100 kPa

	Solid	Liquid	Gas		Liquid
$\Delta_f H°$/kJ mol^{-1}:		-281.6		d:	0.7748 g/mL
$S°$/J mol^{-1}K^{-1}:				η:	1.31 mPa s
C_p/J mol^{-1}K^{-1}:				k:	
COMMENTS:					

Name: Trichloroacetaldehyde   CAS RN: 75-87-6
Synonyms: Chloral   Merck No.: 9538
          Trichloroethanal   Mol. Wt.: 147.387
Mol. Form.: C$_2$HCl$_3$O

PHYSICAL CONSTANTS

| T_m: -57.5°C (215.6 K) | T_c: | μ: |
| T_b: 97.8°C (370.9 K) | P_c: | IP: |

TRANSITION PROPERTIES

$\Delta_{fus}H$ (T_m):   Vapor pressure (0°C): 1.80 kPa
$\Delta_{vap}H$ (T_b):   Vapor pressure (25°C): 6.66 kPa
$\Delta_{vap}H$ (25°C): 39.6 kJ/mol   Vapor pressure (100°C): 108 kPa

PROPERTIES AT 25° AND 100 kPa

	Solid	Liquid	Gas		Liquid
$\Delta_f H°$/kJ mol^{-1}:		-236.2	-196.6	d:	1.505 g/mL
$S°$/J mol^{-1}K^{-1}:				η:	
C_p/J mol^{-1}K^{-1}:		151.0		k:	
COMMENTS:					

Name: Trichloroacetic acid
Synonym: Trichloroethanoic acid

CAS RN: 76-03-9
Merck No.: 9539
Mol. Wt.: 163.387

Mol. Form.: $C_2HCl_3O_2$

PHYSICAL CONSTANTS

T_m: 57.5°C (330.6 K)
T_b: 196.5°C (469.6 K)

T_c:
P_c:

μ:
IP:

TRANSITION PROPERTIES

$\Delta_{fus}H$ (T_m): 5.88 kJ/mol
$\Delta_{vap}H$ (T_b):
$\Delta_{vap}H$ (25°C):

Vapor pressure (0°C):
Vapor pressure (25°C):
Vapor pressure (100°C): 2.47 kPa

PROPERTIES AT 25° AND 100 kPa

	Solid	Liquid	Gas	Solid
$\Delta_f H°$/kJ mol^{-1}:	-503.3			d:
$S°$/J mol^{-1}K^{-1}:				η: N/A
C_p/J mol^{-1}K^{-1}:				k:
COMMENTS:				

Name: Trichloroacetonitrile
Synonym: Trichloroethanenitrile

CAS RN: 545-06-2
Merck No.: 9540
Mol. Wt.: 144.387

Mol. Form.: C_2Cl_3N

PHYSICAL CONSTANTS

T_m: -42°C (231 K)
T_b: 85.7°C (358.8 K)

T_c:
P_c:

μ:
IP:

TRANSITION PROPERTIES

$\Delta_{fus}H$ (T_m):
$\Delta_{vap}H$ (T_b):
$\Delta_{vap}H$ (25°C):

Vapor pressure (0°C):
Vapor pressure (25°C): 9.89 kPa
Vapor pressure (100°C):

PROPERTIES AT 25° AND 100 kPa

	Solid	Liquid	Gas	Liquid
$\Delta_f H°$/kJ mol^{-1}:				d: 1.4403 g/mL
$S°$/J mol^{-1}K^{-1}:			336.6	η:
C_p/J mol^{-1}K^{-1}:			96.1	k:
COMMENTS: Highly toxic				

Name: Trichloroacetyl chloride
Synonym: Trichloroethanoyl chloride

CAS RN: 76-02-8
Merck No.:
Mol. Wt.: 181.832

Mol. Form.: C_2Cl_4O

PHYSICAL CONSTANTS

T_m:
T_b: 117.95°C (391.10 K)

T_c:
P_c:

μ:
IP: 11.00 eV

TRANSITION PROPERTIES

$\Delta_{fus}H$ (T_m):
$\Delta_{vap}H$ (T_b):
$\Delta_{vap}H$ (25°C): 41.0 kJ/mol

Vapor pressure (0°C):
Vapor pressure (25°C): 2.77 kPa
Vapor pressure (100°C): 59.2 kPa

PROPERTIES AT 25° AND 100 kPa

	Solid	Liquid	Gas	Liquid
$\Delta_f H°$/kJ mol^{-1}:		-280.8	-239.8	d: 1.613 g/mL
$S°$/J mol^{-1}K^{-1}:				η:
C_p/J mol^{-1}K^{-1}:				k:
COMMENTS: Highly toxic				

Name: 1,2,4-Trichlorobenzene

CAS RN: 120-82-1
Merck No.: 9543
Mol. Wt.: 181.448

Mol. Form.: $C_6H_3Cl_3$

PHYSICAL CONSTANTS

T_m: 17°C (290 K)	T_c:	μ:
T_b: 213.5°C (486.6 K)	P_c:	IP: 9.04 eV

TRANSITION PROPERTIES

$\Delta_{fus}H$ (T_m):	Vapor pressure (0°C):
$\Delta_{vap}H$ (T_b):	Vapor pressure (25°C):
$\Delta_{vap}H$ (25°C):	Vapor pressure (100°C):

PROPERTIES AT 25° AND 100 kPa

	Solid	Liquid	Gas		Liquid
$\Delta_f H°$/kJ mol^{-1}:				d:	1.459 g/mL
$S°$/J mol^{-1}K^{-1}:				η:	
C_p/J mol^{-1}K^{-1}:		194.6		k:	
COMMENTS: TLV=5 ppm; highly toxic					

Name: 1,3,5-Trichlorobenzene

CAS RN: 108-70-3
Merck No.: 9544
Mol. Wt.: 181.448

Mol. Form.: $C_6H_3Cl_3$

PHYSICAL CONSTANTS

T_m: 63.5°C (336.6 K)	T_c:	μ: 0 D
T_b: 208°C (481 K)	P_c:	IP: 9.32 eV

TRANSITION PROPERTIES

$\Delta_{fus}H$ (T_m): 18.20 kJ/mol	Vapor pressure (0°C):
$\Delta_{vap}H$ (T_b):	Vapor pressure (25°C):
$\Delta_{vap}H$ (25°C):	Vapor pressure (100°C):

PROPERTIES AT 25° AND 100 kPa

	Solid	Liquid	Gas		Solid
$\Delta_f H°$/kJ mol^{-1}:				d:	1.66 g/mL
$S°$/J mol^{-1}K^{-1}:				η:	N/A
C_p/J mol^{-1}K^{-1}:				k:	
COMMENTS:					

Name: 1,1,1-Trichloroethane
Synonym: Methyl chloroform

CAS RN: 71-55-6
Merck No.: 9549
Mol. Wt.: 133.404

Mol. Form.: $C_2H_3Cl_3$

PHYSICAL CONSTANTS

T_m: -30.4°C (242.7 K)	T_c: 272°C (545 K)	μ: 1.755 D
T_b: 74.09°C (347.24 K)	P_c: 4.30 MPa	IP: 11.00 eV

TRANSITION PROPERTIES

$\Delta_{fus}H$ (T_m): 2.73 kJ/mol	Vapor pressure (0°C): 4.80 kPa
$\Delta_{vap}H$ (T_b): 29.86 kJ/mol	Vapor pressure (25°C): 16.5 kPa
$\Delta_{vap}H$ (25°C): 32.50 kJ/mol	Vapor pressure (100°C): 211 kPa

PROPERTIES AT 25° AND 100 kPa

	Solid	Liquid	Gas		Liquid
$\Delta_f H°$/kJ mol^{-1}:		-177.4	-144.6	d:	1.3303 g/mL
$S°$/J mol^{-1}K^{-1}:		227.4	323.1	η:	0.793 mPa s
C_p/J mol^{-1}K^{-1}:		144.3	93.3	k:	0.101 W/m K
COMMENTS: TLV=350 ppm; highly toxic					

Name: 1,1,2-Trichloroethane
Synonym: Vinyl trichloride

Mol. Form.: $C_2H_3Cl_3$

CAS RN: 79-00-5
Merck No.: 9550
Mol. Wt.: 133.404

PHYSICAL CONSTANTS

T_m: -36.6°C (236.5 K)
T_b: 113.8°C (386.9 K)

T_c:
P_c:

μ:
IP: 11.00 eV

TRANSITION PROPERTIES

$\Delta_{fus}H$ (T_m): 11.54 kJ/mol
$\Delta_{vap}H$ (T_b): 34.82 kJ/mol
$\Delta_{vap}H$ (25°C): 40.24 kJ/mol

Vapor pressure (0°C):
Vapor pressure (25°C): 3.10 kPa
Vapor pressure (100°C): 66.6 kPa

PROPERTIES AT 25° AND 100 kPa

	Solid	Liquid	Gas		Liquid
$\Delta_f H°$/kJ mol^{-1}:		-191.5	-151.2		d: 1.4346 g/mL
$S°$/J mol^{-1}K^{-1}:		232.6	337.2		η: 1.10 mPa s
C_p/J mol^{-1}K^{-1}:		150.9	89.0		k:

COMMENTS: TLV=10 ppm; carcinogen; highly toxic

Name: Trichloroethylene
Synonym: Trichloroethene

Mol. Form.: C_2HCl_3

CAS RN: 79-01-6
Merck No.: 9552
Mol. Wt.: 131.388

PHYSICAL CONSTANTS

T_m: -84.75°C (188.40 K)
T_b: 87.21°C (360.36 K)

T_c: 271.1°C (544.2 K)
P_c: 5.02 MPa

μ:
IP: 9.47 eV

TRANSITION PROPERTIES

$\Delta_{fus}H$ (T_m):
$\Delta_{vap}H$ (T_b): 31.40 kJ/mol
$\Delta_{vap}H$ (25°C): 34.54 kJ/mol

Vapor pressure (0°C): 2.70 kPa
Vapor pressure (25°C): 9.50 kPa
Vapor pressure (100°C):

PROPERTIES AT 25° AND 100 kPa

	Solid	Liquid	Gas		Liquid
$\Delta_f H°$/kJ mol^{-1}:		-43.6	-8.1		d: 1.4578 g/mL
$S°$/J mol^{-1}K^{-1}:		228.4	324.8		η: 0.545 mPa s
C_p/J mol^{-1}K^{-1}:		124.4	80.3		k: 0.116 W/m K

COMMENTS: TLV=50 ppm; carcinogen; highly toxic

Name: Trichlorofluoromethane
Synonyms: Refrigerant 11
         CFC-11
Mol. Form.: CCl_3F

CAS RN: 75-69-4
Merck No.: 9553
Mol. Wt.: 137.368

PHYSICAL CONSTANTS

T_m: -111.11°C (162.04 K)
T_b: 23.75°C (296.90 K)

T_c: 198.1°C (471.2 K)
P_c: 4.41 MPa

μ: 0.46 D
IP: 11.77 eV

TRANSITION PROPERTIES

$\Delta_{fus}H$ (T_m): 6.90 kJ/mol
$\Delta_{vap}H$ (T_b): 25.06 kJ/mol
$\Delta_{vap}H$ (25°C): 25.02 kJ/mol

Vapor pressure (0°C): 40.3 kPa
Vapor pressure (25°C): 106 kPa
Vapor pressure (100°C): 824 kPa

PROPERTIES AT 25° AND 100 kPa

	Solid	Liquid	Gas		Gas
$\Delta_f H°$/kJ mol^{-1}:		-301.3	-268.3		d: 5.615 g/L
$S°$/J mol^{-1}K^{-1}:		225.4			η:
C_p/J mol^{-1}K^{-1}:		121.6	78.1		k:

COMMENTS: TLV=1000 ppm; highly toxic

Name: Trichloromethane   CAS RN: 67-66-3
Synonyms: Chloroform   Merck No.: 2141
        Refrigerant 20   Mol. Wt.: 119.377
Mol. Form.: $CHCl_3$

PHYSICAL CONSTANTS

T_m: -63.6°C (209.5 K) T_c: 263.3°C (536.4 K)   μ: 1.04 D
T_b: 61.17°C (334.32 K) P_c: 5.47 MPa   IP: 11.37 eV

TRANSITION PROPERTIES

$\Delta_{fus}H$ (T_m): 8.80 kJ/mol   Vapor pressure (0°C): 8.02 kPa
$\Delta_{vap}H$ (T_b): 29.24 kJ/mol   Vapor pressure (25°C): 26.2 kPa
$\Delta_{vap}H$ (25°C): 31.28 kJ/mol   Vapor pressure (100°C): 308 kPa

PROPERTIES AT 25° AND 100 kPa

	Solid	Liquid	Gas		Liquid
$\Delta_f H°$/kJ mol^{-1}:		-134.5	-103.1		d: 1.480 g/mL
$S°$/J mol^{-1}K^{-1}:		201.7	295.7		η: 0.537 mPa s
C_p/J mol^{-1}K^{-1}:		114.2	65.7		k: 0.117 W/m K

COMMENTS: TLV=10 ppm; carcinogen; highly toxic

Name: (Trichloromethyl)benzene   CAS RN: 98-07-7
Synonym: Benzotrichloride   Merck No.: 1120
                                          Mol. Wt.: 195.475

Mol. Form.: $C_7H_5Cl_3$

PHYSICAL CONSTANTS

T_m: -5°C (268 K) T_c: μ:
T_b: 221°C (494 K) P_c: IP: 9.60 eV

TRANSITION PROPERTIES

$\Delta_{fus}H$ (T_m):   Vapor pressure (0°C):
$\Delta_{vap}H$ (T_b):   Vapor pressure (25°C):
$\Delta_{vap}H$ (25°C):   Vapor pressure (100°C):

PROPERTIES AT 25° AND 100 kPa

	Solid	Liquid	Gas		Liquid
$\Delta_f H°$/kJ mol^{-1}:					d: 1.365 g/mL
$S°$/J mol^{-1}K^{-1}:					η:
C_p/J mol^{-1}K^{-1}:					k:

COMMENTS: Carcinogen; highly toxic

Name: 2,4,6-Trichlorophenol   CAS RN: 88-06-2
Synonym: Phenachlor   Merck No.: 9556
                                          Mol. Wt.: 197.447

Mol. Form.: $C_6H_3Cl_3O$

PHYSICAL CONSTANTS

T_m: 69°C (342 K) T_c: μ:
T_b: 246°C (519 K) P_c: IP:

TRANSITION PROPERTIES

$\Delta_{fus}H$ (T_m):   Vapor pressure (0°C):
$\Delta_{vap}H$ (T_b):   Vapor pressure (25°C):
$\Delta_{vap}H$ (25°C):   Vapor pressure (100°C): 0.495 kPa

PROPERTIES AT 25° AND 100 kPa

	Solid	Liquid	Gas		Solid
$\Delta_f H°$/kJ mol^{-1}:					d: 1.490 g/mL
$S°$/J mol^{-1}K^{-1}:					η: N/A
C_p/J mol^{-1}K^{-1}:					k:

COMMENTS: Carcinogen; highly toxic

Name: 1,2,3-Trichloropropane
Synonyms: Allyl trichloride
          Trichlorohydrin
Mol. Form.: $C_3H_5Cl_3$

CAS RN: 96-18-4
Merck No.:
Mol. Wt.: 147.431

PHYSICAL CONSTANTS

T_m: -14.7°C (258.4 K)
T_b: 157°C (430 K)

T_c:
P_c:

μ:
IP:

TRANSITION PROPERTIES

$\Delta_{fus}H$ (T_m):
$\Delta_{vap}H$ (T_b): 37.12 kJ/mol
$\Delta_{vap}H$ (25°C): 46.94 kJ/mol

Vapor pressure (0°C):
Vapor pressure (25°C): 0.492 kPa
Vapor pressure (100°C): 17.8 kPa

PROPERTIES AT 25° AND 100 kPa

	Solid	Liquid	Gas		Liquid
$\Delta_f H°$/kJ mol^{-1}:		-230.6	-182.9		d: 1.382 g/mL
$S°$/J mol^{-1}K^{-1}:					η: 2.23 mPa s
C_p/J mol^{-1}K^{-1}:		183.6			k:

COMMENTS: Highly toxic

Name: Trichlorosilane
Synonym: Silicochloroform

Mol. Form.: Cl_3HSi

CAS RN: 10025-78-2
Merck No.: 9559
Mol. Wt.: 135.452

PHYSICAL CONSTANTS

T_m: -128.2°C (144.9 K)
T_b: 33°C (306 K)

T_c: 206°C (479 K)
P_c:

μ: 0.86 D
IP: 11.70 eV

TRANSITION PROPERTIES

$\Delta_{fus}H$ (T_m):
$\Delta_{vap}H$ (T_b):
$\Delta_{vap}H$ (25°C): 25.70 kJ/mol

Vapor pressure (0°C):
Vapor pressure (25°C):
Vapor pressure (100°C):

PROPERTIES AT 25° AND 100 kPa

	Solid	Liquid	Gas		Liquid
$\Delta_f H°$/kJ mol^{-1}:		-539.3	-513.0		d: 1.3313 g/mL
$S°$/J mol^{-1}K^{-1}:		227.6	313.9		η: 0.326 mPa s
C_p/J mol^{-1}K^{-1}:			75.8		k:

COMMENTS: Highly toxic; very flammable

Name: 1,1,2-Trichlorotrifluoroethane
Synonyms: Refrigerant 113
          CFC-113
Mol. Form.: $C_2Cl_3F_3$

CAS RN: 76-13-1
Merck No.:
Mol. Wt.: 187.375

PHYSICAL CONSTANTS

T_m: -35°C (238 K)
T_b: 47.7°C (320.8 K)

T_c: 214.2°C (487.3 K)
P_c: 3.42 MPa

μ:
IP: 11.99 eV

TRANSITION PROPERTIES

$\Delta_{fus}H$ (T_m): 2.47 kJ/mol
$\Delta_{vap}H$ (T_b): 27.04 kJ/mol
$\Delta_{vap}H$ (25°C): 28.40 kJ/mol

Vapor pressure (0°C): 16.3 kPa
Vapor pressure (25°C): 46.4 kPa
Vapor pressure (100°C): 437 kPa

PROPERTIES AT 25° AND 100 kPa

	Solid	Liquid	Gas		Liquid
$\Delta_f H°$/kJ mol^{-1}:		-805.8	-777.3		d: 1.5642 g/mL
$S°$/J mol^{-1}K^{-1}:					η: 0.656 mPa s
C_p/J mol^{-1}K^{-1}:		170.1			k:

COMMENTS: TLV=1000 ppm; highly toxic

Name: Tridecane		CAS RN: 629-50-5
		Merck No.:
		Mol. Wt.: 184.365

Mol. Form.: $C_{13}H_{28}$

PHYSICAL CONSTANTS

T_m: -5.39°C (267.76 K)	T_c: 403°C (676 K)	μ:
T_b: 235.47°C (508.62 K)	P_c: 1.71 MPa	IP:

TRANSITION PROPERTIES

$\Delta_{fus}H$ (T_m): 28.50 kJ/mol	Vapor pressure (0°C):
$\Delta_{vap}H$ (T_b): 45.65 kJ/mol	Vapor pressure (25°C): 0.005 kPa
$\Delta_{vap}H$ (25°C): 66.43 kJ/mol	Vapor pressure (100°C):

PROPERTIES AT 25° AND 100 kPa

	Solid	Liquid	Gas		Liquid
$\Delta_f H°$/kJ mol^{-1}:				d:	0.7527 g/mL
$S°$/J mol^{-1}K^{-1}:				η:	1.72 mPa s
C_p/J mol^{-1}K^{-1}:		406.7		k:	0.137 W/m K
COMMENTS:					

Name: 1-Tridecanol	CAS RN: 112-70-9
Synonym: Tridecyl alcohol	Merck No.:
	Mol. Wt.: 200.365

Mol. Form.: $C_{13}H_{28}O$

PHYSICAL CONSTANTS

T_m: 32.5°C (305.6 K)	T_c: 461°C (734 K)	μ:
T_b:	P_c: 1.935 MPa	IP:

TRANSITION PROPERTIES

$\Delta_{fus}H$ (T_m):	Vapor pressure (0°C):
$\Delta_{vap}H$ (T_b):	Vapor pressure (25°C):
$\Delta_{vap}H$ (25°C):	Vapor pressure (100°C):

PROPERTIES AT 25° AND 100 kPa

	Solid	Liquid	Gas		Solid
$\Delta_f H°$/kJ mol^{-1}:	-599.4			d:	
$S°$/J mol^{-1}K^{-1}:				η:	N/A
C_p/J mol^{-1}K^{-1}:				k:	
COMMENTS:					

Name: Triethanolamine	CAS RN: 102-71-6
Synonym: Tris(2-hydroxyethyl)amine	Merck No.: 9581
	Mol. Wt.: 149.190

Mol. Form.: $C_6H_{15}NO_3$

PHYSICAL CONSTANTS

T_m: 20.5°C (293.6 K)	T_c:	μ:
T_b: 335.4°C (608.5 K)	P_c:	IP: 7.90 eV

TRANSITION PROPERTIES

$\Delta_{fus}H$ (T_m):	Vapor pressure (0°C):
$\Delta_{vap}H$ (T_b):	Vapor pressure (25°C):
$\Delta_{vap}H$ (25°C):	Vapor pressure (100°C):

PROPERTIES AT 25° AND 100 kPa

	Solid	Liquid	Gas		Liquid
$\Delta_f H°$/kJ mol^{-1}:				d:	1.1205 g/mL
$S°$/J mol^{-1}K^{-1}:				η:	609 mPa s
C_p/J mol^{-1}K^{-1}:		389.0		k:	
COMMENTS: TLV=0.5 ppm					

Name: Triethylamine
Synonym: *N,N*-Diethylethanamine

Mol. Form.: $C_6H_{15}N$

CAS RN: 121-44-8
Merck No.: 9582
Mol. Wt.: 101.192

PHYSICAL CONSTANTS

T_m: -114.7°C (158.4 K)	T_c: 262.5°C (535.6 K)	μ: 0.66 D
T_b: 89°C (362 K)	P_c: 3.032 MPa	IP: 7.50 eV

TRANSITION PROPERTIES

$\Delta_{fus}H$ (T_m):
$\Delta_{vap}H$ (T_b): 31.01 kJ/mol
$\Delta_{vap}H$ (25°C): 34.84 kJ/mol

Vapor pressure (0°C):
Vapor pressure (25°C): 7.70 kPa
Vapor pressure (100°C):

PROPERTIES AT 25° AND 100 kPa

	Solid	Liquid	Gas		Liquid
$\Delta_f H°$/kJ mol^{-1}:		-127.7	-92.8		d: 0.7230 g/mL
$S°$/J mol^{-1}K^{-1}:					η: 0.347 mPa s
C_p/J mol^{-1}K^{-1}:		219.9			k:

COMMENTS: TLV=10 ppm; flammable

Name: Triethylene glycol
Synonym: 1,2-Bis(2-hydroxyethoxy)ethane

Mol. Form.: $C_6H_{14}O_4$

CAS RN: 112-27-6
Merck No.: 9585
Mol. Wt.: 150.175

PHYSICAL CONSTANTS

T_m: -7°C (266 K)	T_c:	μ:
T_b: 285°C (558 K)	P_c:	IP:

TRANSITION PROPERTIES

$\Delta_{fus}H$ (T_m):
$\Delta_{vap}H$ (T_b): 71.40 kJ/mol
$\Delta_{vap}H$ (25°C): 79.2 kJ/mol

Vapor pressure (0°C):
Vapor pressure (25°C):
Vapor pressure (100°C):

PROPERTIES AT 25° AND 100 kPa

	Solid	Liquid	Gas		Liquid
$\Delta_f H°$/kJ mol^{-1}:		-804.2	-725.0		d: 1.12 g/mL
$S°$/J mol^{-1}K^{-1}:					η:
C_p/J mol^{-1}K^{-1}:		327.6			k:

COMMENTS:

Name: Trifluoroacetic acid
Synonym: Trifluoroethanoic acid

Mol. Form.: $C_2HF_3O_2$

CAS RN: 76-05-1
Merck No.: 9595
Mol. Wt.: 114.024

PHYSICAL CONSTANTS

T_m: -15.25°C (257.90 K)	T_c: 218.2°C (491.3 K)	μ: 2.28 D
T_b: 73°C (346 K)	P_c: 3.258 MPa	IP: 11.46 eV

TRANSITION PROPERTIES

$\Delta_{fus}H$ (T_m):
$\Delta_{vap}H$ (T_b): 33.26 kJ/mol
$\Delta_{vap}H$ (25°C): 38.5 kJ/mol

Vapor pressure (0°C): 4.00 kPa
Vapor pressure (25°C): 15.1 kPa
Vapor pressure (100°C):

PROPERTIES AT 25° AND 100 kPa

	Solid	Liquid	Gas		Liquid
$\Delta_f H°$/kJ mol^{-1}:		-1069.9	-1031.4		d: 1.485 g/mL
$S°$/J mol^{-1}K^{-1}:					η: 0.808 mPa s
C_p/J mol^{-1}K^{-1}:					k:

COMMENTS:

Name: 1,1,1-Trifluoroethane
Synonyms: Methylfluoroform
             Refrigerant 143a
Mol. Form.: $C_2H_3F_3$

CAS RN: 420-46-2
Merck No.:
Mol. Wt.: 84.041

PHYSICAL CONSTANTS

T_m: -111.3°C (161.8 K) T_c: 73.2°C (346.3 K)   μ: 2.347 D
T_b: -47.55°C (225.60 K) P_c: 3.76 MPa   IP: 12.90 eV

TRANSITION PROPERTIES

$\Delta_{fus}H$ (T_m): 6.19 kJ/mol   Vapor pressure (0°C):
$\Delta_{vap}H$ (T_b):   Vapor pressure (25°C):
$\Delta_{vap}H$ (25°C):   Vapor pressure (100°C): N/A

PROPERTIES AT 25° AND 100 kPa

	Solid	Liquid	Gas		Gas
$\Delta_f H°$/kJ mol^{-1}:			-744.6		d: 3.435 g/L
$S°$/J mol^{-1}K^{-1}:			279.9		η:
C_p/J mol^{-1}K^{-1}:			78.2		k:
COMMENTS:					

Name: Trifluoromethane
Synonyms: Fluoroform
             Refrigerant 23
Mol. Form.: CHF_3

CAS RN: 75-46-7
Merck No.: 4102
Mol. Wt.: 70.014

PHYSICAL CONSTANTS

T_m: -155.18°C (117.97 K) T_c: 26.2°C (299.3 K)   μ: 1.651 D
T_b: -82.1°C (191.0 K) P_c: 4.858 MPa   IP: 13.86 eV

TRANSITION PROPERTIES

$\Delta_{fus}H$ (T_m):   Vapor pressure (0°C):
$\Delta_{vap}H$ (T_b): 18.4 kJ/mol   Vapor pressure (25°C):
$\Delta_{vap}H$ (25°C):   Vapor pressure (100°C): N/A

PROPERTIES AT 25° AND 100 kPa

	Solid	Liquid	Gas		Gas
$\Delta_f H°$/kJ mol^{-1}:			-695.4		d: 2.862 g/L
$S°$/J mol^{-1}K^{-1}:			259.7		η:
C_p/J mol^{-1}K^{-1}:			51.0		k:
COMMENTS:					

Name: (Trifluoromethyl)benzene
Synonym: Benzotrifluoride

Mol. Form.: $C_7H_5F_3$

CAS RN: 98-08-8
Merck No.: 1121
Mol. Wt.: 146.112

PHYSICAL CONSTANTS

T_m: -29.1°C (244.0 K) T_c: μ: 2.86 D
T_b: 102.1°C (375.2 K) P_c: IP: 9.69 eV

TRANSITION PROPERTIES

$\Delta_{fus}H$ (T_m): 13.46 kJ/mol   Vapor pressure (0°C):
$\Delta_{vap}H$ (T_b): 32.63 kJ/mol   Vapor pressure (25°C): 5.14 kPa
$\Delta_{vap}H$ (25°C): 37.60 kJ/mol   Vapor pressure (100°C): 95.4 kPa

PROPERTIES AT 25° AND 100 kPa

	Solid	Liquid	Gas		Liquid
$\Delta_f H°$/kJ mol^{-1}:					d: 1.181 g/mL
$S°$/J mol^{-1}K^{-1}:					η: 0.54 mPa s
C_p/J mol^{-1}K^{-1}:		188.4			k:
COMMENTS: Highly toxic; flammable					

Name: Triiodomethane
Synonym: Iodoform

Mol. Form.: CHI_3

CAS RN: 75-47-8
Merck No.: 4926
Mol. Wt.: 393.732

PHYSICAL CONSTANTS

T_m: 119°C (392 K)	T_c:	μ:
T_b: 218°C (491 K)	P_c:	IP: 9.25 eV

TRANSITION PROPERTIES

$\Delta_{fus}H$ (T_m):
$\Delta_{vap}H$ (T_b):
$\Delta_{vap}H$ (25°C):

Vapor pressure (0°C):
Vapor pressure (25°C):
Vapor pressure (100°C): 0.030 kPa

PROPERTIES AT 25° AND 100 kPa

	Solid	Liquid	Gas		Solid
$\Delta_f H°$/kJ mol^{-1}:	141.0			d:	4.008 g/mL
$S°$/J mol^{-1}K^{-1}:			356.2	η:	N/A
C_p/J mol^{-1}K^{-1}:			75.0	k:	
COMMENTS: TLV=0.6 ppm					

Name: Trimellitic anhydride
Synonym: 1,2,4-Benzenetricarboxylic anhydride

Mol. Form.: $C_9H_4O_5$

CAS RN: 552-30-7
Merck No.: 9617
Mol. Wt.: 192.128

PHYSICAL CONSTANTS

T_m: 162°C (435 K)	T_c:	μ:
T_b:	P_c:	IP:

TRANSITION PROPERTIES

$\Delta_{fus}H$ (T_m):
$\Delta_{vap}H$ (T_b):
$\Delta_{vap}H$ (25°C):

Vapor pressure (0°C):
Vapor pressure (25°C):
Vapor pressure (100°C):

PROPERTIES AT 25° AND 100 kPa

	Solid	Liquid	Gas		Solid
$\Delta_f H°$/kJ mol^{-1}:				d:	
$S°$/J mol^{-1}K^{-1}:				η:	N/A
C_p/J mol^{-1}K^{-1}:				k:	
COMMENTS:					

Name: Trimethylamine
Synonym: N,N-Dimethylmethanamine

Mol. Form.: C_3H_9N

CAS RN: 75-50-3
Merck No.: 9625
Mol. Wt.: 59.111

PHYSICAL CONSTANTS

T_m: -117.1°C (156.0 K)	T_c: 159.64°C (432.79 K)	μ: 0.612 D
T_b: 2.87°C (276.02 K)	P_c: 4.087 MPa	IP: 7.82 eV

TRANSITION PROPERTIES

$\Delta_{fus}H$ (T_m): 6.55 kJ/mol
$\Delta_{vap}H$ (T_b): 22.94 kJ/mol
$\Delta_{vap}H$ (25°C): 21.66 kJ/mol

Vapor pressure (0°C): 90.7 kPa
Vapor pressure (25°C): 215 kPa
Vapor pressure (100°C):

PROPERTIES AT 25° AND 100 kPa

	Solid	Liquid	Gas		Gas
$\Delta_f H°$/kJ mol^{-1}:		-45.7	-23.7	d:	2.416 g/L
$S°$/J mol^{-1}K^{-1}:		208.5	287.1	η:	
C_p/J mol^{-1}K^{-1}:		137.9	91.8	k:	
COMMENTS: TLV=5 ppm; very flammable					

Name: 1,2,3-Trimethylbenzene
Synonym: Hemimellitene

Mol. Form.: C_9H_{12}

CAS RN: 526-73-8
Merck No.:
Mol. Wt.: 120.194

PHYSICAL CONSTANTS

T_m: -25.4°C (247.7 K)
T_b: 176.12°C (449.27 K)

T_c: 391.32°C (664.47 K)
P_c: 3.454 MPa

μ:
IP: 8.42 eV

TRANSITION PROPERTIES

$\Delta_{fus}H$ (T_m): 8.37 kJ/mol
$\Delta_{vap}H$ (T_b):
$\Delta_{vap}H$ (25°C): 49.05 kJ/mol

Vapor pressure (0°C):
Vapor pressure (25°C):
Vapor pressure (100°C): 9.44 kPa

PROPERTIES AT 25° AND 100 kPa

	Solid	Liquid	Gas	Liquid
$\Delta_f H°$/kJ mol^{-1}:		-58.5	-9.5	d: 0.890 g/mL
$S°$/J mol^{-1}K^{-1}:		267.9		η:
C_p/J mol^{-1}K^{-1}:		216.4		k:
COMMENTS:				

Name: 1,2,4-Trimethylbenzene
Synonym: Pseudocumene

Mol. Form.: C_9H_{12}

CAS RN: 95-63-6
Merck No.: 7929
Mol. Wt.: 120.194

PHYSICAL CONSTANTS

T_m: -43.8°C (229.3 K)
T_b: 169.38°C (442.53 K)

T_c: 376.02°C (649.17 K)
P_c: 3.232 MPa

μ:
IP: 8.27 eV

TRANSITION PROPERTIES

$\Delta_{fus}H$ (T_m): 3.76 kJ/mol
$\Delta_{vap}H$ (T_b):
$\Delta_{vap}H$ (25°C): 47.93 kJ/mol

Vapor pressure (0°C):
Vapor pressure (25°C): 0.300 kPa
Vapor pressure (100°C): 11.7 kPa

PROPERTIES AT 25° AND 100 kPa

	Solid	Liquid	Gas	Liquid
$\Delta_f H°$/kJ mol^{-1}:		-61.8	-13.8	d: 0.8723 g/mL
$S°$/J mol^{-1}K^{-1}:				η:
C_p/J mol^{-1}K^{-1}:		215.0		k:
COMMENTS: TLV=25 ppm; highly toxic				

Name: Trimethylborate
Synonyms: Boron trimethoxide
       Methyl borate
Mol. Form.: $C_3H_9BO_3$

CAS RN: 121-43-7
Merck No.: 9626
Mol. Wt.: 103.914

PHYSICAL CONSTANTS

T_m: -29.3°C (243.8 K)
T_b: 67.5°C (340.6 K)

T_c: 228.6°C (501.7 K)
P_c: 3.59 MPa

μ:
IP: 10.00 eV

TRANSITION PROPERTIES

$\Delta_{fus}H$ (T_m):
$\Delta_{vap}H$ (T_b):
$\Delta_{vap}H$ (25°C):

Vapor pressure (0°C):
Vapor pressure (25°C): 17.2 kPa
Vapor pressure (100°C):

PROPERTIES AT 25° AND 100 kPa

	Solid	Liquid	Gas	Liquid
$\Delta_f H°$/kJ mol^{-1}:				d: 0.915 g/mL
$S°$/J mol^{-1}K^{-1}:				η:
C_p/J mol^{-1}K^{-1}:				k:
COMMENTS: Flammable				

Name: 2,2,3-Trimethylbutane

Synonym: Triptane

Mol. Form.: C_7H_{16}

CAS RN: 464-06-2

Merck No.:

Mol. Wt.: 100.204

PHYSICAL CONSTANTS

T_m: -25°C (248 K)

T_b: 80.86°C (354.01 K)

T_c: 258.1°C (531.2 K)

P_c: 2.954 MPa

μ:

IP:

TRANSITION PROPERTIES

$\Delta_{fus}H$ (T_m): 2.20 kJ/mol

$\Delta_{vap}H$ (T_b): 28.90 kJ/mol

$\Delta_{vap}H$ (25°C): 32.05 kJ/mol

Vapor pressure (0°C): 9.13 kPa

Vapor pressure (25°C): 28.5 kPa

Vapor pressure (100°C): 172 kPa

PROPERTIES AT 25° AND 100 kPa

	Solid	Liquid	Gas		Liquid
$\Delta_f H°$/kJ mol^{-1}:		-236.5	-204.5		d: 0.6859 g/mL
$S°$/J mol^{-1}K^{-1}:		292.2			η:
C_p/J mol^{-1}K^{-1}:		213.5			k:
COMMENTS: Flammable					

Name: 2,2,5-Trimethylhexane

Mol. Form.: C_9H_{20}

CAS RN: 3522-94-9

Merck No.:

Mol. Wt.: 128.258

PHYSICAL CONSTANTS

T_m: -105.76°C (167.39 K)

T_b: 124.09°C (397.24 K)

T_c: 295°C (568 K)

P_c:

μ:

IP:

TRANSITION PROPERTIES

$\Delta_{fus}H$ (T_m): 6.19 kJ/mol

$\Delta_{vap}H$ (T_b): 33.65 kJ/mol

$\Delta_{vap}H$ (25°C): 40.16 kJ/mol

Vapor pressure (0°C):

Vapor pressure (25°C): 2.21 kPa

Vapor pressure (100°C): 50.0 kPa

PROPERTIES AT 25° AND 100 kPa

	Solid	Liquid	Gas		Liquid
$\Delta_f H°$/kJ mol^{-1}:		-293.3			d: 0.7032 g/mL
$S°$/J mol^{-1}K^{-1}:					η:
C_p/J mol^{-1}K^{-1}:					k:
COMMENTS: Flammable					

Name: 2,2,3-Trimethylpentane

Synonym: 2-*tert*-Butylbutane

Mol. Form.: C_8H_{18}

CAS RN: 564-02-3

Merck No.:

Mol. Wt.: 114.231

PHYSICAL CONSTANTS

T_m: -112.26°C (160.89 K)

T_b: 110°C (383 K)

T_c: 290.4°C (563.5 K)

P_c: 2.730 MPa

μ:

IP:

TRANSITION PROPERTIES

$\Delta_{fus}H$ (T_m): 8.62 kJ/mol

$\Delta_{vap}H$ (T_b): 31.94 kJ/mol

$\Delta_{vap}H$ (25°C): 36.91 kJ/mol

Vapor pressure (0°C):

Vapor pressure (25°C): 4.30 kPa

Vapor pressure (100°C): 76.5 kPa

PROPERTIES AT 25° AND 100 kPa

	Solid	Liquid	Gas		Liquid
$\Delta_f H°$/kJ mol^{-1}:		-256.9	-220.0		d: 0.7121 g/mL
$S°$/J mol^{-1}K^{-1}:					η: 0.6 mPa s
C_p/J mol^{-1}K^{-1}:		245.5			k:
COMMENTS: Flammable					

Name: 2,2,4-Trimethylpentane

Synonym: Isooctane

Mol. Form.: C_8H_{18}

CAS RN: 540-84-1

Merck No.: 5079

Mol. Wt.: 114.231

PHYSICAL CONSTANTS

T_m: -107.3°C (165.8 K)

T_b: 99.22°C (372.37 K)

T_c: 270.9°C (544.0 K)

P_c: 2.568 MPa

μ:

IP: 9.86 eV

TRANSITION PROPERTIES

$\Delta_{fus}H$ (T_m): 9.04 kJ/mol

$\Delta_{vap}H$ (T_b): 30.79 kJ/mol

$\Delta_{vap}H$ (25°C): 35.14 kJ/mol

Vapor pressure (0°C):

Vapor pressure (25°C): 6.50 kPa

Vapor pressure (100°C): 103 kPa

PROPERTIES AT 25° AND 100 kPa

	Solid	Liquid	Gas	Liquid
$\Delta_f H°$/kJ mol⁻¹:		-259.2	-224.0	d: 0.6878 g/mL
$S°$/J mol⁻¹K⁻¹:				η: 0.47 mPa s
C_p/J mol⁻¹K⁻¹:		239.1		k:

COMMENTS: Flammable

Name: 2,3,3-Trimethylpentane

Mol. Form.: C_8H_{18}

CAS RN: 560-21-4

Merck No.:

Mol. Wt.: 114.231

PHYSICAL CONSTANTS

T_m: -100.93°C (172.22 K)

T_b: 114.8°C (387.9 K)

T_c: 300.5°C (573.6 K)

P_c: 2.820 MPa

μ:

IP:

TRANSITION PROPERTIES

$\Delta_{fus}H$ (T_m):

$\Delta_{vap}H$ (T_b): 32.12 kJ/mol

$\Delta_{vap}H$ (25°C): 37.27 kJ/mol

Vapor pressure (0°C):

Vapor pressure (25°C): 3.60 kPa

Vapor pressure (100°C): 66.6 kPa

PROPERTIES AT 25° AND 100 kPa

	Solid	Liquid	Gas	Liquid
$\Delta_f H°$/kJ mol⁻¹:		-253.5	-216.3	d: 0.7223 g/mL
$S°$/J mol⁻¹K⁻¹:				η:
C_p/J mol⁻¹K⁻¹:		245.6		k:

COMMENTS: Flammable

Name: 2,3,4-Trimethylpentane

Mol. Form.: C_8H_{18}

CAS RN: 565-75-3

Merck No.:

Mol. Wt.: 114.231

PHYSICAL CONSTANTS

T_m: -109.2°C (163.9 K)

T_b: 113.5°C (386.6 K)

T_c: 293.4°C (566.5 K)

P_c: 2.730 MPa

μ:

IP:

TRANSITION PROPERTIES

$\Delta_{fus}H$ (T_m):

$\Delta_{vap}H$ (T_b): 32.36 kJ/mol

$\Delta_{vap}H$ (25°C): 37.75 kJ/mol

Vapor pressure (0°C):

Vapor pressure (25°C):

Vapor pressure (100°C): 68.7 kPa

PROPERTIES AT 25° AND 100 kPa

	Solid	Liquid	Gas	Liquid
$\Delta_f H°$/kJ mol⁻¹:		-255.0	-217.3	d: 0.7150 g/mL
$S°$/J mol⁻¹K⁻¹:		329.3		η:
C_p/J mol⁻¹K⁻¹:		247.3		k:

COMMENTS: Flammable

Name: Trinitroglycerol
Synonyms: Nitroglycerin
             1,2,3-Propanetriol trinitrate
Mol. Form.: $C_3H_5N_3O_9$

CAS RN: 55-63-0
Merck No.: 6528
Mol. Wt.: 227.088

PHYSICAL CONSTANTS

T_m: 13°C (286 K) T_c: μ:
T_b: P_c: IP:

TRANSITION PROPERTIES

$\Delta_{fus}H$ (T_m): 21.87 kJ/mol   Vapor pressure (0°C):
$\Delta_{vap}H$ (T_b):   Vapor pressure (25°C):
$\Delta_{vap}H$ (25°C): 100.0 kJ/mol   Vapor pressure (100°C):

PROPERTIES AT 25° AND 100 kPa

	Solid	Liquid	Gas		Liquid
$\Delta_f H°$/kJ mol^{-1}:		-370.9	-270.9		d: 1.592 g/mL
$S°$/J mol^{-1}K^{-1}:					η:
C_p/J mol^{-1}K^{-1}:					k:

COMMENTS: Highly toxic; explosive

Name: 1,3,5-Trioxane
Synonyms: Formaldehyde trimer
             Trioxymethylene
Mol. Form.: $C_3H_6O_3$

CAS RN: 110-88-3
Merck No.: 9646
Mol. Wt.: 90.079

PHYSICAL CONSTANTS

T_m: 60.2°C (333.3 K) T_c: μ: 2.08 D
T_b: 114.5°C (387.6 K) P_c: IP: 10.30 eV

TRANSITION PROPERTIES

$\Delta_{fus}H$ (T_m): 15.11 kJ/mol   Vapor pressure (0°C):
$\Delta_{vap}H$ (T_b):   Vapor pressure (25°C):
$\Delta_{vap}H$ (25°C):   Vapor pressure (100°C): 64.0 kPa

PROPERTIES AT 25° AND 100 kPa

	Solid	Liquid	Gas		Solid
$\Delta_f H°$/kJ mol^{-1}:	-522.5		-465.9		d:
$S°$/J mol^{-1}K^{-1}:	133.0				η: N/A
C_p/J mol^{-1}K^{-1}:	111.4				k:

COMMENTS:

Name: Trisilane
Synonym: Trisilicon octahydride

Mol. Form.: H_8Si_3

CAS RN: 7783-26-8
Merck No.: 9668
Mol. Wt.: 92.320

PHYSICAL CONSTANTS

T_m: -117.4°C (155.7 K) T_c: μ:
T_b: 53°C (326 K) P_c: IP: 9.20 eV

TRANSITION PROPERTIES

$\Delta_{fus}H$ (T_m):   Vapor pressure (0°C):
$\Delta_{vap}H$ (T_b): 28.50 kJ/mol   Vapor pressure (25°C):
$\Delta_{vap}H$ (25°C): 28.4 kJ/mol   Vapor pressure (100°C):

PROPERTIES AT 25° AND 100 kPa

	Solid	Liquid	Gas		Liquid
$\Delta_f H°$/kJ mol^{-1}:		92.5	120.9		d: 0.7 g/mL
$S°$/J mol^{-1}K^{-1}:					η:
C_p/J mol^{-1}K^{-1}:					k:

COMMENTS: Flammable

Name: Tungsten   CAS RN: 7440-33-7
                                                                  Merck No.: 9722
                                                                  Mol. Wt.: 183.840
Mol. Form.: W

PHYSICAL CONSTANTS

T_m: 3422°C (3695 K) T_c: μ:
T_b: 5555°C (5828 K) P_c: IP: 7.98 eV

TRANSITION PROPERTIES

$\Delta_{fus}H$ (T_m): 52.31 kJ/mol   Vapor pressure (0°C): N/A
$\Delta_{vap}H$ (T_b):   Vapor pressure (25°C): N/A
$\Delta_{vap}H$ (25°C):   Vapor pressure (100°C): N/A

PROPERTIES AT 25° AND 100 kPa

	Solid	Liquid	Gas		Solid
$\Delta_fH°$/kJ mol^{-1}:	0.0		849.4	d: 19.3 g/mL
$S°$/J mol^{-1}K^{-1}:	32.6		174.0	η: N/A
C_p/J mol^{-1}K^{-1}:	24.3		21.3	k: 173 W/m K
COMMENTS:				

Name: Tungsten fluoride (WF$_6$)   CAS RN: 7783-82-6
Synonyms: Tungsten hexafluoride   Merck No.: 9723
          Tungsten(VI) fluoride   Mol. Wt.: 297.830
Mol. Form.: F$_6$W

PHYSICAL CONSTANTS

T_m: 2.3°C (275.4 K) T_c: 171°C (444 K)   μ:
T_b: 17°C (290 K) P_c: 4.34 MPa   IP:

TRANSITION PROPERTIES

$\Delta_{fus}H$ (T_m): 4.10 kJ/mol   Vapor pressure (0°C):
$\Delta_{vap}H$ (T_b): 27.00 kJ/mol   Vapor pressure (25°C):
$\Delta_{vap}H$ (25°C): 26.60 kJ/mol   Vapor pressure (100°C):

PROPERTIES AT 25° AND 100 kPa

	Solid	Liquid	Gas		Gas
$\Delta_fH°$/kJ mol^{-1}:		-1747.7	-1721.7	d: 12.173 g/L
$S°$/J mol^{-1}K^{-1}:		251.5	341.1	η:
C_p/J mol^{-1}K^{-1}:			119.0	k:
COMMENTS:				

Name: Undecane   CAS RN: 1120-21-4
                                                                  Merck No.:
                                                                  Mol. Wt.: 156.312
Mol. Form.: C$_{11}$H$_{24}$

PHYSICAL CONSTANTS

T_m: -25.6°C (247.5 K) T_c: 365.70°C (638.85 K)   μ:
T_b: 195.93°C (469.08 K) P_c: 1.955 MPa   IP: 9.56 eV

TRANSITION PROPERTIES

$\Delta_{fus}H$ (T_m): 22.32 kJ/mol   Vapor pressure (0°C):
$\Delta_{vap}H$ (T_b): 42.5 kJ/mol   Vapor pressure (25°C):
$\Delta_{vap}H$ (25°C): 56.43 kJ/mol   Vapor pressure (100°C): 4.38 kPa

PROPERTIES AT 25° AND 100 kPa

	Solid	Liquid	Gas		Liquid
$\Delta_fH°$/kJ mol^{-1}:		-327.2	-270.9	d: 0.7365 g/mL
$S°$/J mol^{-1}K^{-1}:				η: 1.10 mPa s
C_p/J mol^{-1}K^{-1}:		344.9		k: 0.140 W/m K
COMMENTS:				

Name: Uranium

CAS RN: 7440-61-1
Merck No.: 9766
Mol. Wt.: 238.029

Mol. Form.: U

PHYSICAL CONSTANTS

T_m: 1135°C (1408 K) T_c: μ:
T_b: 4131°C (4404 K) P_c: IP: 6.19 eV

TRANSITION PROPERTIES

$\Delta_{fus}H$ (T_m): 9.14 kJ/mol   Vapor pressure (0°C): N/A
$\Delta_{vap}H$ (T_b):   Vapor pressure (25°C): N/A
$\Delta_{vap}H$ (25°C):   Vapor pressure (100°C): N/A

PROPERTIES AT 25° AND 100 kPa

	Solid	Liquid	Gas	Solid
$\Delta_f H°$/kJ mol^{-1}:	0.0		533.0	d: 19.1 g/mL
$S°$/J mol^{-1}K^{-1}:	50.2		199.8	η: N/A
C_p/J mol^{-1}K^{-1}:	27.7		23.7	k: 27.5 W/m K

COMMENTS: Highly toxic

Name: Uranium fluoride (UF$_4$)
Synonyms: Uranium tetrafluoride
          Uranium(IV) fluoride
Mol. Form.: F$_4$U

CAS RN: 10049-14-6
Merck No.: 9771
Mol. Wt.: 314.023

PHYSICAL CONSTANTS

T_m: 1036°C (1309 K) T_c: μ:
T_b: 1417°C (1690 K) P_c: IP:

TRANSITION PROPERTIES

$\Delta_{fus}H$ (T_m):   Vapor pressure (0°C): N/A
$\Delta_{vap}H$ (T_b):   Vapor pressure (25°C): N/A
$\Delta_{vap}H$ (25°C):   Vapor pressure (100°C): N/A

PROPERTIES AT 25° AND 100 kPa

	Solid	Liquid	Gas	Solid
$\Delta_f H°$/kJ mol^{-1}:	-1914.2		-1598.7	d: 6.7 g/mL
$S°$/J mol^{-1}K^{-1}:	151.7		368.0	η: N/A
C_p/J mol^{-1}K^{-1}:	116.0		91.2	k:

COMMENTS: Highly toxic

Name: Uranium fluoride (UF$_6$)
Synonyms: Uranium hexafluoride
          Uranium(VI) fluoride
Mol. Form.: F$_6$U

CAS RN: 7783-81-5
Merck No.: 9768
Mol. Wt.: 352.019

PHYSICAL CONSTANTS

T_t: 64.05°C (337.20 K)* T_c: 232.7°C (505.8 K)   μ:
T_s: 56.5°C (329.6 K)† P_c: 4.66 MPa   IP: 14.00 eV

TRANSITION PROPERTIES

$\Delta_{fus}H$ (T_m): 19.19 kJ/mol   Vapor pressure (0°C):
$\Delta_{vap}H$ (T_b):   Vapor pressure (25°C):
$\Delta_{vap}H$ (25°C):   Vapor pressure (100°C):

PROPERTIES AT 25° AND 100 kPa

	Solid	Liquid	Gas	Solid
$\Delta_f H°$/kJ mol^{-1}:	-2197.0		-2147.4	d: 5.09 g/mL
$S°$/J mol^{-1}K^{-1}:	227.6		377.9	η: N/A
C_p/J mol^{-1}K^{-1}:	166.8		129.6	k:

COMMENTS: Highly toxic. *Triple point. †Sublimation point.

Name: Urea   CAS RN: 57-13-6
Synonym: Carbamide   Merck No.: 9781
                                                                                   Mol. Wt.: 60.056
Mol. Form.: CH_4N_2O

PHYSICAL CONSTANTS

T_m: 132.7°C (405.8 K) T_c: μ:
T_b: P_c: IP: 9.70 eV

TRANSITION PROPERTIES

$\Delta_{fus}H$ (T_m):   Vapor pressure (0°C):
$\Delta_{vap}H$ (T_b):   Vapor pressure (25°C):
$\Delta_{vap}H$ (25°C):   Vapor pressure (100°C):

PROPERTIES AT 25° AND 100 kPa

	Solid	Liquid	Gas		Solid
$\Delta_f H°$/kJ mol^{-1}:	-333.6		-245.8		d: 1.32 g/mL
$S°$/J mol^{-1}K^{-1}:					η: N/A
C_p/J mol^{-1}K^{-1}:					k:
COMMENTS:					

Name: Vanadium   CAS RN: 7440-62-2
                                                                                   Merck No.: 9823
                                                                                   Mol. Wt.: 50.942
Mol. Form.: V

PHYSICAL CONSTANTS

T_m: 1910°C (2183 K) T_c: μ:
T_b: 3407°C (3680 K) P_c: IP: 6.75 eV

TRANSITION PROPERTIES

$\Delta_{fus}H$ (T_m): 21.50 kJ/mol   Vapor pressure (0°C): N/A
$\Delta_{vap}H$ (T_b):   Vapor pressure (25°C): N/A
$\Delta_{vap}H$ (25°C):   Vapor pressure (100°C): N/A

PROPERTIES AT 25° AND 100 kPa

	Solid	Liquid	Gas		Solid
$\Delta_f H°$/kJ mol^{-1}:	0.0		514.2		d: 6.0 g/mL
$S°$/J mol^{-1}K^{-1}:	28.9		182.3		η: N/A
C_p/J mol^{-1}K^{-1}:	24.9		26.0		k: 30.7 W/m K
COMMENTS:					

Name: Vanadium chloride (VCl_4)   CAS RN: 7632-51-1
Synonyms: Vanadium tetrachloride   Merck No.:
          Vanadium(IV) chloride   Mol. Wt.: 192.752
Mol. Form.: Cl_4V

PHYSICAL CONSTANTS

T_m: -25.7°C (247.4 K) T_c: μ:
T_b: 152°C (425 K) P_c: IP: 9.20 eV

TRANSITION PROPERTIES

$\Delta_{fus}H$ (T_m): 2.30 kJ/mol   Vapor pressure (0°C):
$\Delta_{vap}H$ (T_b): 41.40 kJ/mol   Vapor pressure (25°C):
$\Delta_{vap}H$ (25°C): 42.50 kJ/mol   Vapor pressure (100°C):

PROPERTIES AT 25° AND 100 kPa

	Solid	Liquid	Gas		Liquid
$\Delta_f H°$/kJ mol^{-1}:		-569.4	-525.5		d:
$S°$/J mol^{-1}K^{-1}:		255.0	362.4		η:
C_p/J mol^{-1}K^{-1}:			96.2		k:
COMMENTS:					

Name: Vanadium fluoride (VF$_5$)
Synonyms: Vanadium pentafluoride
        Vanadium(V) fluoride
Mol. Form.: F$_5$V

CAS RN: 7783-72-4
Merck No.: 9825
Mol. Wt.: 145.934

PHYSICAL CONSTANTS

T_m: 19.5°C (292.6 K)
T_b: 48.3°C (321.4 K)

T_c:
P_c:

μ:
IP:

TRANSITION PROPERTIES

$\Delta_{fus}H$ (T_m): 49.96 kJ/mol
$\Delta_{vap}H$ (T_b): 44.52 kJ/mol
$\Delta_{vap}H$ (25°C): 46.4 kJ/mol

Vapor pressure (0°C):
Vapor pressure (25°C):
Vapor pressure (100°C):

PROPERTIES AT 25° AND 100 kPa

	Solid	Liquid	Gas		Liquid
$\Delta_f H°$/kJ mol^{-1}:		-1480.3	-1433.9		d: 2.50 g/mL
$S°$/J mol^{-1}K^{-1}:		175.7	320.9		η:
C_p/J mol^{-1}K^{-1}:			98.6		k:
COMMENTS:					

Name: Vanadium oxide (V$_2$O$_5$)
Synonyms: Divanadium pentoxide
        Vanadium(V) oxide
Mol. Form.: O$_5$V$_2$

CAS RN: 1314-62-1
Merck No.: 9826
Mol. Wt.: 181.880

PHYSICAL CONSTANTS

T_m: 670°C (943 K)
T_b: 1800°C (2073 K)

T_c:
P_c:

μ:
IP:

TRANSITION PROPERTIES

$\Delta_{fus}H$ (T_m): 64.50 kJ/mol
$\Delta_{vap}H$ (T_b):
$\Delta_{vap}H$ (25°C):

Vapor pressure (0°C): N/A
Vapor pressure (25°C): N/A
Vapor pressure (100°C): N/A

PROPERTIES AT 25° AND 100 kPa

	Solid	Liquid	Gas		Solid
$\Delta_f H°$/kJ mol^{-1}:	-1550.6				d: 3.35 g/mL
$S°$/J mol^{-1}K^{-1}:	131.0				η: N/A
C_p/J mol^{-1}K^{-1}:	127.7				k:
COMMENTS: Highly toxic					

Name: Vanadium oxytrichloride
Synonym: Vanadyl trichloride

Mol. Form.: Cl$_3$OV

CAS RN: 7727-18-6
Merck No.: 9834
Mol. Wt.: 173.299

PHYSICAL CONSTANTS

T_m: -79°C (194 K)
T_b: 127°C (400 K)

T_c:
P_c:

μ:
IP: 11.60 eV

TRANSITION PROPERTIES

$\Delta_{fus}H$ (T_m):
$\Delta_{vap}H$ (T_b): 36.78 kJ/mol
$\Delta_{vap}H$ (25°C): 39.1 kJ/mol

Vapor pressure (0°C):
Vapor pressure (25°C):
Vapor pressure (100°C):

PROPERTIES AT 25° AND 100 kPa

	Solid	Liquid	Gas		Liquid
$\Delta_f H°$/kJ mol^{-1}:		-734.7	-695.6		d:
$S°$/J mol^{-1}K^{-1}:		244.3	344.3		η:
C_p/J mol^{-1}K^{-1}:			89.9		k:
COMMENTS:					

Name: Vinyl acetate

Synonym: Ethenyl acetate

Mol. Form.: $C_4H_6O_2$

CAS RN: 108-05-4

Merck No.: 9896

Mol. Wt.: 86.090

PHYSICAL CONSTANTS

T_m: -93.2°C (179.9 K)

T_b: 72.5°C (345.6 K)

T_c:

P_c:

μ:

IP: 9.19 eV

TRANSITION PROPERTIES

$\Delta_{fus}H$ (T_m):

$\Delta_{vap}H$ (T_b): 34.55 kJ/mol

$\Delta_{vap}H$ (25°C): 37.20 kJ/mol

Vapor pressure (0°C):

Vapor pressure (25°C): 15.4 kPa

Vapor pressure (100°C):

PROPERTIES AT 25° AND 100 kPa

	Solid	Liquid	Gas		Liquid
$\Delta_f H°$/kJ mol^{-1}:		-280.1	-314.9		d: 0.927 g/mL
$S°$/J mol^{-1}K^{-1}:					η:
C_p/J mol^{-1}K^{-1}:					k:

COMMENTS: TLV=10 ppm; carcinogen; flammable

Name: Water

Mol. Form.: H_2O

CAS RN: 7732-18-5

Merck No.: 9951

Mol. Wt.: 18.015

PHYSICAL CONSTANTS

T_m: 0.00°C (273.15 K)

T_b: 100.0°C (373.15 K)

T_c: 373.99°C (647.14 K)

P_c: 22.06 MPa

μ: 1.854 D

IP: 12.61 eV

TRANSITION PROPERTIES

$\Delta_{fus}H$ (T_m): 6.01 kJ/mol

$\Delta_{vap}H$ (T_b): 40.65 kJ/mol

$\Delta_{vap}H$ (25°C): 43.98 kJ/mol

Vapor pressure (0°C): 0.611 kPa

Vapor pressure (25°C): 3.17 kPa

Vapor pressure (100°C): 101 kPa

PROPERTIES AT 25° AND 100 kPa

	Solid	Liquid	Gas		Liquid
$\Delta_f H°$/kJ mol^{-1}:		-285.8	-241.8		d: 0.9970 g/mL
$S°$/J mol^{-1}K^{-1}:		70.0	188.8		η: 0.893 mPa s
C_p/J mol^{-1}K^{-1}:		75.3	33.6		k: 0.6071 W/m K

COMMENTS:

Name: Xenon

Mol. Form.: Xe

CAS RN: 7440-63-3

Merck No.: 9981

Mol. Wt.: 131.290

PHYSICAL CONSTANTS

T_m: -111.75°C (161.40 K)

T_b: -108.04°C (165.11 K)

T_c: 16.58°C (289.73 K)

P_c: 5.84 MPa

μ: 0 D

IP: 12.13 eV

TRANSITION PROPERTIES

$\Delta_{fus}H$ (T_m): 1.81 kJ/mol

$\Delta_{vap}H$ (T_b): 12.62 kJ/mol

$\Delta_{vap}H$ (25°C):

Vapor pressure (0°C): 4138 kPa

Vapor pressure (25°C): N/A

Vapor pressure (100°C): N/A

PROPERTIES AT 25° AND 100 kPa

	Solid	Liquid	Gas		Gas
$\Delta_f H°$/kJ mol^{-1}:			0.0		d: 5.366 g/L
$S°$/J mol^{-1}K^{-1}:			169.7		η: 23.1 μPa s
C_p/J mol^{-1}K^{-1}:			20.8		k: 0.0055 W/m K

COMMENTS:

Name: *o*-Xylene
Synonym: 1,2-Dimethylbenzene

Mol. Form.: C_8H_{10}

CAS RN: 95-47-6
Merck No.: 9988
Mol. Wt.: 106.167

PHYSICAL CONSTANTS

T_m: -25.2°C (247.9 K) T_c: 357.2°C (630.3 K)   μ: 0.640 D
T_b: 144.5°C (417.6 K) P_c: 3.730 MPa   IP: 8.56 eV

TRANSITION PROPERTIES

$\Delta_{fus}H$ (T_m): 13.61 kJ/mol   Vapor pressure (0°C):
$\Delta_{vap}H$ (T_b): 36.24 kJ/mol   Vapor pressure (25°C): 0.880 kPa
$\Delta_{vap}H$ (25°C): 43.43 kJ/mol   Vapor pressure (100°C): 26.5 kPa

PROPERTIES AT 25° AND 100 kPa

	Solid	Liquid	Gas		Liquid
$\Delta_f H°$/kJ mol^{-1}:		-24.4	19.1		*d*: 0.8759 g/mL
$S°$/J mol^{-1}K^{-1}:					η: 0.760 mPa s
C_p/J mol^{-1}K^{-1}:		186.1			*k*: 0.131 W/m K

COMMENTS: TLV=100 ppm; highly toxic; flammable

Name: *m*-Xylene
Synonym: 1,3-Dimethylbenzene

Mol. Form.: C_8H_{10}

CAS RN: 108-38-3
Merck No.: 9988
Mol. Wt.: 106.167

PHYSICAL CONSTANTS

T_m: -47.8°C (225.3 K) T_c: 343.90°C (617.05 K)   μ:
T_b: 139.12°C (412.27 K) P_c: 3.535 MPa   IP: 8.56 eV

TRANSITION PROPERTIES

$\Delta_{fus}H$ (T_m): 11.55 kJ/mol   Vapor pressure (0°C):
$\Delta_{vap}H$ (T_b): 35.66 kJ/mol   Vapor pressure (25°C): 1.13 kPa
$\Delta_{vap}H$ (25°C): 42.65 kJ/mol   Vapor pressure (100°C): 31.1 kPa

PROPERTIES AT 25° AND 100 kPa

	Solid	Liquid	Gas		Liquid
$\Delta_f H°$/kJ mol^{-1}:		-25.4	17.3		*d*: 0.8601 g/mL
$S°$/J mol^{-1}K^{-1}:					η: 0.581 mPa s
C_p/J mol^{-1}K^{-1}:		183.0			*k*: 0.130 W/m K

COMMENTS: TLV=100 ppm; highly toxic; flammable

Name: *p*-Xylene
Synonym: 1,4-Dimethylbenzene

Mol. Form.: C_8H_{10}

CAS RN: 106-42-3
Merck No.: 9988
Mol. Wt.: 106.167

PHYSICAL CONSTANTS

T_m: 13.2°C (286.3 K) T_c: 343.1°C (616.2 K)   μ: 0 D
T_b: 138.37°C (411.52 K) P_c: 3.511 MPa   IP: 8.44 eV

TRANSITION PROPERTIES

$\Delta_{fus}H$ (T_m): 16.81 kJ/mol   Vapor pressure (0°C):
$\Delta_{vap}H$ (T_b): 35.67 kJ/mol   Vapor pressure (25°C): 1.19 kPa
$\Delta_{vap}H$ (25°C): 42.40 kJ/mol   Vapor pressure (100°C): 32.1 kPa

PROPERTIES AT 25° AND 100 kPa

	Solid	Liquid	Gas		Liquid
$\Delta_f H°$/kJ mol^{-1}:		-24.4	18.0		*d*: 0.8566 g/mL
$S°$/J mol^{-1}K^{-1}:					η: 0.603 mPa s
C_p/J mol^{-1}K^{-1}:		181.5			*k*: 0.130 W/m K

COMMENTS: TLV=100 ppm; highly toxic; flammable

Name: 2,3-Xylenol   CAS RN: 526-75-0
Synonym: 2,3-Dimethylphenol   Merck No.: 9989
                                                                 Mol. Wt.: 122.167
Mol. Form.: $C_8H_{10}O$

PHYSICAL CONSTANTS

T_m: 72.8°C (345.9 K) T_c: 449.7°C (722.8 K)   μ:
T_b: 216.92°C (490.07 K) P_c: IP: 8.26 eV

TRANSITION PROPERTIES

$\Delta_{fus}H$ (T_m): 21.02 kJ/mol   Vapor pressure (0°C):
$\Delta_{vap}H$ (T_b):   Vapor pressure (25°C):
$\Delta_{vap}H$ (25°C):   Vapor pressure (100°C):

PROPERTIES AT 25° AND 100 kPa

	Solid	Liquid	Gas		Solid
$\Delta_f H°$/kJ mol^{-1}:	-241.1		-157.2		d: 0.942 g/mL
$S°$/J mol^{-1}K^{-1}:					η: N/A
C_p/J mol^{-1}K^{-1}:					k:
COMMENTS:					

Name: 2,4-Xylenol   CAS RN: 105-67-9
Synonym: 2,4-Dimethylphenol   Merck No.: 9989
                                                                 Mol. Wt.: 122.167
Mol. Form.: $C_8H_{10}O$

PHYSICAL CONSTANTS

T_m: 24.53°C (297.68 K) T_c: 434.5°C (707.6 K)   μ:
T_b: 210.98°C (484.13 K) P_c: IP: 8.00 eV

TRANSITION PROPERTIES

$\Delta_{fus}H$ (T_m):   Vapor pressure (0°C):
$\Delta_{vap}H$ (T_b): 47.14 kJ/mol   Vapor pressure (25°C): 0.022 kPa
$\Delta_{vap}H$ (25°C): 64.96 kJ/mol   Vapor pressure (100°C): 1.93 kPa

PROPERTIES AT 25° AND 100 kPa

	Solid	Liquid	Gas		Liquid
$\Delta_f H°$/kJ mol^{-1}:		-228.7	-162.9		d: 0.980 g/mL
$S°$/J mol^{-1}K^{-1}:					η:
C_p/J mol^{-1}K^{-1}:					k:
COMMENTS: Highly toxic					

Name: 2,5-Xylenol   CAS RN: 95-87-4
Synonym: 2,5-Dimethylphenol   Merck No.: 9989
                                                                 Mol. Wt.: 122.167
Mol. Form.: $C_8H_{10}O$

PHYSICAL CONSTANTS

T_m: 74.8°C (347.9 K) T_c: 433.8°C (706.9 K)   μ:
T_b: 211.18°C (484.33 K) P_c: IP:

TRANSITION PROPERTIES

$\Delta_{fus}H$ (T_m): 23.38 kJ/mol   Vapor pressure (0°C):
$\Delta_{vap}H$ (T_b): 46.94 kJ/mol   Vapor pressure (25°C): 0.022 kPa
$\Delta_{vap}H$ (25°C): 59.96 kJ/mol   Vapor pressure (100°C): 1.96 kPa

PROPERTIES AT 25° AND 100 kPa

	Solid	Liquid	Gas		Solid
$\Delta_f H°$/kJ mol^{-1}:	-246.6		-161.6		d: 0.975 g/mL
$S°$/J mol^{-1}K^{-1}:					η: N/A
C_p/J mol^{-1}K^{-1}:					k:
COMMENTS:					

Name: 2,6-Xylenol
Synonym: 2,6-Dimethylphenol

Mol. Form.: $C_8H_{10}O$

CAS RN: 576-26-1
Merck No.: 9989
Mol. Wt.: 122.167

PHYSICAL CONSTANTS

T_m: 45.7°C (318.8 K)
T_b: 201.07°C (474.22 K)

T_c: 427.9°C (701.0 K)
P_c:

μ:
IP: 8.05 eV

TRANSITION PROPERTIES

$\Delta_{fus}H$ (T_m): 18.90 kJ/mol
$\Delta_{vap}H$ (T_b): 44.52 kJ/mol
$\Delta_{vap}H$ (25°C): 75.31 kJ/mol

Vapor pressure (0°C):
Vapor pressure (25°C): 0.019 kPa
Vapor pressure (100°C): 3.41 kPa

PROPERTIES AT 25° AND 100 kPa

	Solid	Liquid	Gas		Solid
$\Delta_f H°$/kJ mol^{-1}:	-237.4		-161.8		d: 0.891 g/mL
$S°$/J mol^{-1}K^{-1}:					η: N/A
C_p/J mol^{-1}K^{-1}:					k:
COMMENTS:					

Name: 3,4-Xylenol
Synonym: 3,4-Dimethylphenol

Mol. Form.: $C_8H_{10}O$

CAS RN: 95-65-8
Merck No.: 9989
Mol. Wt.: 122.167

PHYSICAL CONSTANTS

T_m: 60.8°C (333.9 K)
T_b: 227°C (500 K)

T_c: 456.7°C (729.8 K)
P_c:

μ:
IP: 8.09 eV

TRANSITION PROPERTIES

$\Delta_{fus}H$ (T_m): 18.13 kJ/mol
$\Delta_{vap}H$ (T_b): 49.67 kJ/mol
$\Delta_{vap}H$ (25°C): 85.03 kJ/mol

Vapor pressure (0°C):
Vapor pressure (25°C): 0.005 kPa
Vapor pressure (100°C):

PROPERTIES AT 25° AND 100 kPa

	Solid	Liquid	Gas		Solid
$\Delta_f H°$/kJ mol^{-1}:	-242.3		-156.6		d: 1.012 g/mL
$S°$/J mol^{-1}K^{-1}:					η: N/A
C_p/J mol^{-1}K^{-1}:					k:
COMMENTS:					

Name: 3,5-Xylenol
Synonym: 3,5-Dimethylphenol

Mol. Form.: $C_8H_{10}O$

CAS RN: 108-68-9
Merck No.: 9989
Mol. Wt.: 122.167

PHYSICAL CONSTANTS

T_m: 63.6°C (336.7 K)
T_b: 221.74°C (494.89 K)

T_c: 442.5°C (715.6 K)
P_c:

μ:
IP:

TRANSITION PROPERTIES

$\Delta_{fus}H$ (T_m): 18.00 kJ/mol
$\Delta_{vap}H$ (T_b): 49.31 kJ/mol
$\Delta_{vap}H$ (25°C): 82.01 kJ/mol

Vapor pressure (0°C):
Vapor pressure (25°C): 0.007 kPa
Vapor pressure (100°C): 1.10 kPa

PROPERTIES AT 25° AND 100 kPa

	Solid	Liquid	Gas		Solid
$\Delta_f H°$/kJ mol^{-1}:	-244.4		-161.5		d: 0.984 g/mL
$S°$/J mol^{-1}K^{-1}:					η: N/A
C_p/J mol^{-1}K^{-1}:					k:
COMMENTS:					

Name: Yttrium

Mol. Form.: Y

CAS RN: 7440-65-5
Merck No.: 10015
Mol. Wt.: 88.906

PHYSICAL CONSTANTS

T_m: 1526°C (1799 K)
T_b: 3336°C (3609 K)

T_c:
P_c:

μ:
IP: 6.22 eV

TRANSITION PROPERTIES

$\Delta_{fus}H$ (T_m): 11.40 kJ/mol
$\Delta_{vap}H$ (T_b):
$\Delta_{vap}H$ (25°C):

Vapor pressure (0°C): N/A
Vapor pressure (25°C): N/A
Vapor pressure (100°C): N/A

PROPERTIES AT 25° AND 100 kPa

	Solid	Liquid	Gas		Solid
$\Delta_f H°$/kJ mol^{-1}:	0.0		421.3	d:	4.47 g/mL
$S°$/J mol^{-1}K^{-1}:	44.4		179.5	η:	N/A
C_p/J mol^{-1}K^{-1}:	26.5		25.9	k:	17.2 W/m K
COMMENTS:					

Name: Zinc

Mol. Form.: Zn

CAS RN: 7440-66-6
Merck No.: 10025
Mol. Wt.: 65.390

PHYSICAL CONSTANTS

T_m: 419.53°C (692.68 K)
T_b: 907°C (1180 K)

T_c:
P_c:

μ:
IP: 9.39 eV

TRANSITION PROPERTIES

$\Delta_{fus}H$ (T_m): 7.32 kJ/mol
$\Delta_{vap}H$ (T_b):
$\Delta_{vap}H$ (25°C):

Vapor pressure (0°C): N/A
Vapor pressure (25°C): N/A
Vapor pressure (100°C): N/A

PROPERTIES AT 25° AND 100 kPa

	Solid	Liquid	Gas		Solid
$\Delta_f H°$/kJ mol^{-1}:	0.0		130.4	d:	7.14 g/mL
$S°$/J mol^{-1}K^{-1}:	41.6		161.0	η:	N/A
C_p/J mol^{-1}K^{-1}:	25.4		20.8	k:	116 W/m K
COMMENTS:					

Name: Zinc bromide
Synonym: Zinc dibromide

Mol. Form.: Br₂Zn

CAS RN: 7699-45-8
Merck No.: 10028
Mol. Wt.: 225.198

PHYSICAL CONSTANTS

T_m: 394°C (667 K)
T_b: 697°C (970 K)

T_c:
P_c:

μ:
IP:

TRANSITION PROPERTIES

$\Delta_{fus}H$ (T_m): 16.70 kJ/mol
$\Delta_{vap}H$ (T_b): 118.00 kJ/mol
$\Delta_{vap}H$ (25°C):

Vapor pressure (0°C): N/A
Vapor pressure (25°C): N/A
Vapor pressure (100°C): N/A

PROPERTIES AT 25° AND 100 kPa

	Solid	Liquid	Gas		Solid
$\Delta_f H°$/kJ mol^{-1}:	-328.7			d:	4.5 g/mL
$S°$/J mol^{-1}K^{-1}:	138.5			η:	N/A
C_p/J mol^{-1}K^{-1}:				k:	
COMMENTS:					

Name: Zinc chloride
Synonym: Zinc dichloride

Mol. Form.: Cl_2Zn

CAS RN: 7646-85-7
Merck No.: 10031
Mol. Wt.: 136.295

PHYSICAL CONSTANTS

T_m: 290°C (563 K)
T_b: 732°C (1005 K)

T_c:
P_c:

μ:
IP:

TRANSITION PROPERTIES

$\Delta_{fus}H$ (T_m):
$\Delta_{vap}H$ (T_b): 126.00 kJ/mol
$\Delta_{vap}H$ (25°C):

Vapor pressure (0°C): N/A
Vapor pressure (25°C): N/A
Vapor pressure (100°C): N/A

PROPERTIES AT 25° AND 100 kPa

	Solid	Liquid	Gas		Solid
$\Delta_f H°$/kJ mol^{-1}:	-415.1		-266.1		d: 2.907 g/mL
$S°$/J mol^{-1}K^{-1}:	111.5				η: N/A
C_p/J mol^{-1}K^{-1}:	71.3				k:
COMMENTS:					

Name: Zinc fluoride
Synonym: Zinc difluoride

Mol. Form.: F_2Zn

CAS RN: 7783-49-5
Merck No.: 10035
Mol. Wt.: 103.387

PHYSICAL CONSTANTS

T_m: 872°C (1145 K)
T_b: 1500°C (1773 K)

T_c:
P_c:

μ:
IP:

TRANSITION PROPERTIES

$\Delta_{fus}H$ (T_m):
$\Delta_{vap}H$ (T_b): 190.10 kJ/mol
$\Delta_{vap}H$ (25°C):

Vapor pressure (0°C): N/A
Vapor pressure (25°C): N/A
Vapor pressure (100°C): N/A

PROPERTIES AT 25° AND 100 kPa

	Solid	Liquid	Gas		Solid
$\Delta_f H°$/kJ mol^{-1}:	-764.4				d: 4.9 g/mL
$S°$/J mol^{-1}K^{-1}:	73.7				η: N/A
C_p/J mol^{-1}K^{-1}:	65.7				k:
COMMENTS: Highly toxic					

Name: Zinc iodide
Synonym: Zinc diiodide

Mol. Form.: I_2Zn

CAS RN: 10139-47-6
Merck No.: 10040
Mol. Wt.: 319.199

PHYSICAL CONSTANTS

T_m: 446°C (719 K)
T_b: 625°C (898 K)

T_c:
P_c:

μ:
IP:

TRANSITION PROPERTIES

$\Delta_{fus}H$ (T_m):
$\Delta_{vap}H$ (T_b):
$\Delta_{vap}H$ (25°C):

Vapor pressure (0°C): N/A
Vapor pressure (25°C): N/A
Vapor pressure (100°C): N/A

PROPERTIES AT 25° AND 100 kPa

	Solid	Liquid	Gas		Solid
$\Delta_f H°$/kJ mol^{-1}:	-208.0				d: 4.74 g/mL
$S°$/J mol^{-1}K^{-1}:	161.1				η: N/A
C_p/J mol^{-1}K^{-1}:					k:
COMMENTS:					

Name: Zinc oxide   CAS RN: 1314-13-2
                                                                           Merck No.: 10050
                                                                           Mol. Wt.: 81.389

Mol. Form.: OZn

PHYSICAL CONSTANTS

T_m: 1975°C (2248 K)	T_c:	μ:
T_b:	P_c:	IP:

TRANSITION PROPERTIES

$\Delta_{fus}H$ (T_m): 52.30 kJ/mol   Vapor pressure (0°C): N/A
$\Delta_{vap}H$ (T_b):   Vapor pressure (25°C): N/A
$\Delta_{vap}H$ (25°C):   Vapor pressure (100°C): N/A

PROPERTIES AT 25° AND 100 kPa

	Solid	Liquid	Gas		Solid
$\Delta_f H°$/kJ mol^{-1}:	-350.5			d:	5.6 g/mL
$S°$/J mol^{-1}K^{-1}:	43.7			η:	N/A
C_p/J mol^{-1}K^{-1}:	40.3			k:	
COMMENTS:					

Name: Zirconium   CAS RN: 7440-67-7
                                                                            Merck No.: 10076
                                                                            Mol. Wt.: 91.224

Mol. Form.: Zr

PHYSICAL CONSTANTS

T_m: 1855°C (2127.85 K)	T_c:	μ:
T_b: 4409°C (4682 K)	P_c:	IP: 6.63 eV

TRANSITION PROPERTIES

$\Delta_{fus}H$ (T_m): 21.00 kJ/mol   Vapor pressure (0°C): N/A
$\Delta_{vap}H$ (T_b):   Vapor pressure (25°C): N/A
$\Delta_{vap}H$ (25°C):   Vapor pressure (100°C): N/A

PROPERTIES AT 25° AND 100 kPa

	Solid	Liquid	Gas		Solid
$\Delta_f H°$/kJ mol^{-1}:	0.0		608.8	d:	6.52 g/mL
$S°$/J mol^{-1}K^{-1}:	39.0		181.4	η:	N/A
C_p/J mol^{-1}K^{-1}:	25.4		26.7	k:	22.7 W/m K
COMMENTS:					

Name: Zirconium chloride (ZrCl$_4$)   CAS RN: 10026-11-6
Synonyms: Zirconium tetrachloride   Merck No.: 10077
                      Zirconium(IV) chloride   Mol. Wt.: 233.035
Mol. Form.: Cl$_4$Zr

PHYSICAL CONSTANTS

T_m: 437°C (710 K)	T_c: 505°C (778 K)	μ:
T_b:	P_c: 5.77 MPa	IP: 11.20 eV

TRANSITION PROPERTIES

$\Delta_{fus}H$ (T_m): 50.00 kJ/mol   Vapor pressure (0°C): N/A
$\Delta_{vap}H$ (T_b):   Vapor pressure (25°C): N/A
$\Delta_{vap}H$ (25°C):   Vapor pressure (100°C): N/A

PROPERTIES AT 25° AND 100 kPa

	Solid	Liquid	Gas		Solid
$\Delta_f H°$/kJ mol^{-1}:	-980.5			d:	2.80 g/mL
$S°$/J mol^{-1}K^{-1}:	181.6			η:	N/A
C_p/J mol^{-1}K^{-1}:	119.8			k:	
COMMENTS:					

NAME/SYNONYM INDEX

This index includes the primary names and major synonyms given in the tables. It is alphabetized without regard to locant numbers, spaces, and prefixes such as *cis*, *trans*, *sec*, *tert*, etc. The letter a, b, or c following the page number indicates that the entry appears at the top, middle, or bottom of the page, respectively.

NAME/SYNONYM INDEX (continued)

NAME/SYNONYM INDEX (continued)

NAME/SYNONYM INDEX (continued)

NAME/SYNONYM INDEX (continued)

NAME/SYNONYM INDEX (continued)

NAME/SYNONYM INDEX (continued)

NAME/SYNONYM INDEX (continued)

NAME/SYNONYM INDEX (continued)

NAME/SYNONYM INDEX (continued)

NAME/SYNONYM INDEX (continued)

NAME/SYNONYM INDEX (continued)

NAME/SYNONYM INDEX (continued)

NAME/SYNONYM INDEX (continued)

NAME/SYNONYM INDEX (continued)

NAME/SYNONYM INDEX (continued)

NAME/SYNONYM INDEX (continued)

NAME/SYNONYM INDEX (continued)

NAME/SYNONYM INDEX (continued)

MOLECULAR FORMULA INDEX

This index is given in a modified Hill order in which all substances not containing carbon appear first, followed by those that do contain carbon. For each carbon number, the compounds with no hydrogen are listed first, followed by those with one hydrogen, two hydrogens, etc.

MOLECULAR FORMULA INDEX (continued)

MOLECULAR FORMULA INDEX (continued)

MOLECULAR FORMULA INDEX (continued)

MOLECULAR FORMULA INDEX (continued)

MOLECULAR FORMULA INDEX (continued)

MOLECULAR FORMULA INDEX (continued)

MOLECULAR FORMULA INDEX (continued)

MOLECULAR FORMULA INDEX (continued)

MOLECULAR FORMULA INDEX (continued)

CHEMICAL ABSTRACTS SERVICE REGISTRY NUMBER INDEX

CHEMICAL ABSTRACTS SERVICE REGISTRY NUMBER INDEX
(continued)

CHEMICAL ABSTRACTS SERVICE REGISTRY NUMBER INDEX
(continued)

CHEMICAL ABSTRACTS SERVICE REGISTRY NUMBER INDEX
(continued)

CHEMICAL ABSTRACTS SERVICE REGISTRY NUMBER INDEX
(continued)

CHEMICAL ABSTRACTS SERVICE REGISTRY NUMBER INDEX
(continued)